ANALYSIS AND DESIGN OF DIGITAL CIRCUITS AND COMPUTER SYSTEMS

ANALYSIS AND DESIGN OF DIGITAL CIRCUITS AND COMPUTER SYSTEMS

PAUL M. CHIRLIAN

Professor of Electrical Engineering
Stevens Institute of Technology

Matrix Publishers, Inc., Champaign, Illinois

10 9 8 7 6 5 4 3 2 1

Library of Congress catalog card number: 76-6727

Matrix Publishers, Inc.
Champaign, IL 61820

ISBN: 0-916460-03-7

This book was prepared by typewriter composition.
Illustrations by Scientific Illustrators.
In-house Editor was Merl K. Miller.
Printed by KOWA Graphics.

To Barbara, Lisa and Peter

Contents

4 COMBINATIONAL LOGIC-SWITCHING ALGEBRA 65

5 COMBINATIONAL LOGIC CIRCUITS 143

6 SEQUENTIAL CIRCUITS 197

7 MEMORIES 289

APPENDIX — SEMICONDUCTOR DEVICES 541

Preface

This introductory book's objective is to provide the reader with the basic ideas and tools needed to analyze and design digital circuits and computer systems. Since this is a text for use in a basic course, the fundamental ideas of switching circuits and their design is stressed. However, it is felt that, after completing such a first course, the student should have an understanding of a complete digital computer or a computer system. Therefore, such things as elementary computer organization and machine language are discussed. This latter discussion is by no means exhaustive since it would not be appropriate for an elementary course taken by all students. However, sufficient detail is included to give the reader an understanding of how a computer or computer system operates.

The discussion of such things as number systems, chage of radix, and elementary binary arithmetic are made very detailed. This is done so that, if desired, this material need not be covered in class but may just be assigned for home reading.

It is assumed that the readers of this book have studied some very elementary circuit material. However, they need not have any other prerequisites.

The block diagram approach can be used to discuss switching circuits and computers. However, this does not provide a full understanding of computer operation. Consequently, the electronic circuit aspects of switching circuits are discussed in Chapter Five. In addition to basic ideas, computation of such things as switching levels and noise immunity are also considered there. The block diagram approach is used in most of the remainder of the text so that the discussions of Chapter Five can be omitted or deferred.

Chapter Five assumes that the student is familiar with the basic ideas of semiconductor devices. If this is not the case, then the material presented in the Appendix will provide him with the necessary background.

The first chapter is an introductory one which is designed to provide the student with a brief picture of digital device usage.

In Chapter Two we consider number systems and simple binary arithmetic operations.

The elementary ideas of logic and basic gates are discussed in Chapter Three.

In Chapter Four, minimization techniques for combinational circuits are studied. The topics covered include basic ideas of switching algebra, minimization using maps, prime implicants, the Quine-McCluskey method, minimization of multiple output networks, and use of decimal notation in prime implicant tables.

We have discussed logic gates and the mathematical means of minimizing them. Now, in the fifth chapter, we consider the circuits from the viewpoint of an Electrical Engineer. We start with a discussion of logic circuits. First, the idea of logic families is presented and the various criteria for their choice are considered. Next, all the commonly used transistor logic families are discussed. Following this, we consider the MOS logic families. Finally, all the logic families are compared. Again, these devices are studied from the viewpoint of the Electrical Engineer.

In the sixth chapter, we start with a general discussion of sequential circuits. Next we discuss the various flip-flops. Synchronization and triggering are considered in detail. The basic tools needed to analyze synchronous circuits are discussed followed by a discussion of registers. A detailed discussion of the design of synchronous sequential circuits is presented. Counters are considered here. The basic ideas of asynchronous sequential circuit are discussed, and the problem of hazards and races are considered. The design of asynchronous sequential circuits is then presented.

After developing the basic tools, some of the components of the digital computer are discussed. In Chapter Seven we consider memories. Random access memories are considered in great detail. This discussion includes core, film, transistor and MOS memories. Next, auxiliary memories are discussed. Finally, read only memories and decoders are studied.

In Chapter Eight, the ideas of binary codes are discussed. First we consider BCD codes, then alphanumeric codes. Next, reflected codes are considered. Finally we study error detecting and correcting codes.

In Chapter Nine, binary arithmetic is discussed. We start with a consideration of modular arithmetic. Next, complementary arithmetic is studied. Floating point arithmetic is discussed. Precision and accuracy are considered in detail.

Computer arithmetic is presented in Chapter Ten. The actual computer implementation of addition, subtraction, multiplicatior and division are presented. BCD conversion is also considered here. Circuits which implement these operations are explained. The chapter concludes with a complete discussion of the accumulator.

In Chapter Eleven the complete digital computer is considered. We start with a general discussion of the components of a computer. This is followed by a consideration of information storage and transmission. The complete computer is then discussed. A simple machine language is developed and simple programs are written. The object here is not to teach programming but to relate the machine language to the actual operation of the computer. Assembler and higher level languages are then discussed, again, not to teach programming, but to provide a basic understanding of what these are.

In the twelfth chapter, we consider some digital circuits. There is no attempt at completeness here. The purpose of the chapter is to provide some basic ideas and to discuss the elements of computer interfacing. Operational amplifiers are discussed. Next, comparators and sample and hold circuits are presented. These are then used in a discussion of digital to analog and analog to digital conversion. Astable and monostable multivibrators are considered next. The use of these components in fabricating computer interfacing is then discussed.

The Appendix discusses the basic ideas of semiconductor devices. This is done so that the reader who has not been exposed to this material previously can understand the electronic circuits. This Appendix contains a brief discussion of semiconductor physics, the p-n junction diode, the transistor, the field effect transistor, switching in semiconductor devices and integrated circuits. All necessary background material is provided in this appendix.

Many varied problems are given at the end of each chapter.

The author would like to express his gratitude to his colleagues Prof. Alfred C. Gilmore, Jr. and Otto C. Boelens for the numerous and lengthy discussions held with him during the writing of the book. They provided much help and inspiration. The author would also like to thank his colleagues Profs. Preston R. Clement, Emil C. Neu, and Gerald J.Herskowitz for their helpful comments.

Much loving gratitude and heartfelt thanks are again due my wife, Barbara, who not only provided me with continuous encouragement and saw to it that my time was free from interruption, but who also typed the rough draft and final draft of the manuscript and corrected the punctuation and grammar of the copy during a time when she was busily engaged in a time consuming project of her own.

Paul M. Chirlian

1

Introduction

This book will discuss digital systems emphasizing the details of the analysis and design of the elements of these systems. Although emphasis will be placed on the digital computer, other digital systems will also be treated.

Digital systems are often studied by considering them to be made up of blocks which perform logical operations. In addition to this block diagram approach, we shall also discuss the electronic circuits within the blocks.

In this first chapter, we shall discuss some general ideas of digital systems.

1.1 A GENERAL DISCUSSION OF DIGITAL DEVICES

The digital computer can perform many operations. It can, in seconds, solve equations which could take years to calculate by hand. An entire industrial process can be controlled by a digital computer. For instance, special sensors could determine the temperature, pressure, acidity, etc., in a chemical process. In an appropriate form, this information is then supplied to a digital computer. Calculations on this data are then performed. As a result of these calculations, electrical signals, which control the process, are generated. For instance, in response to these signals, heaters could be turned on or off, or additional chemicals added.

When a radar signal is reflected from a distant planet (or object) it is often very weak and obscured by noise. Ordinary techniques cannot be used to detect this signal. However, a digital computer can process it and make the radar signal detectable. Computers can also be used to monitor a hospital recovery room. We have only considered several of the host of tasks that can be performed by a digital computer. In order to perform such complex operations, the digital computer must contain very many circuits. In fact, each computer is made up of the interconnection of thousands of relatively simple circuits. Thus, we must study the individual circuits before considering how they are interconnected to form a complete computer.

Integrated circuit techniques can be used to produce simple digital computers which are small in size. Such computers, called *microprocessors*, are used to perform many tasks. For instance, a microprocessor in an oscilloscope can be used to obtain a reading of the true rms value of a signal. Actually, microprocessors can be used for many of the previously discussed control tasks.

The earliest digital computers used relays as their fundamental elements. Note that a relay is simply a switch controlled by an electromagnet. These switches were either on or off(i.e., either open or closed). Present digital computers have replaced the relays with electronic devices. However, these still act as controlled switches which are either on or off. In the next section we shall discuss how these on or off conditions give rise to a signal which has two levels that are called *zero* and *one*.

In a digital system, all signals are made up of combinations of such zeros and ones. In contrast, in an *analog* system, the signals can vary continuously. For instance, if the number 1.23 is computed by an analog computer, then 1.23 volts might actually appear at the output. In a digital computer, the number 1.23 would be represented by a sequence of zeros and ones. In addition, the output of an analog system can vary continuously while that of a digital one can only vary in discrete steps. The digital system seems more complex and, indeed, it often is. However, it has many advantages. A digital system can be extremely accurate. Many systems can provide greater than 16 place accuracy. Such accuracy is impossible with an analog system, where the limit of precision is usually 3 or 4 significant figures. In addition, analog systems often change their characteristics with time, temperature, etc. (or drift) while digital systems do not. Digital systems can often be more flexible. Thus, in spite of their complexity, digital computers are very widely used for a great variety of operations. In fact, digital systems have replaced analog systems in very many operations. One reason for this is that the use of integrated circuits now allows the extremely large circuits needed for digital computers to be reliably built in small volumes at relatively low costs.

All very large computers, small desk calculators, and most intermediate size computers are digital in nature. Analog computers are still built, but today, most of them are special purpose in nature. In addition, digital systems are now used in areas which were once the province of analog systems. Let us consider some of these. The ordinary voltmeter is an analog device. The accuracy of such instruments is, at best, 1 percent of the *full* scale value. (Thus, they tend to be inaccurate at low scale readings.) On the other hand, digital voltmeters are built which are capable of a fraction of a percent accuracy over almost their entire range. (We shall discuss such voltmeters in Chapter 12.) Digital voltmeters also have the advantage that the voltage is presented as an easily read number(i.e., 346.2 volts appears as the numbers, 346.2). Thus, it is easy to read and no interpolation is required. Another example of a digital device replacing an analog one is the electric filter, which selects one signal and rejects all others. Usually, this is an analog system. (Such filters are used to tune a radio to one station while rejecting the others.) However, in many complex applications, simple digital computers are programmed to perform the mathematical operation of filtering. In these and other applications, analog systems are being replaced by digital ones. Of course, because of their relative simplicity, there are many circumstances where analog systems will continue to be used. However, digital systems are becoming more and more important.

Often, we must work with both analog and digital systems in a single device. For intance, in a digital voltmeter, the input signal, the voltage, is a continuous signal. It must be converted to a digital representation. Similarly, in a digital filter, the input signal is continuous. It must be converted to a digital representation, which is processed digitally, and then converted back to a continuous signal. Devices called analog to digital converters and digital to analog converters are used for such purposes. We shall discuss these in Chapter 12. (At times, the combination of a digital and an analog system is called a *hybrid system*.)

1.2 SOME FUNDAMENTALS OF COMPUTERS

Digital computers always manipulate numbers. For instance, if a nonlinear differential equation is to be solved, then the computer is programmed to implement an appropriate numerical analysis procedure. The program instructs the computer to perform many arithmetic operations (addition, subtraction, multiplication, and division)on a set of input data(i.e., numbers). These operations result in the solution. Similarly, if a digital computer is used as a digital filter, a sequence of numbers, obtained from an analog to digital converter is supplied to the computer. The appropriate arithmetic operations are then performed. These result in a sequence of numbers proportional to the output signal. Digital to analog conversion constructs an analog signal from this sequence. The operations performed by the computer are governed by a series of instructions called a *program*, which is supplied by the user. Although the basic operations of the computer are arithmetic in nature, these can be combined to approximate other mathematical operations such as differentiation and integration.

A computer consists of several basic parts. There is a *control unit* which receives the instructions of the program and directs the appropriate computer operations. The *central processor unit* performs the operations of addition, subtraction, multiplication and division plus some additional operations. The *memory* is used to store both numbers and programs. There are also *input-output* devices such as card readers, teletypes and printers which are used by the human operators to supply the computer with the data and instructions and which, conversely, are used by the computer to print out the calculated values.

All the data handled by the computer is in numeric form. At times, the computer performs non-numeric procedures such as the alphabetizing of a list of names. However, all such information is converted into numerical information using a suitable *code*.

Thus, the storage and manipulation of numbers is of fundamental importance in the study of digital computers. For this reason, we shall now study some fundamental ideas of number systems. We all learned and used the *decimal number* system based on the ten digits 0,1,2,3,4,5,6,7,8, and 9. Such a number system probably developed because humans have ten fingers. Thus, ten was used for convenience. When mechanical adding machines and desk calculators were built, the decimal system was carried over to them easily since gears with ten or multiples of ten teeth were used. However, electronic devices have replaced almost all the mechanical ones, not only in large computers but also in simple adding machines and desk calculators. It is not desirable to use the decimal system in such electronic devices. We shall illustrate this discussion with the simple electronic circuit of Fig.1-1a. The box represents an idealized electronic device; it could approximate a transistor. A plot of output voltage versus input voltage for this device is shown in Fig. 1-1b. Let us consider this characteristic.

If the input voltage v_1 is less than v_{1c}, then the output circuit of the device acts as an open circuit and $v_2=V_{CC}$. In this

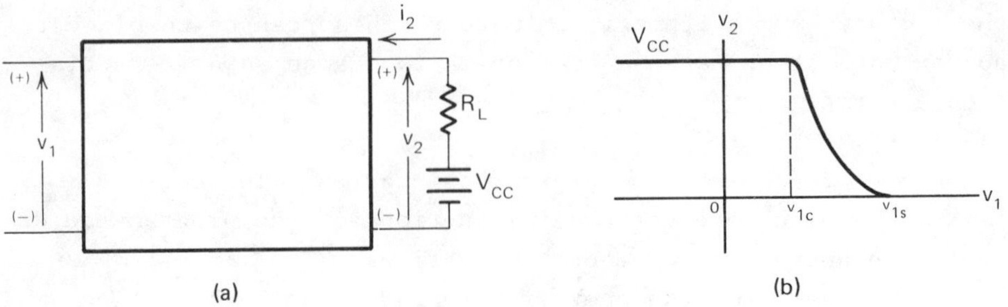

Figure 1-1. (a) An idealized electronic device; (b) the transfer characteristics of this circuit.

condition, the output current $i_2=0$ and the device is said to be *cut off*. If we want to be sure that the device is cut off, then we can make v_1 less than v_{1c}.

If v_1 is greater than v_{1s}, then the output circuit of the device acts as a short circuit and $v_2=0$. In this case, the device is said to be *saturated*. Again, if we want to be sure that the device is saturated, then we can make v_1 greater than v_{1s}.

The cut off and saturation values can be established easily and readily. These voltage levels are relatively far apart so that they can easily be distinguished. For instance, the saturation voltage may range between 0.5 and 0.1 volt while a cut off voltage could range between 9 and 10 volts. Note that the cut off values for each device need not be the same. They must only fall within some specified range. A similar statement can be made for the saturation values. Thus, the digital computer can be extremely reliable since a small shift in levels will still result in voltages which lie in their specified ranges and, thus, be interpreted correctly. Hence, there are two levels (ranges of levels) that are used. If the cut off and saturation levels represent numbers, then there are only two digits available. As before, these are called *zero* and *one*. This is called a *binary number system*. Thus, in digital computers, we work with arithmetic systems that use only the two digits zero and one. These are called *binary digits* and are often called *bits*. Note that this does *not* mean that the data cannot be entered in the usual decimal form. It is converted to the binary form within the computer. Thus, we are restricted to a *binary number system*.

Another reason for operating the active devices at either saturation or cut off is efficiency. At saturation, the current is high but the voltage across the device is extremely small; at cut off, the voltage is large but the current is extremely small. Either of these conditions usually results in small power dissipation. This is very important, especially when many electronic components are contained on one small chip. Actual electronic circuits will be discussed in subsequent chapters of this book. Now that we have established the reasons for using a binary number system, we shall consider this number system in detail in the next chapter.

BIBLIOGRAPHY

Booth, T.L., "Digital Networks and Computer Systems", Chap.1, John Wiley and Sons, Inc., New York, 1971

Hill,F,J. and Peterson,G.R., "Digital Systems Hardware Organization and Design", Chap.1, John Wiley and Sons, Inc., New York 1973

Peatman, J.B., "The Design of Digital Systems", Chap.1, McGraw-Hill Book Co., New York, 1972

PROBLEMS

1-1. Discuss the number of signals used in a digital computer.

1-2. Compare digital and analog systems.

1-3. Discuss the basic elements of a digital computer.

1-4. Why is the binary number system used in a digital computer?

1-5. Discuss the need for codes in a digital system.

Radix 10 Decimal	Radix 2 Binary	Radix 8 Octal	Radix 16 Hexadecimal
31	11111	37	1F
32	100000	40	20
33	100001	41	21
34	100010	42	22
35	100011	43	23
36	100100	44	24

If we expect a number to have any meaning, then we must know what radix is used. Often, it is clear what radix is used. However, if this is not the case, then a subscript will be appended to indicate the radix. For instance, see Table 2-1, we can write

$$8_{10} = 1000_2 = 10_8 = 8_{16} \tag{2-1}$$

The subscript is always written in radix 10. When it is clear which is meant, the subscripts will be omitted from binary and decimal numbers.

Now let us discuss number systems. We shall again start with radix ten since it is the most familiar system. We shall then extend the results to an arbitrary radix. At the start, we shall consider whole numbers and then generalize to numbers with fractional parts. Suppose we write the number 165_{10}. This number actually represents the following:

$$165_{10} = 1(10^2) + 6(10^1) + 5(10^0) \tag{2-2}$$

where all the numbers are written in radix ten. That is, the number 165_{10} represents 1 one hundreds, 6 tens and 5 ones. Let us generalize this. Suppose that we have a number:

$$a_n a_{n-1} a_{n-2} \cdots a_2 a_1 a_0 \tag{2-3}$$

where the a_j represent digits of the base 10 number system.

For instance, in the previous case, we had the number 165, where

$a_0 = 5$

$a_1 = 6$

$a_2 = 1$

We can represent the number of (2-3) by the equation

$$a_n a_{n-1} \ldots a_2 a_1 a_0 = a_n 10^n + a_{n-1} 10^{n-1} + \ldots + a_2 10^2 + a_1 10^1 + a_0 10^0 \qquad (2\text{-}4)$$

Thus, each digit represents the number of times that the appropriate power of ten appears in the summation.

Now let us consider numbers with fractional parts. For example

$$256.123_{10} = 2(10^2) + 5(10^1) + 6(10^0) + 1(10^{-1}) + 2(10^{-2}) + 3(10^{-3})$$

The decimal point is used to separate the positive powers of ten from the negative ones. Therefore, in general, we can write

$$(a_n a_{n-1} \ldots a_2 a_1 a_0 . a_{-1} a_{-2} \ldots a_{-k})_{10} = a_n 10^n + a_{n-1} 10^{n-1} + \ldots + a_2 10^2 + a_1 10^1 + a_0 10^0 + a_{-1} 10^{-1} + a_{-2} 1^{-2} + \ldots + a_{-k} 10^{-k} \qquad (2\text{-}5)$$

In a decimal system, we call the point that separates the positive and negative powers of ten, the decimal point. The general name given to this point is the *radix point*. Note that in binary systems, the radix point is called the *binary point*; in octal systems, the radix point is called the *octal point*, etc.

Now let us consider the expression of a number in a general radix. We shall express the general radix in terms of radix ten, since we are most familiar with radix ten. Integers will be considered first. Then, following Eq. (2-4), we have

$$(a_n a_{n-1} \ldots a_2 a_1 a_0)_r = a_n r^n + a_{n-1} r^{n-1} + \ldots + a_2 r^2 + a_1 r^1 + a_0 r^0 \qquad (2\text{-}6)$$

Note that the numbers to the right of the equals sign are expressed to base ten. We shall discuss some examples of this. (Table 2-1 can be used to verify some of the results.)

$$37_8 = 3_{10}(8_{10}{}^1) + 7_{10}(8_{10}{}^0) = 24_{10} + 7_{10} = 31_{10}$$

$$1E_{16} = 1_{10}(16_{10}{}^0) + 14_{10}(10_{10}{}^0) = 30_{10}$$

$$100100_2 = 1(2_{10}{}^5) + 0(2_{10}{}^4) + 0(2_{10}{}^3) + 1(2_{10}{}^2) + 0(2_{10}{}^1) + 0(2_{10}{}^0) = 32 + 4 = 36_{10}$$

Now let us include the radix point. Following the example of Eq. (2-5), we obtain

$$(a_n a_{n-1} \ldots a_2 a_1 a_0 . a_{-1} a_{-2} \ldots a_{-k})_r = a_n r^n + a_{n-1} r^{n-1} + \ldots + a_2 r^2 + a_1 r^1 + a_0 r^0 + a_{-1} r^{-1} + a_{-2} r^{-2} + \ldots + a_{-k} r^{-k} \qquad (2\text{-}7)$$

where we have expressed the numbers to the right of the equals sign in decimal radix. We shall illustrate this with some examples.

$$127.34_8 = 1(8^2) + 2(8^1) + 7(8^0) + 3(8^{-1}) + 4(8^{-2}) = 87.4375_{10}$$

$$101.101_2 = 1(2^2) + 0(2^1) + 1(2^0) + 1(2^{-1}) + 0(2^{-2}) + 1(2^{-3}) = 5.625_{10}$$

$$EA2.AB3_{16} = 14(16^2) + 10(16^1) + 2(16^0) + 10(16^{-1}) + 11(16^{-2}) + 3(16^{-3}) = 3746.668701_{10}$$

The examples that we have just performed illustrate the conversion from an arbitrary radix to radix ten. Let us now illustrate the conversion from radix ten to an arbitrary radix. Suppose that we have the decimal number

$$N_{10} = (d_n d_{n-1} \dots d_2 d_1 d_0)_{10}$$

and we want to convert it to a binary system number. Thus, we have

$$N_{10} = b_k 2^k + b_{k-1} 2^{k-1} + \dots + b_2 2^2 + b_1 2^1 + b_0 2^0 \qquad (2\text{-}8)$$

We must solve for all the unknown b_j. Note that, in this case, the unknowns are equal to either 0 or 1. There are various procedures that can be used here. The technique that we shall discuss is a convenient one to use. Divide both sides of Eq.(2-8) by 2. This yields

$$N_{10}/2 = b_k 2^{k-1} + b_{k-1} 2^{k-2} + \dots + b_2 2 + b_1 + b_0/2 \qquad (2\text{-}9)$$

The quotient of the division is

$$Q_1 = b_k 2^{k-1} + b_{k-1} 2^{k-2} + \dots + b_2 2 + b_1 \qquad (2\text{-}10)$$

while the remainder is the *rightmost* binary digit. Now divide the quotient Q_1 by 2. This yields

$$Q_1/2 = b_k 2^{k-2} + b_{k-1} 2^{k-3} + b_2 + b_1/2 \qquad (2\text{-}11)$$

The quotient is

$$Q_2 = b_k 2^{k-2} + b_{k-1} 2^{k-3} + b_2 \qquad (2\text{-}12)$$

while the remainder is

$$R_2 = b_1$$

which is the second binary digit. Proceding in this way we can obtain all the binary digits. Let us illustrate this with an example. Let us obtain the binary equivalent of 47_{10}

$$\frac{47_{10}}{2} = 23 + 1/2$$

Thus, $R_1 = 1$

$Q_1 = 23$

Dividing by 2, we obtain

$$\frac{Q_1}{2} = 11 + 1/2$$

Therefore,

$Q_2 = 11$

$R_2 = 1$

Proceding similarly, we have

$Q_2/2 = 5 + 1/2$

$R_3 = 1$

$Q_3 = 5$

$Q_3/2 = 2 + 1/2$

$R_4 = 1$

$Q_4 = 2$

$Q_4/2 = 1$

$R_5 = 0$

$Q_5 = 1$

$Q_5/2 = 1/2$

$Q_6 = 0$

$R_6 = 1$

Thus, the procedure has terminated. We now use all the remainders to obtain the binary digits. Remember that the first remainder yields the rightmost digit. Thus, we have

$$47_{10} = 101111_2$$

Now let us consider the conversion of numbers with fractional parts. A number can always be considered to be the sum of its whole and fractional parts. Thus, we can use the procedure just discussed to obtain the conversion of the whole part. We shall now discuss the conversion of the part to the right of the radix point. We can write

$$F_{10} = (0.d_{-1}d_{-2}\ldots d_{-m})_{10} = (0.b_{-1}b_{-2}\ldots)_2 \qquad (2\text{-}13)$$

That is, we have equated the decimal and binary numbers. This represents the equation

$$d_{-1}10^{-1} + d_{-2}10^{-2} + \ldots + d_{-m}10^{-m} = b_{-1}2^{-1} + b_{-2}2^{-2} + \ldots \qquad (2\text{-}14)$$

Note that we have represented a decimal fraction of finite length by a binary fraction, which may be of infinite length. We shall consider this subsequently.

Now multiply both sides of Eq. (2-14) by 2. This yields

$$2F_{10} = b_{-1} + b_{-2}2^{-1} + b_{-3}2^{-2} + \ldots \qquad (2\text{-}15)$$

The nonfractional part of this expression represents the first digit to the right of the binary point. The fractional part of Eq.(2-15) is then

$$F_2 = b_{-2}2^{-1} + b_{-3}2^{-2} + \ldots \qquad (2\text{-}16)$$

Multiplying this by two, we have

$$2F_2 = b_{-2} + b_{-3}2^{-1} + \ldots \qquad (2\text{-}17)$$

Again, the nonfractional part represents the next digit. We can then procede in this way to obtain all successive digits. Let us illustrate this with some examples.

Let us obtain the binary equivalent of

0.25_{10}

Multiplying by 2, we have

$2(0.25) = 0.5$

There is no whole part. Thus,

$b_{-1} = 0$

Multiplying again, we have
$2(0.5) = 1$

The whole part is 1. Hence,

$b_{-2} = 1$

Removing the whole part, we are left with zero. Hence, the procedure terminates. Then,

$$0.25_{10} = 0.01_2 \qquad (2\text{-}18)$$

Let us consider another example. Obtain the binary equivalent of

0.426_{10}

Multiplying by 2 yields

$2(0.426) = 0.852$

Thus, $b_{-1} = 0$. Proceeding, we have

$2(0.852) = 1.704$

Then,

$b_{-2} = 1$

Removing the whole part and multiplying, we obtain

$2(0.704) = 1.408$

Hence,

$b_{-3} = 1$

Continuing, we have

$2(0.408) = 0.816$

$b_{-4} = 0$

$2(0.816) = 1.632$

$b_{-5} = 1$

$2(0.632) = 1.264$

$b_{-6} = 1$

$2(0.264) = 0.528$

$b_{-7} = 0$

Hence, the result is

$$0.426_{10} = 0.0110110\ldots\ldots_2 \qquad (2\text{-}19)$$

Note that the procedure did not terminate. At times, an infinite number of binary digits must be used to *exactly* represent a decimal fraction. Humans and computers can only work with a finite number of digits. Thus, in the case of fractions, the conversion from one radix to another often results in an error.

In the following example

$$0.426_{10} = 0.01101_2 + \varepsilon \tag{2-20}$$

ε represents the error.

Let us estimate the size of this error. In radix 2, we can write the fractional part *exactly* using the infinite series

$$F_{10} = \sum_{n=1}^{\infty} b_{-n}2^{-n} \tag{2-21}$$

When only a finite number of binary digits are used, we represent this by a finite sum plus an error. Thus,

$$F_{10} = \sum_{n=1}^{N} b_{-n}2^{-n} + \varepsilon \tag{2-22}$$

where the error is equal to the terms in the summation which have been left out. Hence,

$$\varepsilon = \sum_{n=N+1}^{\infty} b_{-n}2^{-n} \tag{2-23}$$

The b_{-n} are either zeros or ones. In this case, this series will converge rapidly. In fact, it can be shown that

$$|\varepsilon| \leq 2^{-N} \tag{2-24}$$

Thus, we need only carry the fraction out until the 2^{-N} becomes less than some desired value. For instance, in Eq.(2-20)

$$\varepsilon < 2^{-5} = 0.03125_{10}$$

If this is not a sufficiently small error, then additional terms should be taken. For instance, if we write

$$0.246_{10} = 0.0110110$$

then

$$\varepsilon < 0.0078125_{10}$$

In general, an error will occur when numbers are converted from one radix to another. This is called a *roundoff error*. (We shall subsequently discuss other causes of roundoff errors.)

Let us illustrate this discussion with an additional example. Let us express

47.426_{10}

in binary form. We shall use a shorthand form here to represent the division. The whole part is 47. Then,

$2|\underline{47}$

$2|\underline{23} + 1/2$

$2|\underline{11} + 1/2$

$2|\underline{5} + 1/2$

$2|\underline{2} + 1/2$

$2|\underline{1} + 0/2$

$0 + 1/2$

Thus, $47_{10} = 101111_2$. Then, to obtain the fractional part, we perform the following calculations

$2(0.426) = 0.852$

$2(0.852) = 1.704$

$2(0.704) = 1.408$

$2(0.408) = 0.816$

$2(0.816) = 1.632$

$2(0.632) = 1.264$

.
.
.

Hence,

$47.426_{10} \approx 101111.011011_2$

We have illustrated conversion from radix 10 to radix 2. The procedures for conversion from base 10 to other bases follow this one. The radix r replaces the 2 in the previous procedures Let us illustrate this with an example. Let us express 47.426_{10} in octal. We again work with the whole and fractional parts separately. Thus,

$8 | \underline{47}$

$8 | \underline{5} + 7/8$

$0 + 5/8$

Hence,

$$47_{10} = 57_8$$

Now, consider the fractional part

$8(.426) = 3.408$

$8(.408) = 3.264$

$8(.264) = 2.112$

$8(.112) = 0.896$

$8(.896) = 7.168$

$\vdots$

Therefore,

$$47.426_{10} \approx 57.33207_8$$

The conversion between arbitrary radices follows the same general ideas. It is somewhat more difficult to perform the computations in this case since we are not used to performing arithmetic in other bases.

Conversion between binary and octal or hexadecimal is especially easy. Let us write a binary number and show how it can be converted to an octal number.

$$1010111_2 = 1(2^6) + 0(2^5) + 1(2^4) + 0(2^3) + 1(2^2) + 1(2^1) + 1(2^0)$$

Note that the terms to the right of the equals sign are expressed in decimal. Now let us rewrite this as 001010111_2. Note that we have added zeros to the left which does not change the number. Then,

$$001010111_2 = 0(2^8) + 0(2^7) + 1(2^6) + 0(2^5) + 1(2^4) + 0(2^3) + 1(2^2) + 1(2^1) + 1(2^0)$$

Note that we have added zeros so as to cause nine terms to appear. Now we can write

$$1010111_2 = \{0(2^2) + 0(2^1) + 1(2^0)\}8^2 + \{0(2^2) + 1(2^1) + 0(2^0)\}8^1 + \{1(2^2) + 1(2^1) + 1(2^0)\}8^0$$

Let us express each bracketed term as a binary number.

$$001010111_2 = \{001_2\}{8_{10}}^2 + \{010_2\}{8_{10}}^1 + \{111_2\}{8_{10}}^0$$

The bracketed digits are simply the digits of the binary number. The right hand side of this expression is in "octal form". That is, if we convert the binary coefficients to decimal numbers, we obtain the"summation form" of octal representation. (Note that, for numbers between 0 and 7, the decimal and octal representations are the same.) Hence, we have

$$001010111_2 = 127_8$$

We can use this procedure for any binary-octal conversion of an integer. That is, break up the binary number into groups of three digits, adding zeros at the left if necessary. Start at the binary point and convert each binary group into a decimal number. This yields the equivalent octal number. Let us consider another example.

$$1010101_2 = 001\ \ 010\ \ 101_2 = 125_8$$

Now let us consider a binary number with fractional parts.

$$0.1011_2 = 1(2^{-1}) + 0(2^{-2}) + 1(2^{-3}) + 1(2^{-4}) + 0(2^{-5}) + 0(2^{-6})$$

Note that we have added zeros to the right so that there are six digits following the binary point. Then, manipulating, we have

$$0.1011_2 = \left[1(2^2) + 0(2^1) + 1(2^0)\right]8^{-1} + \left[1(2^2) + 0(2^1) + 0(2^0)\right]8^{-2}$$

Here, we have put this in an "octal form". Thus, we can write

$$0.1011_2 = 0.54_8$$

Again, we break up the binary digits into groups of three adding zeros at the right if necessary and express each group in decimal form. Each group then represents the corresponding octal digit. Note that *there is no roundoff error* when numbers are converted from binary to octal form. Let us consider another example. Let us convert

$$1101101.10101_2$$

to octal. First add zeros so that there are six digits to the right of the binary point. Then, proceeding as before, we obtain

$$1\ \ 101\ \ 101.\ \ 101\ \ 010_2 = 155.52_8$$

The converse procedure can be used to convert from octal to binary . That is, each octal digit is expressed as a three digit binary number. These are all combined in sequence to obtain the binary number. For instance,

$$155.52 = 001\ \ 101\ \ 101.\ \ 101\ \ 010_2$$

Note that the addition of trailing zeros to a binary number can give a false sense of precision. For instance, consider the number

0.1011_2

To convert it to octal, we write it as

0.101100_2

When a decimal number is written in this way, it implies that the last two digits are *known* (e.g., if we write 0.1011, this could represent 0.101100 or 0.101101, etc.) Thus,

0.54_8

implies six digits of precision in the binary number whereas there are only four digits of precision. Thus, a false precision has been obtained.

The conversion of binary to hexadecimal is essentially the same as that for binary to octal, except that groups of four binary digits are used. For instance

$$1101101.10101_2 = 0110\ \ 1101.\ \ 1010\ \ 1000_2$$

$$= 6D.A8_{16}$$

Of course, the converse procedure is used to convert the hexadecimal number to binary. Note that no roundoff error occurs when there is a conversion from binary to hexadecimal and vice versa.

After gaining some experience, the conversions from binary to octal or hexadecimal and vice versa can be done by inspection. (The conversion is always simple whenever the radix is a power of 2.) This simplicity of conversion is a reason for the importance of the octal and hexadecimal systems for humans working with computers. As we have discussed, numbers are usually worked with and stored in binary form. Often, *people* operating computers must inspect the binary numbers to study the computer's operation. However, numbers in radix 2 generally have many more digits than their octal or hexadecimal equivalents and people find them tedious to work with. For the above discussed reasons, octal or hexadecimal representations are often used by *people* working with computers. Note that conversions could be made from the binary to the decimal system. However, this is more complex and would be subject to roundoff error.

2.2 A FUNDAMENTAL DISCUSSION OF SOME BASIC ARITHMETIC OPERATIONS

In this section we shall consider some fundamental aspects of addition using **binary numbers**. In Chapter Nine we shall consider arithmetic operations in greater detail. In general, the rules of arithmetic apply independent of the radix. We shall specifically consider how they are applied to binary arithmetic. In addition, we will also discuss some aspects of arithmetic resulting from constraints imposed by digital computer circuitry.

Let us start by considering the addition of two numbers. We shall use whole numbers here. However, the discussion also applies to numbers with fractional parts. We begin with examples in base ten since these are familiar to use. Consider the addition

$$\begin{array}{r} 12734 \\ +\ \underline{26251} \\ 38985 \end{array}$$

Each column is added and this produces the result. In the chosen example, the sum of each column is less then the radix (10). If this is not true, then we must "carry a number" into the next column to the left. For instance, consider the addition

$$\begin{array}{r} 14978 \\ +\ \underline{36415} \\ 51393 \end{array}$$

The first column's addition is 8 + 5 = 13. The three is written down as the answer while the remaining 10 is carried over to the next column by adding 1 to that column. This procedure is repeated, where necessary, in all successive columns. (Note that 1000 is carried into the fourth column, etc.)

The same procedure is used if the radix is different from ten. For instance, let us consider octal addition

$$\begin{array}{r} 1247_8 \\ +\ 2510_8 \\ \hline 3757_8 \end{array}$$

The reader can verify that this arithmetic will "check" if the octal numbers are converted to decimal numbers. Now let us consider octal addition with carrying (Table 2-1 will be helpful here.)

$$\begin{array}{r} 111 \\ 1365_8 \\ +\ 4627_8 \\ \hline 6214_8 \end{array}$$

Note the first column: $5_8 + 7_8 = 14_8$. The 4 is written in the answer and the $10_8(8_{10})$ is carried to the next column. Note that octal addition may seem awkward. However, this is only because we are not as familiar with the "octal addition table" as we are with the base 10 addition table.

Let us discuss binary addition. It follows the previous ideas. As before, we shall start with an example which does not require carrying.

$$\begin{array}{r} 10101_2 \\ +\ 01000_2 \\ \hline 11101_2 \end{array}$$

Consider the following example which does require carrying.

$$\begin{array}{r} 111 \\ 10101_2 \\ +\ 00111_2 \\ \hline 11100_2 \end{array} \tag{2-25}$$

Inspect the first column. Here we have $1_2 + 1_2 = 10_2$. Again, we write the 0 in the answer and carry the $10_2(2_{10})$ to the next column. Thus, addition in the binary system is very similar to that in the decimal system.

Now let us consider addition of numbers with fractional parts. Again, this addition follows the rules for "ordinary" addition.

For example

$$\begin{array}{r} 10110.1101_2 \\ +\ 00101.1101_2 \\ \hline 11100.1010_2 \end{array} \qquad (2\text{-}26)$$

Thus, binary addition can be readily accomplished.

A problem occurs when numbers are added in a computer which does not occur when we add numbers using a pencil and paper. Consider the following addition

$$\begin{array}{r} 10110_2 \\ +\ 11101_2 \\ \hline 110011_2 \end{array}$$

The two numbers added each have five digits. Each binary digit is called a *bit*. Thus, each number has five bits. However, a carry takes place in the leftmost addition. Thus, the resulting answer has six bits. When humans add, this extra digit presents no problem. However, in computers, numbers are either stored in devices called *registers*, or are added and stored in devices called *accumulators*. The storage portion of an accumulator can be called its register. A register or accumulator can only store a fixed number of bits. If the binary number has more bits than can be stored, errors often result. To understand these, we shall discuss some fundamental ideas of storage here.

A register which can store a binary number of 8 bits is represented in Fig. 2-1. (We shall consider the electronic circuits that perform the storage in Chapters 5 and 6.) Note that the right hand digit is called the *least significant digit* since it has the smallest numerical value. Similarly, the leftmost digit is called the *most significant figure* since it has the largest numerical value. For the time being, we shall restrict ourselves to integer numbers. In Chapter 9 we shall generalize these results.

In a register, the numbers stored are the least significant digits. For instance, if we use the register of Fig. 2-1, to store the addition

$$\begin{array}{r} 11010110_2 \\ \underline{11011001_2} \\ 110101111_2 \end{array}$$

then the number stored in the register would be

$$10101111_2$$

which is equivalent to 175_{10} rather than 431_{10}. This is a substantial error and is called an *overflow*. Thus, the computer designer and user, to avoid error, must have a thorough understanding of number storage. In Chapter 9 we shall discuss how these disadvantages can, at times, be used to advantage.

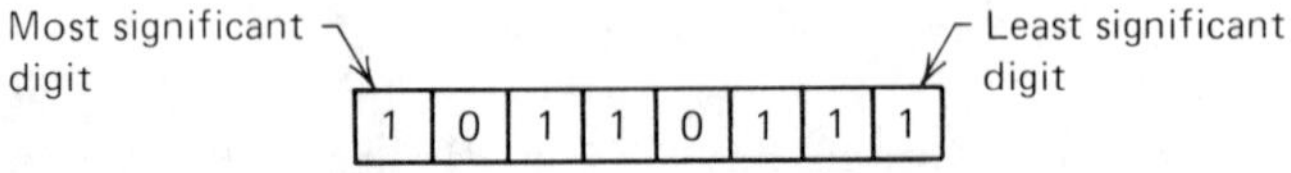

Figure 2-1. A diagrammatic representation of a register storing 10110111_2.

BIBLIOGRAPHY

Booth, T.L., "Digital Networks and Computer Systems", Chap. 2, John Wiley and Sons, Inc., New York, 1971

Hill, F.J., and Peterson, G.R., "Introduction to Switching Theory and Logical Design", Second Ed., Chap. 2, John Wiley and Sons, Inc., New York, 1974

Mano, M.M., "Computer Logic Design", Chap. 1, Prentice Hall, Inc., Englewood Cliffs, N.J. 1972

O'Malley, J., "Introduction to the Digital Computer", Chap. 1, Holt, Rinehart and Winston, Inc., New York, 1970

PROBLEMS

2-1. Obtain the values of the numbers from 0_{10} to 36_{10} in base 5.

2-2. Convert the following numbers from base 2 to base 10.

101, 101000, 11010110, 1110011101

2-3. Convert the following numbers from base 8 to base 10.

127, 4567, 7772, 1467

2-4. Convert the following numbers from base 16 to base 10.

37, 1000, 1A3, ABCD

2-5. Repeat Prob. 2-2 for

101.101, 10110.11010, 101.0110101

2-6. Repeat Prob. 2-3 for

127.1, 4567.0714, 1775.01245, 11.0121741

2-7. Repeat Prob. 2-4 for

37.A1, 1A3.101AB, BCA101.AC29B

2-8. Convert the following numbers from base 10 to base 2.

26, 98, 147, 1679, 24873

2-9. Repeat Prob. 2-8 but now write the numbers in octal.

2-10. Repeat Prob. 2-8 but now write the numbers in hexadecimal.

2-11. Convert the following numbers from base 10 to binary. Estimate any error.

26.123, 98.476, 147.001, 1246.8012, 24.16793

2-12. Repeat Prob. 2-11 but now write the numbers in octal.

2-13. Repeat Prob. 2-11 but now write the numbers in hexadecimal.

2-14. Convert the following binary numbers to octal.

101.10101, 101.0101, 101.0100101

2-15. Repeat Prob. 2-14 but now write the numbers in hexadecimal.

2-16. Convert the following octal numbers to binary

127, 1243.107, 721.7461

2-17. Convert the following hexadecimal numbers to binary.

AB1, ABC.01A, 1A7.96BAE

2-18. Perform the following addition. The numbers are in binary.

$$\begin{array}{r} 101101.0101 \\ \underline{101010.0100} \end{array}$$

2-19. Perform the addition A+B where

$$A = 7261.427_8$$

$$B = 674.3165_8$$

2-20. Repeat Prob. 2-19 for

$$A = CB2.3A_{16}$$

$$B = AB9.90A_{16}$$

2-21. Repeat Prob. 2-19 for

$$A = 1011.1101_2$$

$$B = AB.ACE_{16}$$

3

Introduction to Logical Operations

In a digital computer, we deal with binary numbers where each binary digit can have one (and only one) of two values, 0 or 1. In mathematical logic, we consider propositions that are either *true* or *false*. Thus, we can define logical variables that have either the value true or the value false. Since mathematical logic and digital computers are both two variable systems, much of the mathematics that has been developed over the years for logical variables can be applied to the analysis and design of digital systems. In this chapter we shall introduce some fundamental ideas of logical operations as applied to digital computers, and give some examples which illustrate these applications. In the next chapter, we shall formalize these ideas and develop design procedures.

3.1 BASIC LOGICAL OPERATIONS

To introduce logical operations, let us consider a simple binary addition of two single digit numbers A and B. Suppose that we write this in the following way:

$$\begin{array}{r} a_1 \\ \underline{b_1} \\ c_1 s_1 \end{array} \tag{3-1}$$

where a_1,b_1,c_1 and s_1 all represent binary digits(i.e., either 0 or 1). For instance, if A=1, $a_1=1$ and if B=1, then $b_1=1$. In this case, we have

$$\begin{array}{r} 1 \\ \underline{+\ 1} \\ 10 \end{array}$$

In this case, $s_1=0$ and $c_1=1$. We can now write the following statements. If a_1 and b_1 are both 1, then $s_1=0$ and $c_1=1$. If a_1 and b_1 are both 0, then $s_1=0$ and $c_1=0$. If either a_1 or b_1, but not both, are 1, then $s_1=1$ and $c_1=0$. These statements have their analogs in mathematical logic. In that logic there are also two variables, called *true* and *false*. Thus, the above statements could be written in terms of logic as: if a_1 and b_1 are both true, s_1 is false and c_1 is true. If a_1 and b_1 are both false, then s_1 and c_1 are both false. If either a_1 or b_1, but not both, are true, then s_1 is true and c_1 is false.

We have illustrated a simple addition as a logical operation. Actually, we can represent almost all computer operations by logical statements, although some may be considerably more complex than the simple one we have discussed.

Let us now formalize the previous discussion. In addition, we shall introduce the concept of circuits which electrically perform the logical operations. That is, we shall consider operations or circuits which have a number of inputs which can be either 0 or 1(true or false). These inputs produce an output in accordance with some logical rules. For instance, we previously discussed a set of logical rules which resulted in simple binary addition. In this section we shall introduce *combinational logic* and the corresponding circuit which are called *combinational circuits*. In these, the present values, or logical inputs, zeros and ones, produce the output. There is no memory associated with these circuits. In Chapter 6 we shall consider *sequential logic* and *sequential* circuits. Here memory is involved. That is, the output is not only a function of the present value of the inputs, but also of the past values of the inputs.

Let us start by defining a *binary variable*. It is simply a variable which can take on the values 0 or 1. For instance, in (3-1) a_1, b_1, c_1 and s_1 are binary variables. Now consider a switch which is closed if a particular binary variable is a 1 and open if that binary variable is a 0. Such a switch is illustrated in Fig. 3-1.Such switches can be implemented electronically using semiconductor devices. In their earliest forms they were *relays*, that is, switches which were opened and closed magnetically(by the current in a solenoid)

Figure 3-1. A switch which is open if A = 0 and closed if A = 1.

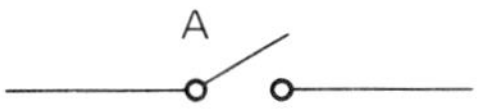

We shall now work with the circuit of Fig. 3-2. Here we assume that the output V_o is a 1 if V_o=V and is a 0 if V_o=0. Note that in an actual circuit, the designer can specify the voltage levels that correspond to a 0 or a 1. They do not have to be zero volts and one volt. In Fig. 3-2 V_o will be a 1 if A or B are 1, and, in addition, if C is also 1. For instance, the following combinations result in V_o's being a 1.

$$A = 1, B = 0, C = 1 \qquad (3\text{-}2a)$$

$$A = 0, B = 1, C = 1 \qquad (3\text{-}2b)$$

$$A = 1, B = 1, C = 1 \qquad (3\text{-}2c)$$

Also, the following combinations of variables result in V_o's being a zero.

$$A = 0,\ B = 0,\ C = 0 \tag{3-3a}$$

$$A = 0,\ B = 0,\ C = 1 \tag{3-3b}$$

$$A = 1,\ B = 0,\ C = 0 \tag{3-3c}$$

$$A = 0,\ B = 1,\ C = 0 \tag{3-3d}$$

$$A = 1,\ B = 1,\ C = 0 \tag{3-3e}$$

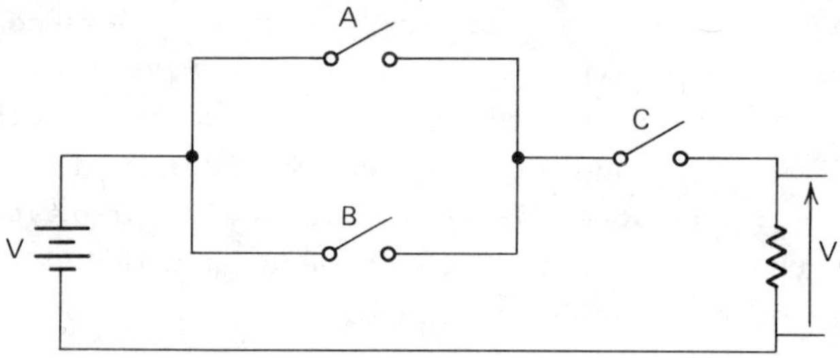

Figure 3-2. A circuit containing three switches.

Figure 3-2 is a combinational logic circuit and Eqs.(3-2) and (3-3) characterize it. Complex logic circuits are characterized by many such equations. Thus, it is desirable to write them in a compact form. This is done by constructing a table which is called a *truth table*. Here each line represents one combination of variables. For instance, the truth table for Fig. 3-2 is

TRUTH TABLE FOR FIGURE 3.2

A	B	C	V_o
0	0	0	0
0	0	1	0
0	1	0	0
0	1	1	1
1	0	0	0
1	0	1	1
1	1	0	0
1	1	1	1

Note that we list all the independent binary variables on one side of a vertical line and the dependent binary variable(s) on the other. In this truth table, the first line represents Eq. (3-3a). To do things in an orderly fashion, the independent binary variables are written in order of increasing binary value. That is, if the right hand side of the truth table is considered to be a binary number, then each row is one more(in value)than the preceding one. In this way the output for *every possible set of inputs is given.* Let us consider an additional example.

We shall write the truth table for the logic circuit of Fig. 3-3.

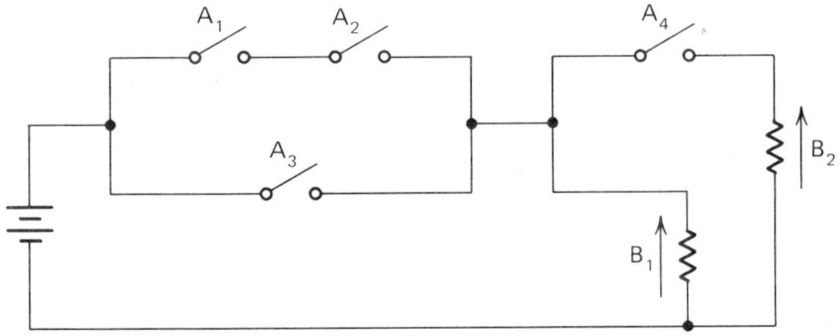

Figure 3-3. A more complex switching circuit.

TRUTH TABLE FOR FIGURE 3.3

A_1	A_2	A_3	A_4	B_1	B_2
0	0	0	0	0	0
0	0	0	1	0	0
0	0	1	0	1	0
0	0	1	1	1	1
0	1	0	0	0	0
0	1	0	1	0	0
0	1	1	0	1	0
0	1	1	1	1	1
1	0	0	0	0	0
1	0	0	1	0	0
1	0	1	0	1	0
1	0	1	1	1	1
1	1	0	0	1	0
1	1	0	1	1	1
1	1	1	0	1	0
1	1	1	1	1	1

We have discussed some elementary ideas of logical operations. In the next section these ideas shall be formalized.

3.2 BASIC COMBINATIONAL LOGIC OPERATIONS

All of the operations of combinational logic circuits can be expressed in terms of some fundamental logic operations. These are used to characterize and study logic circuits. Let us consider these fundamental operations.

The AND Operation

If we have two logical variables A_1 and A_2, the AND operation gives a value of 1, if , and only if, A_1 and A_2 are both 1. It gives a value of 0 otherwise. This is indicated by the switch circuit of Fig. 3-4. Note that B will be 1 if A_1 *and* A_2 are 1. It will be 0 otherwise. We indicate the AND operation by a dot (·) between the variables. This is the same as an ordinary multiplication sign. Thus, for Fig. 3-4, we have

$$B = A_1 . A_2 \tag{3-4a}$$

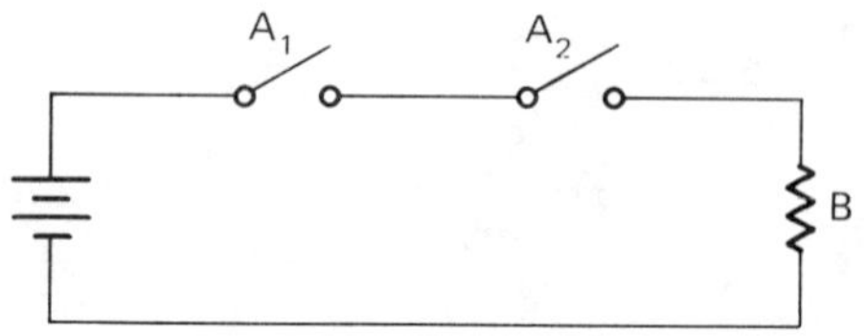

Figure 3-4. A simple AND switch circuit.

The truth table for the AND operation is

TRUTH TABLE FOR EQUATION 3-4a

A_1	A_2	B
0	0	0
0	1	0
1	0	0
1	1	1

At times, as in the case of ordinary multiplication, the dot is omitted. Thus, we could write

$$B = A_1A_2 \qquad (3\text{-}4b)$$

An alternate notation that is used for the AND operation is

$$B = A_1 \wedge A_2 \qquad (3\text{-}4c)$$

In general, we shall use the notation of Eq. (3-4a) or that of (3-4b).

We have illustrated the AND function with two independent variables. However, there can be more. For instance, in Fig. 3-4, if we had three switches in series, the relation would be

$$B = A_1 \cdot A_2 \cdot A_3 \qquad (3\text{-}5)$$

In this case, B would be 1 if, and only if, A_1 *and* A_2 *and* A_3 were *all* 1's.

In modern digital computers, semiconductor circuits perform logical operations such as the AND operation. Such circuits are called *gates*. If the AND operation is performed by a gate, then it is called an AND *gate*. Many gates can be incorporated on a single silicon chip. A computer may have the interconnection of thousands of gates. To simplify the wiring diagrams, special symbols are used to represent the various gates. The AND gate symbol is illustrated in Fig. 3-5. Two input and three input AND gates are shown in Figs. 3-5a and 3-5b, respectively. Note that many different symbols for gates are in general use. Representative ones will be used in this text.

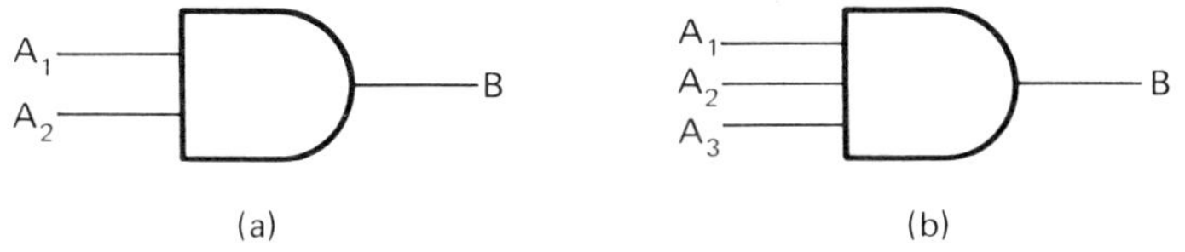

Figure 3-5. Symbols for AND gates. (a) Two input; (b) three input.

The "OR" Operation

A switch circuit which illustrates this operation is shown in Fig. 3-6. Note that if *either or both* of the independent variables A_1 and A_2 are 1, then B=1. If both A_1 and A_2 are 0, then B=0. The OR operation is indicated by a + sign. Thus, the logical equation for the switch circuit of Fig. 3-6 is

$$B = A_1 + A_2 \qquad (3\text{-}6a)$$

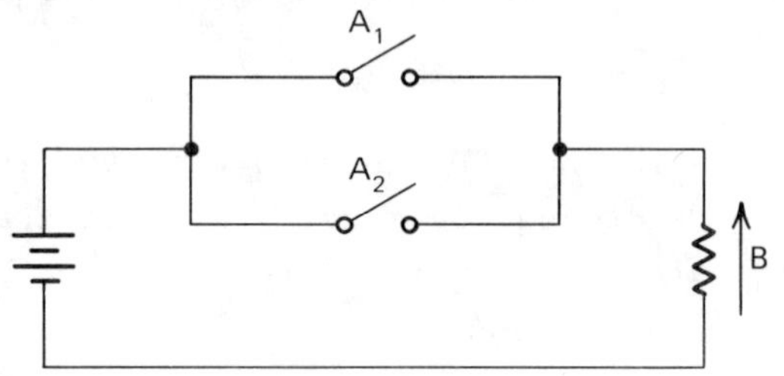

Figure 3-6. A simple OR switch circuit.

An alternate symbolic notation for the OR operation is

$$B = A_1 \vee A_2 \qquad (3\text{-}6b)$$

The truth table for the OR operation is

A_1	A_2	B
0	0	0
0	1	1
1	0	1
1	1	1

Symbols for the OR gate are shown in Fig. 3-7. The logical equation that represents Fig. 3-7b is

$$B = A_1 + A_2 + A_3 + A_4 \qquad (3\text{-}7)$$

In this case, B is 1 if any or all of A_1 or A_2 or A_3 or A_4 are 1.

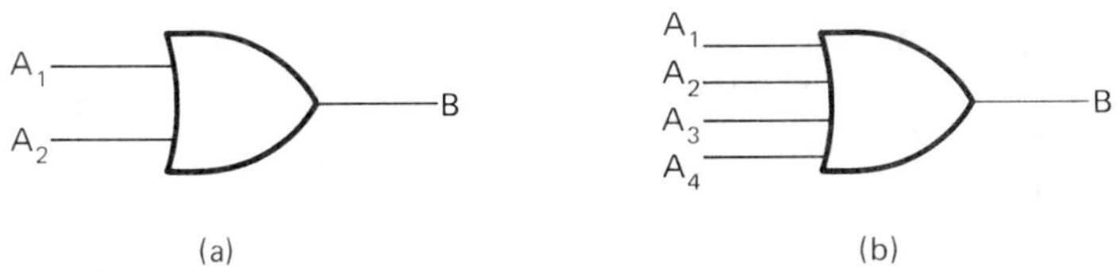

Figure 3-7. Symbols for OR gates. (a) Two input; (b) four input.

We can combine gates to form more complicated logical operations. For instance, consider the circuit of Fig. 3-2. We can represent it by the logical expression

$$V_o = (A+B)\cdot C \tag{3-8}$$

That is, V_o is 1 if A or B are 1, and, in addition, if C is 1 (i.e., if (A+B) and C are both 1). We can write the truth table for this as

A	B	C	A+B	(A+B)·C
0	0	0	0	0
0	0	1	0	0
0	1	0	1	0
0	1	1	1	1
1	0	0	1	0
1	0	1	1	1
1	1	0	1	0
1	1	1	1	1

Thus, by breaking the expression into parts, we can obtain the truth table for complex logical expressions. A representation of Fig. 3-2 using gate symbols is shown in Fig. 3-8.

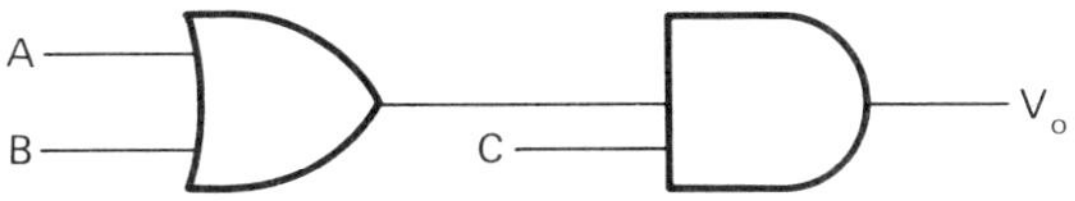

Figure 3-8. An interconnection of gates which results in the logical expression V = (A + B) · C.

Complement – NOT or Inversion

Before we consider the next gate, let us consider an operation of mathematical logic. The complement of a variable A is defined in the following way. If A is 1, then its complement is 0, and if A is 0, then its complement is 1. We shall write the complement of a variable by placing a bar over the variable. Thus, the complement of A is $\bar{A}$. (At times, a prime, i.e., A' is used to represent the complement.) Similarly, the complement of (B+C).A is written as $\overline{(B+C).A}$. The truth table for the complement is given by

A	$\bar{A}$
0	1
1	0

A gate which performs the mathematical operation of taking the complement is called a NOT gate or an inversion gate or an inverter. Its symbol is shown in Fig. 3-9.

Figure 3-9. The symbol for a NOT gate.

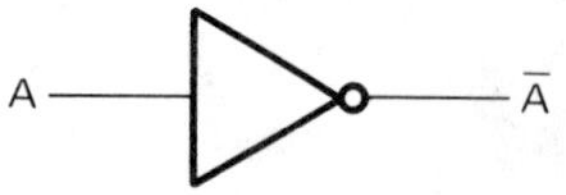

Let us consider some useful symbolism. The small circle in Fig. 3-9 can also be used to represent a complement operation. For instance, in Fig. 3-10, we represent

$$B = A_1 + \overline{A_2}$$

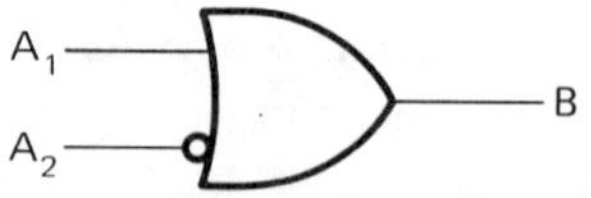

Figure 3-10. A representation of the logic equation $B = A_1 + \overline{A_2}$.

The operation of taking the complement is called a *uniary operation* since only one variable is involved. The operations of AND and OR are called binary operations since two (or more) variables are involved.

The NOR Operation

This consists of taking the complement of the OR operation. That is, the NOR operation is defined by the logical expression

$$B = \overline{A_1 + A_2} \tag{3-10}$$

An alternative name which has been used for the NOR operation is the *Pierce arrow* and an alternative notation is

$$B = A_1 \downarrow A_2 = \overline{A_1 + A_2} \tag{3-11}$$

The truth table for the NOR is

A_1	A_2	B
0	0	1
0	1	0
1	0	0
1	1	0

The symbolic representation for a NOR gate is shown in Fig. 3-11. Note that it consists of the symbol for an OR gate with the "inverting circle" added.

Figure 3-11. A symbol for the NOR gate.

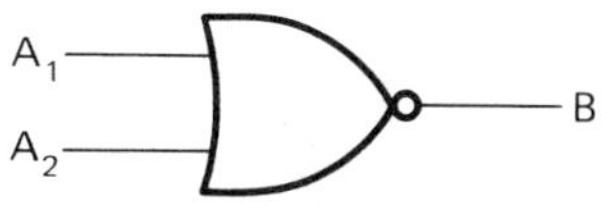

The NAND Operation

We just discussed an operation that was the complement of the OR operation. Now we shall consider one which is the complement of the AND operation. It is called the NAND operation and is defined by

$$B = \overline{A_1 . A_2} \tag{3-12}$$

The NAND operation has also been called the *Sheffer stroke*. An alternate symbolism is

$$B = A_1 \uparrow A_2 = \overline{A_1 . A_2} \tag{3-13}$$

The truth table for the NAND operation is

A_1	A_2	$B = \overline{A_1 . A_2}$
0	0	1
0	1	1
1	0	1
1	1	0

The symbolic representation for the NAND gate is shown in Fig. 3-12. We have shown three inputs here. The logical expression for this is

$$B = \overline{A_1 . A_2 . A_3} \tag{3-14}$$

Figure 3-12. The symbol for a three input NAND gate.

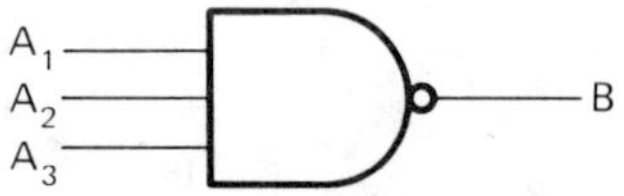

Note that the symbol for the NAND gate is just that for the AND gate followed by the "inverting circle". Naturally, the output of the NAND gate is just the complement of the output of an AND gate.

The Exclusive OR(EXOR) Operation

The EXOR is similar to the OR operation except that the output is a 0 if the inputs are both 1 or both 0. That is, if the inputs are A_1 and A_2 and B is the output, then B is 1 if A_1 or A_2 is 1. However, B is 0 if *both* A_1 and A_2 are 0 or *both* A_1 and A_2 are 1. The symbolic representation for the exclusive OR operation is

$$B = A_1 \oplus A_2$$

The truth table for the exclusive OR(EXOR)operation is

A_1	A_2	B
0	0	0
0	1	1
1	0	1
1	1	0

The symbolic representation for an EXOR gate is shown in Fig. 3-13.

Figure 3-13. An exclusive OR gate (EXOR).

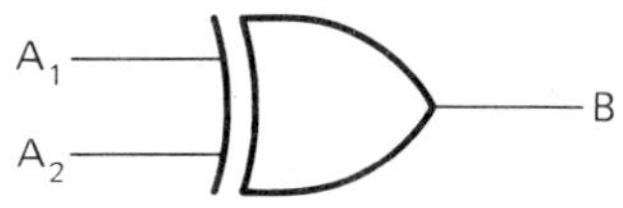

Let us discuss EXOR operations involving more than two variables. This is somewhat more involved than multiple variable AND and OR operations. Let us consider

$$B = A_1 \oplus A_2 \oplus A_3 \tag{3-16a}$$

We can write this as

$$B = (A_1 \oplus A_2) \oplus A_3 \tag{3-16b}$$

That is, we apply the EXOR operations to A_1 and A_2 and then apply the EXOR operation again to the result of the previous operation and A_3. Let us illustrate this with a truth table.

A_1	A_2	A_3	$A_1 \oplus A_2$	$(A_1 \oplus A_2) \oplus A_3$
0	0	0	0	0
0	0	1	0	1
0	1	0	1	1
0	1	1	1	0
1	0	0	1	1
1	0	1	1	0
1	1	0	0	0
1	1	1	0	1

(3-17)

Note that if $A_1=1$, $A_2=1$, and $A_3=1$, then $A_1 \oplus A_2 \oplus A_3 = 1$, and not 0.

We need not place the parentheses as we did in Eq. (3-16b). For instance, all the following are equivalent.

$$A_1 \oplus A_2 \oplus A_3 = (A_1 \oplus A_2 \oplus A_3 = A_1 \oplus (A_2 \oplus A_3) = (A_1 \oplus A_3) \oplus A_2) \tag{3-18}$$

This can be demonstrated by writing the truth table for each of these. The details of this will be left to the reader.

3.3 REPRESENTATION OF GENERAL LOGICAL EXPRESSIONS

In general, we can represent any combinational logical expression using the operations discussed in the last section. Alternatively, we can say that any combinational logic circuit can be built from an interconnection of the gates presented there. In general, we can represent the behavior of a complex interconnection of gates by relations such as Eq. (3-8). Such logical expressions are also called *switching functions*. If a set of operations is such that it can be used to express any switching function, then that set is said to be *functionally complete*.

Actually we do not need all the operations presented here to obtain a functionally complete set. In fact, we can show that many of the operations can be expressed in terms of the others. For instance, suppose we have a set consisting of AND, OR and NOT. We can obtain NAND, NOR and EXOR from these. The first two represent the complement of the AND and OR operations, respectively. Thus, their operations can be obtained by cascading a NOT gate with an AND or OR gate. This is shown in Figs. 3-14a and b. We can demonstrate the equivalence by comparing truth tables. For instance, consider Fig. 3-14b, which is equivalent to the NOR circuit. Let us write its truth table

A_1	A_2	$A_1 + A_2$	$\overline{A_1 + A_2}$
0	0	0	1
0	1	1	0
1	0	1	0
1	1	1	0

If this is compared with the truth table for the NOR gate, it can be seen that the results are the same. Of course, we could have demonstrated this in a more compact way by comparing the output $\overline{A_1+A_2}$ with Eq. (3-10).

Now let us demonstrate that we can replace an exclusive OR gate by an interconnection of AND, OR and NOT gates. Consider the interconnection of gates shown in Fig. 3-14(c). This is equivalent to the EXOR. This can be demonstrated by constructing the truth table for $(A_1+A_2).\overline{A_1A_2}$ and showing that it is

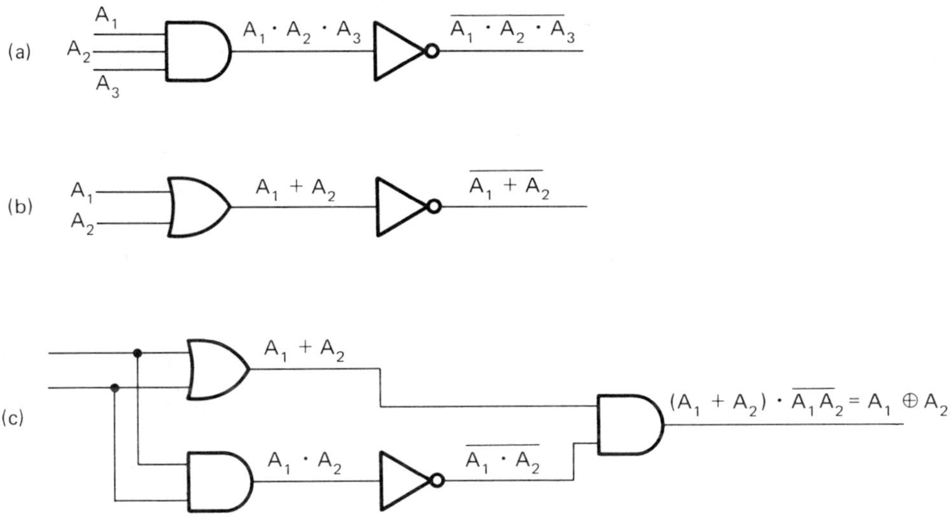

Figure 3-14. Interconnection of AND, OR and NOT gates to obtain other logical operations: (a) NAND; (b) NOR; and (c) EXOR.

equivalent to that for the EXOR. This will be left to the reader. Alternatively, we can use the following reasoning. $(A_1+A_2) \cdot \overline{A_1 A_2}$ represents A_1+A_2 AND $\overline{A_1 \cdot A_2}$. Now A_1+A_2 will be 1 when A_1 or A_2 are 1. In addition, $\overline{A_1 \cdot A_2}$ will be 1 except when A_1 and A_2 are both 1. Thus, $(A_1+A_2) \cdot \overline{A_1 \cdot A_2}$ will be 1 only when A_1 or A_2, but not both, are 1. Thus, we have achieved the exclusive OR. Thus, we have shown that the AND,OR and NOT operations are functionally complete, at least as far as NAND, NOR and EXOR are concerned.

There are other functionally complete sets of operations. For instance, the NOR operation is functionally complete by itself. Let us demonstrate this. We shall do this by showing that we can construct the NOT, OR and AND gates just from interconnections of NOR gates. Since we have just shown that the NOT, OR and AND operations are functionally complete, this will demonstrate that the NOR operation is itself functionally complete.

Consider Fig. 3-15a. This is a NOR gate with both inputs connected together. The truth table for this circuit can be obtained from the previously given truth table for the NOR operation for those cases where $A_1 = A_2$. Using this we obtain the truth table

A	B
0	1
1	0

This is the same as the truth table for the NOT operation.

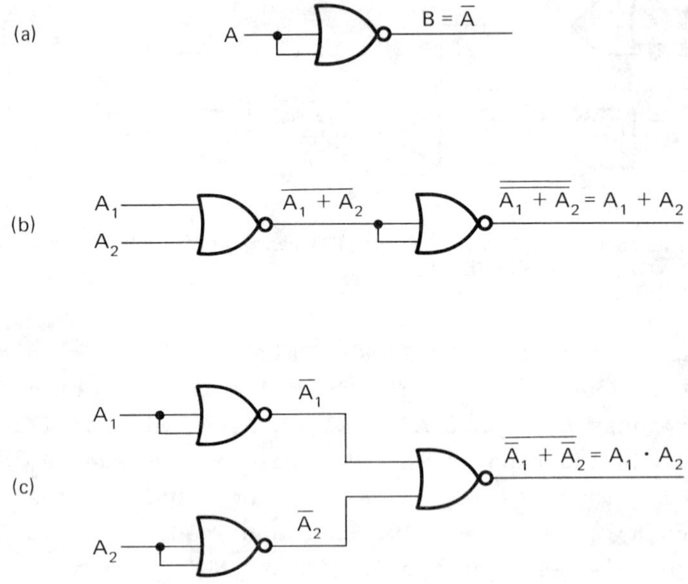

Figure 3-15. Interconnections of NOR gates to achieve other logical operations. (a) NOT; (b) OR; (c) AND.

Now consider Fig. 3-15b. The output of the first NOR gate is $\overline{A_1 + A_2}$ (by definition). We have just shown that the second gate merely takes the complement. The complement of a complement is just the variable itself. That is

$$\bar{\bar{A}} = A \tag{3-19}$$

This can be verified by considering the truth table. Thus, the output of the second gate is $\overline{\overline{A_1 + A_2}} = A_1 + A_2$. Hence, Fig. 3-15b represents an OR gate.

Finally, consider Fig. 3-15c. The output is $\overline{\bar{A}_1+\bar{A}_2}$. This is equivalent to the AND operation. We shall demonstrate this by constructing the truth table

A_1	A_2	$\bar{A}_1$	$\bar{A}_2$	$\bar{A}_1+\bar{A}_2$	$\overline{\bar{A}_1+\bar{A}_2}$	$A_1 \cdot A_2$
0	0	1	1	1	0	0
0	1	1	0	1	0	0
1	0	0	1	1	0	0
1	1	0	0	0	1	1

The last two columns are identical. This demonstrates that

$$A_1 \cdot A_2 = \overline{\bar{A}_1 + \bar{A}_2} \tag{3-20}$$

Therefore, Fig. 3-15c represents an AND gate. Hence, we have demonstrated that the NOR operation is functionally complete.

In a similar way, we can show that the NAND operation is functionally complete. In Fig. 3-16 we illustrate the NOT, AND and OR operations realized using NAND gates. In Fig. 3-16a we obtain a NOT gate. Note that $\overline{A \cdot A}=\bar{A}$. This can be verified by writing the truth table.

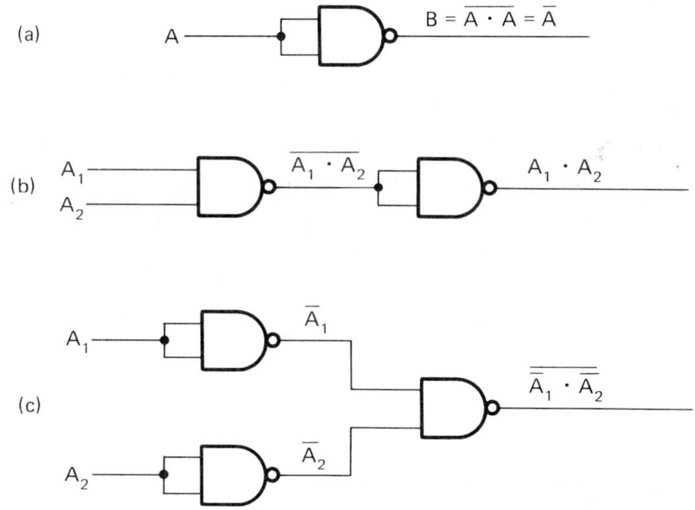

Figure 3-16. Interconnections of NAND gates to achieve other logical operations. (a) NOT; (b) AND; (c) OR.

In Fig. 3-16b we obtain the AND operation. Note that the output of the NAND gate is the complement of the AND operation. Thus, following a NAND gate by a NOT gate produces the desired result.

The output of Fig.3-16c is $\overline{\bar{A}_1 \cdot \bar{A}_2}$. This is equivalent to the OR operation. The following truth table verifies this.

A_1	A_2	$\bar{A}_1$	$\bar{A}_2$	$\bar{A}_1 . \bar{A}_2$	$\overline{\bar{A}_1 . \bar{A}_2}$	A_1+A_2
0	0	1	1	1	0	0
0	1	1	0	0	1	1
1	0	0	1	0	1	1
1	1	0	0	0	1	1

Thus, Fig. 3-16c is an OR gate. Hence, we have demonstrated that the NAND operation is functionally complete.

We have demonstrated functional completeness by realizing NOT, AND and OR operations. Actually, we need only consider certain pairs of these. For instance, NOT and AND are functionally complete. This is true since NAND is functionally complete and NAND can be realized by cascading AND and NOT gates. Similarly, using the NOR operations we can demonstrate that NOT and OR are functionally complete.

When computers are built, many complex logical expressions must be implemented as gate circuits. It is necessary to work with a functionally complete set of gates so that the designer can implement any required logical operation no matter how complex, using standard gates. Each of these functionally complete sets discussed(i.e., AND and NOT;or OR and NOT; or NAND; or NOR) can be used for this purpose. Actually NAND and NOR gates can be simply constructed using semiconductors so they are often used for this purpose. These circuits will be illustrated in Chapter 5.

3.4 SOME EXAMPLES OF LOGIC CIRCUITS

Let us now see how logic circuits can be interconnected to result in some practical operations. First we shall consider an adder circuit which is a circuit whose input is two binary numbers and whose output is a binary number which is the sum of these input numbers. Again, this example will be very elementary. We shall consider actual adder circuit in much greater detail in Chapter 10.

We shall start by discussing a circuit called a *half adder* which adds two one bit binary numbers, a_1 and b_1. Such addition was illustrated in (3-1). The truth table for the half adder is

a_1	b_1	c_1	s_1
0	0	0	0
0	1	0	1
1	0	0	1
1	1	1	0

(3-21)

Note that we have

$$\begin{array}{r} a_1 \\ \underline{b_1} \\ c_1 s_1 \end{array}$$

It is conventional to call s_1 the sum digit and c_1 the carry digit. From the truth table we see that the logical expressions for c_1 and s_1 are

$$s_1 = a_1 \oplus b_1 \tag{3-22a}$$

$$c_1 = a_1 \cdot b_1 \tag{3-22b}$$

We will use AND, OR and NOT gates to construct the half adder. (Note (see Sec. 3-3) that we could implement this using NOR or NAND gates, etc. This is often done in actual computers.) Figure 3-13 can be used for the EXOR gate for s_1 while a single AND gate can be used for c_1. The resulting half adder is shown in Fig. 3-17a. Note that there are two AND circuits that both have the same inputs a_1 and b_1. Thus, their outputs are the same and only one AND circuit is needed. This is illustrated in Fig. 3-17b, where there is one less gate. Here, the leftmost AND circuit not only supplies c_1 but it also supplies the input to the inverter. Thus, Fig. 3-17b represents a simpler or reduced form of Fig. 3-17a.

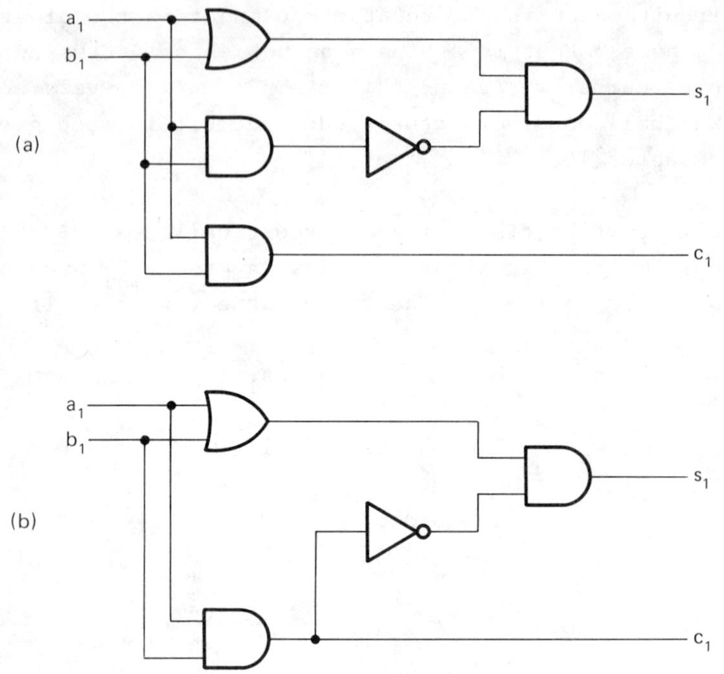

Figure 3-17. (a) A half adder; (b) an implementation that uses one less gate.

This is an example of a multilevel logic circuit. This means that a signal must pass through several gates. (This is true for s_1, but not for c_1.) There is a disadvantage to this. Practical gates take a finite amount of time to respond. For instance, if a signal must pass through two gates successively, the second gate cannot respond until the first one has produced its output. For example, if it takes 10^{-8} seconds for each gate to respond and the signal must pass through two gates, then the response time is $2x10^{-8}$ seconds. In a computer there is an extremely large number of such operations performed when even a simple computer program is executed. Thus, the seemingly small response times of each component can result in minutes, or hours of actual computer time when a lengthy program is run.

When logic circuits are designed, they are often *minimized*. That is, the number of components is reduced. An example of an extremely simple minimization has been illustrated in Fig. 3-17. Minimization is important not only because it reduces the number of components(this can be substantial since there are very many components in a computer)but also because it increases reliability. Reduction in the number of levels will also reduce response time.

The half adder we have studied can be used to add two binary digits. If the number to be added has more than one bit, it may seem as though we need only use a half adder for each digit. This would be true if there were no carrying. However, carrying does occur and the half adder has no provision for the carry bit.

A more complex circuit called the *full adder* resolves this problem. Remember that the adder circuit only works with one bit. Each full adder must work with single bits of A and B and also with a carry bit from the preceding digit. Note that the half adder produces the sum plus a carry bit. The full adder must do this also. Let us consider a simple addition

$$\begin{array}{r} 1\,1\,1\,1 \\ 01011 \\ +\ \underline{00111} \\ 10010 \end{array}$$

Thus, each addition after the first digit must accept a carry digit. The inputs to the adder which adds the "j^{th}" bit will be a_j, b_j, and c_{j-1} where a_j and b_j represent the jth digit of A and B, respectively and c_{j-1} represents the carry from the j-1 column addition. Thus, we have

$$\begin{array}{r} c_{j-1} \\ a_j \\ \underline{b_j} \\ c_j s_j \end{array} \qquad (3\text{-}23)$$

Note that in Eqs.(3-22) and in the preceding comments, we used the notation a_1, b_1, etc. Here we use the more generalized notation a_j, b_j, etc.

Let us write the desired truth table, which is obtained simply by performing the binary addition. For instance, if $c_{j-1}=1$,

$a_j=1$ and $b_j=1$, then the addition yields $c_j=1$, $s_j=1$. Proceeding similarly, we obtain the following truth table

a_j	b_j	c_{j-1}	s_j	c_j
0	0	0	0	0
0	0	1	1	0
0	1	0	1	0
0	1	1	0	1
1	0	0	1	0
1	0	1	0	1
1	1	0	0	1
1	1	1	1	1

(3-24)

In a complex case, it is desirable to obtain a logical expression or switching function(see Sec. 3-3)which characterizes each of the outputs. From this, the gate circuit can be designed. Let us discuss how this can be done.

The value of the logical expression should be a 1 corresponding to each time that a 1 appears in the "output column" of the truth table. Let us consider the first line where $s_j=1$. In this case, $a_j=0$, $b_j=0$ and $c_j=1$. When $a_j=0$, then $\bar{a}_j=1$. Thus, we can write this in compact form as $s_j=1$ if

$$\bar{a}_j.\bar{b}_j.c_{j-1} = 1$$

There are a number of possibilities that result in $s_j=1$. Since the occurrence of *any* one of them causes $s_j=1$, we must combine all these possibilities using the OR relation. Thus, for the truth table (3-24), we have

$$s_j = \bar{a}_j.\bar{b}_j.c_{j-1} + \bar{a}_j.b_j.\bar{c}_{j-1} + a_j.\bar{b}_j.\bar{c}_{j-1} + a_j.b_j.c_{j-1} \qquad (3\text{-}25)$$

Note that any other possible combinations of a_j, b_j and c_{j-1} results in a 0 for s_j. Thus, none of these terms appears in Eq. (3-25). Proceeding similarly, we can obtain the expression for c_j. It is

$$c_j = \bar{a}_j.b_j.c_{j-1} + a_j.\bar{b}_j.c_{j-1} + a_j.b_j.\bar{c}_{j-1} + a_j.b_j.c_{j-1} \qquad (3\text{-}26)$$

We can use these logical expressions to perform a simple switching circuit design. For instance, Eq. (3-25) indicates that we should have an OR gate with four inputs, $\bar{a}_j . \bar{b}_j . c_{j-1}$, $\bar{a}_j . b_j . \bar{c}_{j-1}$, $a_j . \bar{b}_j . \bar{c}_{j-1}$, and $a_j . b_j . c_{j-1}$. Each of these inputs can be obtained from a separate AND gate. The input to the AND circuits are the inputs, or their complements, which can be obtained from a NOT gate. This is shown in the upper portion of Fig. 3-18 where s_j is obtained. In a similar way, the lower portion of Fig. 3-18 implements Eq. (3-26) where c_j is obtained. Note that Eq. (3-25) and Eq. (3-26) have one term in common, $a_j . b_j . c_{j-1}$. Thus, one of the AND gates is used by both parts of the diagram. This represents a saving of one gate.

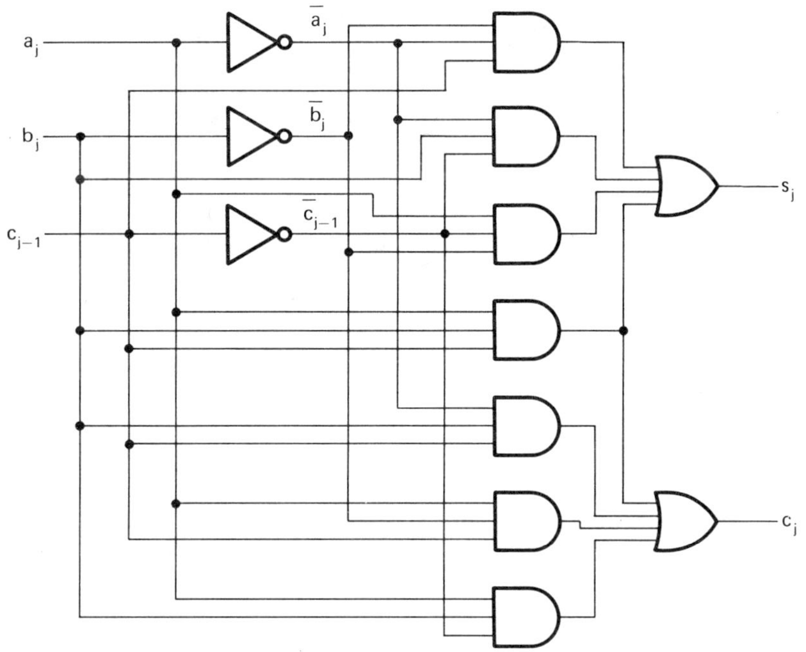

Figure 3-18. A full adder based upon Eqs. (3-25) and (3-26). This is an unreduced form.

Let us consider the procedure that we have used in general terms. We form the truth table. Then, for each of the output variables, we determine which entries produce 1's. Each set of inputs or their complements that produces a 1 represents a set of inputs to an AND gate. The outputs of all the AND gates are supplied to an OR gate. Thus, if N 1's appear in the output column, then N AND gates, one OR gate and the necessary NOT gates are used.

Although this method is simple, it often results in an excessive number of gates. Equations such as (3-25) and (3-26) can often be manipulated to minimize the number of gates required. The general techniques for minimization will be discussed in detail in the next chapter. In this section we shall just consider some examples that illustrate that these minimizations can take place.

We shall start by illustrating a minimization of the full adder that was just discussed. We can write s_j in terms of the exclusive or operation. This is

$$s_j = a_j \oplus b_j \oplus c_{j-1} \qquad (3\text{-}27)$$

This can be seen by comparing the truth table (3-17) for the EXOR operation applied to three variables and the truth table (3-24) for s_j.

Now let us consider the expression for c_j. We can write this as

$$c_j = a_j \cdot b_j + b_j \cdot c_{j-1} + c_{j-1} \cdot a_j \qquad (3\text{-}28)$$

This can be verified by constructing the truth table for Eq.(3-28) and comparing it with the truth table (3-24) applying to c_j. If this is done, it can be seen that they are the same. Thus, we can characterize the full adder by

$$s_j = a_j \oplus b_j \oplus c_{j-1} \qquad (3\text{-}29a)$$

$$c_j = a_j . b_j + b_j . c_{j-1} + c_{j-1} . a_j \qquad (3\text{-}29b)$$

Now let us consider the half adder. From Eqs.(3-22), we have

$$s = a \oplus b \qquad (3\text{-}30a)$$

$$c = a \cdot b \qquad (3\text{-}30b)$$

The half adder performs an EXOR and an AND operation. We can use two half adders and an OR gate to obtain a full adder. Let us demonstrate this. We shall use some simplified diagrams here. In Fig. 3-19a, we show a *block diagram* of a half adder. The inputs and outputs are shown. However, the internal connections of the gates are omitted. This is a "shorthand" type of drawing. Figure 3-17b could represent the internal wiring of the box.

In Fig. 3-19b, we have shown the interconnection of two half adders and an OR gate to produce a full adder. Since the s output of a half adder produces an EXOR, then s_j is of the form of Eq. (3-29a) which is correct. Let us consider c_j. The c output of the second half adder is $(a_j \oplus b_j).c_{j-1}$. Thus, the output of the OR gate is $(a_j \oplus b_j).c_{j-1} + a_j.b_j$. Let us show that this is equivalent to $a_j.b_j + b_j.c_{j-1} + c_{j-1}.a_j$. We shall use the truth table to do this.

a_j	b_j	c_{j-1}	$a_j \oplus b_j$	$(a_j \oplus b_j).c_{j-1}+a_j.b_j$	$a_j.c_{j-1} + b_j.c_{j-1} + a_j.b_j$
0	0	0	0	0	0
0	0	1	0	0	0
0	1	0	1	0	0
0	1	1	1	1	1
1	0	0	1	0	0
1	0	1	1	1	1
1	1	0	0	1	1
1	1	1	0	1	1

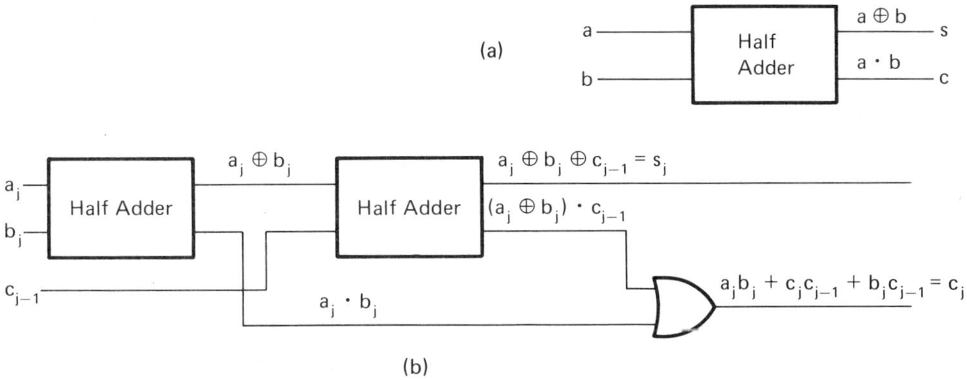

Figure 3-19. (a) A half adder; (b) a full adder composed of two half adders and an OR gate.

Comparing the last two columns, we have

$$(a_j \oplus b_j).c_{j-1} + a_j.b_j = a_j.c_{j-1} + b_j.c_{j-1} + a_j.b_j \qquad (3\text{-}31)$$

Thus, the proper value for c_j is obtained, see Eq. (3-29b).

The half adder of Fig. 3-17b uses four gates. Thus, the full adder of Fig. 3-19b requires a total of nine gates. The full adder of Fig. 3-18 uses 12 gates. This represents a substantial reduction in the number of gates(25%). Digital computers contain an extremely large number of gates. Thus, if we can substantially reduce this number, a considerable saving takes place. In this case, we simply appeared to obtain the reduced form through luck. Again, systematic procedures do exist for reducing the number of gates and they will be considered in the next chapter.

Each full adder adds two one bit numbers plus a carry term. Thus, if we wish to add numbers which have a total of 16 bits, 16 adders are required. The connections of these will be considered in Chapter 10.

Let us consider one further example of a reduction. Figure 3-20 is a block diagram of a device called a *multiplexer*. This particular one has four inputs a_0, a_1, a_2, and a_3, one output b, and two control inputs c_0 and c_1. The multiplexer is a controlled switch which connects the output b to *one* of the inputs in accordance with the digital number supplied by the control. For instance, if $c_0=1$ and $c_1=1(11_2=3_{10})$then b will be connected to a_3. That is, the digital value of b will be equal to that of a_3. Multiplexers find extensive use in digital systems. For instance, suppose that we are working with a digital communication system that has four different output channels and that we want to be able to select one of these channels at a time. If each channel is connected to one input of a multiplexer, then we can electronically select the desired signal by placing the appropriate signal on the control leads.

In digital computers, we often want to vary the interconnections in accordance with the operation being performed. A multiplexer can be used here as a high speed, easily controlled,

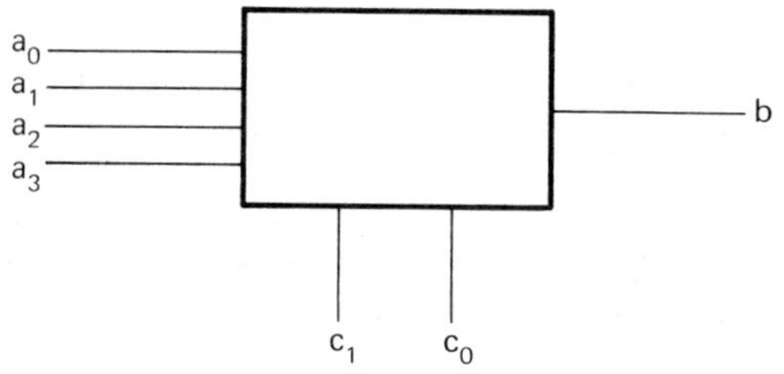

Figure 3-20. The block diagram for a multiplexer.

electronic switch. This adds greatly to the versitility of the computer.

The truth table for this device has 64 entries. For purposes of brevity, we shall only consider part of it.

c_1	c_0	a_0	a_1	a_2	a_3	b
0	0	0	0	0	0	0
0	0	0	0	0	1	0
0	0	0	0	1	0	0
0	0	0	0	1	1	0
0	0	0	1	0	0	0
0	0	0	1	0	1	0
0	0	0	1	1	0	0
0	0	0	1	1	1	0
0	0	1	0	0	0	1
0	0	1	0	0	1	1
0	0	1	0	1	0	1
0	0	1	0	1	1	1
0	0	1	1	0	0	1
0	0	1	1	0	1	1
0	0	1	1	1	0	1
0	0	1	1	1	1	1

In the total table, half(or 32)of the entries will be 1's. (The details of constructing the remainder of the table will be left to the reader.) Thus, if we used the formal procedure discussed earlier in this section, we would require 32 AND gates, four NOT gates and an OR gate. The OR gate would require 32 inputs and each of the AND gates would require 6 inputs. That is, a 32 term switching function can be obtained from the truth table to characterize b.

We shall now show that b can be expressed in terms of a four term switching function. We shall just state this without derivation. Again, the purpose of presenting it here is to show that minimization techniques can be very powerful. We can write b as

$$b = \bar{c}_1 \cdot \bar{c}_0 \cdot a_0 + \bar{c}_1 \cdot c_0 \cdot a_1 + c_1 \, \bar{c}_0 \cdot a_2 + c_1 \cdot c_0 \cdot a_3 \tag{3-32}$$

This equation represents four AND gates. If any of the inputs to an AND gate is 0, then its output will be 0. The control variables or their complements in each term represent the subscript of the input which is to be "connected" to the output. For instance, if $\bar{c}_1 \cdot \bar{c}_0 = 1$, then $c_1 = 0$ and $c_0 = 0 (00_2 = 0_{10})$ and $b = a_0$. Similarly, if $c_1 \cdot \bar{c}_0 = 1$, then $c_1 = 1$, $c_0 = 0 (10_2 = 2_{10})$ and $b = a_2$. Thus, the only AND gate whose output can possibly be 1 is properly selected by the control values. That is, if the number $(c_1 c_0)_2 = d_{10}$, then the output will be given by

$$b = a_d \tag{3-33}$$

Thus, we have achieved the desired multiplex operation. However, only four AND gates rather than 32 are required. The gate diagram for this circuit is shown in Fig. 3-21.

BIBLIOGRAPHY

Booth, T.L., "Digital Networks and Computer Sytems", Chapter 3, John Wiley and Sons, Inc., New York, 1971

Mano, M.M., "Computer Logic Design:", Chap. 1, Prentice Hall, Inc., Englewood Cliffs, N.J., 1972

O'Malley, J., "Introduction to the Digital Computer", Chap. 2, Holt, Rinehart and Winston, Inc., New York, 1972

Peatman, J.B., "The Design of Digital Systems", Chap. 2, McGraw-Hill Book Company, New York, 1972

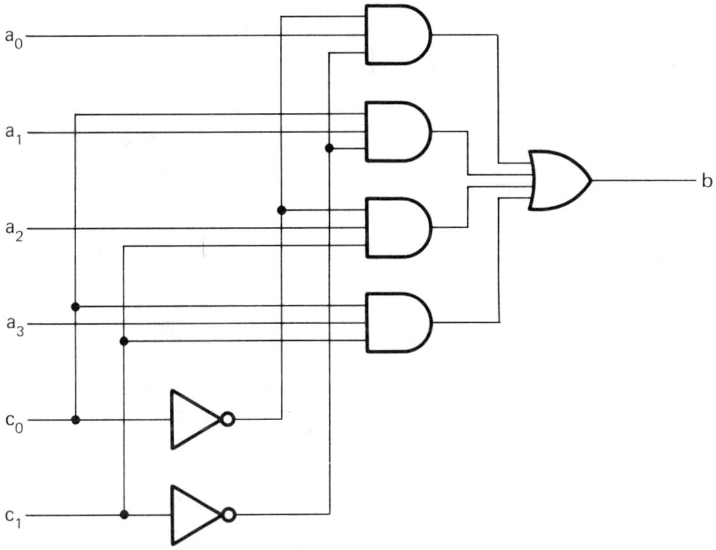

Figure 3-21. A multiplexer.

PROBLEMS

3-1. Write a truth table for the switch circuit of Fig. 3-22.

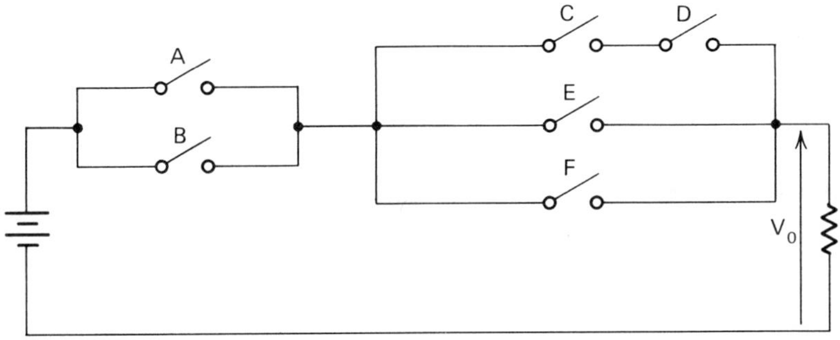

Figure 3-22.

3-2. Write a truth table for the gate circuit of Fig. 3-23.

3-3. Write a logical expression for the gate circuit of Fig. 3-23.

3-4. Repeat Prob. 3-2 for Fig. 3-24.

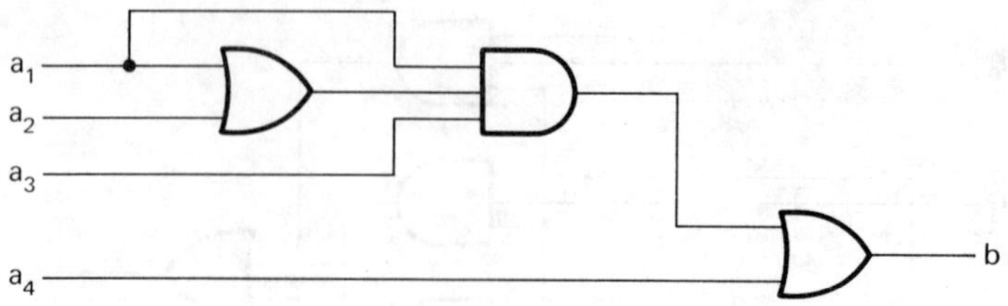

Figure 3-23.

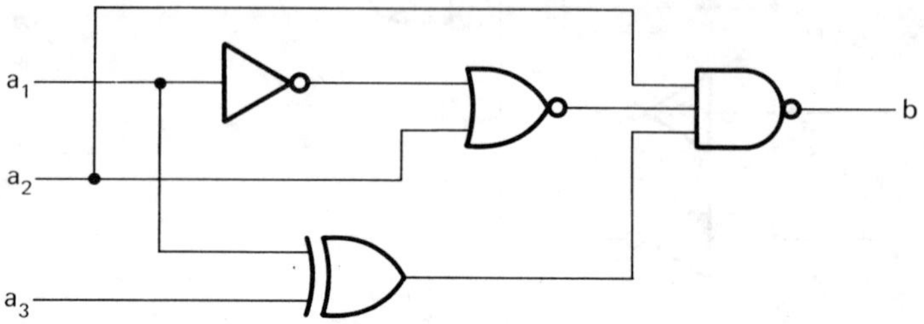

Figure 3-24.

3-5. Repeat Prob. 3-3 for Fig. 3-24.

3-6. Show that the NAND operation is functionally complete.

3-7. Redraw the circuit of Fig. 3-23 using NOR gates only.

3-8. Repeat Prob. 3-7 using NAND gates.

3-9. Obtain a truth table for a half subtractor(i.e., one that does not provide for borrowing). The output should consist of a borrow bit and a difference bit.

3-10. Write a switching function for the truth table of Prob.3-9.

3-11. Obtain a gate circuit using AND, OR and NOT gates for the truth table of Prob. 3-9.

3-12. Obtain a truth table for a full subtractor, which does have provision for borrowing.

3-13. Obtain a switching function for the truth table of Prob. 3-12.

3-14. Obtain a gate circuit for the truth table of Prob. 3-12.

3-15. Redraw Fig. 3-17b using only NOR gates.

3-16. Redraw Fig. 3-17b using only NAND gates.

3-17. Obtain a switching function for a multiplexer with eight inputs.

3-18. Draw a gate circuit for the multiplexer of Prob. 3-17.

4

Combinational Logic-Switching Algebra

In Chapter 3, we introduced some basic ideas of switching algebra and demonstrated the desirability of minimizing switching functions. In this chapter we shall extend these ideas.

The mathematics of logical operations, called *Boolean algebra*, was developed in the 19th century by George Boole. A special branch of this algebra, called *switching algebra*, is applicable to switching circuits. We shall study it here.

4.1 THE BASIC IDEAS OF SWITCHING ALGEBRA

We wish to develop a mathematics which can be applied to switching circuits. In this case, the variables can take on either of two values, namely 0 and 1. In Chapter 3 we also developed the fundamental operations AND, OR, NOT, NAND, NOR and EXOR. It is assumed that the reader is familiar with these operations and all the discussions of Chapter 3.

Consider the switching function for the sum term of a full adder given in Eq. (3-25) as repeated here

$$s = \bar{a}_j.\bar{b}_j.c_{j-1} + \bar{a}_j.b_j.\bar{c}_{j-1} + a_j.\bar{b}_j.\bar{c}_{j-1} + a_j.b_j.c_{j-1} \tag{4-1}$$

We can state that s_j is a function of the variables a_j, b_j, and c_{j-1}. Thus, in a generalized form, we can write

$$s_j = f(a_j, b_j, c_{j-1}) \tag{4-2}$$

This notation is essentially the same as that used in ordinary algebra. A word of caution is in order here. Many of the symbols used in switching algebra are the same as those used in ordinary algebra. However, there can be *substantial differences in the meanings of these symbols*.

In order to develop an algebra we must define some basic operations. This has already been done in Sec. 3-1. Now let us develop some others. In Chapter 3 we demonstrated that certain relations were equivalent by making use of truth tables. For instance, see Eq. (3-20), we showed that

$$A_1.A_2 = \overline{\bar{A}_1 + \bar{A}_2} \tag{4-3}$$

This was done by writing the truth table which contained all possible combinations of the values of A_1 and A_2, for $A_1.A_2$ and $\bar{A}_1+\bar{A}_2$. It was shown that for every possible combination of values of A_1 and A_2, $A_1.A_2$ and $\bar{A}_1+\bar{A}_2$ were the same. Thus, they must be equivalent. This is called a proof by *perfect induction*. Note that, in ordinary algebra, where the variables can take on any values, we cannot use such a proof because there are an infinite number of possible combinations. In switching algebra, proof by perfect induction is possible since each variable can only be one of two values. Thus, we can exhaustively check all possibilities. However, if the number of variables is large, then the truth table becomes very large. For instance, if there are n variables, then there are 2^n entries in the truth table. Thus, if n is greater than 4, proof by perfect induction is tedious.

Just as in the case of ordinary algebra, we can use a *proof by deduction*, (i.e., direct proof). That is, we start with basic postulates and use these to derive results. This avoids the tedium of proof by perfect induction. Let us now develop some basic properties that can be used in proofs.

Some Basic Properties Of Switching Algebra

We shall now develop fundamental operations. Remember that sometimes the symbolism will look like ordinary algebra but the meanings may be very different.

Equivalence

Let us formalize an idea that we have been using. Two switching functions are said to be *logically equivalent* or, simply, *equivalent*, if, and only if they always have the same value (0 or 1)for all possible combinations of the independent binary variables. For instance,

$$f(x,y,z) = g(x,y,z)$$

if f and g have the same values (0 or 1) for all 8 possible combinations of x,y, and z.

Indempotency Laws

$$x + x = x \tag{4-4}$$

This states that if x is 1, then x OR x will be 1 and if x is 0, then x OR x will be 0. Proceeding to prove this by perfect induction, we have

x	x + x
0	0
1	1

Note that Eq. (4-4) would not be true for ordinary algebra.

The second indempotency law is

$$x \cdot x = x \tag{4-5}$$

Again, this can be proven by perfect induction. The details will be left to the reader.

Zero and One Associated With A Variable

Often, when we deal with several variables, we know that one of them is 0 or 1. In such cases, the following relations can prove helpful.

$$0 \cdot x = 0 \qquad (4\text{-}6a)$$

$$1 \cdot x = x \qquad (4\text{-}6b)$$

$$0 + x = x \qquad (4\text{-}6c)$$

$$1 + x = 1 \qquad (4\text{-}6d)$$

These may be verified by perfect induction or by the application of logic. For instance, for Eq. (4-6a) 0 AND x will always be 0 since one of the terms is always 0.

Use Of Parentheses

We use parentheses in switching algebra in the same way as in orindary algebra. For instance,

$$(x + y) \cdot (a + b)$$

means to first take x OR y and then take a OR b. Finally, apply the AND operation to the result of both operations.

The hierarchy, or order, of operations is the same as in ordinary algebra. For instance, $x + y \cdot z$ means $x + (y \cdot z)$.

Involution

The complement of a complement is the variable itself. That is

$$\overline{(\bar{x})} = x \qquad (4\text{-}7)$$

Note that if we complement a variable, we change it from a 1 to a 0 or vice versa. Doing this twice must result in the original variable.

Complementation Laws

$$x + \bar{x} = 1 \qquad (4\text{-}8a)$$

$$x \cdot \bar{x} = 0 \qquad (4\text{-}8b)$$

If we consider x and $\bar{x}$, one will always be 1 and the other 0. Then, Eqs.(4-8) follow.

Commutative Laws

$$x + y = y + x \qquad (4\text{-}9a)$$

$$x \cdot y = y \cdot x \qquad (4\text{-}9b)$$

Associative Laws

$$x + (y+z) = (x+y) + z = (x+z) + y \qquad (4\text{-}10a)$$

$$x \cdot (y \cdot z) = (x \cdot y) \cdot z = y \cdot (x \cdot z) \qquad (4\text{-}10b)$$

These laws can be proven by perfect induction. Note that, in this case, they are the same as the rules of ordinary algebra.

Distributive Laws

$$x \cdot (y+z) = x \cdot y + x \cdot z \qquad (4\text{-}11a)$$

$$x + y \cdot z = (x+y) \cdot (x+z) \qquad (4\text{-}11b)$$

Again, perfect induction can be used. For the second equation, we can also write

$$(x+y) \cdot (x+z) = x \cdot x + x \cdot z + y \cdot x + y \cdot z \qquad (4\text{-}12)$$

This can be demonstrated by perfect induction. We can also demonstrate the reduction of Eq. (4-12) after developing some additional identities.

Absorption

This states

$$x + x \cdot y = x \qquad (4\text{-}13)$$

Let us demonstrate this. From Eqs. (4-6b) and (4-11a), we have

$$x + x \cdot y = x \cdot 1 + x \cdot y = x \cdot (1+y)$$

Then, using Eqs. (4-6d) and (4-6b), we have

$$x \cdot (1+y) = x$$

Thus, we have demonstrated Eq. (4-13).

Now consider Eq. (4-12). Applying Eqs. (4-5) and (4-11a), we obtain

$$(x+y) \cdot (x+z) = x + x \cdot (y+z) + y \cdot z$$

Substituting Eq. (4-13) (replace y+z by a single variable) we have Eq. (4-11b). Thus, we have verified Eq. (4-12).

There are two other forms of absorption. One is

$$x + \bar{x} \cdot y = x + y \qquad (4\text{-}14)$$

To prove this we use Eq. (4-11b). Hence,

$$x + \bar{x} \cdot y = (x + \bar{x}) \cdot (x + y)$$

Substituting Eqs. (4-8a) and (4-6b), we obtain Eq. (4-14).

Finally, we have

$$x \cdot (\bar{x} + y) = x \cdot y \qquad (4\text{-}15)$$

To prove this we can write

$$x \cdot (\bar{x} + y) = x \cdot \bar{x} + x \cdot y \qquad (4\text{-}16)$$

Substituting Eqs. (4-8b) and (4-6c) we obtain Eq. (4-15)

This concludes our discussion of some very fundamental operations. There are some additional operations which can be obtained. However, before doing this, let us see how these simple operations can be used to simplify some switching functions. Suppose we have the switching function

$$f(x,y,z) = x + x \cdot y + x \cdot z + x \cdot (y \cdot \bar{z} + \bar{x}) \tag{4-17}$$

Using Eq. (4-13) we have

$$x + x \cdot y = x$$

Thus,

$$f(x,y,z) = x + x \cdot z + x \cdot (y \cdot \bar{z} + \bar{x}) \tag{4-18a}$$

Again, applying Eq. (4-13), we obtain

$$f(x,y,z) = x + x \cdot (y \cdot \bar{z} + \bar{x})$$

Again, using Eq. (4-13) and remembering that $y \cdot \bar{z} + \bar{x}$ can be thought of as a single variable, we have

$$f(x,y,z) = x \tag{4-18b}$$

This represents a considerable reduction.

De Morgan's Theorems

We shall now consider two theorems that will considerably strengthen our ability to reduce switching functions. They are

$$\overline{x + y} = \bar{x} \cdot \bar{y} \tag{4-19}$$

and

$$\overline{x \cdot y} = \bar{x} + \bar{y} \tag{4-20}$$

Note that when the bar extends over the entire function, it means to evaluate the function and *then* take the complement.

These theorems are proved using perfect induction. For instance, consider Eq. (4-20)

x	y	$x \cdot y$	$\overline{x \cdot y}$	$\bar{x} + \bar{y}$
0	0	0	1	1
0	1	0	1	1
1	0	0	1	1
1	1	1	0	0

The last two columns of the truth table are identical. Thus, Eq. (4-20) is proven.

Equations (4-19) and (4-20), which are called *DeMorgan's theorems*, have been stated for two variables. However, they can be extended to any number of variables. For instance, consider $\overline{x + y + z}$. Let $y + z = a$. Thus, we have $\overline{x + a} = \bar{x} \cdot \bar{a} = \bar{x} \cdot \overline{y + z}$. Again, using Eq. (4-19) we obtain

$$\overline{x + y + z} = \bar{x} \cdot \bar{y} \cdot \bar{z} \tag{4-21}$$

Proceeding similarly, we can derive

$$\overline{x \cdot y \cdot z} = \bar{x} + \bar{y} + \bar{z} \tag{4-22}$$

We can use these theorems with mixed AND and OR operations. For instance, suppose we wish the complement of $x + y \cdot z$. Let $y \cdot z = a$. Thus,

$$\overline{x + (y \cdot z)} = \overline{x + a} = \bar{x} \cdot \bar{a} = \bar{x} \cdot (\overline{y \cdot z})$$

Using Eq. (4-20), we have

$$\overline{x + (y \cdot z)} = \bar{x} \cdot (\bar{y} + \bar{z}) \tag{4-23}$$

If we continue in this fashion, we obtain a general expression for DeMorgan's theorem which states: *The complement of any switching function can be obtained by replacing each variable by its complement, replacing each* AND *operation by an* OR *operation and each* OR *operation by an* AND *operation.* (Note that, where they are present, 0's are replaced by 1's and vice versa.)

Equation (4-23) illustrates this. Let us consider another example.

$$\overline{(x + \bar{y}) \cdot (z \cdot a)} = (\bar{x} \cdot y) + (\bar{z} + \bar{a}) \tag{4-24}$$

Now let us illustrate how a simple reduction can make use of DeMorgan's theorems. Suppose

$$F(x,y,z) = x \cdot \{[x + (y \cdot z)] + [\bar{x} \cdot (\bar{y} + \bar{z})]\} \tag{4-25}$$

Consider the terms within the braces. Using DeMorgan's theorem, we have

$$\overline{x + (y \cdot z)} = \bar{x} \cdot (\bar{y} + \bar{z})$$

Then, substituting Eq.(4-8a), we have

$$[x + (y \cdot z)] + [\bar{x} \cdot (\bar{y} + \bar{z})] = 1$$

Substituting Eq.(4-6b) we obtain

$$F(x,y,z) = x \tag{4-26}$$

This is considerably simpler than Eq.(4-25).

There are two other general relations, known as *Shannon's theorems*, that sometimes prove helpful. These are

$$f(x,y,z,\ldots) = x \cdot f(1,y,z,\ldots) + \bar{x} \cdot f(0,y,z,\ldots) \tag{4-27}$$

and

$$f(x,y,z,\ldots) = [x + f(0,y,z,\ldots)] \cdot [\bar{x} + f(1,y,z,\ldots)] \tag{4-28}$$

Let us clarify the notation. $f(x,y,z,\ldots)$ is a switching function with the variables x, y and z and others. $f(1,y,z,\ldots)$ is the same function where every x is replaced by a 1.

Perfect induction can be used to demonstrate Eqs.(4-27) and (4-28). For instance, in Eq.(4-27), if x = 1, then

$$f(1,y,z,\ldots) = 1 \cdot f(1,y,z,\ldots) + 0 \cdot f(0,y,z,\ldots)$$

Substituting Eqs.(4-6), we have

$$f(1,y,z,\ldots) = f(1,y,z,\ldots)$$

as it should. Now consider that $x = 0$. Then,

$$f(0,y,z,\ldots) = 0 \cdot f(1,y,z,\ldots) + 1 \cdot f(0,y,z,\ldots)$$

Thus,

$$f(0,y,z,\ldots) = f(0,y,z,\ldots)$$

as it should. We have demonstrated that Eq.(4-27) is valid for $x = 1$ and all possible $y,z,\ldots$ and it is valid for $x = 0$ and all possible $y,z,\ldots$. Thus, the relation is always valid. We can prove Eq.(4-28) in a similar way. The details will be left to the reader.

Canonical Forms

There are several forms that can be used to express any switching functions. These are general. However, they are often not in minimal form. Such generalized forms are called *canonical forms*. To study them, we require two fundamental definitions. We shall now consider the first of these. Suppose that we have switching functions of several variables such as $f(x,y,z)$. Now consider a logical expression which contains one each of the three variables or their complements related by the AND operation. These are

$$\begin{aligned} &x \cdot y \cdot z \\ &x \cdot y \cdot \bar{z} \\ &x \cdot \bar{y} \cdot z \\ &x \cdot \bar{y} \cdot \bar{z} \\ &\bar{x} \cdot y \cdot z \\ &\bar{x} \cdot y \cdot \bar{z} \end{aligned} \tag{4-29}$$

$\bar{x} \cdot \bar{y} \cdot z$

$\bar{x} \cdot \bar{y} \cdot \bar{z}$

These terms are called *minterms* or *canonical products.* Note that if there are three binary variables, then there are eight possible minterms. In general, if there are n binary variables, there are 2^n possible minterms.

If we call the AND operation a *logical product* or, simply, a *product*, then the minterms are the products of all the variables *or* their complements. They are called minterms since each minterm will be zero except for one combination of the binary variables. For instance, $x \cdot y \cdot \bar{z}$ is zero except if x=1, y=1, and z=0. The other seven combinations of x,y, and z all result in $x \cdot y \cdot \bar{z} = 0$.

Any switching function of these variables can be expressed as an OR combination of minterms such as those in (4-29). For instance, Eq.(4-1) represents a function that is an OR combination of minterms.

Before proceeding, let us discuss some convenient terminology. If two logical expressions are related by an OR operation, then they are said to be a *logical sum*, or simply a *sum.* For instance, A + B is the sum of A and B. Similarly, the AND operation is said to be a logical product or, simply a product. For instance, A · B is said to be the product of A and B.

Thus, we can state that any logical function can be expressed as a sum of minterms. Actually, see Sec. 3-4, we have demonstrated this by showing that a logical expression could be obtained from a truth table by summing the minterms which correspond to a 1 for the value of the function. It is assumed that the reader is familiar with this procedure. For instance, suppose that we have the truth table,

x	y	z	A
0	0	0	1
0	0	1	0
0	1	0	1
0	1	1	0
1	0	0	1
1	0	1	1
1	1	0	0
1	1	1	1

(4-30)

The logical expression for A is then

$$A = \bar{x} \cdot \bar{y} \cdot \bar{z} + \bar{x} \cdot y \cdot \bar{z} + x \cdot \bar{y} \cdot \bar{z} + x \cdot \bar{y} \cdot z + x \cdot y \cdot z \qquad (4\text{-}31)$$

That is, A is the sum of all minterms that result in its being 1.

Since *any* switching function can be expressed as a sum of minterms, such general expressions are *canonical* forms. In particular, such an expression is called a *canonical sum of products form* or a *disjunctive canonical form*.

Let us discuss some notation which will be of help in describing canonical forms. A *normal product* is the product of variables(either complemented or uncomplemented). A *normal sum* or a *disjunctive normal form* is a sum of normal products. Thus, a canonical sum is a normal sum. However, the converse need not be true. For instance,

$$F(x,y,z) = x \cdot y + \bar{x} \cdot y \cdot z + x \cdot \bar{z}$$

is a normal sum, but it is *not* a canonical sum since $x \cdot y$ and $x \cdot \bar{z}$ are not minterms. Note that $x \cdot y$ and $x \cdot \bar{z}$ do not contain all three variables.

Now let us define another quantity called a *maxterm*, or *canonical sum*. Again, let us illustrate this with a logical expression of three variables $f(x,y,z)$. Now consider logical expressions which are the sum(OR operations)of each of the three variables or their complements. These are(for these variables)

$x + y + z$

$x + y + \bar{z}$

$x + \bar{y} + z$

$x + \bar{y} + \bar{z}$ (4-32)

$\bar{x} + y + z$

$\bar{x} + y + \bar{z}$

$\bar{x} + \bar{y} + z$

$\bar{x} + \bar{y} + \bar{z}$

These are called maxterms since each will be one for all possible combinations of x,y, and z except 1. For instance, $x + \bar{y} + z$ is 1 except when x=0, y=1 and z=0.

We shall now proceed with a development which will show that any switching function can be expressed in terms of maxterms. We start by expressing the complement of the variable $\bar{A}$ in terms of minterms. This can be obtained by adding all the minterms for which A=0(i.e., $\bar{A}$=1). For instance, using the truth table (4-30), we obtain

$$\bar{A} = \bar{x} \cdot \bar{y} \cdot z + \bar{x} \cdot y \cdot z + x \cdot y \cdot \bar{z} \qquad (4\text{-}33)$$

Now take the complement of both sides of the expression. Using DeMorgan's theorem we obtain

$$A = (x+y+\bar{z}) \cdot (x+\bar{y}+\bar{z}) \cdot (\bar{x}+\bar{y}+z) \qquad (4\text{-}34)$$

Note that $\bar{\bar{A}} = A$. Thus, we have expressed A as a product of maxterms.

This procedure can be used in general. We can always express the complement of the output as a sum of minterms. Taking the complement of a minterm we obtain a maxterm. Thus, any switching function can be expressed as a product of maxterms. Thus, the product of maxterms is also a canonical form, which is called a canonical *product of sums form* or a *conjunctive canonical form*.

In Sec. 3-4 we demonstrated that a switching function consisting of the sum of n minterms could be realized by the interconnection of n AND gates and an OR gate, with a suitable number of inverters. In a similar way we see that a switching function consisting of the product of m maxterms can be realized using m OR gates and one AND gate. These two realizations do not necessarily use the same number of gates. For instance, the realization of A using Eq. (4-31) requires 4 AND gates and 1 OR gate, while the realization using Eq. (4-34) requires 3 OR gates and 1 AND gate. In general, if there are more 1's than 0's in the output column of the truth table, then the realization using maxterms will require fewer gates than the realization using minterms. Similarly, if there are more 0's than 1's, the realization using minterms will have fewer gates. However, there may often be other realizations which use far fewer gates than either of these.

Minterms and maxterms are often listed in terms of a short hand notation. In Table 4-1, we shall illustrate this notation assuming three independent variables.

TABLE 4-1
MINTERMS AND MAXTERMS WHEN THERE ARE THREE SWITCHING VARIABLES

x_1	x_2	x_3	Minterm	Designation	Maxterm	Designation
0	0	0	$\bar{x}_1 \cdot \bar{x}_2 \cdot \bar{x}_3$	m_0	$x_1+x_2+x_3$	M_0
0	0	1	$\bar{x}_1 \cdot \bar{x}_2 \cdot x_3$	m_1	$x_1+x_2+\bar{x}_3$	M_1
0	1	0	$\bar{x}_1 \cdot x_2 \cdot \bar{x}_3$	m_2	$x_1+\bar{x}_2+x_3$	M_2
0	1	1	$\bar{x}_1 \cdot x_2 \cdot x_3$	m_3	$x_1+\bar{x}_2+\bar{x}_3$	M_3
1	0	0	$x_1 \cdot \bar{x}_2 \cdot \bar{x}_3$	m_4	$\bar{x}_1+x_2+x_3$	M_4
1	0	1	$x_1 \cdot \bar{x}_2 \cdot x_3$	m_5	$\bar{x}_1+x_2+\bar{x}_3$	M_5
1	1	0	$x_1 \cdot x_2 \cdot \bar{x}_3$	m_6	$\bar{x}_1+\bar{x}_2+x_3$	M_6
1	1	1	$x_1 \cdot x_2 \cdot x_3$	m_7	$\bar{x}_1+\bar{x}_2+\bar{x}_3$	M_7

The switching variables are listed in the usual order. In the minterm column we write the minterm which is 1 when the switching variables have the designated values. The minterm is designated by a subscripted lower case m. The subscript is in base 10 and is equal to the base 2 representation of the variable.

For instance, $\bar{x}_1 \cdot x_2 \cdot x_3$ is designated m_3 since $3_{10}=011_2$ and this minterm is 1 when $x_1=0$, $x_2=1$ and $x_3=1$. Similarly, m_4 corresponds to $x_1 \cdot \bar{x}_2 \cdot \bar{x}_3$ since this is 1 when $x_1=1$, $x_2=0$ and $x_3=0$. Note $4_{10}=100_2$. Note that there is some ambiguity here. We arbitrarily listed the independent variables in the order x_1, x_2, x_3. Other orders could have been used(e.g., x_3, x_2, x_1).

A minterm is 1 only for one of the 2^n possible combinations of the independent variables. Thus, the designation of the minterms is based on their being 1. Conversely, a maxterm is 0 for only one of the 2^n combinations of the binary variables. Hence, we designate the maxterms on the basis of their being 0. For instance, M_5 designated $\bar{x}_1+x_2+\bar{x}_3$; $5_{10}=101_2$ and $\bar{x}_1+x_2+\bar{x}_3 = 0$ when $x_1=1$, $x_2=0$, and $x_3=1$. Note that the corresponding minterms and maxterms are complements. That is

$$m_j = \bar{M}_j \tag{4-35}$$

Suppose that we have a switching function expressed as

$$f(x_1,x_2,x_3) = \bar{x}_1 \cdot x_2 \cdot \bar{x}_3 + x_1 \cdot \bar{x}_2 \cdot \bar{x}_3 + x_1 \cdot x_2 \cdot x_3$$

We can write this as

$$f(x_1,x_2,x_3) = m_2 + m_4 + m_7$$

A short hand notation that is sometimes used is

$$f(x_1,x_2,x_3) = \sum (2,4,7)$$

We can use a similar notation with maxterms. For instance, suppose

$$g(x_1,x_2,x_3) = (x_1+x_2+x_3) \cdot (x_1+\bar{x}_2+x_3) \cdot (\bar{x}_1+\bar{x}_2+\bar{x}_3)$$

We can write this as

$$g(x_1,x_2,x_3) = M_0 \cdot M_2 \cdot M_7$$

In short hand notation, this is

$$g(x_1,x_2,x_3) = \pi\ (0,2,7)$$

Expression Of An Arbitrary Switching Function In Canonical Form

We have seen how an arbitrary switching function can be expressed as a sum of minterms or a product of maxterms. These forms can be generated easily from the truth table. Thus, if we are given an arbitrary switching function and we want to express it in canonical form, we need only construct the truth table and obtain the function. However, this is often tedious and there are other procedures which can be used to obtain a canonical form which is equivalent to a given switching function. Let us illustrate this with an example. We shall express the function

$$A(x,y,z) = x + \bar{y} + z \qquad (4\text{-}36)$$

as a sum of minterms. Each term should be a product of the three independent variables x,y, and z, or their complements. Consider each term. For instance, the first is x. Now $x\cdot 1=x$; also $y+\bar{y}=1$. Thus, we can write

$$x = x \cdot (y+\bar{y}) = x \cdot y + x \cdot \bar{y}$$

Now repeat this procedure with z. Then,

$$x = (x\cdot y + x\cdot\bar{y}) \cdot (z+\bar{z}) = x\cdot y\cdot z+x\cdot y\cdot\bar{z}+x\cdot\bar{y}\cdot z+x\cdot\bar{y}\cdot\bar{z}$$

Hence, we have expressed x as a sum of minterms. The second term in Eq.(4-36) is missing an x term. Thus, we write

$$\bar{y}\cdot z = \bar{y}\cdot z(x+\bar{x}) = \bar{y}\cdot z\cdot x + \bar{y}\cdot z\cdot\bar{x} = x\cdot\bar{y}\cdot z + \bar{x}\cdot\bar{y}\cdot z$$

Combining terms, we have

$$A(x,y,z) = x\cdot y\cdot z + x\cdot y\cdot\bar{z} + x\cdot\bar{y}\cdot z + x\cdot\bar{y}\cdot\bar{z} + \bar{x}\cdot\bar{y}\cdot z \qquad (4\text{-}37)$$

There are several terms which are repeated. Using Eq.(4-4), we obtain

$$A(x,y,z) = x\cdot y\cdot z + x\cdot y\cdot\bar{z} + x\cdot\bar{y}\cdot z + x\cdot\bar{y}\cdot\bar{z} + \bar{x}\cdot\bar{y}\cdot z$$

This type of procedure can be used in general. That is, after manipulation, a switching function can be expressed as a sum of terms. Next, the "missing" variables can be included by means of the above procedure.

If we want to express a function as a product of maxterms, we can obtain the complement of the given function, express it as a product of minterms and, finally, take the complement of the resulting expression. For instance, from Eq. (4-36), we have

$$\bar{A} = \bar{x} \cdot (y+\bar{z}) = \bar{x} \cdot y + \bar{x} \cdot \bar{z}$$

We can write this as

$$\bar{A} = \bar{x} \cdot y \cdot (z+\bar{z}) + \bar{x} \cdot \bar{z}(y+\bar{y})$$

or

$$A = \bar{x} \cdot y \cdot z + \bar{x} \cdot y \cdot \bar{z} + \bar{x} \cdot y \cdot \bar{z} + \bar{x} \cdot \bar{y} \cdot \bar{z}$$

Eliminating the repeated terms, we have

$$\bar{A} = \bar{x} \cdot y \cdot z + \bar{x} \cdot y \cdot \bar{z} + \bar{x} \cdot \bar{y} \cdot \bar{z}$$

Taking the complement of both sides of this expression, we obtain

$$A = (x+\bar{y}+\bar{z}) \cdot (x+\bar{y}+z) \cdot (x+y+z) \qquad (4\text{-}38a)$$

There is an alternative procedure. Let us illustrate it with an example. We shall again use Eq. (4-36).

$$A(x,y,z) = x + \bar{y} \cdot z$$

Now express this as a product

$$A(x,y,z) = (x+\bar{y}) \cdot (x+z) \qquad (4\text{-}38b)$$

Note that $(x+\bar{y}) \cdot (x+z) = x + x \cdot \bar{y} + \bar{y} \cdot z + z \cdot x$. However, see Eq. (4-13), $x+x \cdot \bar{y}+x \cdot z = x$. Thus, Eqs. (4-38a) and (4-38b) are equivalent. We can always make use of suitable manipulations, which may be tedious, to express a switching function as a product, such as Eq. (4-38b). Now examine each parentheses of the product to determine the missing terms. These can be added making use of Eq. (4-8b). Thus,

$$A(x,y,z) = (x+\bar{y}+z\cdot\bar{z}) \cdot (x+z+y\cdot\bar{y})$$

Now consider the first term

$$(x+\bar{y}+z\cdot\bar{z}) = (x+\bar{y}+z) \cdot (x+\bar{y}+\bar{z})$$

This can be demonstrated by carrying out the operations and using Eq. (4-13). Thus, we can write

$$A(x,y,z) = (x+\bar{y}+z) \cdot (x+\bar{y}+\bar{z}) \cdot (x+y+z) \cdot (x+\bar{y}+z)$$

Using Eq. (4-5) to eliminate the repeated terms, we have

$$A(x,y,z) = (x+\bar{y}+z) \cdot (x+\bar{y}+\bar{z}) \cdot (x+y+z) \tag{4-39}$$

Uniqueness Of Canonical Forms

The canonical form of representation of a switching function is unique. For instance, if a function is represented as a sum of minterms, then any sum of minterms which represents this function would contain *exactly* the same minterms. We can prove this by contradiction. Suppose that the same switching function could be represented by two different sums of minterms. Each minterm is 1 only for *one* of the 2^n possible combinations of the independent binary variables. Thus, there must be one set of binary variables which causes one sum of minterms to be 1 while the other must be 0. However, a function cannot be 0 and 1 simultaneously. Thus, the assumption that sums of different minterms could represent the same function is false. Thus, this demonstrates the uniqueness. A similar proof can be used to demonstrate the uniqueness of the product of maxterms.

Duality

After considering the developments of this section, one fact appears. For each relation, there is an equivalent one where the AND and OR operations have been interchanged and the 0's and 1's have been interchanged. For instance, in this section, we developed

$x + 0 = x$ and $x \cdot 1 = x$

$x + x = x$ and $x \cdot x = x$

$\overline{x+y} = \bar{x} \cdot \bar{y}$ and $\overline{x \cdot y} = \bar{x} + \bar{y}$

$x+x \cdot y = x$ and $x \cdot (x+y) = x$

Each relation in the second column can be obtained from the first by interchanging 0's and 1's and $\cdot$ and +.

This principle can be carried throughout all switching algebra. That is, if a relation is valid, then its *dual*, obtained by replacing all 0's by 1's, 1's by 0's, OR by AND and AND by OR is also valid.

Let us illustrate this with one further example. In Eq.(4-23) we showed that

$$\overline{x+(y \cdot z)} = \bar{x} \cdot (\bar{y}+\bar{z}) \qquad (4\text{-}40)$$

Then, the following will also be valid

$$\overline{x \cdot (y+z)} = \bar{x}+(\bar{y} \cdot \bar{z}) \qquad (4\text{-}41)$$

Note that duals are not, in general, equal. For instance, $\overline{x+(y \cdot z)} \neq \overline{x \cdot (y+z)}$.

We have used the $(\cdot)$ to indicate the AND operation. However, it is conventional and convenient to omit it if no ambiguity is present. In the remainder of this book we shall do so. Thus, we shall write

$$x \cdot y = xy \qquad (4\text{-}42)$$

Logical Expressions From Specifications

In Chapter 3 we discussed the need for logical expressions in terms of computer circuits such as adders. At times, small special purpose computers are built to perform some simple tasks. Examples are elevator controls, change making mechanisms, simple process controls, etc. Let us illustrate how a logical expression can be generated from the specifications. For instance, suppose that entrance to an honors program is to be determined by passing or failing three tests. The first two, T_1 and T_2 cover specialized areas and the student is only required to know one of the subjects. The third is general and all students are required to pass it. Thus, to enter the honors program, the student must pass T_1 or T_2 or both, and T_3. Thus, the truth table is

T_1	T_2	T_3	P
0	0	0	0
0	0	1	0
0	1	0	0
0	1	1	1
1	0	0	0
1	0	1	1
1	1	0	0
1	1	1	1

(4-43)

where 1 represents pass and 0 represents fail. If P is 1, the student enters the program. We want to design a simple gate circuit which will determine if a student enters the program. Using the truth table, we can write P as a sum of minterms. This is

$$P = \bar{T}_1T_2T_3 + T_1\bar{T}_2T_3 + T_1T_2T_3 \qquad (4\text{-}44)$$

This realization requires 3 AND gates and one OR gate and 2 NOT gates. However, we can simplify the relation. The following steps can be used

$$P = (T_1+\bar{T}_1)T_2T_3 + T_1\bar{T}_2T_3$$

$$P = T_2T_3 + T_1\bar{T}_2T_3$$

$$P = T_3(T_2+T_1\bar{T}_2)$$

$$P = T_3(T_1+T_2)$$

This can be realized using one OR gate and one AND gate. No NOT gates are required. The realization is shown in Fig. 4-1.

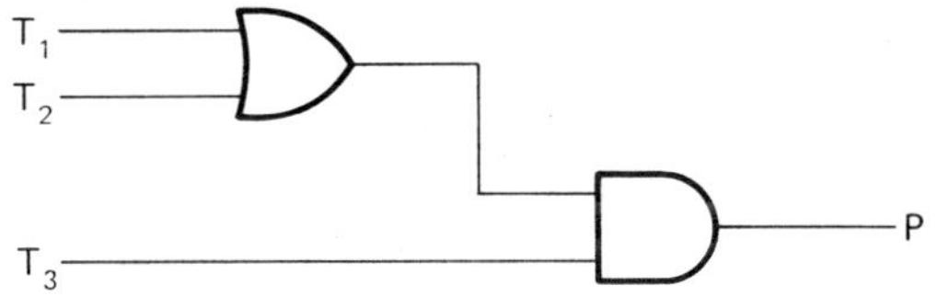

Figure 4-1. The gate circuit realization of Eq. (4-44).

4.2 KARNAUGH MAPS MINIMIZATION USING THE MAP METHOD

In Sec. 4-1, we illustrated several minimizations of switching functions. However, the unsystematic procedures used there become very tedious when relatively large logical expressions are encountered. In this section we shall discuss a map which is a symbolic representation of the truth table that will provide a systematic means of reducing the function. The map we use is called a *Karnaugh map* and is a two dimensional plot of a truth table. It allows us to make use of the relations

$$x(y+\bar{y}) = x$$

and

$$x + \bar{x}y = x + y$$

in an organized way.

We shall start by considering the example of a switching function of two variables. For instance,

$$f(x,y) = x\bar{y} + xy + \bar{x}y \tag{4-45}$$

Its Karnaugh map is shown in Fig. 4-2a. The variables x and y are listed along the side of the map. Ones are put in the appropriate boxes. These boxes are called *cells*. For instance, the lower right cell corresponds to x=1,y=1. A 1 is put there since xy is a minterms of f(x,y). Similarly, the upper right hand cell corresponds to x=0,y=1. A 1 is put there since $\bar{x}y$ is also a minterm which appears in f. Thus, to obtain the complete Karnaugh map, we expand the function in the canonical form using the sum of minterms and then put a 1 in each cell which corresponds to a minterm. These are called 1-cells. The blank cells actually should have 0's in them, i.e., 0-cells, since the minterms which correspond to them do not appear in the expression for the function. However, the zeros are usually omitted to avoid cluttering the map. Note that each cell corresponds to a value of the function with a particular set of variables. This is illustrated in Fig. 4-2b.

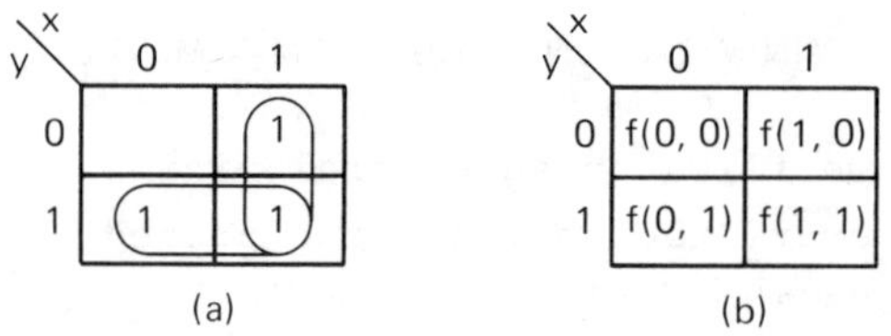

Figure 4-2. (a) A two variable Karnaugh map for the switching function of Eq. (4-45); (b) the function corresponding to the input variables.

Now consider the Karnaugh map. If we move horizontally from one cell to an adjacent cell then one, and only one, of the two variables of the corresponding minterms changes. For instance, the two lower boxes correspond to $x\bar{y}$ and xy. In general, any two adjacent boxes will be of this form. (We shall see how this extends to more than two independent variables subsequently.) Now

$$x\bar{y} + xy = x(y+\bar{y}) = x \tag{4-46}$$

Thus, if 1's appear in adjacent horizontal boxes, the two minterms can be combined into a single, simpler term. This is indicated on the Karnaugh map by encircling the variables, see Fig.4-2a.

In a similar way, if two 1's are adjacent vertically (one above the other) only one variable of the corresponding minterm will change(to its complement). For instance, in the right column, we have

$$\bar{x}y + xy = (\bar{x}+x)y = y \qquad (4\text{-}47)$$

(See Eq. (4-8a)) Thus, if two 1's are adjacent vertically, the corresponding minterms can be combined into a simpler expression. We indicate this also by encircling the appropriate 1-cells. (Note that we can combine 1-cells that are adjacent horizontally or adjacent vertically, but *not* diagonally.)

Proceeding with the simplification, we can write

$$f(x,y) = x + \bar{x}y = y + x\bar{y} \qquad (4\text{-}48)$$

Using Eq. (4-14) we can write either of these as

$$f(x,y) = x + y \qquad (4\text{-}49)$$

Now, let us use the Karnaugh map to interpret this. One circled pair represents x, the other circled pair represents y. Their combination is the sum of the two x+y. Thus, we have used the Karnaugh map to systematically reduce the expression. We shall illustrate this further subsequently.

We have illustrated some simple ideas using a two variable Karnaugh map. Now let us extend these to additional variables. The Karnaugh map is always a two dimensional map. Thus, each cell may have to represent more than two variables. For instance, a three variable Karnaugh map is shown in Fig. 4-3. The upper 0's and 1's represent the values y and z. The leftmost number of each pair represents y, the other z, since y and z are written in this order above the map. Note the order of the numbers along the top of the map. It is not in the order of ordinary digital numbers but is such that adjacent numbers differ by only one binary digit. Note that the Karnaugh map is useful for minimization because cells which are adjacent (horizontally or vertically)represent a change of just one variable to its complement. Thus, as we have seen, 1's in adjacent boxes represent minterms which can be combined to yield just a single term in the switching function. Ordering the binary numbers so that adjacent numbers differ by only one binary digit accomplishes this.

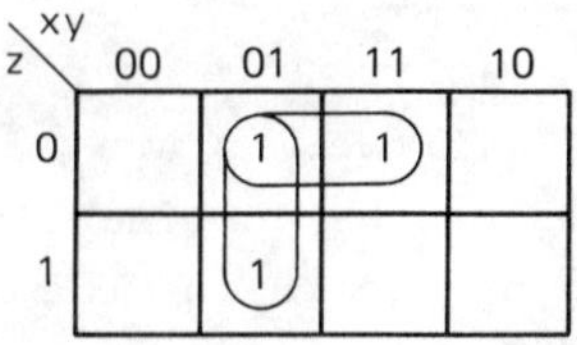

Figure 4-3. A three variable Karnaugh map for the switching function of Eq. (4-50).

In Chapter 8 we shall discuss general procedures for encoding of numbers so that successive numbers differ by only one binary digit. Such encoding is called a *Gray code* or a *reflected code*. Here, we need not consider the general details of Gray coding. The appropriate ordering of the binary digits of the Karnaugh map will simply be given in the text.

Let us illustrate this using the Karnaugh map of Fig. 4-3 as an example. The switching function that corresponds to this map is

$$f(x,y,z) = \bar{x}y\bar{z} + xy\bar{z} + \bar{x}yz \qquad (4\text{-}50)$$

The two 1's in the second column represent

$$\bar{x}y\bar{z} + \bar{x}yz = \bar{x}y(z+\bar{z}) = \bar{x}y \qquad (4\text{-}51a)$$

As in the case of the two variable Karnaugh map, a pair of vertically adjacent 1's results in the elimination of a variable. Now consider the two adjacent horizontal 1's. They represent

$$\bar{x}y\bar{z} + xy\bar{z} = (\bar{x}+x)(y\bar{z}) = y\bar{z} \qquad (4\text{-}51b)$$

Thus, adjacent 1-cells represent a potential for eliminating one variable. Combining Eqs. (4-51a) and (4-50), we have

$$f(x,y,z) = \bar{x}y + xy\bar{z} = y(\bar{x}+x\bar{z}) = y(\bar{x}+\bar{z}) = \bar{x}y + y\bar{z} \qquad (4\text{-}52)$$

As in the case of the two variable map, this can be obtained by adding Eqs.(4-51a) and (4-51b). That is, the value of the function is the sum of the values of the encircled pairs. This is because the sum of all the encircled pairs is just the sum of all the minterms.

Let us consider some additional examples. The switching function for the Karnaugh map of Fig. 4-4 is

$$f(x,y,z) = \bar{x}y\bar{z} + \bar{x}yz + xy\bar{z} + xyz \tag{4-53}$$

We can simplify this as

$$f(x,y,z) = (\bar{x}+x)y\bar{z} + (\bar{x}+x)yz = y\bar{z} + yz$$

$$f(x,y,z) = y \tag{4-54}$$

Now consider the circles drawn around all four 1-cells. We could break this up into four loops, each enclosing a pair of horizontal or vertical 1-cells. From what we have seen previously, a horizontal or vertical loop indicates a pair of minterms which can be combined into a single term since they differ only in the complementation of one variable. Because of the Gray coding of the variable locations, we would expect any group of variables enclosed in a group(such as a square)to contain similarly related terms. Thus, such groupings should contain candidates for reduction. This is illustrated by Eqs,(4-53) and (4-54). Note that we can write Eq. (4-53) as

$$f(x,y,z) = (\bar{x}\bar{z} + \bar{x}z + x\bar{z} + xz)y \tag{4-55}$$

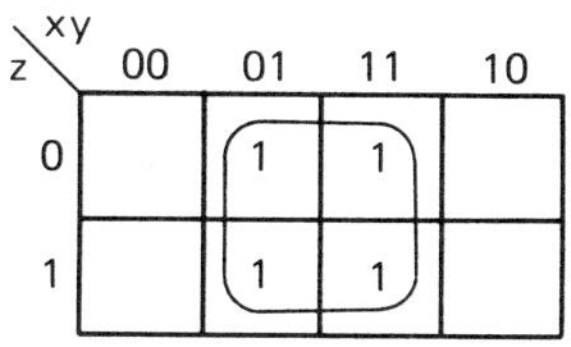

Figure 4-4. A three variable Karnaugh map for the switching function of Eq. (4-53).

The term in the parentheses represents 1. Thus, the reduction of Eq. (4-54) results. The grouping on the Karnaugh map systematically yields this minimization. We shall consider several additional examples and then obtain a direct procedure for determining the reduced function is a very simple way.

Let us consider the Karnaugh map of Fig. 4-5. The switching function for this map is

$$f(x,y,z) = \bar{x}\bar{y}z + \bar{x}\bar{y}\bar{z} + x\bar{y}z + x\bar{y}\bar{z} \tag{4-56}$$

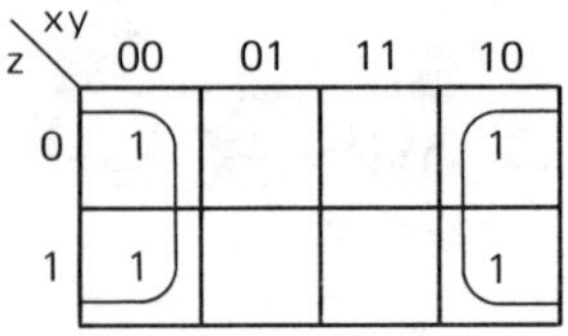

Figure 4-5. A three variable Karnaugh map for the switching function of Eq. (4-56).

Consider the terms

$$\bar{x}\bar{y}\bar{z} + x\bar{y}\bar{z} = \bar{y}\bar{z}(\bar{x}+x) = \bar{y}\bar{z} \tag{4-57}$$

This indicates that the Karnaugh map locations which correspond to these minterms should be adjacent. However, they are on opposite sides of the map in the upper right hand and upper left hand corners, respectively. Thus, they are not physically adjacent. To resolve this difficulty, we consider that the right and left edges of the map are in contact. That is, consider it as a cylinder, not as a plane. This resolves the difficulty since these squares are now adjacent horizontally. Proceeding in this way, we can encircle the square formed by four 1-cells.

Let us see if Eq. (4-56) can be reduced as we might expect. We can write Eq. (4-56) as

$$f(x,y,z) = \bar{x}\bar{y}(z+\bar{z}) + x\bar{y}(z+\bar{z}) = \bar{x}\bar{y} + x\bar{y}$$

Hence,

$$f(x,y,z) = \bar{y} \tag{4-58}$$

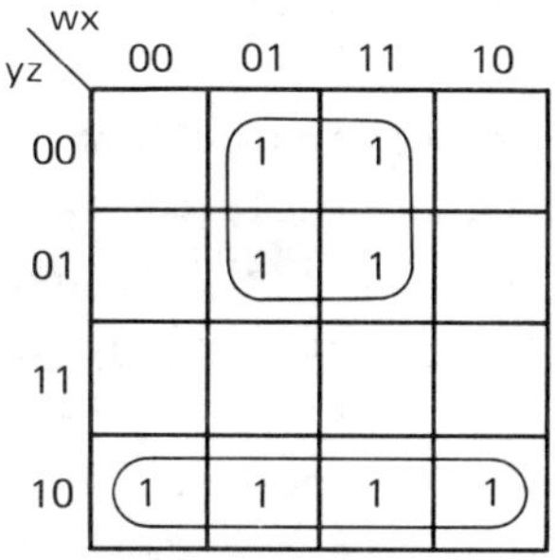

Figure 4-6. A four variable Karnaugh map for the switching function of Eq. (4-59).

Now let us extend our results to a four variable Karnaugh map. Such a map is illustrated in Fig. 4-6. This represents the switching function

$$f(w,x,y,z) = \bar{w}x\bar{y}\bar{z} + wx\bar{y}\bar{z} + \bar{w}x\bar{y}z + wx\bar{y}z + \bar{w}\bar{x}y\bar{z} + \bar{w}xy\bar{z} + wxy\bar{z} + w\bar{x}y\bar{z} \qquad (4\text{-}59)$$

Again notice that any pair of 1-cells that are adjacent either horizontally or vertically results in a pair of minterms that differ only in one variable and it is complemented in one and uncomplemented in the other. Thus, we should expect the same type of minimization here. Now consider the two groupings isolated in the map. First consider the upper square. This represents the minterm sum

$$\alpha_1 = \bar{w}x\bar{y}\bar{z} + wx\bar{y}\bar{z} + \bar{w}x\bar{y}z + wx\bar{y}z = x\bar{y}\bar{z} + x\bar{y}z = x\bar{y} \qquad (4\text{-}60)$$

Thus, we again see that the groupings indicate where considerable reduction can occur. Now consider the sum of minterms represented by the encircled lower rectangle. This is

$$\alpha_2 = \bar{w}\bar{x}y\bar{z} + \bar{w}xy\bar{z} + wxy\bar{z} + w\bar{x}y\bar{z} = y\bar{z} \qquad (4\text{-}61)$$

Again, a grouping on the Karnaugh map leads to a considerable reduction. Note that these reductions occur because of the complement relation in one variable in each of the minterms. Thus we can reduce Eq. (4-59) to

$$f(w,x,y,z) = \alpha_1 + \alpha_2 = x\bar{y} + y\bar{z} \qquad (4\text{-}62)$$

We shall systematize these reductions subsequently.

Let us now introduce some terminology. As we have discussed, each square, representing a single minterm of the Karnaugh map, is called a *cell*. If two cells are adjacent horizontally or vertically, they are simply said to be *adjacent*. In general, if two cells represent minterms which differ by only one variable, (i.e., it is complemented in one and not in the other) they are said to be adjacent. Thus, if two cells containing 1's, i.e., 1-cells, are adjacent, a variable can be eliminated from a pair of minterms. A grouping of adjacent 1-cells is defined as a *subcube*. The subcube is said to cover the cells containing the ones. For instance, Fig. 4-6 has two subcubes, each of which covers four cells.

yz \ wx	00	01	11	10
00	m_0	m_4	m_{12}	m_8
01	m_1	m_5	m_{13}	m_9
11	m_3	m_7	m_{15}	m_{11}
10	m_2	m_6	m_{14}	m_{10}

(a)

yz \ wx	00	01	11	10
00	0	4	12	8
01	1	5	13	9
11	3	7	15	11
10	2	6	14	10

(b)

Figure 4-7. (a) A four variable Karnaugh map showing a designation in terms of minterms; (b) an alternative form of this representation.

Let us also discuss some additional notation which will enable us to discuss these maps more easily. The cells of a Karnaugh map are often identified by the minterms which they define. For instance, for three variables, minterms are designated in Table 4-1. Remember that the subscript, when converted from decimal to binary, identifies the value of the independent variables which result in 1 for the minterm. Figure 4-7 indicates these values for a four variable Karnaugh map. As an example, consider m_5. Then $5_{10} = 0101$. Hence, $m_5=1$ if $w=0$, $x=1$, $y=0$ and $z=1$. This represents $\bar{w}x\bar{y}z$. Thus, the subscript of the minterm can be obtained by writing the binary number using the values of w,x,y and z as digits. (i.e., they are 1's for uncomplemented variables and 0 for the complemented ones.)

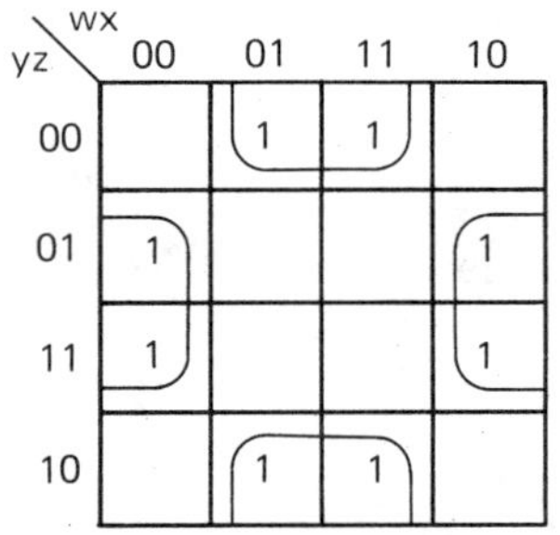

Figure 4-8. A four variable Karnaugh map for the switching function of Eq. (4-63).

Now we shall consider some further reductions of four variable maps. Let us work with the map of Fig. 4-8. It represents the switching function

$$f(w,x,y,z) = \bar{w}x\bar{y}\bar{z} + wx\bar{y}\bar{z} + \bar{w}xy\bar{z} + wxy\bar{z} + \bar{w}\bar{x}\bar{y}z + w\bar{x}\bar{y}z + \bar{w}\bar{x}yz + w\bar{x}yz \tag{4-63}$$

Terms on the left and right edges of the map are adjacent. This corresponds to the three variable map. In a similar way, the corresponding cells on the top and bottom edges are also adjacent. Thus, we have encircled two subcubes of four cells each. Consider the subcube which consists of the upper and lower sections, i.e., $m_4+m_{12}+m_6+m_{14}$, see Fig. 4-7. Let us see how it reduces. The minterms which correspond to it are

$$\alpha_1 = \bar{w}x\bar{y}\bar{z} + wx\bar{y}\bar{z} + \bar{w}xy\bar{z} + wxy\bar{z} \tag{4-64a}$$

This can be simplified in the following way

$$\alpha_1 = x\bar{y}\bar{z} + xy\bar{z} = x\bar{z} \tag{4-64b}$$

Similarly, the minterms which correspond to the remaining subcube, i.e., $m_1+m_3+m_9+m_{11}$, are

$$\alpha_2 = \bar{w}\bar{x}\bar{y}z + w\bar{x}\bar{y}z + \bar{w}\bar{x}yz + w\bar{x}yz \tag{4-65a}$$

This can be simplified in the following way

$$\alpha_2 = \bar{w}\bar{x}z + w\bar{x}z = \bar{x}z \tag{4-65b}$$

Thus, Eq. (4-63) can be expressed as

$$f(w,x,y,z) = x\bar{z} + \bar{x}z \tag{4-66}$$

Of course, this is the sum of the "functions" corresponding to each of the subcubes.

Before proceeding with some more complex functions, let us make some general statements based upon what we have observed. Each subcube of 1-cells can be expressed as a single product of the independent variables. It is conventional to call variables represented by letters(e.g., $x, y, \bar{z}$, etc.) *literals*. *Thus, each subcube can be represented by a single product of literals.* When a subcube consists of two adjacent cells, one variable can be eliminated from the sum of the corresponding minterms. (i.e., the one which changes to its complement). In the examples, we have seen(e.g.,Figs. 4-4, 4-5, 4-6, and 4-8)that if a subcube covers four cells, then two variables can be eliminated. This actually is general. If a map has n variables, then a subcube covering 2^k adjacent cells can be represented by the product of n-k literals. Then, *a subcube covering 2^k adjacent cells can be represented by a single product of n-k literals*. Thus, for a four variable Karnaugh map, we can write the following table.

TABLE 4-2
NUMBER OF LITERALS IN A SINGLE PRODUCT NEEDED TO REPRESENT THE 1-CELLS COVERED BY A SUBCUBE IN A FOUR VARIABLE KARNAUGH MAP

Number of 1-cells covered by a subcube	Number of literals in corresponding term
1	4
2	3
4	2
8	1
16	0

Note that if the entire Karnaugh map consists of 1's, then the function is a 1 and no literals are needed. Thus, it is desirable to use subcubes as large as possible. We can then write the product for each subcube and simplify the expression. Actually, we save this step. The minimized expression for any subcube can be obtained simply by inspecting the subcube. The expression for the subcube is just the *product of all literals whose values do not change in the subcube*. If the value of the literal is always 1, then the literal appears in the product. If the value of the literal is always 0, then the complement of the literal appears in the product. This comes about because the literals which change appear both complemented and uncomplemented. When the function is reduced, these literals are eliminated. Let us illustrate this with some examples. In Fig. 4-4, the only variable in the subcube that does not change is z. Thus, this subcube(and complete Karnaugh map)can be represented by z. This is verified by Eq.(4-54).

In Fig. 4-8, there are two subcubes. The first is m_4, m_{12}, m_6 and m_{14}. The two variables that do not change are x and $\bar{z}$. (In the subcube, the value of x is always 1 and that of z is always 0.) Thus, this subcube represents $x\bar{z}$. The second subcube consists of minterms m_1, m_3, m_9 and m_{11}. Again, the variables that do not change are x and z. However, in this case, we have x=0 and z=1. Thus, the subcube can be represented by $\bar{x}z$. The complete function is given by $x\bar{z}+\bar{x}z$. This is verified by Eq.(4-66). Thus, once the subcubes are identified, the minimized function can be written by inspection.

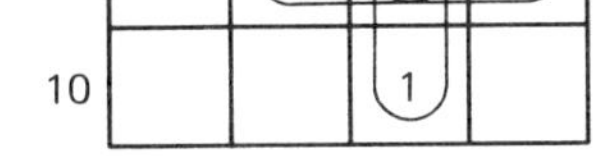

Figure 4-9. A four variable Karnaugh map for the switching function of Eq. (4-67).

Let us consider an example of one more minimization. Then we shall formalize some rules. The Karnaugh map is given in Fig. 4-9. We have indicated four subcubes. (One has only one cell.)

We shall write the switching function in terms of the subcubes. It is

$$
\begin{aligned}
f(w,x,y,z) = {} & \bar{w}\bar{x}\bar{y}\bar{z} \\
& + \bar{w}x\bar{y}z + wx\bar{y}z + \bar{w}xyz + wxyz \\
& + wx\bar{y}\bar{z} + wx\bar{y}z + wyxz + wxy\bar{z} \qquad (4\text{-}67) \\
& + wxyz + w\bar{x}yz
\end{aligned}
$$

Note that we have written the minterms corresponding to each subcube on a separate line. We have repeated terms. For instance, the minterm wxyz appears three times. This is allowable since A+A=A. It may seem that doing this is contrary to the idea of minimization. Actually, it helps minimization. For instance, see Table 4-2, increasing the size of the subcube results in simpler functions for it. Also, consider Figs. 4-3 and 4-4. Figure 4-4 has one more minterm but it reduces to a much simpler expression. That is, Eq. (4-54) is simpler than Eq.(4-52). It is for reasons such as these that we repeat minterms. That is, we form subcubes which are as large as possible. Of course, we do not use more subcubes than necessary.

Now let us consider the reduction of Eq. (4-67). Let us write

$$f(w,x,y,z) = \alpha_1 + \alpha_2 + \alpha_3 + \alpha_4 \qquad (4\text{-}68)$$

where α_1, α_2, α_3 and α_4 represent the sums of the minterms for the four subcubes. Thus,

$$\alpha_1 = \bar{w}\bar{x}\bar{y}\bar{z} \qquad (4\text{-}69)$$

This is one term subcube and cannot be reduced further.

$$\alpha_2 = \bar{w}x\bar{y}z + wx\bar{y}z + \bar{w}xyz + wxyz \qquad (4\text{-}70a)$$

Using the rule for obtaining the value of a subcube(or just reducing)Eq.(4-70a) results in

$$\alpha_2 = xz \qquad (4\text{-}70b)$$

Two variables have been eliminated. Similarly,

$$\alpha_3 = wx\bar{y}\bar{z} + wx\bar{y}z + wxyz + wxy\bar{z} = wx \tag{4-71}$$

$$\alpha_4 = wxyz + w\bar{x}yz = wyz \tag{4-72}$$

Note that each subcube has reduced ao a single product term and that the number of terms in the subcube decreases with the size of the subcube, see Table 4-2. Again, this substantiates our observations.

The general procedure is to form subcubes which are as large as possible. Let us state some rules which will help the formulation. The basic idea of the rules is to eliminate those cells that cannot be used in the larger subcubes. The remaining cells can then be used in larger subcubes. The following steps are used:

1)Find all single 1-cells that *cannot be made part of a larger subcube*. Encircle each one. These are single cell subcubes. For instance, in Fig. 4-9, the cell m_0 is such a subcube.

2)Next find all cells with 1's that can be made part of a two cell, but no larger, subcube. Encircle each pair. For instance, in Fig. 4-9, the cells m_{15} and m_{11} form such a subcube.

3)Next find all cells with 1's that can be made part of a four, but no larger, cell subcube. Encircle each group of four. For instance, in Fig. 4-9, the cells m_5. m_3, m_7 and m_{15} constitute one such set and m_{12}, m_{13}, m_{15}, and m_{14} constitute another set. A subcube should not be picked if *all* of its cells have been covered in an earlier step.

4)Repeat this procedure using groups of eight. Then, repeat using groups of 16. If the Karnaugh map has more than four variables, keep repeating this procedure doubling the size of the subcubes in each step.

5)When all possible subcubes are covered, stop. At this point, all 1's on the Karnaugh map should be in subcubes. Now examine the Karnaugh map to see if any subcubes can be removed and still have all the 1's covered by subcubes. If this can be done, then these "superfluous" subcubes should be removed. Now the procedure is completed.

Remember that no subcube is picked if it can be completely made part of a larger subcube. However, individual cells of one subcube may be part of larger subcubes. Thus, this procedure attempts to cover all the 1-cells with the largest possible subcubes, and the smallest number of subcubes. An illustration of this procedure as applied to the Karnaugh map of Fig. 4-9 is shown in Fig. 4-10.

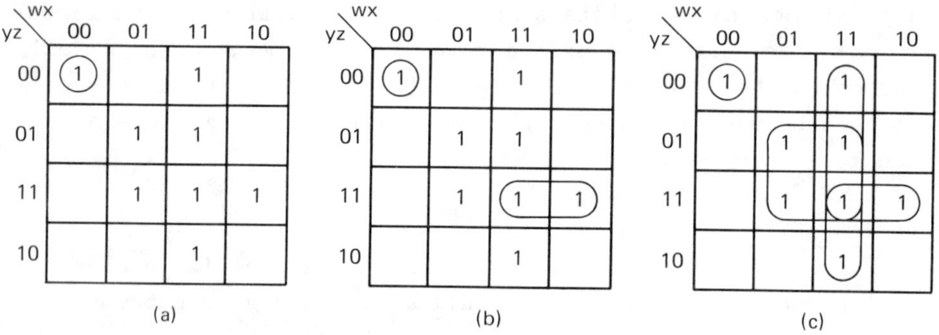

Figure 4-10. An illustration of the minimization procedure used in Fig. 4-9. (a) Karnaugh map after step 1; (b) Karnaugh map after step 2; (c) Karnaugh map after step 3.

At times, the procedure will lead to a unique minimization. That is, only one set of subcubes will be generated. However, this is not always the case. In Fig. 4-11 we illustrate an example where two different minimizations are possible. The subcubes indicated by the dashed lines are not part of the minimization. All of the subcubes, including those indicated by the dashed lines would be generated by the procedure. (The subcubes of Figs. 4-11a and b are the same when all subcubes are considered.) *However, there is one excess subcube*. In Figs. 4-11a and b, the minimization is completed without using the subcube indicated by dashed lines. Note that a different excess subcube is removed in each figure. Thus, two different minimizations result.

Note that all possible subcubes need not be used. For instance, in Fig. 4-11, four subcubes (each with 4 cells) were

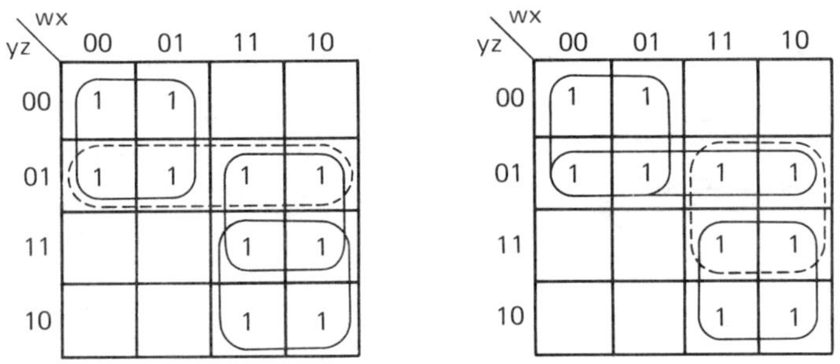

(a) (b)

Figure 4-11. Two different minimizations of the same Karnaugh map.

picked. However, only three subcubes were used in each of the two minimizations.

Karnaugh Maps With More Than Four Variables

The size of a Karnaugh map increases as 2^n where n is the number of literals. Thus, for large values of n, they become unwieldy. In fact, Karnaugh maps are rarely used if the number of variables exceeds 7. The convenience of determining subcubes is diminished if there are more than four literals.

Let us illustrate the use of a Karnaugh map if there are six variables. Consider Fig. 4-12. There, a map is shown with the minterm designations.

xyz \ uvw	000	001	011	010	110	111	101	100
000	m_0	m_8	m_{24}	m_{16}	m_{48}	m_{56}	m_{40}	m_{32}
001	m_1	m_9	m_{25}	m_{17}	m_{49}	m_{57}	m_{41}	m_{33}
011	m_3	m_{11}	m_{27}	m_{19}	m_{51}	m_{59}	m_{43}	m_{35}
010	m_2	m_{10}	m_{26}	m_{18}	m_{50}	m_{58}	m_{42}	m_{34}
110	m_6	m_{14}	m_{30}	m_{22}	m_{54}	m_{62}	m_{46}	m_{38}
111	m_7	m_{15}	m_{31}	m_{23}	m_{55}	m_{63}	m_{47}	m_{39}
101	m_5	m_{13}	m_{29}	m_{21}	m_{53}	m_{61}	m_{45}	m_{37}
100	m_4	m_{12}	m_{28}	m_{20}	m_{52}	m_{60}	m_{44}	m_{36}

Figure 4-12. A six variable Karnaugh map.

Because of the Gray coding, all of the previous discussions apply. For instance, suppose that we had a subcube consisting of m_{25}, m_{17}, m_{27} and m_{29}, see Fig. 4-12. This could be represented by the term $\bar{u}v\bar{x}z$. Note that these are the unchanging literals in the subcube. In a similar way, we can simplify any subcube.

A problem arises here. There are subcubes whose cells are not physically adjacent to each other. For instance, consider

$$m_{26} + m_{58} = \bar{u}vw\bar{x}y\bar{z} + uvw\bar{x}y\bar{z} \tag{4-73}$$

There is only one change of variable. Thus, these cells are said to be adjacent even though they are not physically adjacent. Thus, we cannot just rely on cell's being physically adjacent to define subcubes in the higher ordered Karnaugh map. However, there is a simplification here. Observe the heavy lines drawn in the Karnaugh map. Cells which have mirror symmetry about this line will differ in only one variable. Thus, these cells will also be adjacent cells. This is the case for $m_{26}+m_{58}$. As an additional example, consider a subcube which covers four cells. It consists of $m_{27} + m_{59} + m_{31} + m_{63}$. The same type of reflected symmetry existed in the other Karnaugh maps that we studied. For instance, Figs. 4-8 and 4-5 indicate adjacent cells that were not physically adjacent. In the case of four or fewer variables on Karnaugh maps, it was convenient to think of the edges as touching. However, the mirror symmetry discussed here also applies.

If we have only a five variable Karnaugh map, then it is similar to the upper half of the six variable map. A seven variable Karnaugh map is indicated in Fig. 4-13. Note that there are now three vertical "reflecting lines" as well as one horizontal one. Subcubes not only cover physically adjacent cells, but also cells which have symmetry about the reflecting lines. Two such four cell subcubes are shown. Using the previously developed rules, we can express the function represented by this as

$$f(t,u,v,w,x,y,z) = \bar{u}\bar{v}w\bar{y}z + tvw\bar{y}z \tag{4-74}$$

Note that a simplified function can easily be obtained once the subcubes are identified.

xyz \ tuvw	0000	0001	0011	0010	0110	0111	0101	0100	1100	1101	1111	1110	1010	1011	1001	1000
000																
001		1									1			1	1	
011																
010																
110																
111																
101		1									1			1	1	
100																

Figure 4-13. A seven variable Karnaugh map.

Don't Care Conditions

We have thus far assumed that we care about every possible cell in the Karnaugh map. That is, we assume that the 1's on the Karnaugh map indicate when the function should be 1 and that it should be 0 for all other combinations of the literals. This is often the case, but it may not always be so. For instance, suppose that we want to represent the alphabet by a code of 5 binary digits. There are 26 letters. However, 5 binary digits can represent 32 different quantities. Thus, there must be 7 combinations of binary digits which are never used. That is, in a properly working system, those values would never be generated. Thus, we "don't care" what the function is when they occur(since they never occur). These are called *don't care conditions*

Don't care conditions also arise if a function is actually not a function of one particular variable but, for convenience, or to conform with other functions in the system, we list it as a function of that variable. For instance, consider Eqs.(4-56) and (4-58). Here we have

$$f(x,y,z) = x\bar{y}\bar{z} + \bar{x}\bar{y}\bar{z} + xy\bar{z} + \bar{x}y\bar{z} = \bar{z} \qquad (4\text{-}75a)$$

Here we don't care about the values of x and y. We can write this the following way

$$f(x,y,z) = f(d,d,z) \qquad (4\text{-}75b)$$

The d is read either as 0 or 1.

yz \ wx	00	01	11	10
00				d
01		1	d	
11		d	1	
10	d			

Figure 4-14. A four variable Karnaugh map indicating a don't care condition.

If there are don't care minterms, then these can be assumed to be either 0 or 1 on the Karnaugh map. The value is chosen so that there is a minimum number of subcubes and that they are as large as possible. For instance, suppose that we have the Karnaugh map given by Fig. 4-14. If we consider that the d=0, then there are two one cell subcubes and the switching function is

$$f(w,x,y,z) = \bar{w}x\bar{y}z + wxyz \tag{4-76}$$

However, if the d's corresponding to m_7 and m_{13} are considered to be 1's and the d's equivalent to m_2 and m_8 are considered to be 0's, then we have

$$f(w,x,y,z) = xz \tag{4-77}$$

Note that for all cells but the "don't care" ones, both Eqs. (4-76) and (4-77) yield the same results.

KARNAUGH MAPS USING MAXTERMS

We have based the previous development on simplification of a sum of minterms. Actually we could proceed in a dual way and work with the product of maxterms. In this case, we work with the 0's on the Karnaugh map. Now, it is convenient to mark the 0-cells and leave the 1-cells blank. Note that the Karnaugh map for a function is unique. It does not change if we work with minterms or maxterms.

Let us consider an example. We shall use the Karnaugh map of Fig. 4-15. This is *exactly* the same as the Karnaugh map of Fig. 4-6. We have indicated the 0's instead of the 1's as was done in Fig. 4-6. Actually, a complete Karnaugh map could show both 0's and 1's. However, this would be too cluttered. We can easily obtain the function as a product of maxterms by pretending that the 0's are 1's and then expanding the complement as a sum of minterms and, finally taking the complement of the result. Doing this yields

$$\begin{aligned}\bar{f}(w,x,y,z) = {} & \bar{w}\bar{x}\bar{y}\bar{z} + \bar{w}\bar{x}\bar{y}z + w\bar{x}\bar{y}\bar{z} + w\bar{x}\bar{y}z \\ & + \bar{w}\bar{x}yz + \bar{w}xyz + wxyz + w\bar{x}yz \end{aligned} \tag{4-78}$$

Taking the complement, we obtain

$$f(w,x,y,z) = (w+x+y+z)(w+x+y+\bar{z})(\bar{w}+x+y+z)(\bar{w}+x+y+\bar{z})$$
$$(w+x+\bar{y}+\bar{z})(w+\bar{x}+\bar{y}+\bar{z})(\bar{w}+\bar{x}+\bar{y}+\bar{z})(\bar{w}+x+\bar{y}+\bar{z}) \qquad (4\text{-}79)$$

We could have minimized $\bar{f}$ before we took the complement. Remember that when we obtain $\bar{f}$, we consider that the 0's are 1's. Thus, we form subcubes as before. These are shown in Fig. 4-15. Thus,

$$\bar{f}(w,x,y,z) = \bar{x}\bar{y} + yz \qquad (4\text{-}80)$$

Then, taking the complement yields

$$f(w,x,y,z) = (x+y)(\bar{y}+\bar{z}) \qquad (4\text{-}81)$$

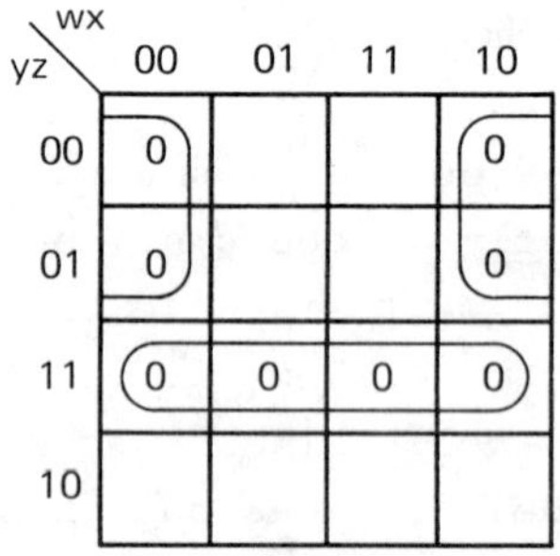

Figure 4-15. A four variable Karnaugh map which is identical to Fig. 4-6.

After a little experience is gained with this procedure, it will not be necessary to go through the steps of taking the complement. The following procedure is used. The subcubes are determined using the same rules except that the 0's take the place of the 1's in the previous development. Each subcube represents a sum of literals. As before, the literals which appear are those whose values do not change in the minterms corresponding to the cells of the subcube. However, the complemented literals of the maxterms are now represented by 1's and the uncomplemented literals are represented by 0's. Finally, the product of all the sums (each one representing a subcube) is taken. Thus, this procedure is the dual of the one using minterms.

4.3 PRIME IMPLICANTS—QUINE MC CLUSKEY METHOD

The Karnaugh map minimization technique is very easy to use if the number of literals is four or less. However, if there are more than seven literals, it becomes unwieldly. In this section we shall discuss another procedure which is easier to use when the number of literals is large. Before discussing the procedure, we must consider some fundamental definitions.

Prime Implicants

We want to minimize a switching function. To aid us we shall now discuss some definitions.

A switching function f is said to *imply* another switching function g if g is 1 whenever f is 1. Let us consider an example of this. Suppose

$$f = xy \tag{4-82a}$$

$$g = xy + \bar{x}z + yz \tag{4-82b}$$

Whenever f is 1, g will also be 1. Note that f and g are not equivalent since g may be 1 when f is zero. f is said to be an *implicant* of g.

If f is an implicant of g, then g is said to *cover* f. That is, g covers f if g is 1 whenever f is 1. For instance, see Eq. (4-82), g covers f. (Note that g may be 1 when f is 0.)

A *prime implicant* is a special form of implicant. It must be a product of literals and, also, it must be such that if *any* literal is deleted from the product, the resulting term is no longer an implicant. For instance, consider Eqs.(4-82), f is a prime implicant of g since (1)f is an implicant of g; (2)it is a product of literals and, (3)if any literal is deleted from f, it will no longer be an implicant. That is, x alone is not an implicant and y alone is not an implicant. Consider another example

$$f_1 = xyz \tag{4-83}$$

is an implicant of g, since whenever $\hat{f_1}=1$, g=1. However, we can delete the z from f_1 and still have an implicant of g. Thus, f_1 is not a prime implicant of g.

If f is an implicant of g, then when f=1, g=1. However, if f=0, g may be either 0 or 1. If we combine the sums of enough implicants such as

$$A = f_1 + f_2 + \dots + f_n \tag{4-84}$$

then we may have the situation that A=1 whenever g=1. If g=0, then all its implicants must be 0. Thus, A=0 whenever g=0. Hence, we can then state

$$g = A \tag{4-85}$$

In this case, we have expressed g as a sum of its implicants. It is always possible to find such a sum. For example, a switching function can always be expressed as a sum of minterms and each minterm is an implicant. If we are interested in minimizing the function, then each implicant used in the expansion should be a prime implicant, since it does not have any unneeded literals. Simply expanding a function in terms of prime implicants does not necessarily produce a minimal function. If, however, the appropriate prime implicants are used, the function will be minimal. We shall now demonstrate this and discuss some ideas which will enable us to obtain minimal functions.

We can use Karnaugh maps to determine prime implicants and to expand the given function in terms of these implicants. We shall do this since it will provide some insight into the process. However, we shall also develop techniques that do not require Karnaugh maps since this is our objective here. Consider the minimal functions obtained from the subcubes of the Karnaugh map. These were sums of products, each product representing a single subcube. Each product is a prime implicant. Note that when each product in the sum was 1, the function was 1, and that no variable can be omitted from a product since(we assume that) the minterms covered by each subcube have been expressed in the minimal form discussed in the last section. For example, consider Fig. 4-16. Each subcube represents a prime implicant.

These are

$\bar{w}\bar{x}\bar{z}$, wxz, xy and $y\bar{z}$

Of course, the sum of these prime implicants gives the function.

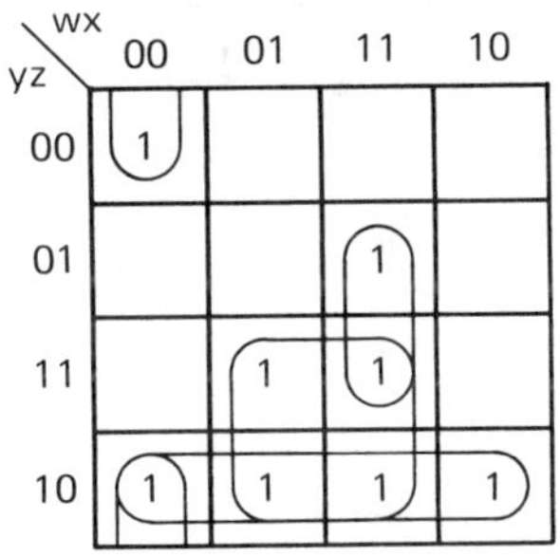

Figure 4-16. A four variable Karnaugh map showing an optimum set of prime implicants.

If a prime implicant covers *at least* one minterm which is *not covered by any other prime implicant*, it is called an *essential prime implicant*. All of the prime implicants shown in Fig. 4-16 are essential prime implicants. However, we shall not always work only with essential prime implicants. We shall illustrate this subsequently.

Now condider the Karnaugh map of Fig. 4-17. Note that four prime implicants are shown there. However, one consisting of $m_7 + m_{15} + m_6 + m_{14}$ contains only minterms which have been included in the other prime implicants. Thus, this prime implicant need not be included in the sum for the function. Such a prime implicant, *all* of whose minterms are contained in essential prime implicants, is called an *absolutely eliminable prime implicant*. Now let us develop a procedure for minimizing switching functions without using Karnaugh maps.

Quine–Mc Cluskey Simplification

We can state some general rules for minimizing functions using prime implicants.

The basic idea is to express the function as a sum of products. Thus, (1) determine the prime implicants; (2) then determine the essential prime implicants. If all the minterms are

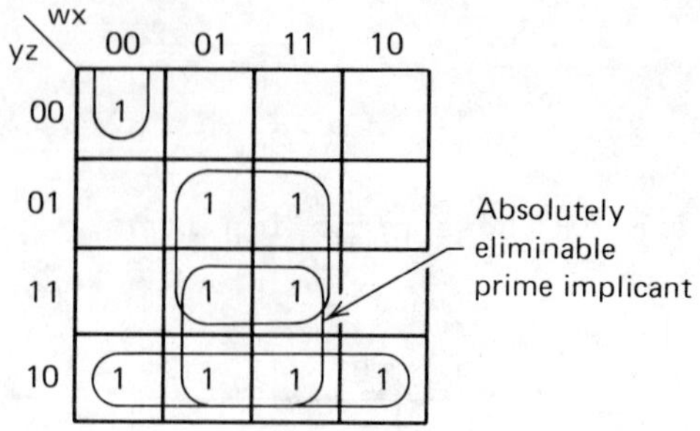

Figure 4-17. A four variable Karnaugh map.

covered by these, then the function has been minimized. If all the minterms have not been covered, then (3) add additional prime implicants, in some minimal way, until all the minterms are covered.

These rules will yield a reduced function. However, it is not obvious how to implement them when the function is complex. We shall develop a procedure which will enable us to do this. For instance, we must determine the essential prime implicants for step 2 and we must choose any additional prime implicants in step 3 in a minimal way. The Karnaugh map provides a procedure for doing this. Since we do not want to use maps, a tabular procedure which is suitable for a large number of variables, and which is also suitable for computer implementation, will be developed.

The first step is to identify all the prime implicants. We assume that we start with a canonical sum of products form(i.e., a sum of minterms). If we used a Karnaugh map, we would identify subcubes. Let us consider what is done there so that we can develop a tabular procedure. Adjacent terms on the Karnaugh map differ in only one literal and that literal is complemented in one term and uncomplemented in the other. Thus, we can consider that a two cell subcube is formed by eliminating these literals. A four cell subcube can be considered to result from the simplification of two adjacent two cell subcubes. This procedure can be repeated for higher ordered subcubes.

Let us now discuss a tabular procedure that can be used to implement the process of forming prime implicants or of finding subcubes. We start with the function as a canonical sum of minterms. For instance, for the Karnaugh map of Fig. 4-16, we have

$$f(w,x,y,z) = \bar{w}\bar{x}\bar{y}\bar{z} + \bar{w}\bar{x}y\bar{z} + wx\bar{y}z + wxyz$$

$$+ \bar{w}xyz + \bar{w}xy\bar{z} + wxy\bar{z} + w\bar{x}y\bar{z} \qquad (4\text{-}86)$$

We want to determine in some systematic way if minterms are adjacent to each other. We use the following procedure. Write each minterm using a coded notation. Replace each uncomplemented variable by a 1 and each complemented variable by a 0. Thus we have

$\bar{w}\bar{x}\bar{y}\bar{z} \rightarrow 0000$

$\bar{w}x\bar{y}z \rightarrow 0101$

$\bar{w}xyz \rightarrow 0111$

etc.

We now form a table containing the digital representation of all the minterms. The table is arranged in groups. The first group contains those minterms with no 1's. The second group contains those minterms with exactly one 1, the third group contains those minterms with exactly two 1's, etc. Thus, for Eq. (4-86), the table is

TABLE 4-3
CODED REPRESENTATION OF THE MINTERMS OF EQ. (4-86)

Minterm	Coded Representation
$\bar{w}\bar{x}\bar{y}\bar{z}$	0000 No 1's
$\bar{w}\bar{x}y\bar{z}$	0010 One 1
$\bar{w}xy\bar{z}$	0110
$w\bar{x}y\bar{z}$	1010 Two 1's
$\bar{w}xyz$	0111
$wx\bar{y}z$	1101
$wxy\bar{z}$	1110 Three 1's
$wxyz$	1111 Four 1's

In Table 4-3 we have written the minterms. This is usually not done; it was just done for illustrative purposes here.

Now suppose that we want to combine two minterms. They must be identical except for one variable, which must be complemented in one and uncomplemented in the other. Thus, if we are to combine two minterms, their coded representation must be identical except for one column. There must be a 1 in one and a 0 in the other. We have grouped the minterms in order of the number of 1's. Thus, starting with the first group (no 1's) if we want to see if a minterm can combine with another, we must compare it with each minterm in the *next* group. Let us consider this and rewrite Table 4-3 to illustrate it.

TABLE 4-4
PRIME IMPLICANTS FOR EQ. (4-86)

	First reduction	Second reduction
w x y z	w x y z	w x y z
0 0 0 0 ✓	0 0 - 0	
0 0 1 0 ✓	0 - 1 0 ✓	- - 1 0
0 1 1 0 ✓	- 0 1 0 ✓	- 1 1 -
1 0 1 0 ✓	0 1 1 - ✓	
0 1 1 1 ✓	- 1 1 0 ✓	
1 1 0 1	1 - 1 0 ✓	
1 1 1 0 ✓	- 1 1 1 ✓	
1 1 1 1 ✓	1 1 - 1	
	1 1 1 - ✓	

Consider 0000 and 0010. They differ in the third column, Thus, their sum can be represented by 00-0. That is, $\bar{w}\bar{x}\bar{y}\bar{z} + \bar{w}\bar{x}y\bar{z} = \bar{w}\bar{x}\bar{z}$. The dash indicates which variable is missing. It is important to keep track of this. For instance, 0-00 $\rightarrow \bar{w}\bar{y}\bar{z}$ is different from 00-0 $\rightarrow \bar{w}\bar{x}\bar{z}$.

The difference between two combined terms must be that the 1's of the lower group coincide exactly with the 1's of the upper group and that the lower group has an additional 1.

When we combine two rows, we put check marks on those rows. This means that these minterms have been combined. Thus, they are not prime implicants and are not needed in the expression. Note that even if a row has been checked off(because of combination with an upper group term)it is still compared with a row below it. This is because a minterm can be in more than one subcube.

After all the eliminations have been completed, the column entitled "first reduction" is obtained. It is possible that there can be reduction of those terms. (Note that Fig. 4-16 shows this because there are subcubes covering four cells and the first reduction only combines pairs of cells. Thus, we would now expect further reduction involving these pairs.)

Now we repeat the reduction using the second column. Note that because of the ordering of the rows, when a combination occurs, the number of 1's in the reduced element will be the same as the number of 1's that were in the original upper element. Thus, the second column will be ordered as before.

Examine the two terms in the second column. Each has been obtained from two reductions. For instance, 0-10 + 1-10 = --10 and also, -010 + -110 = --10 .

If two variables are to combine, they must only differ in one column and this difference must be a 0 in the upper row and a 1 in the lower row. Then, any dashes(-)must coincide. Actually, the dashes are easy to check visually and their presence speeds the process.

After the second reduction is performed, the process is repeated if possible. In this example, we must stop after the second reduction.

W hen the process is completed, all the unchecked terms are prime implicants. Thus, we can write

$$f(w,x,y,z) = \bar{w}\bar{x}\bar{z} + wxz + y\bar{z} + xy \qquad (4\text{-}87)$$

The sum of prime implicants may not be in minimum form since the tabular minimization makes use of the relation $a(b+\bar{b})=a$, but it does not use $a+ab=a$. We must now test to see if the function is in minimum form or if some prime implicants can be eliminated.

We shall again use a tabular type procedure to be systematic. Now we shall use a two dimensional chart listing the minterms along one axis and the prime implicants along the other. In this way we can see which prime implicants cover the needed minterms. Thus, any duplicating prime implicants can be eliminated. Let us illustrate this by continuing the example. In Fig. 4-18 we have drawn the prime implicant choice chart. The minterms which are covered by a prime implicant can be determined by considering all possible values of the blank. For instance, if we have the prime implicant 00-0, then it covers the minterms 0010 and 0000. That is, $\bar{w}\bar{x}\bar{z} = \bar{w}\bar{x}y\bar{z} + \bar{w}\bar{x}\bar{y}\bar{z}$. In a similar way, --10 covers the minterms 0010 and 1010 and 0110 and 1110. That is

$$y\bar{z} = \bar{w}\bar{x}y\bar{z} + w\bar{x}y\bar{z} + \bar{w}xy\bar{z} + wxy\bar{z}$$

We put check marks on the table to indicate which minterms are covered by the corresponding prime implicants.

Prime implicants \ Minterms	0000	0010	0110	1010	0111	1101	1110	1111
√ 00−0	√	√						
√ 11−1						√		√
√ −−10		√	√	√			√	
√ −11−			√		√		√	√

Figure 4-18. A chart used to determine the optimum set of prime implicants from Table 4-4.

Next we determine the essential prime implicants, and put a check mark next to them. This is done in the following way. Examine each column. If a column has only one check then that minterm is covered by only one prime implicant. Thus, this must be an essential prime implicant.

After all the essential prime implicants are determined, we examine the columns to determine if there are any minterms that are not covered. If there are, then these must be covered by some, or all, of the remaining prime implicants. This may involve a trial and error procedure to determine a minimal realization. Let us consider some rules which can be used to help in this choice.

If *all* of one prime implicant's checked columns also have checks from the essential prime implicants, then there is no reason to ever use that prime implicant. In fact, it is an absolutely eliminable prime implicant. In Fig. 4-18 there are no such prime implicants.

The number of (-)'s in the prime implicant representation is analogous to the size of a subcube. If there are no (-)'s, then that prime implicant will cover one minterm. If there is one (-), then that prime implicant will cover two minterms. If there are two (-)'s, then that prime implicant will cover four minterms. In general, we can state that if there are k dashes (-), that the prime implicant will conver 2^k minterms.

The above rules can aid in the selection of the prime implicants needed to cover the minterms which remain after the essential prime implicants have been chosen.

In the examples we considered, the essential prime implicants cover all the minterms. Indeed, all the prime implicants were essential. Thus, using the checked prime implicants in Fig. 4-18, we obtain

$$f(w,x,y,z) = \bar{w}\bar{x}\bar{z} + wxz + y\bar{z} + xy \qquad (4\text{-}88)$$

In this case we did not reduce the function using the chart. At the end of this section we shall consider examples where we do.

Don't Care Conditions

If there are don't care conditions, see Sec. 4-2, they can aid in the reduction of the expression. The following procedure is used. The don't care minterms are treated as 1's when the reduction table is formed. This is done since extra terms here might result in additional combinations, and their resultant reductions. This is equivalent to saying that extra 1's on the

Karnaugh map could result in some larger subcubes. However, in forming the chart that determines which prime implicants to use, *the don't care minterms are omitted*. Note that their inclusion could cause prime implicants to be added to the expansion. Since we don't care about these minterms, we do not include them there. This is equivalent to setting only those don't care minterms, which cause larger subcubes to be formed, equal to 1 and setting the other don't care minterms equal to 0.

As an example, let us assume that, in Fig. 4-16, $m_{12} = wx\bar{y}\bar{z}$ is a don't care minterm. The new Karnaugh map is given in Fig. 4-19. To Table 4-4, we would then add a row 1100. The prime implicant determining table becomes

TABLE 4-5
PRIME IMPLICANTS FOR FIG. 4-19

w x y z	First reduction w x y z	Second reduction
0 0 0 0 ✓	0 0 - 0	
0 0 1 0 ✓	0 - 1 0 ✓	- - 1 0
1 1 0 0 ✓	- 0 1 0 ✓	
0 1 1 0 ✓	1 1 0 - ✓	1 1 - -
1 0 1 0 ✓	1 1 - 0 ✓	- 1 1 -
0 1 1 1 ✓	0 1 1 - ✓	
1 1 0 1 ✓	- 1 1 0 ✓	
1 1 1 0 ✓	1 - 1 0 ✓	
1 1 1 1 ✓	- 1 1 1 ✓	
	1 1 - 1 ✓	
	1 1 1 - ✓	

If we compare this with Table 4-4, we see that one of the prime implicants has become simpler. That is, 11-1 has been replaced by 11--.

The prime implicant determining chart is shown in Fig. 4-20. Note that the 1100 minterm is *not* included. Again we have put checks next to the essential prime implicants. All minterms have been covered by the essential prime implicants, which are the only prime implicants. Thus, the minimized expression is

$$f(wxyz) = \bar{w}\bar{x}\bar{z} + y\bar{z} + wx + xy \tag{4-89}$$

Thus, we have taken advantage of the don't care conditions to reduce the expression somewhat.

yz \ wx	00	01	11	10
00	1		d	
01			1	
11		1	1	
10	1	1	1	1

Figure 4-19. A Karnaugh map with a don't care condition.

Prime implicants \ Minterms	0000	0010	0110	1010	0111	1101	1110	1111
√ 00—0	√	√						
√ ——10		√	√	√			√	
√ 11——						√	√	√
√ —11—			√		√		√	√

Figure 4-20. A chart used to obtain the set of optimum prime implicants from Table 4-5.

Note that the third row has only three checks even though the prime implicant has two dashes. This is becuase one of the checks covers a don't care condition which is omitted from the table.

More Complex Switching Functions

The previous examples considered minimization where the minimization chart was simple in that all the prime implicants were essential ones. However, there are cases where the chart may require study, even after the essential prime implicants are identified. Let us consider such an example.

Suppose that we have the function

$$f(w,x,y,z) = \bar{w}\bar{x}\bar{y}\bar{z} + \bar{w}x\bar{y}\bar{z} + w\bar{x}\bar{y}\bar{z} + \bar{w}xy\bar{z} + w\bar{x}\bar{y}z + wx\bar{y}\bar{z}$$

$$+ \bar{w}xyz + w\bar{x}yz + wxy\bar{z} + wxyz \qquad (4\text{-}90)$$

Its prime implicant determining table is determined as before. Thus, we have

TABLE 4-6
PRIME IMPLICANTS FOR EQ. (4-90)

	First reduction	Second reduction
w x y z	w x y z	w x y z
0 0 0 0 ✓	0 - 0 0 ✓	- - 0 0
0 1 0 0 ✓	- 0 0 0 ✓	- 1 - 0
1 0 0 0 ✓	0 1 - 0 ✓	- 1 1 -
0 1 1 0 ✓	- 1 0 0 ✓	
1 0 0 1 ✓	1 0 0 -	
1 1 0 0 ✓	1 - 0 0 ✓	
0 1 1 1 ✓	0 1 1 - ✓	
1 0 1 1 ✓	- 1 1 0 ✓	
1 1 1 0 ✓	1 0 - 1	
1 1 1 1 ✓	1 1 - 0 ✓	
	- 1 1 1 ✓	
	1 - 1 1	
	1 1 1 - ✓	

Now form the prime implicant chart. This is indicated in Fig. 4-21. In this more complex case, we proceed in the following way. As before, we determine the essential prime implicants and check them. Next consider the minterms covered by each essential prime implicant and check them. Place these checks along the top of the chart. Now examine the unchecked prime implicants. If any of them cover only checked minterms, they are absolutely eliminable prime implicants. Place a cross next to them. The prime implicant -1-0 is absolutely eliminable.

The remaining unchecked or uncrossed prime implicants must cover the unchecked minterms in an optimum way. It is often convenient not to use the original chart to determine this, since it is cluttered. Thus, we redraw the chart using only the unchecked minterms and the unchecked or uncrossed prime implicants. This is shown in Fig. 4-22. From this we see that the two minterms will be covered either by 100- + 1-11 or by

Prime implicants \ Minterms	√ 0000	√ 0100	√ 1000	√ 0110	1001	√ 1100	√ 0111	1011	√ 1110	√ 1111
100—			√		√					
10—1					√			√		
1—11								√		√
√ ——00	√	√	√			√				
× —1—0		√		√		√			√	
√ —11—				√			√		√	√

Figure 4-21. A chart used in the first step to obtain the optimum prime implicants from Table 4-6.

10-1. The second set (10-1)is the minimum one. Thus, we place an asterisk (*) next to it. Then, the reduced function consists of the sum of the checked minterms of Fig. 4-21 and the minterms marked with an asterisk in Fig. 4-22. Hence,

$$f(w,x,y,z) = \bar{y}\bar{z} + xy + w\bar{x}z \tag{4-91}$$

	Prime implicants \ Minterms	1001	1011
a	100—	√	
b	* 10—1	√	√
c	1—11		√

Figure 4-22. Reduced chart used in the second step to obtain the optimum prime implicants from Table 4-6.

In examining Fig. 4-22, we obtained two sets of prime implicants which could be used to cover the minterms. If the chart is large, it may be tedious to identify such sets. The following procedure can be used. Assume that each prime implicant is a variable. These are labeled a,b, and c, in Fig. 4-22. Now consider each column. The term corresponding to the first column will be 1 if a+b is 1. Similarly, the term corresponding to the second column will be 1 if b+c is 1. Then, we can write a function F which will be 1 if both columns are one. The function is

$$F = (a+b)(b+c) \tag{4-92}$$

This represents the prime implicants which cover the minterms. To form the function we examine each column and form a sum consisting of those variables which have a check in that column. Next we take the product of all such sums. Thus, the minterms are all covered. Now let us simplify the resulting expression. Using Eq. (4-11b) we obtain

$$F = b + ac \tag{4-93}$$

Each of the terms in the sum correspond to a set of prime implicants which will cover the desired minterms. Thus, the sets of prime implicants are

10-1

or

100- and 1-11

Of course, thse are the same as the previously determined sets and the minimal expression is the one determined previously.

There are other reduction techniques that are, at times, less tedious. However, the one discussed here will always find all possible coverings.

4.4 MULTIPLE OUTPUT NETWORKS

Thus far we have considered the minimization of a single switching function. However, there are occasions where we have more than one function of the same variables. For instance, consider the full adder of Fig. 3-18. Here the inputs were two variables a_1 and b_1. There were two outputs s_1 and c_1. The switching functions for these were given by Eqs.(3-25) and (3-26) where

$$s_j(a_j,b_j,c_{j-1}) = \bar{a}_j\bar{b}_jc_{j-1} + \bar{a}_jb_j\bar{c}_{j-1} + a_j\bar{b}_j\bar{c}_{j-1} + a_jb_jc_{j-1} \quad (4\text{-}94a)$$

$$c_j(a_j,b_j,c_{j-1}) = \bar{a}_jb_jc_{j-1} + a_j\bar{b}_jc_{j-1} + a_jb_j\bar{c}_{j-1} + a_jb_jc_{j-1} \quad (4\text{-}94b)$$

The most straightfoward procedure for minimizing these functions would be to consider each one individually. However, this often will not yield the simplest network. For instance, even in the unminimized form, we see that both functions have a minterm in common($a_jb_jc_{j-1}$). Thus, one gate could realize this minterm for both functions. This saving of a gate by sharing it with both functions is illustrated in Fig. 3-18.

One procedure that can be used is to reduce each individual function separately. Then, examine the reduced functions to see if they have any prime implicants in common. Only one gate should be used to realize each set of common prime implicants. Thus, some saving results. This is not a general procedure, as we shall subsequently discuss. However, it often yields satisfactory results. Let us illustrate it.

Consider the two functions whose Karnaugh maps are given in Fig. 4-23. If each is minimized separately, we obtain

$$f_1(w,x,y,z) = xyz + wyz \quad (4\text{-}95a)$$

and

$$f_2(w,x,y,z) = \bar{w}xz + xyz \quad (4\text{-}95b)$$

These can share an AND gate, the one whose output is xyz. This can be easily seen since both maps have a common subcube. The gate circuit realization is illustrated in Fig. 4-24. The drawing of these circuits has been simplified. We assume that the complement is available. Also, we have not drawn common bus bars for the independent variables but have just labeled the gate inputs. It is assumed that all the x inputs are connected together, etc. This reduces the clutter of the diagrams and makes them easier to read. As discussed, Fig. 4-24b has one less gate than does Fig. 4-24a.

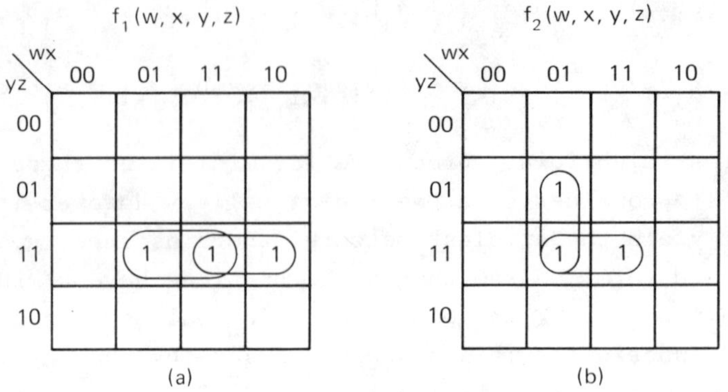

Figure 4-23. Karnaugh maps for two functions of the same variables: (a) $f_1(w, x, y, z)$; (b) $f_2(w, x, y, z)$.

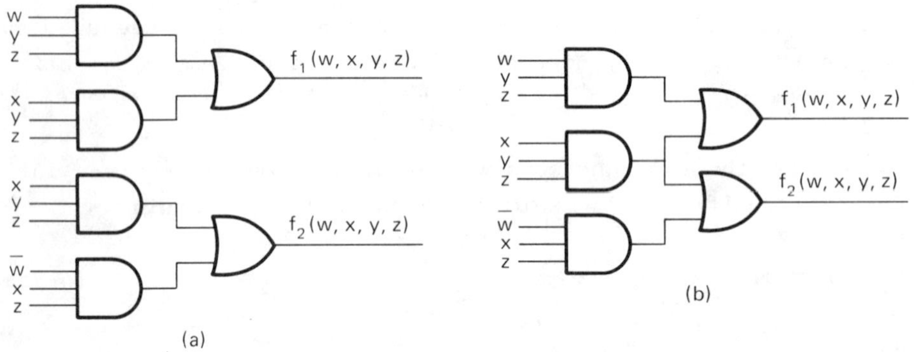

Figure 4-24. (a) A realization of Eqs. (4-95); (b) another realization that uses one less gate.

Often, when we have multiple output functions, we can obtain reductions other than these simple ones. Let us illustrate this and then determine a procedure for implementing this reduction. Consider the Karnaugh maps of Fig. 4-25. The values of these functions are

$$f_1(w,x,y,z) = xyz + wyz \tag{4-96a}$$

and

$$f_2(w,x,y,z) = \bar{w}xz + x\bar{y}z \tag{4-96b}$$

These have no prime implicants in common. However, a reduction can still be made. We can write

$$f_1(w,x,y,z) = \bar{w}xyz + wyz \tag{4-97a}$$

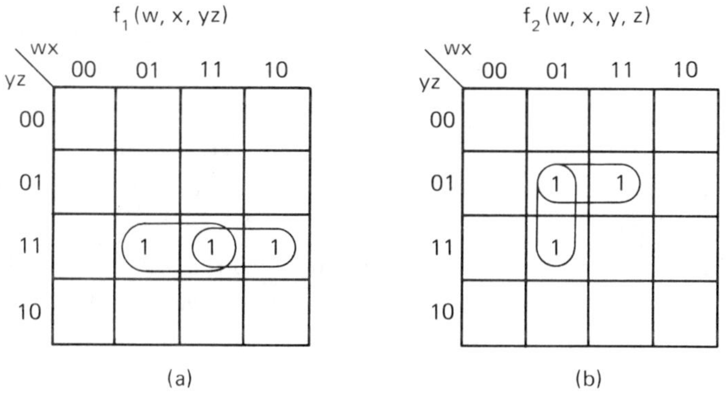

Figure 4-25. Karnaugh maps of two functions of the same variables: (a) $f_1(w, x, y, z)$; (b) $f_2(w, x, y, z)$.

Note that

$$\bar{w}xyz + wyz = yz(\bar{w}x+w) = yz(w+x) = xyz + wyz$$

We can also write

$$f_2(w,x,y,z) = \bar{w}xyz + x\bar{y}z \tag{4-97b}$$

Note that

$$\bar{w}xyz + x\bar{y}z = xz(\bar{w}y+\bar{y}) = xz(\bar{w}+\bar{y}) = \bar{w}xz + x\bar{y}z$$

Now, f_1 and f_2 have the minterms $\bar{w}xyz$ in common. Thus, both functions can be realized using three AND gates instead of four. Note that $\bar{w}xyz$ is an implicant of both f_1 and f_2. However, it is a prime implicant of f_1f_2.

If the switching function is complex, we may not be able to easily identify modifications such as that which resulted in Eq. (4-97). The following procedure will enable us to systematically reduce the functions.

Suppose that we want to realize several functions of the same variables; $f_1(w,x,y,z,\ldots)$, $f_2(w,x,y,z,\ldots)$, and $f_3(w,x,y,z,\ldots)$. Now consider their product $f_1f_2f_3$. If b is a prime implicant of the product, then it also will be an implicant of each of f_1, f_2, and f_3, since when b=1, $f_1=1$, $f_2=1$, and $f_3=1$. Thus, when we minimize, we need not only consider the prime implicants of f_1, f_2, and f_3, but we can also consider the prime implicants of f_1f_2, f_1f_3, f_2f_3, and of $f_1f_2f_3$. Note that if we can find prime implicants of the product of two, or more, functions, then these will be implicants of those functions. We can realize these with common gates which will probably result in a saving of gates. Thus, a minimization procedure would be to consider the functions and all combinations of their products when obtaining the prime implicants. A chart type of simplification can then be used to obtain an optimum network.

A word of caution is in order here. When we combine minterms, we must keep track of which functions they originated from. In general, we should keep the minterms for f_1f_2, f_1f_3, f_2f_3 and $f_1f_2f_3$ separate and obtain the prime implicants for each separately. Then, when f_1 is realized, the prime implicants from f_1, f_1f_2, f_1f_3, and $f_1f_2f_3$ can be used in its realization. *However, the others* (i.e., those from f_2, f_3 and f_2f_3) *cannot be used* for realizing f_1. A simple example will illustrate this. Suppose that $f_1=x$ and $f_2=\bar{x}$. $\bar{x}$ is a prime implicant of f_2. However, it cannot be used in realizing f_1.

Let us now consider an example. Suppose that

$$f_1(wxyz) = m_0 + m_1 + m_5 + m_9 \tag{4-98a}$$

$$f_2(wxyz) = m_0 + m_1 + m_7 + m_{10} \tag{4-98b}$$

$$f_3(wxyz) = m_0 + m_1 + m_5 + m_7 \tag{4-98c}$$

where we have used the notation of Fig. 4-7. Then we form the prime implicant determining table.

TABLE 4-7
PRIME IMPLICANTS FOR EQS. (4-98)

f_1	0000 ✓	000- ×
	0001 ✓	0-01 ×
	0101 ✓	-001
	1001 ✓	
f_2	0000 ✓	000- ×
	0001 ✓	
	1010	
	0111 ×	
f_3	0000 ✓	000- ×
	0001 ✓	0-01 ×
	0101 ✓	01-1
	0111 ✓	
f_1f_2	0000 ✓	000- ×
	0001 ✓	
f_2f_3	0000 ✓	000- ×
	0001 ✓	
	0111	
f_1f_3	0000 ✓	000- ×
	0001 ✓	0-01
	0101 ✓	
$f_1f_2f_3$	0000 ✓	000-
	0001 ✓	

(Note that we can find the canonical sum of minterms for the product of two functions simply be determining their minterms in common.)

We form the table for f_1, f_2, f_3, f_1f_2, f_2f_3, f_1f_3, and $f_1f_2f_3$ as though they were independent functions. That is, we form the prime implicants of each without regard to the others. Now examine the table to see if there is any duplication of prime implicants. For instance, f_1f_3 has a prime implicant 0-01. This can be used for both f_1 and f_3. Both f_1 and f_3, individually, have 0-01 as a prime implicant. It is desirable to use the f_1f_3 prime implicant since this represents gates in common. Thus, we place × marks next to the 0-01's in f_1 and f_3. This means that we shall not use them. Instead, we shall use the 0-01 from f_1f_3. Similarly, we can eliminate the 000- from f_1, f_2 and f_3, since it appears in f_1f_2, f_2f_3, and f_1f_3. It can be eliminated from these since it appears in $f_1f_2f_3$. In general, when there are repeated prime implicants, we keep the one in the term which has the largest number of "factors". Of course, we can only eliminate a prime implicant for a function if it also appears as a prime implicant of a function which is the product of the *original* function and other functions.

Prime implicants \ Minterms	f_1				f_2				f_3			
	√ 0000	√ 0001	√ 0101	√ 1001	√ 0000	√ 0001	√ 1010	√ 0111	√ 0000	√ 0001	√ 0101	√ 0111
√f_1 { −001		√		√								
√f_2 { 1010							√					
×f_3 { 01−1											√	√
√f_2f_3 { 0111								√				√
√f_3f_1 { 0−01		√	√							√	√	
√$f_1f_2f_3$ { 000−	√	√			√	√			√	√		

Figure 4-26. A chart used to obtain the optimum prime implicants from Table 4-7.

Now we form the chart used to obtain the optimum prime implicants, see Fig. 4-26. Note that only the appropriate boxes are checked. For instance, the prime implicants of f_1 only cover f_1, the prime implicants of f_2f_3 only cover f_2 and f_3. Next we check off the essential prime implicants. Then, the covered minterms are checked. Now we look for absolutely eliminable prime implicants. In this example, 01-1 for f_3 is an absolutely eliminable prime implicant. Since all the remaining prime implicants are essential prime implicants, there is no need to draw a reduced chart. To obtain each function we add all the prime implicants used in covering the function. Hence,

$$f_1(w,x,y,z) = \bar{w}\bar{x}\bar{y} + \bar{w}\bar{y}z + \bar{x}\bar{y}z \tag{4-99a}$$

$$f_2(w,x,y,z) = \bar{w}\bar{x}\bar{y} + \bar{w}xyz + w\bar{x}y\bar{z} \tag{4-99b}$$

$$f_3(w,x,y,z) = \bar{w}\bar{x}\bar{y} + \bar{w}\bar{y}z + \bar{w}xyz \tag{4-99c}$$

A realization of this is shown in Fig. 4-27.

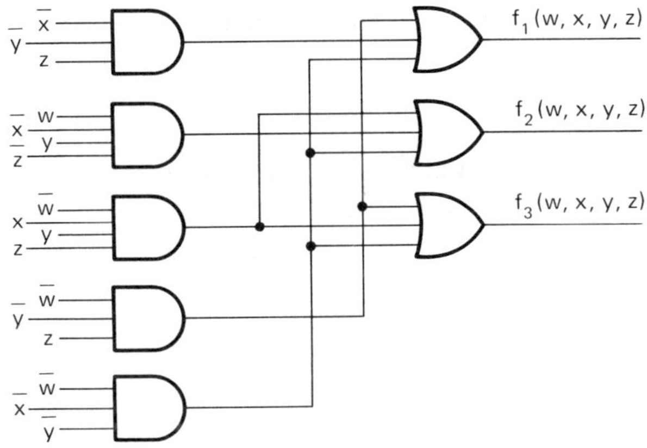

Figure 4-27. A gate circuit realization of eqs. (4-99).

When tables such as Table 4-7 are formed, there is much duplication of minterms. Condensed forms of the tables can be used where a tag is added to the minterms indicating to which function(s) they belong. Then, minterms do not have to be repeated. Such condensed forms are given in the McClusky reference cited in the Bibliography at the end of this chapter. It is not used here since it is felt that the noncondensed form of table is more appropriate for beginners to use. The ideas are the same in both cases.

4.5 USE OF DECIMAL NOTATION IN PRIME IMPLICANT TABLES

When we formed the tables to determine the prime implicants, a binary code was used to represent the minterms. It is more compact to use the decimal(or octal)representation of these numbers, in the tables. For instance

$$w\bar{x}yz \rightarrow 1011 \rightarrow 11_{10}$$

When we combine two minterms, the upper minterm is combined with the one below it which has exactly one more 1. All the 1's of the upper minterm exactly coincide with 1's of the lower minterm. Thus, on a decimal basis, two minterms that combine will be in adjacent groups and the lower minterm will be exactly

$$2^n \quad n = 0, 1, 2, \ldots$$

more than the lower one.

Let us illustrate this. As an example, we shall work with Table 4-4.

TABLE 4-8
A MODIFICATION OF TABLE 4-4 USING DECIMAL NUMBERS

wxyz		First reduction	Second reduction
0000	0 ✓	0,2(2)	
0010	2 ✓	2,6(4) ✓	2,6,10,14(4,8)
0110	6 ✓	2,10(8) ✓	
1010	10✓	6,7(1) ✓	6,7,14,15(1,8)
0111	7 ✓	6,14(8) ✓	
1101	13	10,14(4) ✓	
1110	14✓	13,15(2)	
1111	15✓	7,15(8) ✓	
		14,15(1) ✓	

Consider the first column. We have shown both the decimal and binary number there. In the first group, the difference between 0 and 2 is 2^1. Thus, in the next column we write 0,2 to represent the combination of the minterms. We also write (2). The difference between the two decimal numbers indicates the column where the 0 and 1 differ in the two minterms. Thus, it supplies the same information as the dash did in the previous table.

We proceed similarly, for instance, 2 and 6 combine with a difference of 4. As usual, we check the combined terms. Again, lines are drawn to separate the new groups. Remember that when a new group is formed, the number of 1's is the same as the number of 1's in the upper(original)group.

Now consider the second reduction. If terms in adjacent groups are to combine, then the dashes(i.e., the number in parentheses) must be the same. Hence, 0,2(2) cannot combine with any number in the next group. However, 2,6(4) can possibly combine with 10,14(4) in the next group since both have (4). We must now check each difference between both sets of corresponding numbers. They must be the same and the lower group must be 2^n more than the upper group. In this case, the difference is 10-2 = 14-6 = 8 = 2^3. Thus, the combination can be made. The resulting term is written in the second reduction as

2, 6, 10, 14(4,8)

Note that 2,10(8) and 6,14(8) combine as 2,6,10,14(8,4) which is the same as 2,6,10,14(4,8). Note that (4,8) is equivalent to (8,4) since they indicate dashes in the same position. We can proceed in this way if further reductions were possible.

The unchecked terms represent the prime implicants. They are, in decimal code:

0,2(2)

13,15(2)

2,6,10,14(4,8)

6,7,14,15(1,8)

Now let us express these in terms of literals. 0,2(2) represents either 0 or 2 expressed in binary with the 2 column missing. That is, 00-0. Note that $0_{10} = 0000_2$ and $2_{10} = 0010_2$. Thus, this represents $\bar{w}\bar{x}\bar{z}$.

Similarly, 2,6,10,14(4,8) can be obtained from the binary representation of any of 2,6,10,14 with the 4 and 8 columns deleted.

$2_{10} = 0010_2$

$6_{10} = 0110_2$

$10_{10} = 1010_2$

$14_{10} = 1110_2$

Deleting the 4 and 8 columns, we have --10 → $y\bar{z}$. Proceeding in this way, we obtain

$$f = \bar{w}\bar{x}\bar{z} + wxz + y\bar{z} + xy$$

The decimal notation is a more compact way of representing the minterm determining tables.

4.6 USE OF NAND GATES AND NOR GATES IN THE REALIZATION OF SWITCHING FUNCTIONS

We have, thus far, discussed minimization procedures which realized almost all switching functions using AND, OR and NOT gates. We shall see, in Chapter 5, that it is often best to use NAND (or NOR) gates exclusively. In this section we shall consider procedures for applying the previously discussed minimization techniques to realizations using NAND(or NOR) gates.

Realization Using NAND Gates

In Sec. 3-3 we demonstrated that the NAND gate was a functionally complete set. Thus, we could realize the AND, OR and NOT functions using only NAND gates as shown in Fig. 3-16. These representations could simply replace all the AND, OR and NOT gates in any realization. *However, this would result in an excessive number of gates being used.* We shall now discuss a procedure which will not, in general, result in an increase in the number of gates used. Thus, we take advantage of the previously discussed minimization techniques.

We shall start by obtaining an equivalent representation for the NAND gate. From Eq. (4-20), we have

$$\overline{xy} = \bar{x} + \bar{y} \tag{4-100}$$

The left hand side of this equation is the NAND operation. The right hand side of Eq. (4-100) represents the OR operation applied to the complement of the variables. These two equivalent forms of the NAND operation are shown in Fig. 4-28. We shall call the representation of Fig. 4-28b the NOT-OR form, and shall call Fig. 4-28a the conventional form. (Note that the circle indicates that the complement is taken.)

Figure 4-28. Two representations of a NAND gate. (a) conventional form; (b) NOT-OR form.

We shall now show how AND, OR and NOT gates can be replaced by NAND gates without changing the switching functions. A sum of products realization shall be used here. Consider Fig. 4-29a which is the AND, OR, NOT realization of the sum of products

$$f(x,y,z) = x\bar{y} + y\bar{z} \tag{4-101}$$

We shall demonstrate that Fig. 4-29b, where we have replaced all gates by NAND gates, realizes the same function. The gates which replace the AND gates have been drawn in the conventional form while the NAND gate which replaced the OR gate has been drawn in the NOT-OR form. The NOT-OR representation of the NAND gate is just that. That is, we can consider that(mathematically) the complement is taken, and then the OR operation is performed. If two NOT(complement)operations are performed on a variable, the result is the original variable. Thus, the inputs to the OR portion of the NAND gate C of Fig. 4-29b are the same as the inputs to the OR gate C of Fig. 4-29a. Hence, the outputs must be the same and the networks are equivalent. Therefore, we can make the following general statement. If a network is realized using a sum of products form(using one AND gate for each product and one OR gate to sum the products)then *all* the AND gates and the single OR gate can be replaced by NAND gates, without changing the network function.

Let us consider this. If the sum of products form is used, then a number of AND gates will always drive a single output OR gate. When all of these gates are replaced by NAND gates, a representation such as that of Fig. 4-29b will result. Thus, there will always be a pair of fictitious NOT gates in cascade. Therefore, these "NOT gates" can be mathematically eliminated and, hence, the original switching function will be obtained.

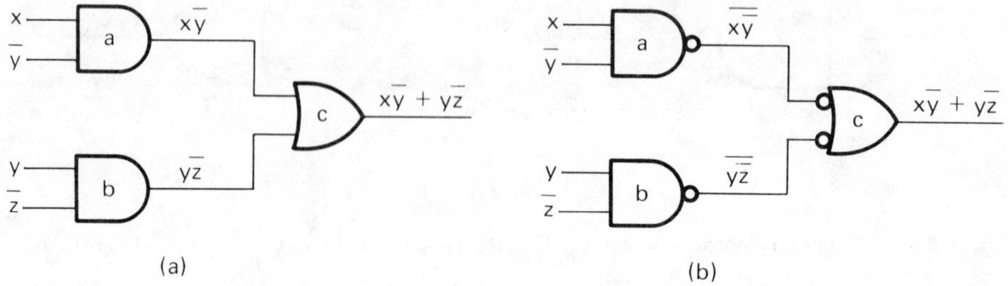

Figure 4-29. Realizations of Eq. (4-100). (a) using sum of products form (AND and OR gates); (b) using NAND gates.

Suppose that we realize a switching network using a product of sums form. For instance, consider Fig. 4-30a. This uses many OR gates (two in this case) driving a single AND gate. Now consider the realization of Fig. 4-30b which has replaced the AND and OR gates by NAND gates. *Note that the input variables are replaced by their complements.* This cancels the effect of the NOT portions of the NOT-OR representation of the NAND gates. Thus, the outputs of the NAND gates a and b are the same as those of the OR gates a and b, respectively. Then, the output of the NAND gate c will be the complement of the desired output. Hence, a NOT gate is added to obtain the desired output. We have not exclusively used NAND gates here. Note, however, that if a NOT gate is not available, then it can be realized using NAND gates, see Fig. 3-16a.

In general, if a network is realized using a product of sums form (using many OR gates and one AND gate) then all the gates can be replaced by NAND gates without changing the switching function provided that all the inputs and the output are complemented. Note that taking the complement increases both the number of gates and the number of levels. Thus, it is usually desirable to use a sum of products realization when NAND gates are used, unless the product of sums form uses many less gates.

Realization Using NOR Gates

The discussion of realization using NOR gates essentially parallels, on a dual basis, the previous discussion of the realiza-

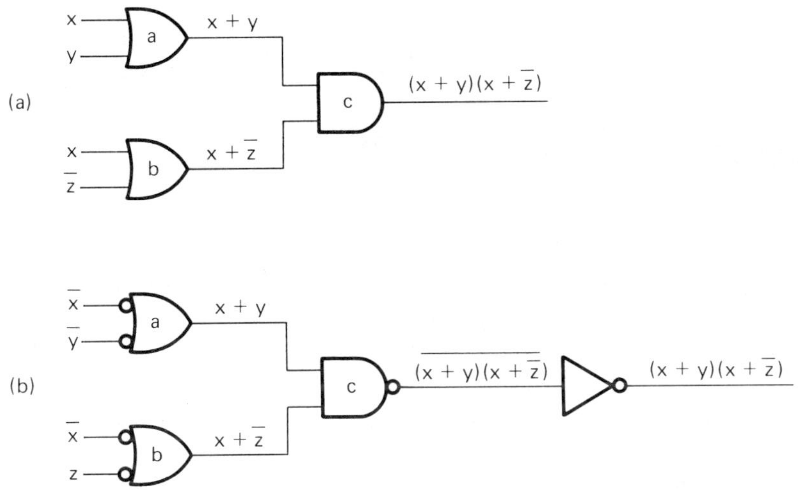

Figure 4-30. (a) A canonical product of sums realization; (b) an equivalent switching network using NAND gates.

tion using NAND gates. Thus, we shall omit some of the details. We start by considering Eq. (4-19).

$$\overline{x+y} = \bar{x}\bar{y} \qquad (4\text{-}102)$$

The left hand side of this expression is just a statement of the NOR operation. Thus, Fig. 4-31a and b are equivalent representations of the NOR operation. We shall call Fig. 4-31a the conventional representation and Fig. 4-31b the NOT-AND representation.

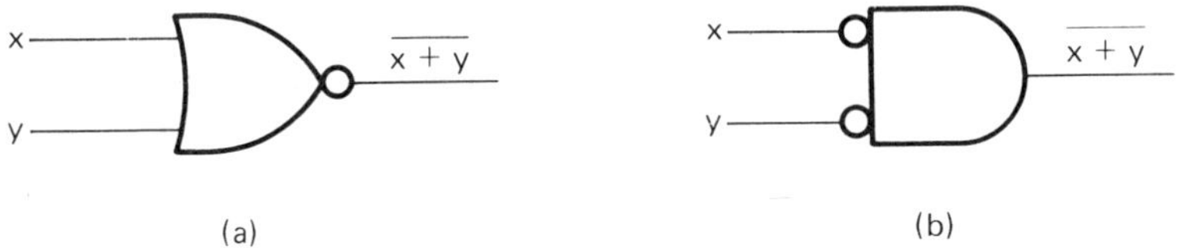

Figure 4-31. Two representations of a NOR gate. (a) conventional form; (b) NOT-AND form.

Suppose that we have a product of sums form such as

$$f(x,y,z) = (x+y)(x+\bar{z}) \qquad (4\text{-}103)$$

Its realization is shown in Fig. 4-32a. Now consider Fig. 4-32b. As before, the cascading of two NOT operations yields the original function. Thus, Figs. 4-32a and b are equivalent.

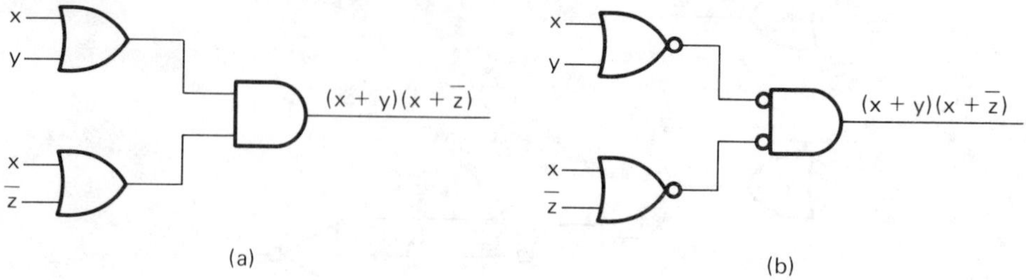

Figure 4-32. Realization of Eq. (4-103). (a) product of sums form using OR and AND gates; (b) using NOR gates.

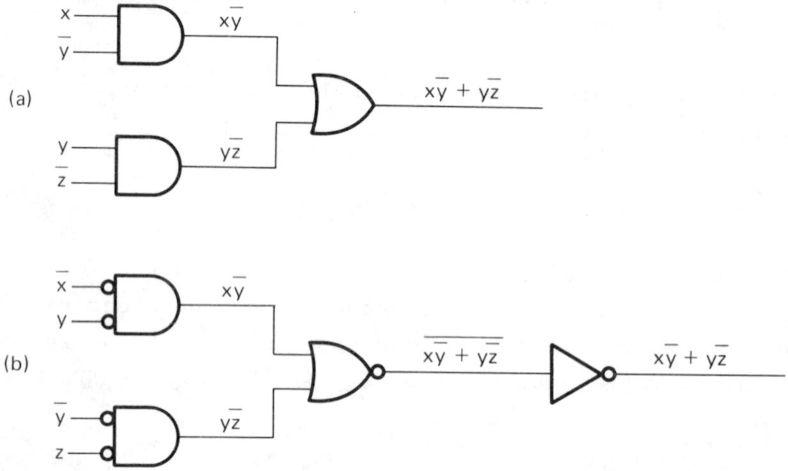

Figure 4-33. (a) A sum of products realization; (b) an equivalent switching network using NOR gates.

Thus, we can state: If a network is realized using a product of sums form (using many OR gates driving a single AND gate) then all OR and AND gates can be replaced by NOR gates without changing the network function.

When the sum of products form is used, we can replace all AND and OR gates by NOR gates, provided that *all* the inputs and the outputs are complemented. For instance, Figs. 4-33a and b have equivalent switching functions. Note that the sum of products form realization using OR gates will use more gates and levels than the product of sums form would. Thus, if an OR gate realization is to be used, the product of sums form is more desirable, unless the sum of products form uses many less gates.

We have considered examples containing single output networks in this section. However, all the ideas carry directly over to multiple output networks.

BIBLIOGRAPHY

Booth, T.L., "Digital Networks and Computer Systems", Chapters 3, 5, and 6, John Wiley and Sons, Inc.,New York, 1971

Kohavi, Z., "Switching and Finite Automatic Theory", Chapters 3 and 4, McGraw-Hill Book Co., New York,1970

Mano, M.M., "Computer Logic Design", Chapters 2 and 3, Prentice Hall, Inc., Englewood Cliffs, N.J., 1972

McCluskey, E.J., :Introduction to the Theory of Switching Circuits", McGraw-Hill Book Co., New York, 1965

Rhyne, V.T., "Fundamentals of Digital System Design", Chapters 2 and 4, Prentice Hall, Inc., Englewood Cliffs, N.J., 1973

PROBLEMS

4-1. Discuss proofs by perfect induction and by deduction. Also tell why proofs by perfect induction are used with switching algebra but never with ordinary algebra.

4-2. Which of the following switching functions are equivalent to $(x+y)\cdot x$?

$$x\cdot y\cdot z + y\cdot z + x\cdot y\cdot z + \bar{y}\cdot z\cdot x$$

$$(x+z)\cdot(y+\bar{z})$$

$$(x+z)\cdot(z+x\cdot y)$$

4-3. Simplify the following expression

$$f(x,y,z) = x + x\cdot y + x\cdot y + x\cdot x$$

4-4. Simplify the following expression

$$f(x,y,z) = x\cdot(y+\bar{y}) + y\cdot x\ \bar{x}$$

4-5. Repeat Prob. 4-4 for

$$f(x,y,z) = \bar{x}\cdot(\bar{y}+y)\cdot(x+x)$$

4-6. Use perfect induction to prove the associative laws, see Eq. (4-10).

4-7. Use perfect induction to prove Eq. (4-14).

4-8. Simplify the following switching function

$$f(w,x,y,z) = x+y + x\cdot z + y\cdot z + x\cdot w + y\cdot w$$

4-9. Repeat Prob. 4-8 for

$$f(w,x,y,z) = w\cdot x + (\bar{w}+\bar{x})(y+z)$$

4-10. Find the complement of the following switching function

$$f(x,y,z) = (x+y)\cdot(z+x) + y\cdot z$$

4-11. Repeat Prob. 4-10 for

$$f(w,x,y,z) = w\cdot(x+y\cdot\bar{z}) + x\cdot\bar{y}(\bar{y}+z)$$

4-12. Prove Eq. (4-28)

4-13. A switching function is characterized by the following truth table

x	y	z	f
0	0	0	0
0	0	1	1
0	1	0	1
0	1	1	0
1	0	0	1
1	0	1	0
1	1	0	1
1	1	1	1

Express the switching function as a sum of minterms.

4-14. Express the switching function of Prob. 4-13 as a sum of maxterms.

4-15. Draw a gate circuit diagram which realizes the function of Prob. 4-13.

4-16. Repeat Prob. 4-15 for the function of Prob. 4-14.

4-17. Use the designations of Table 4-1 to express the function of Prob. 4-13.

4-18. Express the following function in a canonical sum of products form

$$f(x,y,z) = x\cdot(y+z) + x\cdot\bar{y}(z+\bar{x}) + y\cdot\bar{x}(z+\bar{x}y)$$

4-19. Repeat Prob. 4-18 for

$$f(w,x,y,z) = w\cdot(x+\bar{y})+(z+\bar{x})(y+\bar{w})+(x+\bar{y})(\bar{w}+z)$$

4-20. Express the function of Prob. 4-18 in a canonical product of sums form.

4-21. Repeat Prob.4-20 for the function of Prob. 4-19.

4-22. A logic circuit is to be built for an automatic change machine for a rapid transit system. The fare is 15 cents. The money deposited can consist of nickels, dimes and quarters. If there is too much deposited, one nickel or one dime is to be returned. The outputs are to consist of a signal for money paid, and signals to the nickel or the dime change releases. Obtain a switching function for the system. Hint: assume that binary variables express the number of nickels, dimes and quarters. For intance, N_1N_2 represents the number of nickels.

4-23. Obtain a Karnaugh map for the switching function

$$f(x,y) = x + x\bar{y}$$

4-24. Repeat Prob. 4-23 for

$$f(x,y) = xy + \bar{x}\bar{y}$$

4-25. Obtain a switching function in a canonical sum of products for the Karnaugh map of Fig. 4-34.

y \ x	0	1
0	1	1
1	1	

Figure 4-34.

4-26. Obtain a Karnaugh map for the switching function

$$f(x,y,z) = \bar{x}y + \bar{x}\bar{y}z + \bar{x}yz + xyz$$

4-27. Obtain a function in a canonical sum of products form for the Karnaugh map of Fig. 4-35.

z \ xy	00	01	11	10
0	1			1
1	1		1	1

Figure 4-35.

4-28. Simplify the switching function of Prob. 4-25, using the Karnaugh map.

4-29. Repeat Prob. 4-28 for the switching function of Fig. 4-29.

4-30. Obtain a Karnaugh map for the switching function

$$f(w,x,y,z) = wxyz + wx\bar{y}z + w\bar{x}yz + w\bar{x}y\bar{z}$$

4-31. Obtain a switching function in canonical sum of products form for the Karnaugh map of Fig. 4-36.

yz \ wx	00	01	11	10
00	1			1
01		1	1	1
11		1	1	1
10	1			1

Figure 4-36.

4-32. Obtain a Karnaugh map for the switching function

$$f(w,x,y,z) = m_0+m_1+m_2+m_4+m_{15}+m_{14}$$

4-33.Repeat Prob. 4-31 for

$$f(w,x,y,z) = m_0+m_4+m_1+m_5+m_3+m_7+m_{15}+m_{11}+m_2+m_6$$

4-34. Use the Karnaugh map to minimize the switching function of Prob. 4-30. Draw a gate circuit that realizes this network.

4-35. Repeat Prob. 4-34 for the switching function of Fig. 4-30.

4-36. Repeat Prob.4-34 for the switching function of Prob. 4-32.

4-37. Repeat Prob. 4-34 for the switching function of Prob. 4-33.

4-38. Obtain the truth table for the switching function of Prob. 4-33.

4-39. Modify Table 4-2 for a six variable Karnaugh map.

4-40. Use the Karnaugh map to minimize the switching function

$$f(w,x,y,z) = m_0+m_8+m_2+m_{10}+m_5+m_{13}+m_7+m_{15}+m_{11}+m_1$$

4-41. Repeat Prob. 4-40 for the switching function

$$f(u,v,w,x,y,z) = m_0+m_4+m_{32}+m_{36}+m_{17}+m_{49}+m_{27}+m_{59}$$

4-42. Repeat Prob. 4-40 for the switching function

$$f(u,v,w,x,y,z) = m_1+m_9+m_{41}+m_{33}+m_5+m_{13}+m_{45}+m_{37}+m_{18}+m_{50}+m_{22}$$
$$+m_{54}+m_{31}+m_{63}+m_{59}+m_{21}+m_{24}+m_{25}+m_{27}+m_{26}+m_{30}$$
$$+m_{31}+m_{29}+m_{28}$$

4-43. Minimize the switching function whose Karnaugh map is given in Fig. 4-37.

yz \ wx	00	01	11	10
00	d			1
01		1	1	d
11		1	1	
10	d	d		

Figure 4-37.

4-44. Use the Karnaugh map to minimize the following switching function

$$f(w,x,y,z) = m_0+m_5+m_{13}$$

where the following positions are "don't care" positions: m_7, m_{15}, m_2, m_8, and m_{12}.

4-45. Use the Karnaugh map to minimize the following switching function.

$$f(u,v,w,x,y,z) = m_0+m_{36}+m_{18}+m_{54}+m_9+m_{34}$$

where the following are "don't care" positions: m_4, m_5, m_{22}, m_2, m_{10}, m_{50}, m_{32}.

4-46. Repeat Prob. 4-45 but now omit the "don't care" terms.

4-47. Use the Karnaugh map to minimize the switching function of Prob. 4-30. Use maxterms and "zeros" in this reduction.

4-48. Repeat Prob. 4-47 for the map of Fig. 4-36.

4-49. Repeat Prob. 4-47 for the switching function of Prob. 4-32.

4-50. Repeat Prob. 4-47 for the switching function of Prob. 4-33.

4-51. Repeat Prob. 4-47 for the switching function of Prob. 4-41.

4-52. Repeat Prob. 4-47 for the switching function of Prob. 4-44.

4-53. Determine an implicant of

$$f(x,y,z) = x + yz + xyz$$

4-54. If $f_1, f_2, \ldots f_n$ are prime implicants of a function g, and if g=0, what are the values of $f_1, f_2 \ldots f_n$?

4-55. Discuss the meaning of essential prime implicants and absolutely eliminable prime implicants.

4-56. Use the Quine McClusky procedure to minimize the switching function of Prob. 4-30. Draw a gate circuit that realizes this function.

4-57. Repeat Prob. 4-56 for the switching function of Fig. 4-36.

4-58. Repeat Prob. 5-46 for the switching function of Prob. 4-32.

4-59. Repeat Prob. 4-56 for the switching function of Prob. 4-33.

4-60. Repeat Prob. 4-56 for the switching function of Prob. 4-40.

4-61. Repeat Prob. 4-56 for the switching function of Prob. 4-41.

4-62. Repeat Prob. 4-56 for the switching function of Prob. 4-42.

4-63. Repeat Prob. 4-56 for the wwitching function of Fig. 4-37.

4-64. Repeat Prob. 4-56 for the switching function of Prob. 4-44.

4-65. Repeat Prob. 4-56 for the switching function of Prob. 4-45.

4-66. Minimize the number of gates used to realize the following switching functions

$$f_1(x,y,z) = m_0+m_1+m_3+m_5$$

$$f_2(x,y,z) = m_0+m_1+m_7$$

4-67. Repeat Prob. 4-66 for the switching function

$$f_1(w,x,y,z) = m_0+m_1+m_5+m_9+m_{15}$$

$$f_2(w,x,y,z) = m_0+m_1+m_9+m_{12}+m_{14}$$

$$f_3(w,x,y,z) = m_0+m_1+m_5+m_{12}+m_{14}+m_{15}$$

4-68. Repeat Prob. 4-67 but now assume that the following are "don't care" positions: in f_1, m_{14}; in f_2, m_5.

4-69. Repeat Prob. 4-56 but now use decimal numbers in the prime implicant determining table.

4-70. Repeat Prob. 4-69 for the switching function of Prob. 4-32.

4-71. Repeat Prob. 4-69 for the switching function of Prob. 4-33.

4-72. Repeat Prob. 4-69 for the switching function of Prob. 4-42.

4-73. Repeat Prob. 4-69 for the switching function of Prob. 4-44.

4-74. Repeat Prob. 4-34 but now use NAND gates to realize the network.

4-75. Repeat Prob.4-74 but now use NOR gates to realize the network.

4-76. Repeat Prob. 4-35 but now use NAND gates to realize the network.

4-77. Repeat Prob. 4-76 but now use NOR gates to realize the network.

4-78. Repeat Prob. 4-74 for the network of Prob.4-66.

4-79. Repeat Prob. 4-75 for the network of Prob.4-66.

5

Combinational Logic Circuits

Thus far, we have discussed switching functions, their minimizations, and their implementation using gate circuits. These gates have been considered as block diagrams. For instance, we show an AND gate in Fig. 3-5a as a box with inputs A_1 and A_2 and output B. We then state that this box performs the logical operation $B=A_1A_2$. However, except for simple switches, we did not discuss how this logic operation would be implemented electronically. In this chapter we shall discuss this implementation using semiconductor circuits. Virtually all computer circuits make use of integrated circuits. Such circuits will be discussed here.

We shall start our discussion by briefly considering semiconductor devices and their switching. Those readers who desire a more thorough (or more basic) discussion are referred to the Appendix where the material discussed in Sec. 5-1 will be explained in detail.

5.1 SEMICONDUCTOR DEVICES

In this section we shall briefly consider the various semiconductor devices that are used in switching circuits. We shall not consider the physics of these devices here but shall just consider them as circuit elements. (A discussion of physical properties is given in Sec. A-1.)

The p-n Junction Diode

This device is one which essentially conducts current in one direction but not in the other. A symbol for the diode and a typical voltage current characteristic is shown in Fig. 5-1 a and b, respectively.

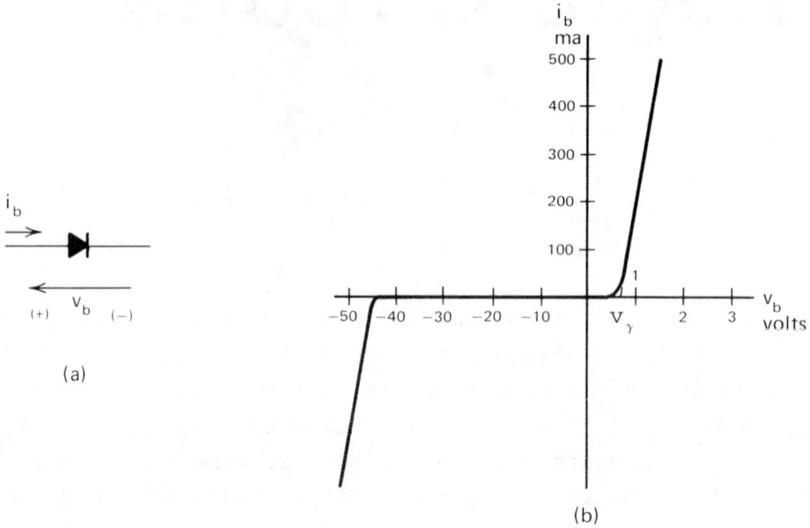

Figure 5-1. The p-n junction diode. (a) its symbol; (b) a typical voltage current characteristic.

Consider Fig. 5-1b. When v_b is positive, the diode conducts. This polarity of v_b is termed *forward bias*, and the diode is said to be conducting in the *forward direction*. When v_b is negative, we have *reverse bias*. Now the current i_b is essentially zero. If a sufficiently large reverse bias voltage is applied, substantial current will result. This is termed *reverse breakdown*. Usually diodes are operated so that the reverse breakdown region is avoided. However, certain diodes called *Zener diodes* make use of reverse breakdown.

When the diode is forward biased, its current is small until the voltage v_γ is reached, see Fig. 5-1. This is called the *turn on voltage*. Often, we approximate the diode characteristics by considering that its current is zero if v_b is less than v_γ and that the diode conducts if v_b is greater than v_γ .

In switching circuits, the polarity of v_b is often switched from positive to negative(or vice versa). This represents switching from a 0 to a 1 or vice versa. If v_b is switched from a positive voltage to a negative voltage, we would expect that i_b would switch from a relatively large current to essentially zero. This happens but it does not occur instantaneously. The time that it takes to switch the diode is called its *switching time*. This is an imporatnt quantity since the speed at which a device switches determines the speed of its gate's response. Since a digital computation involves the switching of very many gates in succession, the switching speed of each device can have a profound effect on the time required by the computer to perform an operation.

In general, the time required to switch a diode off(i.e., v_b switches from positive to negative voltage)is greater than that required to switch it on. When the diode conducts in the forward direction, it is said to be saturated. The more heavily it is saturated(i.e., the greater its forward current)the longer the switching time required to turn it off. When a diode is saturated excess charge is said to be stored.

p-n junction diodes are discussed in greater detail in Sec. A-2. Their switching is discussed in Sec. A-5.

The Junction Transistor

The element which is most commonly used in switching circuits is the junction transistor. In contrast to the diode, a small current can be used to control(i.e., to switch)a much larger current when the transistor is used. This is a very desirable property which can be used to advantage in logic circuits, see Sec.5-3.

Transistors are either p-n-p or n-p-n type. Their symbols and terminology are shown in Fig. 5-2. To describe the operation of the transistor, let us consider the simple n-p-n transistor circuit of Fig. 5-3. The power supply voltage V_{CC} causes v_{CE} to be positive (i.e., the collector is positive with respect to the emitter). Under these conditions, if v_{BE} is positive as well as greater than a fraction of a volt, the collector current i_c will be large and will essentially be limited by the external resistance, R. In this case, the voltage v_{CE} will be small (a fraction of a volt), and the transistor is said to be *saturated*. The minimal value of v_{BE} that produces saturation is called the *base emitter turn on voltage*. A typical value is 0.6 volt.

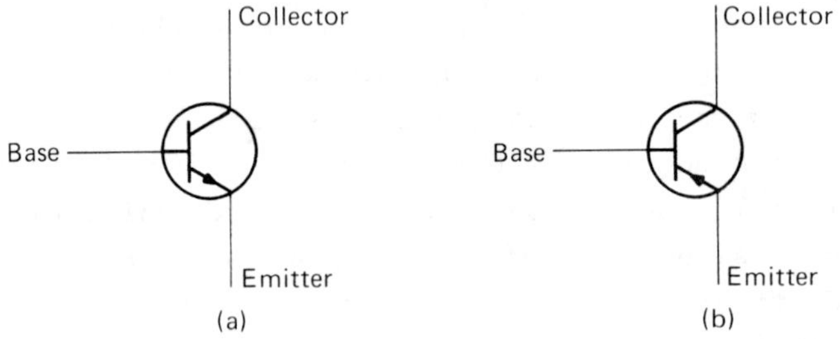

Figure 5-2. Symbols for the junction transistor. (a) n-p-n; (b) p-n-p.

If v_{BE} is negative, the current i_C will essentially be zero. In this case, the transistor is said to be *cutoff*. Often, we can approximate transistor operation by assuming that it is cut off if v_{BE} is less than the base emitter turn on voltage and that it is saturated if v_{BE} is greater than the base emitter turn on voltage. (This assumes that V_{CC} is sufficiently large.)

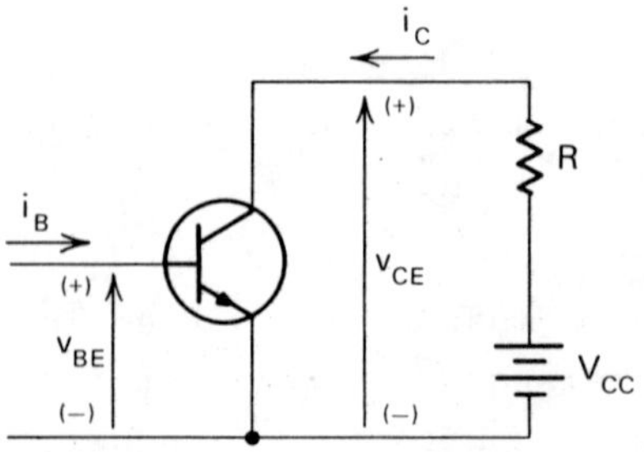

Figure 5-3. A simple transistor circuit.

We have considered the n-p-n transistor. The previous discussion would also apply to the p-n-p transistor if all potentials were reversed.

In switching circuits, a quantity that is of use is the average ratio of the collector current to the base current. This is given by

$$\beta_F = i_C/i_B \tag{5-1}$$

Typical values of β_F are 100 or more.

If, in Fig. 5-3, v_{BE} were switched from a saturation to cut off value, then i_C would fall from its saturation value to essentially zero. However, this does not occur in zero time. In general, the greater the saturation value of i_C, the greater will be the time required to switch the current off. (i.e.,the greater will be the switching time). We shall see, when specific logic circuits are considered, that a compromise must often be made. For instance, increasing the saturation value of i_C increases the reliability of the cicuit which is an advantage. However, increasing the saturation value also increases the switching time, which is a disadvantage. In transistors, as in diodes, we have an excess stored charge as a result of saturation, see Sec. A-5.

When the transistor is switched from cut off to saturation, there is also a finite switching time involved. However, in this case, the switching is faster and the saturation effects do not slow it down.

Junction transistors are discussed in greater detail in Sec. A-3. Their switching is discussed in Sec. A-5.

The Insulated Gate FET-IGFET-MOSFET

Another semiconductor device which finds extensive use in switching circuits is the insulated gate FET(IGFET). It is also called the metal-oxide-silicon FET(MOSFET). These devices can be switched on or off by the application of extremely small currents. These control currents are, in general, very much less than the base currents of junction transistors.

MOSFETs are either n channel or p channel types. Their symbols are shown in Fig. 5-4. The MOSFET acts as a controlled resistance between the drain and source. A signal applied to the gate is capacitively coupled to the semiconductor material between the drain and source. The electric field due to this signal varies the resistance. There is only capacitive coupling between the gate and the bulk of the device and the input(i.e., gate)current is extremely small. MOSFETs used for digital computer applications are usually of a type called enhancement MOSFETs(see Sec. A-4).

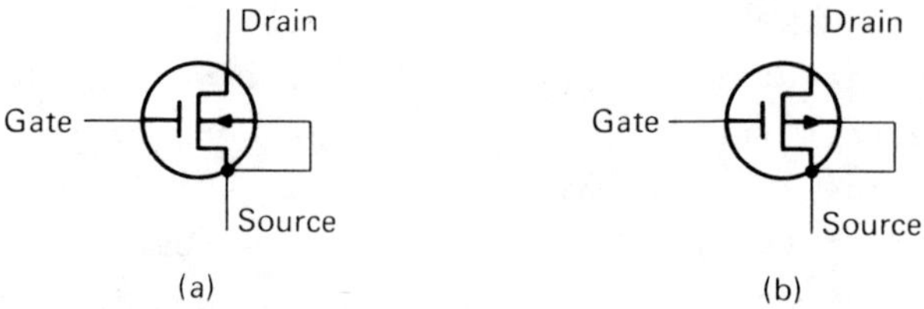

Figure 5-4. Symbols for the MOSFET. (a) n-channel; (b) p-channel.

Let us consider the circuit of Fig. 5-5 to explain the operation of the MOSFET. When v_{GS} is zero or negative, the current i_D will be essentially zero and the MOSFET is said to be *cut off*. When v_{GS} is made positive, i_D increases. If v_{GS} is sufficiently large, the voltage drop v_{DS} becomes very small and the current is essentially limited by the resistance R. This is the *saturation* condition.

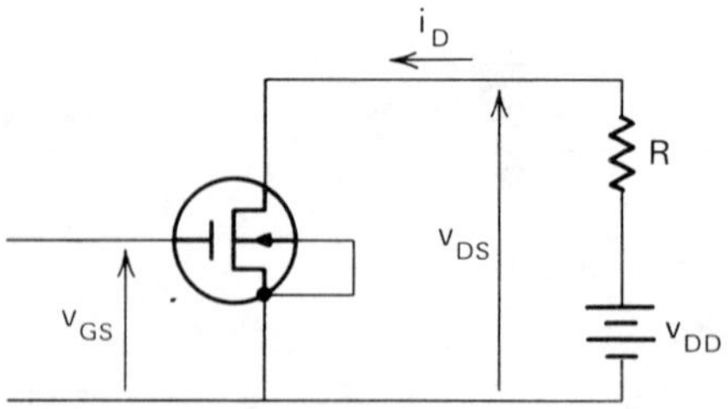

Figure 5-5. A simple MOSFET circuit.

MOSFET devices, in general, can be used in circuits which consume much less power than those circuits which use junction transistors. However, the switching speed of fast junction transistor circuits is considerably greater than that of fast MOSFET circuits.

MOSFETs are discussed in more detail in Sec. A-4.

Integrated Circuits

The modern digital system uses electronic components which are fabricated in such a way that many elements such as transistors, MOSFETs, resistors and capacitors can be incorporated in a single silicon chip. It would not be feasible to construct very large computers if it were not for integrated circuits. In addition, small minicomputers would not be practical without integrated circuits. Integrated circuits can now be fabricated containing thousands of elements on a single chip. This is called *large scale integration*(LSI). Small digital systems can be constructed on several chips using LSI. These *microprocessors* allow simple digital computers to be incorporated in many devices.

A more detailed discussion of integrated circuits is given in Sec. A-6.

5.2 LOGIC FAMILIES

The various combinational logic gates can be made using many different circuit combinations. Some use transistors, others use FETs. Some use devices which are driven into saturation and others do not. It is usual to categorize logic circuits into groups called *logic families*, such as diode-transistor logic, or transistor-transistor logic. In the following sections we shall discuss the commonly used logic families.

Usually, each logic family consists of one or a few basic gates which form a functionally complete set, see Sec. 3-3. These are used to make all switching functions. For instance, all switching functions may be made up of NAND gates. Thus, the NAND gate can be the basic gate of the particular family.

Gate families are compared on the basis of several quantities. One of the most important of these, the propagation delay, is a function of the switching speed. The propagation delay gives the time required for an input at the gate to produce 90% of the final output. Another basis of comparison is the power dissipated by each gate. Let us consider some other standards of comparison.

All electrical devices are subject to extraneous electrical signals called *noise*. This can be caused by stray pick up of other signals, which could even be generated within the computer. For instance, a pulse generated in one part of the computer could be coupled by stray capacitance to another circuit. This is termed *cross talk*. Noise may be generated externally by motors, etc. Finally, when very low level signals are encountered, the noise due to the random motion of electrons can become troublesome. Noise can obscure the signals and make zeros appear as ones and vice versa.

Interference is not the only thing that can cause erroneous signals. Assume that a transistor saturates when its base current is greater than 10 ma. Becuase of variation in components, suppose that the driving circuit only supplies 9 ma. The transistor will not be fully saturated, but will be close to it. The succeeding stage will then be driven by a signal which is not as small as it should be. However, the departure from the desired signal will be small and an error will not result. However, if there are a great many circuits cascaded(i.e., one stage drives the next, etc.)then those small departures can eventually accumulate to produce an erroneous signal. This is also termed noise in a computer. Since the manufacturing tolerances on semiconductor devices are reasonably large, the designer should take such departures into account. For intance, cutoff voltage could be larger than necessary to insure that the device is really cut off. Different logic families have different immunities to all types of noise. This is another criterion used in picking logic families.

Often, one gate drives several other gates. Each gate requires some current to drive it. The driving gate must be capabe of supplying the current to, and maintaining the voltage levels required by, the succeeding gates. This is a function of both the output capabilities of the output gate and of the input requirements(e.g.,impedance)of the driven gates. Usually, for any given logic family, one gate will drive others of the same type. If the output impedance is low, and the input impedance high, then, all other things being equal, many gates can be driven. Conversely, few gates can be driven if the output impedance is high and the input impedance is low. The maximum number of outputs that a gate can drive and still function properly, is called the *fanout*. This is also a criterion used for logic families.

At times, several gates are used to drive a single one. Problems can occur when too many of these gates are connected. The *fan in* is the maximum number of inputs that can be connected to a gate if it is to operate properly.

Additional criteria which determine the choice of logic family are ease of fabrication and the stability of the device once it is fabricated.

In the remainder of this chapter we shall discuss the commonly used logic families and see how the discussions of this section apply to them. Some of the logic families are very often used; others were used at one time but are rarely used now. These will only be discussed briefly.

Now let us present some terminology. Different voltage levels are used to represent 0's and 1's(e.g.,a 1 could be represented by +5 volts and a 0 by 0 volts). The voltages representing 0's and 1's can both be positive, both be negative, or one can be positive and the other negative. If the voltage for a 1 is more positive than the voltage for a 0, then we say that this is *positive logic*. Conversely, if the voltage for a 0 is more positive than that for a 1, we say that this is *negative logic*.

Suppose that we have a circuit whose output voltage v_o is either 5 volts or 0.1 volt. If positive logic is used, then the 5 volt signal represents a 1 while the 0.1 volt signal represents a 0. On the other hand, if negative logic is used, then the 5 volt signal would represent a 0 while the 0.1 volt signal would represent a 1.

Now consider a particular circuit which has two inputs v_A and v_B and an output v_o. Table 5-1 expresses v_o as a function of v_A and v_B.

TABLE 5-1
v_o AS A FUNCTION OF v_A and v_B

v_A	v_B	v_o
0.1	0.1	0.1
0.1	5	0.1
5	0.1	0.1
5	5	5

Suppose that positive logic is used. Then, the circuit of Table 5-1 yields the following truth table.

TABLE 5-2
TRUTH TABLE FOR THE CIRCUIT OF TABLE 5.1 WHEN POSITIVE LOGIC IS USED

A	B	V_o
0	0	0
0	1	0
1	0	0
1	1	1

Thus, this circuit represents an AND gate.

On the other hand, suppose that negative logic is used. Then, the truth table becomes

TABLE 5-3
TRUTH TABLE FOR THE CIRCUIT OF TABLE 5.1 WHEN NEGATIVE LOGIC IS USED

A	B	V_o
0	0	0
0	1	1
1	0	1
1	1	1

Thus, this circuit now represents an OR gate.

In general, a positive logic AND gate is also a negative logic OR gate. Similarly, a positive logic OR gate is also a negative logic AND gate. In the discussion in the remainder of this chapter, we shall assume positive logic when we describe the performance of gates. Remember that if negative logic were used, then the circuits described as AND gates would become OR gates and vice versa. In a similar way, positive logic NAND gates become negative logic NOR gates and vice versa.

We speak of voltages representing 0's and 1's. Actually, these are ranges of voltages. For instance, a 0 could be represented by a range 0 to 0.5 volt, and a 1 by the range 4 to 6 volts.

We shall begin our discussion of logic gate families by considering those which use transistors in combination with resistors and diodes. In Sec. 5-4 we shall consider those logic families that use MOS devices. There are several classes of transistor logic families. We shall consider the most important of them.

5.3 DIODE-TRANSISTOR LOGIC-DTL

One form of logic circuit uses both diodes and transistors, and is thus called *diode-transistor logic*, abbreviated DTL. The basic gates of DTL for positive logic are the NAND gate and the NOR gate. (Remember that each of these is functionally complete, see Sec. 3-3.)

In Fig. 5-6a we have a positive logic DTL,NAND gate. The common lead is marked with a ground symbol. (This also means that this lead may be connected to a common chasis point.) The power supply is connected between the common lead and R_C as shown. The three inputs are between A, B and C and the common lead. To conserve space on the wiring diagram, the common lead is often omitted. Thus, Fig. 5-6b is equivalent to Fig. 5-6a. We shall use this simplified form of schematic diagram in the remainder of the book. Remember that the unidentified sides of a power supply or input voltage, are connected to the common lead. The voltage

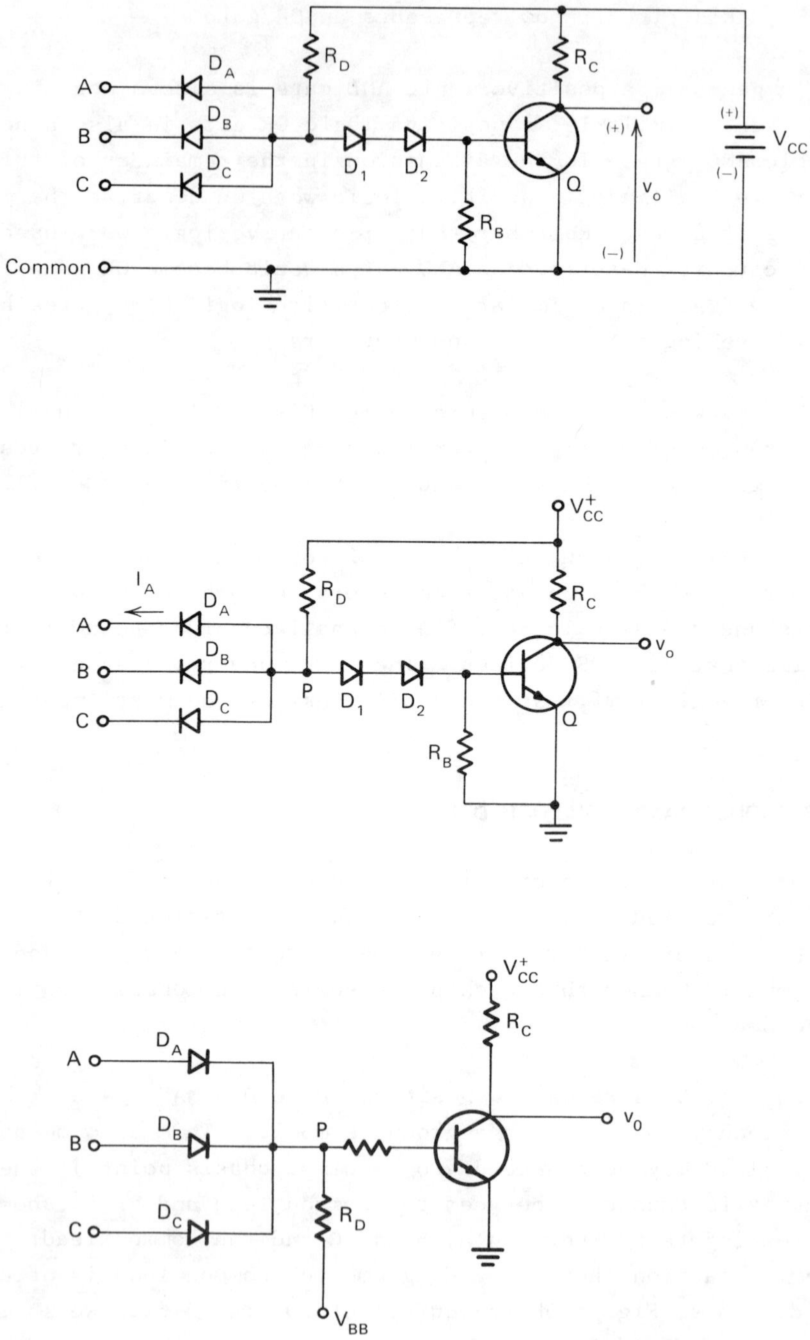

Figure 5-6. (a) a positive logic DTL, NAND gate; (b) a simplified schematic representation of (a); (c) a DTL, NOR gate.

of any point, unless otherwise marked, is referred to the common lead. Figure 5-6c is the schematic of a three input positive logic DTL,NOR gate

At first, let us discuss these gates qualitatively. Consider the gate of Figs. 5-6a or b. We assume positive logic here. Suppose that a 0 is represented by 0 volts and a 1 by 5 volts. Assume that V_{CC}=5 volts. (When we discuss this circuit quantitatively, we shall modify these values slightly. However, for purposes of this qualitative discussion, these values suffice.) Suppose that any of A,B or C represents a 0. The, V_{CC} will cause the corresponding input diode to conduct. Point P will then be at a low voltage. We say that point P is low. The diodes D_1 and D_2 will be reverse biased since the voltage at point P will be less than their cut in voltages. (We shall subsequently discuss the reason for putting two diodes in series.) There will be no base current and the transistor will be essentially cut off. Thus, if any of A, B or C is a 0, the output V_o will be a 1.

Now suppose that A, B and C are 1's. Their diodes will be cut off. Now, consider the series circuit consisting of V_{cc}, R_D, D_1, D_2,and R_B and the emitter-base junction in parallel. The voltage V_{CC} will now forward bias D_1 and D_2 and there will be base current i_B in the transistor Q_1. The values of R_D and R_B are chosen so that i_B will become large enough to produce saturation. Thus, v_o will be a small voltage which represents a 0. Hence, if A AND B AND C represent 1's, then v_o will represent a 0. Thus, this circuit is a NAND gate. That is

$$v_o = \overline{ABC} \tag{5-2}$$

Now consider the circuit of Fig. 5-6c. Again assume that a 0 is zero volts and a 1 is 5 volts. Now suppose that any of A, B or C is a 1. The corresponding input diode (D_A or D_B or D_C)will be forward biased and the voltage at point P will be about +5 volts. The transistor will receive base current, through R_B, which will drive it into saturation. Thus, v_o will represent a 0 if A OR B OR C is a 1.

Now assume that A,B, and C are all 0's. The potential of point P becomes negative because of the negative power supply. Thus, the emitter-base junction is reverse biased and the transistor is cut off. Hence, v_o will represent a 1. Then, if A AND B AND C are 0's, then $v_o = 1$. Note that V_o is the logical representation of the voltage v_o. Then

$$V_o = \overline{A+B+C} \tag{5-3}$$

We have discussed these circuits qualitatively. Now let us consider Fig. 5-6a quantitatively. We shall assume the following typical values for the parameters and the voltage drops.

$V_{CC} = 5$ volts

$R_D = 5000$ ohms

$R_C = 2000$ ohms

$R_B = 4000$ ohms

We also assume that the turn on voltage for the diodes is $V_{\gamma d} =$ 0.6 volts. The voltage across a forward biased diode is a function of the current through it. However, the curve of current versus voltage has a large slope. Thus, since we work in limited current range, we can assume that the voltage drop across a conducting diode is constant. We shall use the value 0.8 volt. The base-emitter turn on voltage for the transistor V_γ=0.6 volt. Under saturation conditions, we have $B_{BE,sat}$=0.7 volt and $V_{CE,sat}$= 0.3 volt.

Now let us discuss the circuit. Suppose that any of A,B or C is low. We assume that the transistor is cut off. We must demonstrate that the voltage and currents justify this. The appropriate input diodes will be forward biased and the voltage across them will be 0.8 volt. Actually, a zero input does not represent zero volts. For instance, the saturation value of v_{CE}=0.3 volt.

It is reasonable to assume that this circuit is driven by a similar one. Thus, a 0 would be represented by 0.3 volt. Hence,

$$v_P = 0.3 + 0.8 = 1.1 \text{ volts}$$

The cut in voltages of D_1 and D_2 are each 0.6 volt. Thus, the voltage across the series interconnection must be 1.2 volts if they are to be turned on. Since it is not, D_1 and D_2 act as an open circuit. Then, $v_{BE} = 0 < V_\gamma$ and the transistor is cut off. Thus, our assumption is valid.

Note that if we used only one diode in place of the series combination of D_1 and D_2, it could conduct and the voltage drop across it would be 0.8 volt. Thus,

$$v_{BE} = 1.1 - 0.8 = 0.3 \text{ volt}$$

This is less than V_γ=0.6 volt. Thus, the transistor would still be cut off.

It seems as though only one diode should be used in place of D_1 and D_2. However, suppose that some noise which is of a positive polarity,and is not large enough to reverse bias D_A(or D_B or D_C), appeared on the input lines. This voltage will add to v_{BE}. If there is only one series diode and v_{BE}, without noise, is 0.3 volt, then any noise voltage greater than $V_\gamma - 0.3 = 0.3$ volt would drive the transistor from cut off. Thus, a noise voltage greater than 0.3 volt would cause problems.

When D_2 is included, if $v_{BE} \geq V_\gamma$, the noise voltage must be such that

$$v_n + 1.1 - V_{\gamma D1} - V_{\gamma D2} \geq V_\gamma$$

or, equivalently,

$$v_n \geq -1.1 + 0.6 + 0.6 + 0.6 = 0.7 \text{ volt}$$

Thus, the second diode has increased the immunity to an extraneous noise pulse by better than two to one. At times, a third diode is added to further increase noise immunity.

We have obtained the conditions for the cutoff of the transistor. In this case,

$v_o = 5$ volts

where we assume that the extremely small cutoff current produces negligible drop in R_C.

Now let us consider that the inputs are all 1s, which means that all input voltages are essentially 5 volts. The diodes D_A, D_B, and D_C will now be cutt off. Let us now assume that the transistor Q is saturated. We must demonstrate that the voltages and currents are such that this is correct. Since Q is saturated

$v_{BE,sat} = 0.7$ volt

and the voltage drop across D_1 and D_2 is 1.6 volts. The input circuit can be modeled as shown in Fig. 5-7. In this case, we have

$$I_B = \frac{V_{CC}-0.7-1.6}{R_D} = \frac{5-0.7-1.6}{5000} = .54 \times 10^{-3}a = 0.54ma \qquad (5\text{-}4)$$

The transistor is essentially a current amplifying device. Thus, we must see that this value of I_B is sufficient as a base current to account for the saturation collector current. The current through R_C is the collector current. We assume here, for the moment, that no current is taken by the output. Thus,

$$I_{C,sat} = \frac{V_{CC}-v_{CE}}{R_C} = \frac{5-0.3}{2000} = 2.35 \times 10^{-3}\ a = 2.35\ ma$$

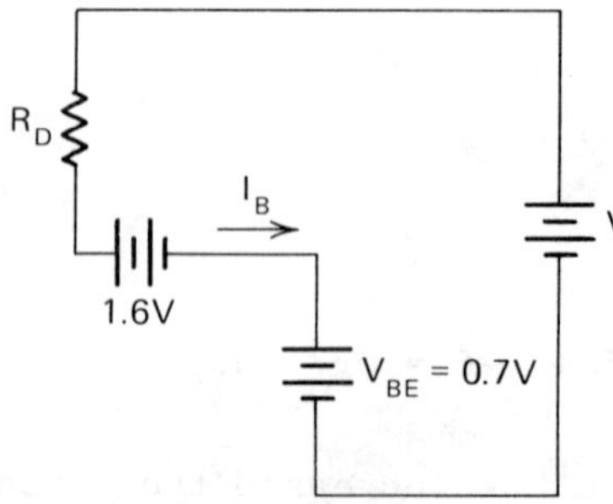

Figure 5-7. A model which can be used to calculate I_B of Fig. 5.6b when all the inputs are 1's.

We defined a β_F which represents, on the average, the collector current divided by the base current. If the $\beta_F I_B$ is not equal to or greater than $I_{C,sat}$, then the value of base current will not produce saturation and the circuit will not operate properly. Thus, we must have

$$\beta_F = \frac{2.35}{.54} = 4.35 \qquad (5\text{-}5)$$

Usually β_F lies between 30 and 300. Thus, saturation will result.

Now let us consider the effects of an extraneous noise signal when all the inputs are 1's. In this case, assume that the noise voltage tends to drive the input voltage towards zero. Problems arise if D_A, D_B, or D_C are driven into conduction. If Q is saturated, $v_{BE,sat}$=0.7 volt and D_1 and D_2 are conducting. Thus,

$$v_P = 0.7 + V_{D1} + V_{D2} = 0.7 + 0.8 + 0.8 = 2.3 \text{ volts}$$

Thus, if D_A or D_B or D_C are to start conducting, the input voltage must drop below

$$2.3 - V_{\gamma d} = 2.3 - 0.6 = 1.7 \text{ volts}$$

since the input voltage is 5 volts. If the noise is to cause trouble, then

$$v_n > 5 - 1.7 = 3.3 \text{ volts}$$

Note that, in this case, the extra diode increases the sensitivity to noise. However, this is a relatively large voltage which is greater than the noise voltage which turns Q on. Thus, it is desirable to include the extra diode.

In our analysis we neglected any output current. That is, we assume that v_o did not drive a succeeding circuit. Actually, this is not the case. Assume that this circuit drives one or more with similar inputs. The current through an input diode is nonzero only when its input voltage represents a 0. Then, I_A, see Fig. 5-6b, is given by

$$I_A = \frac{V_{CC}-V_P}{R_D} \tag{5-6}$$

where we assume that only D_A is conducting. (This will be the worst case.) Note that, in this case, D_1, D_2, D_B and D_C are cut off. Thus, all the current through R_D must pass through D_A. Substituting, we obtain

$$I_A = \frac{5-1.1}{5000} = 0.78 \times 10^{-3} = 0.78 \text{ ma}$$

Now consider Fig. 5-8 which shows the output of the NAND gate. The transistor must not only carry the current through R_C but also that through I_A. This increases the collector current. However, the base current remains the same. Thus, the value of β_F, see Eq. (5-5) must be greater than before. In this case, we have

$$\beta_F = \frac{2.35+.78}{.54} = 5.8$$

This is still a relatively low value of β_F and there should not be any problem involved.

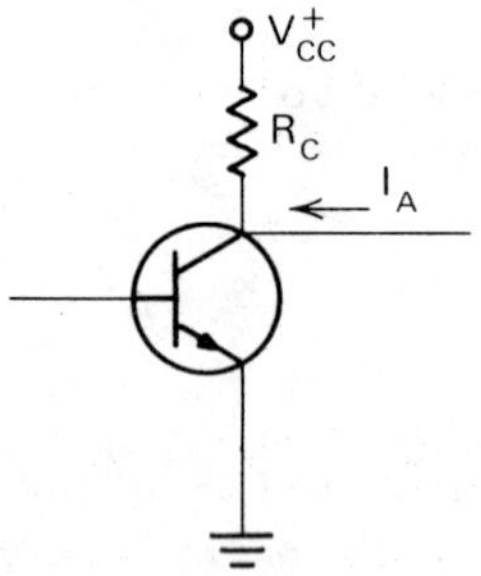

Figure 5-8. A representation of the output of the NAND gate of Fig. 5.6b.

Let us assume that the transistors used have a minimum value of β_F of

$$\beta_{F,min} = 30$$

and determine the fanout of this circuit. That is, what is the maximum number of gates that this circuit can drive? Assume that N gates are driven. Then, the total collector current is

$$I_C = 2.35 \times 10^{-3} + .78 \times 10^{-3} N \tag{5-7}$$

Then,

$$\beta_F \geq \frac{2.35+.78N}{.54} \tag{5-8}$$

If β_F=30, we have

$$N \leq \frac{.54(30)-2.35}{.78} = 17.756$$

Thus, the circuit can drive 17 gates. It is said to have a fanout of 17. The transistor must be capable of supplying the required collector current, which is

$$I_C = (2.35+0.78 \times 17)10^{-3} = 15.61 \times 10^{-3}a = 15.61ma$$

The fanout of this circuit can be greatly increased by replacing the diode D_1 by a transistor. This is shown in Fig.5-9. The emitter-base junction of transistor Q_{D1} takes the place of D_1. However, now the base current of Q is greatly increased by the current amplification of Q_{D1}.

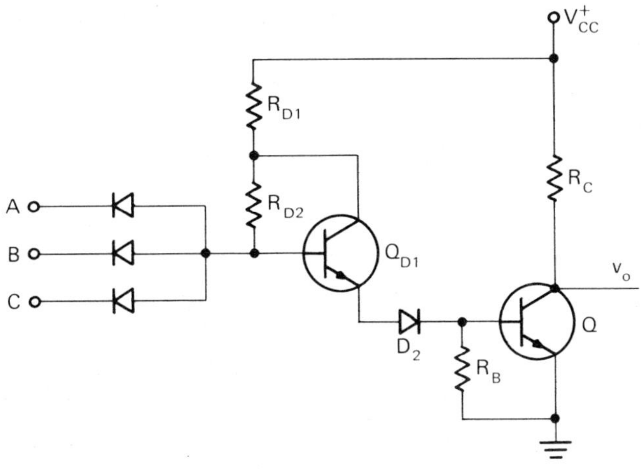

Figure 5-9. A modification of Fig. 5.6b which has increased fanout.

We have analyzed this circuit fairly thoroughly. These techniques can be applied to many other circuits. We shall, therefore, not repeat this analysis in detail.

5.4 HIGH-THRESHOLD LOGIC-HTL

We increased the noise immunity of the DTL gate by adding the second diode D_2, see Fig. 5-6. If the computer is to be used in a very noisy atmosphere, for instance, near heavy electrical machinery, then the circuit of Fig. 5-9 can be used with the diode D_2 replaced by a *Zener diode*. This is a diode which makes use of the reverse breakdown phenomenon illustrated in Fig. 5-1. Note that when a diode is reverse biased, it acts as an open circuit until a certain breakdown voltage is reached. After that, the current increases rapidly. Thus, in a sense, this is like a reverse cut in voltage. As we saw, the addition of D_2 prevented low level noise from turning on the output transistor. The increased voltage drop of the Zener diode increases this noise immunity still further. A typical breakdown voltage for the Zener diode used for this purpose lies in the 6-9 volt range. The power supply voltage is increased to about 15 volts and the resistances are increased to keep the currents in the same range. Much higher noise immunity can be obtained with this circuit. (Note that by increasing the signal levels and, thus, the difference between a 0 and a 1, we increase the noise immunity.) However, the power dissipated is increased and the switching speed reduced. Thus, this circuit is only used when very high noise requires it.

5.5 TRANSISTOR-TRANSISTOR LOGIC-TTL-T^2L

The most commonly used logic circuit at the present time is called *transistor-transistor logic* (TTL). This is because it has the fastest switching of all the logic devices that use saturated devices. In many ways, it is closely related to DTL. The basic (positive logic) gate is the NAND gate. Thus, for negative logic, the basic gate is the NOR gate.

The TTL circuits usually make use of a transistor which has several emitters. The construction of these transistors is shown in Fig. 5-10a. This drawing is not to scale and the base region between the emitter and the collector is really very narrow.

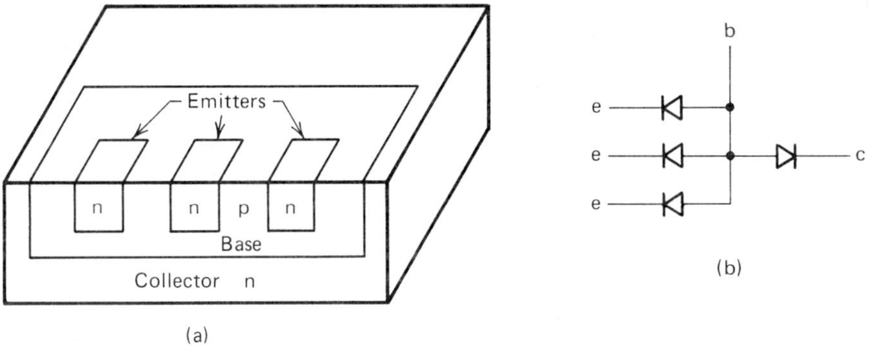

Figure 5-10. (a) A three-emitter transistor; (b) an oversimplified model for this transistor.

Figure 5-10b is an oversimplified model for this transistor showing only the emitter-base and collector-base junction diodes. We have neglected transistor action which we shall see is important to the switching speed of the circuit. However, we shall use the simplified model to present the basic ideas of the logic circuit.

A basic TTL circuit is shown in Fig. 5-11. If we replace the multi-emitter transistor by the simplified model of Fig. 5-10B, then this circuit becomes essentially the same as that of Fig. 5-9. The position of the diode D_2 and the transistor Q_{D1} are interchanged but the operation is essentially the same.

Let us now consider an analysis of this circuit. We shall use the following typical values:

V_{CC} = 5 volts

R_C = 2000 ohms

R_{B1} = 4000 ohms

R_{C1} = 1000 ohms

R_B = 4000 ohms

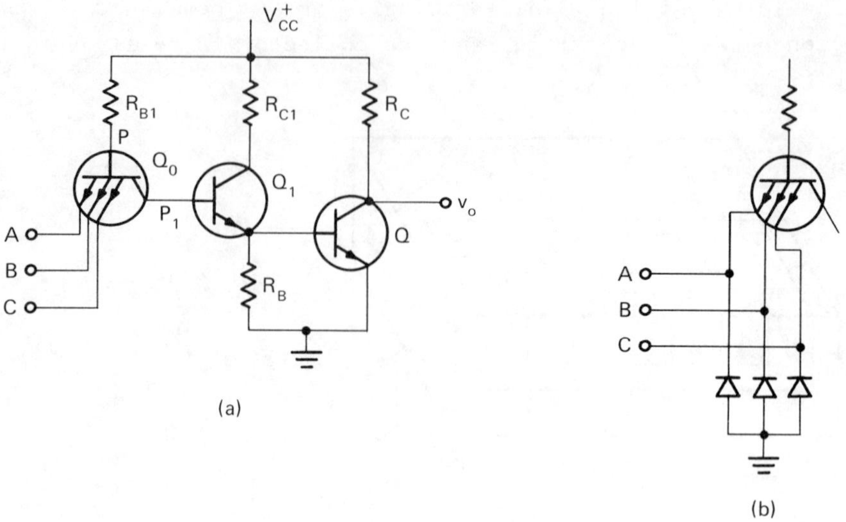

Figure 5-11. (a) A basic positive logic TTL NAND circuit; (b) an actual TTL input circuit.

As before, for the transistor, we assume that $V_\gamma = 0.6$ volt, and, under saturation conditions, $V_{BE,sat}=0.7$ volt and $V_{CE,sat}=0.3$ volt. If all the inputs are 1's, then the "input diodes" will be cut off. Thus, there will be no emitter current in the transistor Q_0. However, we can consider that there are three diodes in series, the base collector junction of the input transistor Q_0 and the base emitter junction of transistors Q_1 and Q. The voltage drop across this junction under saturation conditions is

$$0.7 + 0.7 + 0.3 = 1.7 \text{ volts}$$

Since $V_{CC} = 5$ volts > 1.7 volts, these junctions will be forward biased. Thus, Q_1 and Q will be saturated. Actually, the emitter current of Q_1 will be higher than its base current. This certainly causes transistor Q to saturate. As in the case of the DTL circuit, the transistor action of Q_1 increases the fanout.

Now suppose that one(or more)of the inputs are zero. The input voltage is then 0.3 volt. That(those)emitter-base junction (s)of Q_0 will be forward biased. Thus,

$$v_P = 0.3 + 0.7 = 1.0 \text{ volt} \tag{5-9}$$

This is less than $V_{\gamma 1}+V_{\gamma 2}$ = 1.2 volts. Thus, the collector-base junction of the input transistor will not be turned on. Thus, Q_2 will be off and the v_o rises to 5 volts.

We have thus far ignored the transistor action of the input transistor. Actually, it is important. Suppose that the inputs are switched from a condition where all inputs are 1's to the case where at least one input is a 0. In this case, transistors Q_1 and Q are driven from saturation to cut off. When one of the emitters of the input transistor is forward biased, its collector-base junction is reverse biased. This is because v_P=1.0 volt but $v_{P1} = 2V_{BE,sat}$ = 1.4 volts. Thus, for Q_0, v_{CE}= -0.4 volt. Hence Q_0 is in its active region, see Sec. A-3. The emitter current, thus, is much larger than the base current. This large current tends to drive excess stored charge from Q_1 to Q rapidly, see Sec. A-5. This greatly increases the switching speed. For this reason, the TTL circuit has the fastest speed of any logic that uses saturated transistors.

In an actual integrated circuit, extra diodes are added to the input circuit. These are shown in Fig. 5-11b. These diodes are reverse biased and normally do not affect the operation of the circuit. At times, because of inductance in the transmission lines that interconnect the various circuits, damped sinusoid oscillations are set up. These may lead to the occurrance of relatively large negative voltages, in a positive logic circuit. The added diodes in Fig. 5-11 keep these negative voltages at a very low level, preventing them from causing damage.

The delay caused by excess stored charge is an important factor in slowing the switching speed. If gold doping, see Sec.A-1, is appropriately added during the fabrication process, then the storage time can be greatly reduced. Unfortunately, this is an expensive process which increases the risk of a faulty circuit's being produced. We shall next consider a procedure which almost eliminates the excess stored charge without introducing the problem of gold doping.

5.6 SCHOTTKY DIODE-TRANSISTOR-TRANSISTOR LOGIC-STTL-ST^{2}L

The switching speed of TTL circuits is fast because the current due to transistor action helps to rapidly remove the excess stored charge(see Sec.A-5). The switching speed could be greatly increased if we could prevent the transistor from becoming saturated, in which case, stored charge would not have to be removed. If we could incorporate a device in the circuit that would automatically hold the transistor at the *edge* of saturation, then we would obtain all the advantages of saturated operation(e.g.,reliability, low power compensation)without sacrificing the switching time required to remove the excess stored charge that results when the transistor is heavily saturated.

Let us consider the circuit of Fig. 5-12a. Here we illustrate a single transistor used in a logic gate. A diode is added between the base and collector of the transistor. (We shall explain the special symbol for the diode subsequently.) If we attempt to forward bias the collector-base junction of the transistor, the diode will also be forward biased. This will divert current from the transistor. Thus, the excess stored charge will be reduced. Note that the diode will be reverse biased until saturation voltages are applied so that the diode does not substantially affect the circuit until the "edge of saturation" has been reached. Therefore, this circuit appears to have achieved the desired results.

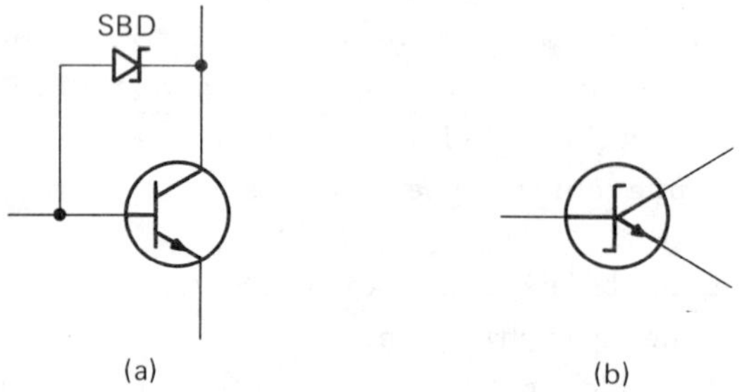

Figure 5-12. (a) A transistor-Schottky barrier diode circuit that improves switching speed; (b) a symbol for the composite transistor.

If the diode is an ordinary p-n junction diode, then the performance of this circuit will not be satisfactory. There will be excess charge stored in the diode when it draws a large forward current. When the transistor is to be cut off, the diode must act as an open circuit. However, until its excess stored charge is eliminated, the diode acts as a low impedance, see

Fig. A-32. Thus, when switching times are considered, the reduced storage time of the transistor will be offset by the increased storage time of the diode and great improvements in switching speed will not be obtained.

There is a diode called a *Schottky-barrier diode* which can be used to remove this difficulty. This diode consists of a metal to doped semiconductor junction, rather than a p-n junction. Such a device can exhibit all the rectifying properties of the p-n junction diode. However, when Schottky diodes are forward biased, they have essentially zero charge storage. Hence, they do not exhibit storage time effects and their switching is very rapid. The diode symbol of Fig. 5-12a is that of the Schottky diode.

In addition, Schottky diodes can be constructed so that their forward bias cut in voltage can be made 0.3 to 0.4 volt lower than that of the forward bias collector-base cut in voltage of a transistor. Hence, in the configuration of Fig. 5-12, most of the collector-base current will be diverted to the diode when the transistor's collector-base junction is forward biased. Thus, during "saturation", the excess stored charge of the transistor will be very low. Thus, the addition of the Schottky barrier diode can significantly improve the switching speed.

When Schottky barrier diodes are incorporated in the already fast TTL circuits, a very rapid circuit results. These are called Schottky diode transistor-transistor logic circuits, which are written as STTL or ST^2L. All of the transistors of the gate, including the multiple emitter input transistor, incorporate Schottky diodes. The transistor Schottky diode combination is represented by a single symbol, see Fig. 5-12b.

Another advantage of these circuits is that they present no great difficulty to fabricate as integrated circuits. In fact, they are only slightly more difficult to fabricate than are ordinary TTL integrated circuits.

5.7 WIRED LOGIC

Usually, complex switching functions are generated by interconnecting the outputs of gates to the inputs of others. However, there are times when the outputs of two or more gates are connected in parrallel. This is called *wired logic*. For instance, suppose two NAND gates have their outputs parallel, see Fig. 5-13. Note that we have shown separate leads for V_{CC} in each output. Actually, they would be connected to a common power supply. A similar statement also applies to the ground leads.

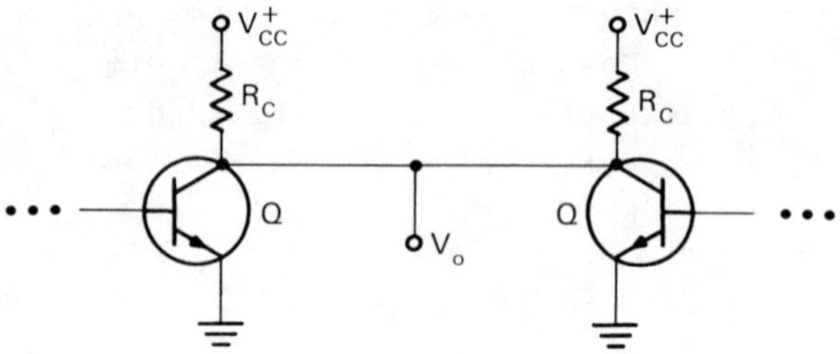

Figure 5-13. Interconnection of output circuits for wired logic.

If either transistor is saturated, v_o will be at a very low voltage, while if both transistors are cut off, v_o will be high.

If either(or both) transistors are saturated, then there will be current through both R_C's. This will be given by

$$I = \frac{V_{CC} - V_{CE,sat}}{R_C} \tag{5-10}$$

Thus, the total current is twice this and each resistor dissipates power. Actually, only one R_C is needed in this case. Integrated circuit gates which are designed for wired logic often have the R_C omitted so that when they are interconnected, only one R_C need be used.

Suppose that the left hand logic circuit is a NAND gate with four inputs X_1, X_2, X_3, and X_4 and the right hand circuit is a NAND gate with three inputs Y_1, Y_2 and Y_3. Then, the output will be a 1 only if both outputs are 1's, that is, if both $\overline{X_1X_2X_3X_4}$ AND $\overline{Y_1Y_2Y_3}$ are 1's. Thus, we have the switching function

$$V_o = \overline{(X_1X_2X_3X_4)}\,\overline{(Y_1Y_2Y_3)}$$

5.8 ACTIVE PULL-UP CIRCUITS

The output stage of the logic circuit we have considered consists of the resistor R_C in series with the transistor Q, see Fig. 5-6. This is called a *passive pull-up circuit*, since the change in voltage v_o can be considered to be brought about by the change in voltage drop across the passive R_C. Now consider that Q is switched from saturation to cutoff. The output voltage across Q rises from its saturation value to its cut off value. There is an effective capacitance C_o between the collector and emitter of the transistor. As the voltage rises, this capacitance must be changed. The time constant of the RC circuit is

$$T = R_C C_o \tag{5-11}$$

The large T is, the longer it takes for the output voltage to reach its "final" value. This limits switching speed. In general, C_o is fixed. We can reduce R_C. However, this will increase the dissipated power under saturation conditions. Note that this power dissipated in R_C is

$$P = \frac{V_{CC} - V_{CE,sat}}{R_C} \tag{5-12}$$

and V_{CC} and $V_{CE,sat}$ are independent of R_C.

An alternate circuit can eliminate the difficulty of increased power. We desire that R_C be small under cut off conditions when the current through it and, hence, the power dissipated by it, is small. Then, C_o can be charged quickly. However, we want R_C to be large when Q is saturated. (Note that C can be discharged through the saturated transistor.) To obtain such a variable R_C, we replace it by a transistor, see Fig. 5-14. Assume that the voltages v_1 and v_2 are "out of phase" so that when v_1 produces saturation of Q, v_2 results in the cut off of Q_P and vice versa. Now, when Q is switched from saturation to cut off, Q_P is saturated. Thus, it acts as a small resistance and the switching can take place rapidly. On the other hand, when Q is saturated, then Q_P is cut off. This causes its collector current to be extremely small. This, in turn, causes the dissipation to be limited. This circuit is called an *active pull-up* because the passive R is replaced by the active transistor Q_P. It is also called a *totem pole* circuit because one transistor is above the other one. This circuit can switch faster, with lower power dissipation, than the passive pull-up circuit.

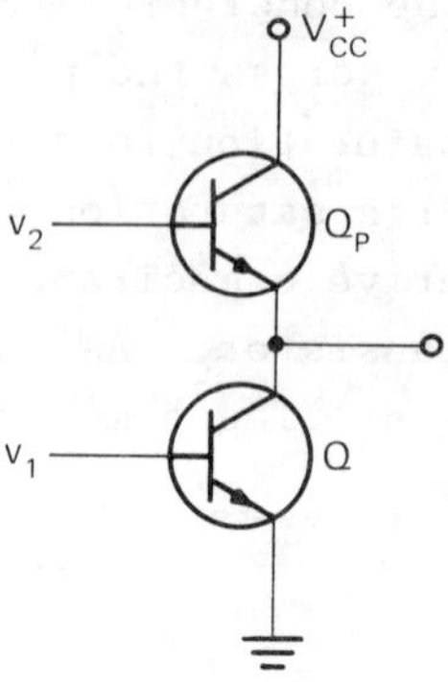

Figure 5-14. A basic active pull-up output circuit.

Now let us discuss the practical circuit of Fig. 5-15, which could be used for the output of Fig. 5-11. When Q_1 is cut off then V_a will be high. Thus, Q_P will be saturated. Since Q_1 is cut off, then so will be Q. When Q_1 is saturated, V_a will be low and Q_P will be cut off and Q will be saturated. Thus, we have achieved the desired operation. The resistor R_1 has a small value and is used to limit current during switching. A transistor switches from cut off to saturation faster than in the reverse direction. Thus, there will be a short transient period of time, near the switching time, when both transistors are saturated. If R_1 were not present, then the power supply would be "short circuited" during this period. R_1 limits the current, reducing both power dissipation and large power supply current pulses which could act as noise in other parts of the computer. A typical value for R_1 is 100 ohms.

The diode D is included to obtain proper bias for the transistor Q_P. Suppose that Q_1 and Q_2 are saturated and that D is not present. Then, using the values for the TTL circuit

$$v_a = v_{CE,sat} + v_{BE,sat} = 0.3 + 0.7 = 1 \text{ volt}$$

$$v_o = 0.3 \text{ volt}$$

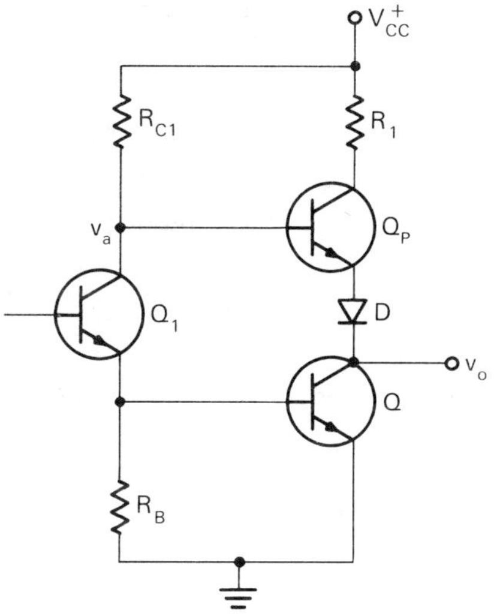

Figure 5-15. A practical active pull-up output circuit.

If the diode D were not present, then for Q_P

$$V_{BE} = v_a - v_o = 1 - 0.3 = 0.7 \text{ volt}$$

This is greater than $V_{\gamma d} = 0.6$ volt, and Q_P would not be cut off.

Now suppose that we include the diode. Its cut in voltage is $V_{\gamma d} = 0.6$ volt. In this case, we require

$$0.6 + 0.6 = 1.2 \text{ volts}$$

if D and Q_P are to conduct. Since 0.7 is less than this, both Q_P and D are cut off. Thus, the diode D must be included.

Now suppose that Q_1 is switched to cut off. Then, Q will be cut off since its $v_{BE}=0$. However, the voltage cannot change instantaneously at the first instant of time, and $v_o=0.3$ volt just after switching. Then, the sum of v_{BE} for Q_P and the voltage across D and that across R_{C1} is

$$5 - 0.3 = 4.7 \text{ volts}$$

This is greater than the sum of V_γ and $V_{\gamma d}$. Hence, both Q_P and D will conduct. The voltage v_o then rapidly rises. The equilibrium condition occurs when Q_P is just at saturation and D is just at cut in. This is

$$v_o = V_{CC} - I_{CE}R_1 - V_{CE,sat} - V_{\gamma d} = 5 - 0.3 - 0.6 = 4.1 \text{ volts}$$

where we have neglected the small drop in R_1.

In general, the current through R_1 will be small since either Q_P or Q will be cut off. The only exception to this is during the transient period when both are saturated. In this case, the current is

$$I = \frac{V_{CC} - 2v_{CE,sat} - V_D}{R_1} = \frac{5-2(0.3)-0.8}{100} = 36 \times 10^{-3} = 36 \text{ ma} \qquad (5\text{-}13)$$

(where V_D is the voltage across the diode). This is a large current. However, it only persists for a short transient time after switching.

Note that this form of active pull up should never be used with wired logic. If, in one circuit, Q_P is saturated and, in the other, Q is saturated, then we have the same 36 ma of current as calculated in Eq. (5-13). However, now this is not just a transient condition. Thus, excessive currents can result.

5.9 RESISTOR-TRANSISTOR LOGIC RTL

Another form of logic circuit which makes use of resistors and transistors is shown in Fig. 5-16. A NOR gate is illustrated there. This is the basic gate of this family for positive logic. If any input is a 1, then its transistor saturates and v_o becomes low. If all inputs are 0's, then the transistors are cut off and v_o is high. Thus, the desired logic function is obtained.

This circuit uses very little space on an integrated circuit chip. It also can operate at low voltage levels. For instance, typical values are $V_{CC} = 3.5$ volts, $R_B = 450$ ohms and $R_C = 700$ ohms.

Let us consider the fan out of this circuit. Suppose that the output wire is connected to two inputs. This would be as shown in Fig. 5-17. In general, transistors do not have identical

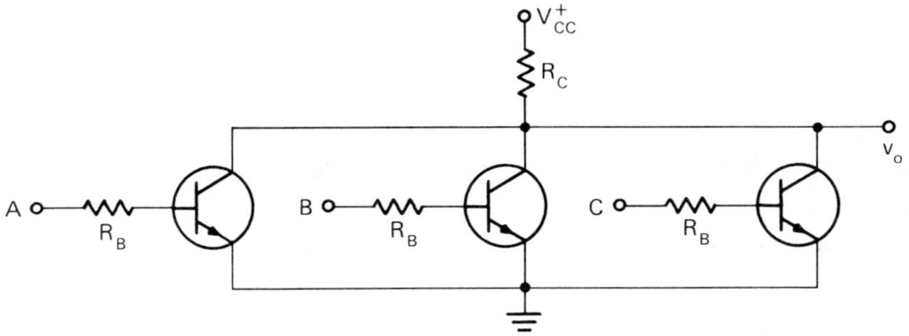

Figure 5-16. A positive logic RTL NOR gate.

characteristics. Thus, if Q_1 has a lower v_{BE} than Q_2, it would take more than its share of base current were it not for R_B. In saturation, i_B is essentially a function of R_B and not of v_{BE}, provided that R_B is large enough(i.e., greater than several hundred ohms). Thus, keeping the two values of R_B nearly equal will insure that a proper division of current results even if the transistors are different.

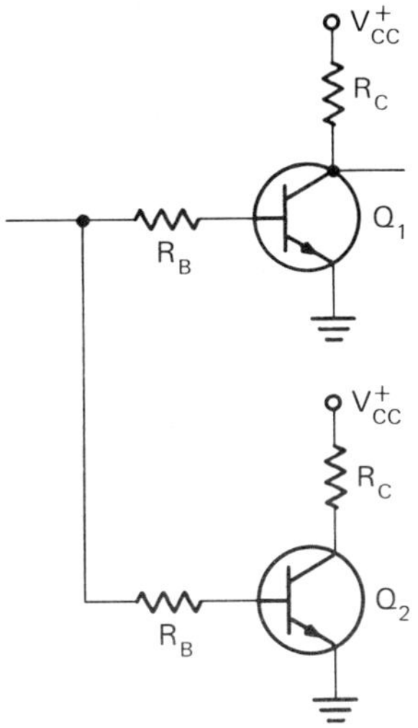

Figure 5-17. An RTL input circuit which would represent a fan out of two for the circuit of Fig. 5-16.

5.10 DIRECT-COUPLED TRANSISTOR LOGIC-DCTL

If we set the value of R_B=0 in Figs. 5-16 and 5-17, the logic circuit obtained is called *direct-coupled transistor logic*. Its advantage over RTL is that the switching speed is increased since the larger base current can remove excess charge faster.

However, DCTL has several disadvantages. Under saturation conditions, there is not any R_B to regulate the base current. Thus, great differences in base current can result. Actually, one transistor can take so much current that the other is not driven into saturation. This is called "*current hogging*". Thus, great care must be taken to make the transistors nearly identical. In addition, the very large values of base currents can result in large amounts of excess charge which tends to offset the previously discussed advantage of this circuit. This circuit can operate at very low voltage levels, which is an advantage. However, using low voltages also means that low noise voltage can produce errors. Other logic circuits are now much more commonly used than is this one.

5.11 EMITTER-COUPLED LOGIC-ECL

The proviously discussed logic circuits all utilized saturated transistors. This makes for reliable operation and low power dissipation since the voltage across the transistor is low when the current through it is high and vice versa. However, saturation slows switching speed. Now let us consider a nonsaturated logic circuit which has the fastest switching speed. However, its power dissipation is relatively high. The basic idea of this circuit is illustrated in Fig. 5-18a. Here a current generator drives the parallel combination of two resistors. For this circuit we have

$$I_1 = \frac{IR_2}{R_1 + R_2} \tag{5-14a}$$

and

$$I_2 = \frac{IR_1}{R_1 + R_2} \tag{5-14b}$$

Now suppose that R_1 is a variable resistor. If $R_1 >> R_2$, then

$$I_1 = IR_2/R_1 \approx 0$$

$$I_2 = I$$

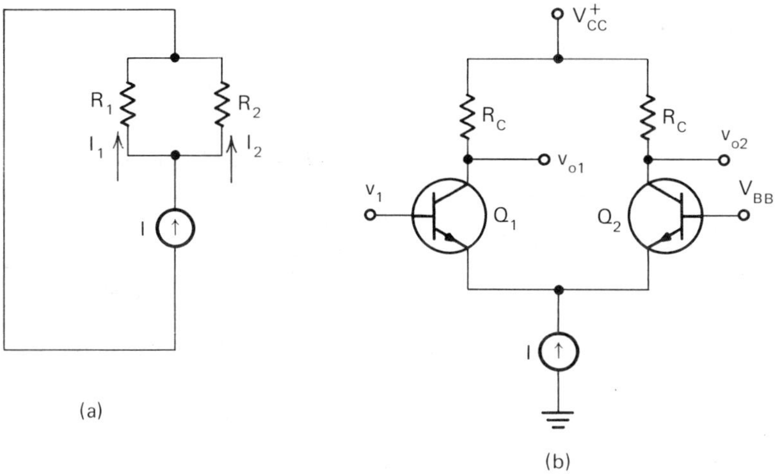

Figure 5-18. (a) A simple current mode model; (b) a transistor implementation of this model.

Thus, essentially all the current passes through R_2. Similarly, if $R_1 << R_2$, then essentially all the current will pass through R_1. Again, we can implement the "variable resistance" using a transistor. This circuit is shown in Fig. 5-18b. The voltage V_{BB} is a fixed voltage while v_1 can be varied. If v_1 is such that Q_1 is cut off, then the current I will be through Q_2. Similarly, if we make v_1 much larger than V_{BB}, but not large enough to cause saturation, then I will be through Q_1 while the current in Q_2 will be very small. We shall also see that the bias of Q_2 also ahifts to aid this action. Since it is the current that is switched in this circuit, the circuit is called *current mode logic CML*. Also, because the two transistors are coupled through their emitter circuits, this is also called *emitter-coupled logic ECL*. The basic gate of this family for positive logic is the OR or NOR gate.

Now let us consider the more practical circuit of Fig. 5-19. Here the one "variable" transistor has been replaced by three so that we have a three input gate. The constant current generator has been replaced by a large resistor in series with a voltage source. If R_E is large enough, the current through it will be essentially a constant.

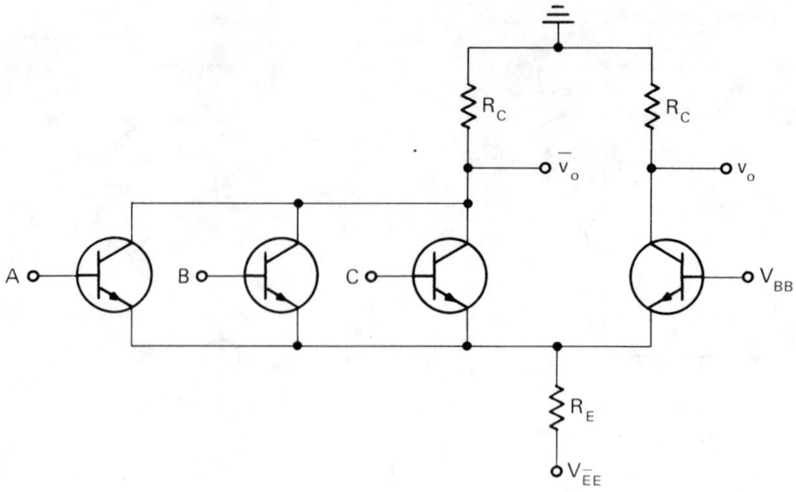

Figure 5-19. A more practical ECL-NOR or OR positive logic gate.

Typically V_{EE} = -5 volts. Now if A or B or C represents a 1, then the corresponding transistor will conduct heavily. Then, $\bar{v}_o$ will represent a 1 and v_o a 0. If all of A, B and C are 0's, then $\bar{v}_o$=0 and v_o=1. Thus, the output v_o gives a NOR gate while $\bar{v}_o$ is its complement, an OR gate. One problem with this circuit is that the voltage at the collector of the transistor differs from the base voltage. Thus, the voltage levels at the output will differ from those at the input. It is desirable to keep the voltages corresponding to 0's and 1's constant throughout the system. This can be accomplished by adding the common collector amplifiers shown in Fig. 5-20. This common collector amplifier has a very low output impedance and, thus, increases the fan out.

Let us analyze the circuit using the following typical values.

V_{EE} = -5.2 volts

R_E = 1000 ohms

R_C = 300 ohms

R_{E1} = 2000 ohms

$V_{BB} = -1.15$ volts

Assume that $v_{BE} = 0.8$ volt for a conducting transistor. Now let us calculate the outputs and demonstrate that they can be produced by equal inputs. Also, the values of R_{C1} must be slightly different from R_C. Let us also obtain the value of R_{C1}.

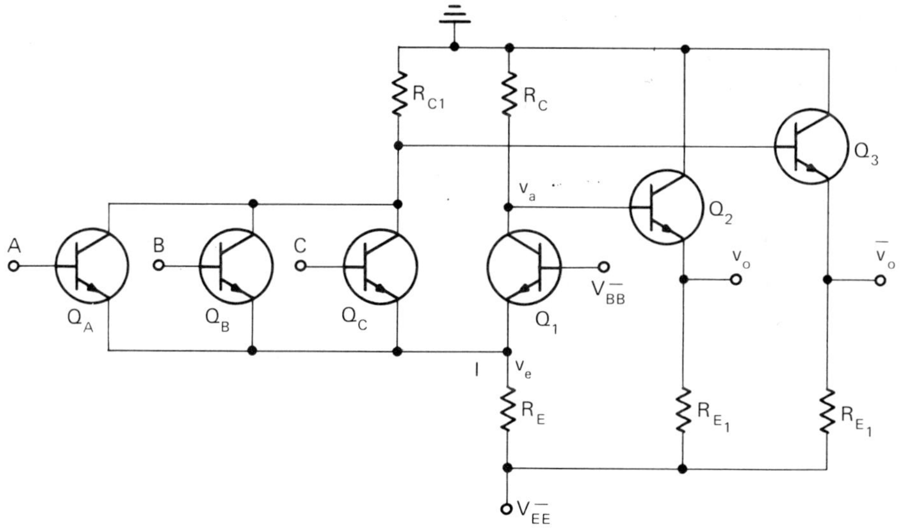

Figure 5-20. A ECL NOR or OR gate with no level shift between input and output.

Assume that all inputs are zero so that Q_A, Q_B, and Q_C are cut off. Then, the current through R_E is

$$I = \frac{V_{ee} - (V_{BE}+1.15)}{R_E} = \frac{5.2-(1.15+0.8)}{1000} = \frac{3.25}{1000} = 3.25 \times 10^{-3}$$

$$= 3.25 \text{ ma} \tag{5-15}$$

Thus, the voltage drop across R_C is IR_C. The output voltage is this minus the base emitter voltage of Q_2. Hence,

$$v_o = -3.25 \times 10^{-3} \times 300 - 0.8 = -1.775 \text{ volts}$$

(We assume that the collector current of Q_1 is much greater than the base current of Q_2.) Now we must show that -1.774 volts at the input will cut off the input transistor. The voltage v_e will be

$$v_e = V_{EE}+IR_E = -5.2 + 3.25 = -1.95 \text{ volts}$$

Thus, the base emitter voltage of the input transistor is

$$v_{BE} = -1.775 - (-1.95) = .175$$

This is less than V_γ for a typical transistor(i.e., 0.6 volt). Thus, the value of -1.775 is a valid one.

Now assume that A, B or C represents a 1. In this case, the circuit will be switched from Q_1. The current through Q_1 is very small. We shall assume it to be zero. The voltage V_{EE} in series with R_{E1} and R_C produces a base current in the transistor Q_2. This, of course, produces collector current. In general, we must investigate the characteristics to determine the collector emitter voltage since this is equal to v_o. Let us assume a typical value for this. Of course, in an actual circuit, the transistor characteristics would have to be verified. This would involve some cutting and trying since R_{E1} is in both the collector and base circuits. A typical value of $v_{EC} = v_o$ would be

$$v_o = -0.8 \text{ volt}$$

Let us see if this results in Q_1's being cut off, if an input signal equal to this is applied. In this case,

$$v_e = -0.8 - v_{BE} = -0.8 - 0.8 = -1.6 \text{ volts}$$

Then, for the transistor Q_1

$$v_{BE} = -1.15-(-1.6) = .45 \text{ volt}$$

This is less than the cut in value. Thus, the transistor Q_1 is cut off. Thus, these voltage levels are allowable. We assumed a 0.8 voltage drop across Q_2. Actually, a wide range of values could be used and the output will be allowable in that an input equal to that output would cut off Q_1.

Now let us determine R_C. In this case, when Q_1 is cut off

$$I = \frac{5.2-1.6}{1000} = 3.6 \times 10^{-3} \tag{5-16}$$

This differs from that given by Eq.(5-15) when Q_1 conducts. If we desire equal voltages and v_o and $\bar{v}_o$, then

$$R_{C1}(3.6) = R_C(3.25)$$

$$R_{C1} = \frac{300 \times 3.25}{3.6} = 271 \text{ ohms.}$$

Note that the difference between the levels for a 0 and for a 1 is small, 1.775-0.8 = .975 in contrast to more than four volts for saturated logic circuits. This reduced voltage difference results in reduced noise immunity for ECL.

The emitter-coupled logic is faster, but dissipates more power and has worse noise immunity than TTL. Thus, each has its advantages.

When very high speed circuits, such as ECL, are used, the circuits are not simply interconnected by wires. Special transmission lines are used for this purpose.

5.12 MOSFET LOGIC FAMILIES-MOS LOGIC FAMILIES

We shall now consider logic families that use MOSFET devices, specifically, those using enhancement MOSFETs. In general, MOS integrated circuits can be fabricated using fewer steps than those using junction transistors. In addition, the isolation problem is reduced and power dissipation can be much less with MOSFET logic. However, the switching speed is much faster using junction transistors.

Basic MOSFET Logic-MOS Logic

A basic MOSFET logic element consists of an FET in series with a resistive pull up element, see Fig. 5-5. However, it is easier to fabricate MOSFETs than resistors in integrated circuits. Thus, an active pull up is used, see Fig. 5-21. This circuit also has other advantages which we shall discuss subsequently. Assume, for the moment, that the upper FET acts as a constant resistance. If A represents a 1 which is a positive signal, then FET F_1 will saturate. This results in v_o's being a small positive voltage output. Typically V_{DD}=10 volts and the level for a 0 is a fraction of a volt. If A represents a 0, then the input bias will be small and the transistor F_1 will be almost cut off. In this case, v_o will be close to V_{DD}. Thus, we have achieved inversion.

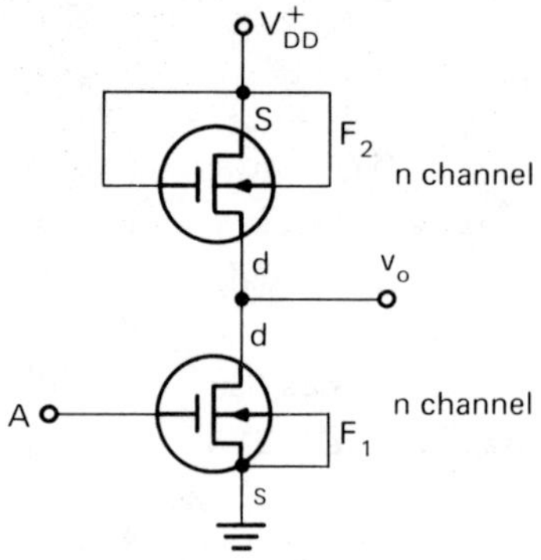

Figure 5-21. A basic positive logic MOSFET inverter.

We have assumed that the upper FET acts as a fixed resistance. Actually, when v_o increases, the value of v_{DS} of F_2 decreases. This tends to reduce the effective resistance of F_2, which causes v_o to become almost equal to V_{DD}. Conversely, when F_1 saturates, the resistance of F_2 increases which brings the value of v_o closer to zero. Thus, the variation of the "resistance" of F_2 is desirable. Subsequently, we shall demonstrate a circuit that vastly increases this desirable effect.

A positive logic NOR gate is shown in Fig. 5-22a. Note that if any input is a 1, the corresponding FET will saturate and v_o will represent a 0. If all inputs are 0's then the FETs F_1, F_2 and F_3 will be cut off and v_o will represent a 1. Similarly, Fig. 5-22b is a positive logic NAND gate. The NOR and NAND are the basic MOS logic gates.

Note the substrate connections. In an integrated circuit, the substrate connection is connected to a region immediately around

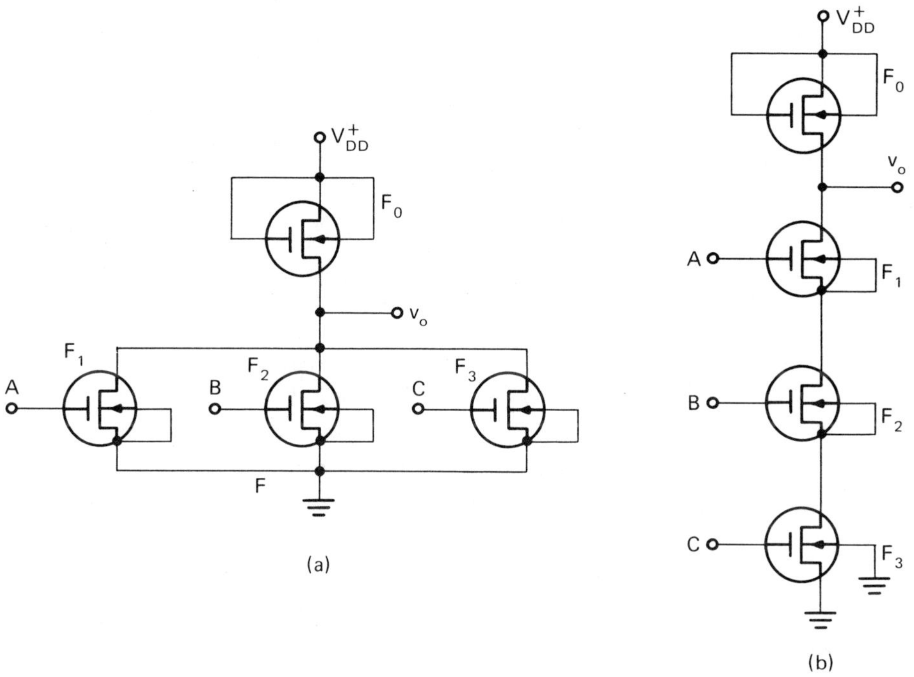

Figure 5-22. MOSFET positive logic gates. (a) NOR; (b) NAND.

the FET in question. The substrate's electrical connections are made to isolate the particular FET by reverse biasing the p-n junction which is between the surrounding region and the remainder of the integrated circuit.

5.13 · COMPLEMENTARY SYMMETRY MOSFET LOGIC-COSMOS LOGIC-CMOS LOGIC

If we use both n-channel and p-channel enhancement MOSFETs in the logic circuit, circuits which draw very little current can be obtained. In Fig. 5-23 we illustrate a basic inverter circuit using this complementary symmetry. Let us assume that a 0 is represented by a small positive voltage essentially equal to zero, and a 1 by a positive voltage essentially equal to V_{DD}, typically 10 volts. Suppose that input A is a 1. Then, F_1 will be saturated while F_2 will be cut off. Note that F_1 is an n-channel device while F_2 is a p-channel device. Thus, the effective resistance of F_2 becomes extremely large while that of F_1 is very small. Thus, $v_o \approx 0$ volts.

Now suppose that A represents a 0. Then, F_1 will be cut off. It may appear as though F_2 would also be cut off. However, the gate to source voltage of F_2 is

$$v_{GS} = v_1 - V_{DD} \tag{5-17}$$

Since V_{DD} is positive and $v_1=0$, then v_{GS} will be negative and equal in magnitude to V_{DD}. Then, F_2 will be saturated. Hence, $v_o = V_{DD}$.

Note that either FET F_1 or F_2 will always be cut off. Thus, the current through the circuit will be extremely small. Hence, this circuit dissipates very little power and subsequently, many circuits can be put on a single chip. The very low power and high packaging density are prime advantages of this circuit.

For a transient period after switching, there will be some current. This is because the various capacitances of the circuit, see Sec. A-4, must be charged or discharged.

Since these are insulated gate FETs, the input impedances of the gates are very high. Thus, the fan out of these circuits is high. Note that there is a capacitance across the input. Thus, if very short pulses are considered, their capacitances will act as a load. Hence, the very high fan out is obtained only when very short pulses are not to be used.

Let us consider some of the details of the operation of this circuit. Just as in the case of the transistor circuits, we shall assume the values of some pertinant parameters for the FET. (These shall be based on the characteristics of the enhancement MOSFET of Fig. A-21.) When the gate to source voltage $v_{GS}=0$, the drain current I_D is a small leakage value. We shall consider this to be both constant and equal to I_γ. In the example we shall use $I_\gamma=100\mu a$. Actually, the current I_D remains at almost I until v_{GS} becomes sufficiently positive. This can be called the *cut in value* for v_{GS}. We shall call this v_γ. Typically $|v_\gamma| = 2$ volts. (Note that we use both n-channel and p-channel FETs in this circuit so that both positive and negative values of v_γ will be used.)

If the external circuit is such that it limits the current to the small value I_γ, then v_{DS} will be a small value which is essentially independent of v_{GS} provided that $v_{GS} > v_\gamma$. We shall assume that this small value is $V_{min} = 0.1$ volt.

Now let us consider the operation of the circuit of Fig. 5-23. We assume that $V_{DD} = 10$ volts. Let us calculate the signal levels of the 0's and 1's. At the start we assume that a 1 is represented by 9 volts or more while a 0 is represented by 1 volt or less. Now suppose that the input is a 0. Then, $V_1 = 1$ volt. Then, v_{GS1} will be less than V_γ and F_1 will be cut off. Now consider F_2. Here

$$v_{GS2} = V_1 - V_{DD} = 1 - 10 = -9 \text{ volts}$$

This is much greater in magnitude than V_γ. Since F_2 is a p-channel FET, it will now act as a low impedance. However, the current I is limited by the fact that F_1 is cut off. Hence,

$$I = 100\ \mu a = 0.1 \text{ ma}$$

Thus, for F_2, $v_{DS2} = 0.1$ volt. Then,

$$v_0 = V_{DD} - v_{DS2} = 9.9 \text{ volts}$$

Thus, a 0 input produces a 1 output. We can consider that the level of a 1 is 9.9 volts. (The value of $V_{DS1} > V_{min}$ since F_1 is cut off and F_2 is not. Thus, most of the voltage drop appears across the channel of F_1.)

Now suppose that the input at A is a 1. Thus, $V_1 = 9.9$ volts. Now F_1 will conduct heavily since $v_{GS1} = V_1 = 9.9$ volts. (Note that F_1 is an n-channel FET.)

For F_2 we have

$$v_{GS2} = V_1 - V_{DD} = 9.9 - 10 = -.1 \text{ volt}$$

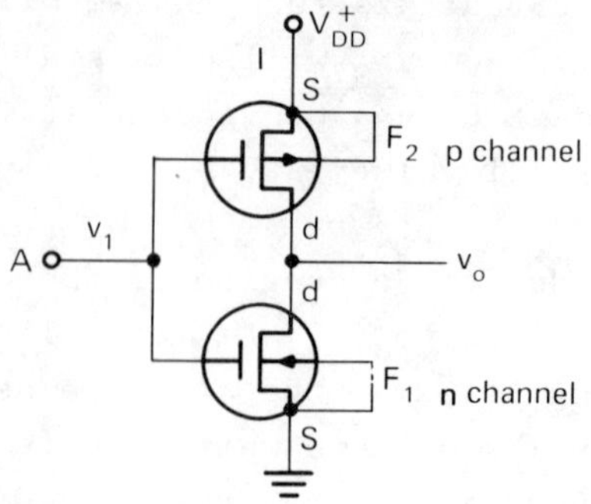

Figure 5-23. A COSMOS positive logic inverter.

This magnitude is less than that of V_γ. F_2 will now be cut off and limit I. Now $v_{DS1} = V_{min}$. Hence,

$$v_0 = 0.1 \text{ volt}$$

Thus, for this circuit, a 1 will be 9.9 volts while a 0 will be 0.1 volt.

A positive logic NOR gate is shown in Fig. 5-24. We can consider pairs of FETs, one in the lower "parallel" section and the other in the upper "series" section. The behavior of each pair is similar to the pair of Fig. 5-23. If any input is a 1, the corresponding parallel lower FET will saturate, while the corresponding series upper FET will be cut off. Thus, the output will be a 0.

If all the inputs are 0, then all the lower FETs will be cut off and all the upper ones will be saturated and the output will be a 1. Thus, this device functions as a NOR gate.

A positive logic NAND gate is illustrated in Fig. 5-25. The operation of this is the converse of Fig. 5-24.

5.14 CLOCKED LOGIC CIRCUITS: DYNAMIC MOS LOGIC

We have considered the basic MOS gates. In Chapter 6 we shall discuss the use of clocked logic. In such circuits, the levels of all the logic circuits are changed synchronously. That is,

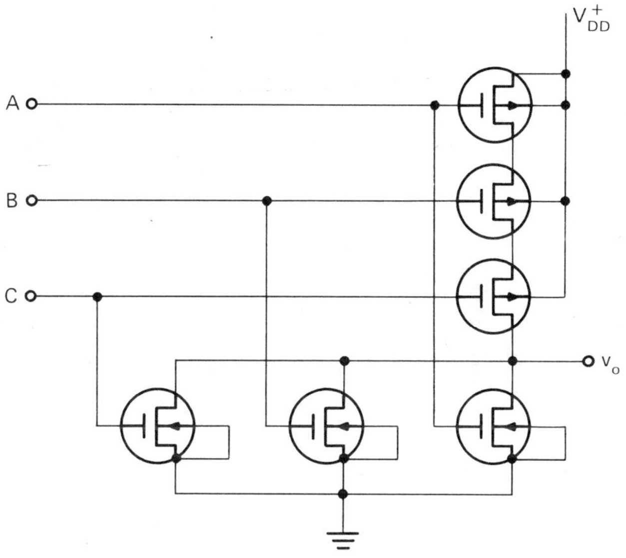

Figure 5-24. A COSMOS positive logic NOR gate.

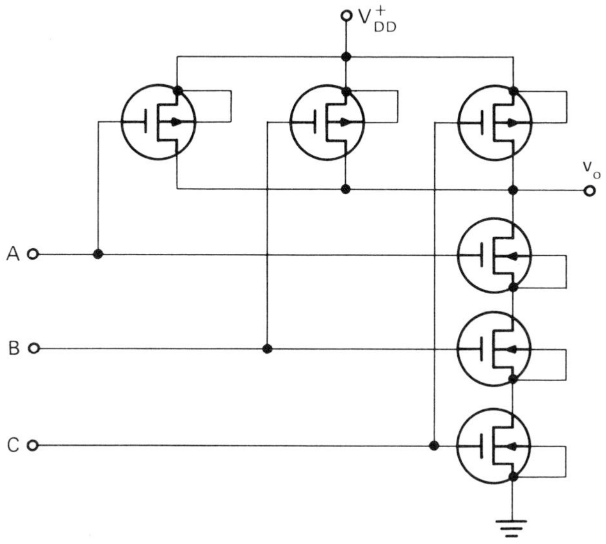

Figure 5-25. A COSMOS positive logic NAND gate.

there are pulses produced by a device called a master clock. The gates can only respond to signals in response to these clock pulses. For instance, the gates' levels can only change when a clock pulse is present. Such circuits are called synchronous sequential circuits. In the next chapter we shall consider these circuits in detail. However, especially when MOS circuits are used, synchronous circuits provide advantages which we shall now discuss. (Note that this material is included here rather than in Chapter 6 since it is essentially a discussion of electronic circuits.)

Let us start by considering the circuit of Fig. 5-21. If the voltage at A switches from a 1 to a 0, the voltage v_0 will not rise up to the 1 level immediately. This is due to stray capacitance such as that which is between the output lead and ground. The voltage v_0 cannot rise to its final value until this stray capacitance becomes charged. (A similar effect occurs when the input is switched from a 1 to a 0.) The switching speed can be increased by increasing the current, since this permits the capacitance to charge or discharge at a faster rate. However, this increases the power dissipation, which is undesirable.

The circuit of Fig. 5-26 tends to reduce the capacitance effects without increasing the power dissipation. This circuit is similar to the inverter of Fig. 5-21 except that now F_3 has been added. The gate of F_3 is driven by the clock pulses CP. Thus, ideally F_3 and, hence, the rest of the circuit, is cut off unless there is a clock pulse present.

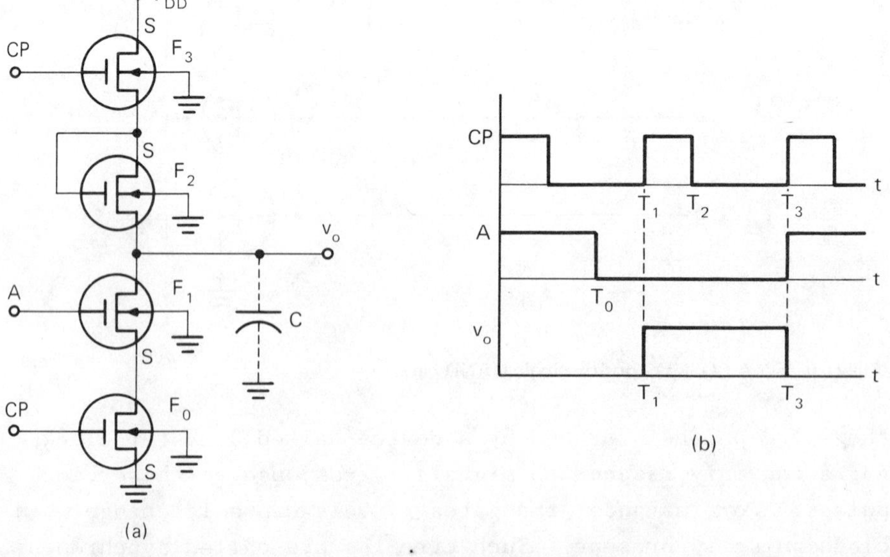

Figure 5-26. (a) A clocked MOSFET inverter; (b) some typical pulses.

Since clock pulses are periodic and are such that their length is only a small fraction of a period, then the circuit will be cut off for much of the time and the power dissipation is considerably reduced. Thus, we can increase the current when the clock pulses are present, and not increase the average power. However, the increased current decreases the switching speed.

Now let us consider the operation of the circuit. Some representative pulses are shown in Fig. 5-26b. (Note that the pulses have been drawn in an idealized fashion, since we have ignored rise and fall times.) Figure 5-26b shows the input A switching from a 1 to a 0 at T_0. However, the output does not change since the clock pulse CP is off and F_3 is cut off. (Note that F_0 is also cut off.) At $t=T_1$, a clock pulse occurs; A is still a 0. Now the output v_o switches to a 1, since F_3 is no longer cut off. At $t=T_2$ the clock signal becomes zero. However, v_o remains constant because of the capacitor C. That is, the charge stored in C causes the voltage to remain constant. This persists until T_3. Now, F_0 is turned on and the capacitor can discharge through the channels of F_1 and F_0. If the input at A is a 0, then the output will become a 1. Similarly, if the input is a 1, the output will be a 0. Thus, the circuit functions as an inverter. Note that F_0 and F_3 only allow v_o to change during clock pulses.

Note that we assumed that the voltage v_o did not change between T_2 and T_3. Actually, the capacitance will discharge slightly. The current through the cut off FETs will not be zero. In addition, there will be leakage through the reverse biased channel-substrate junction. These effects will result in a leakage current of about 1μa, which will slowly discharge the capacitor. However, the repetition rate of the clock pulses is made much shorter than the "time constant" of the leakage circuit. Thus, for all practical purposes, v_o does not change between clock pulses, if the input at A remains a 1. We say that v_o is *refreshed* at each clock pulse since charge will be added to it to bring the voltage back to a 1 level.

We have represented C as a stray output capacitance. Actually, such circuits usually drive similar ones. In such cases, the predominant capacitance is the gate to channel capacitance of the driven circuits.

In the circuit just discussed, the gate capacitance actually stores the signal between clock pulses. This is a dynamic type of condition in that the signals stored as charge on the capacitors will eventually leak off. Thus, we rely on the signal's being refreshed at each successive clock pulse. In some operations, this is undesirable. However, there are many cases where such operation does not present any difficulties.

Now we shall discuss a more practical circuit. Consider the circuit of Fig. 5-27a which uses two clock pulses. Note that the clock pulses are not only applied to the gates but they also act as "power supply voltages". The timing of the clock pulses is shown in Fig. 5-27b. Note that four different sets of pulses are shown while the circuit of Fig. 5-27a only requires two sets of pulses. We shall discuss the reason for this subsequently.

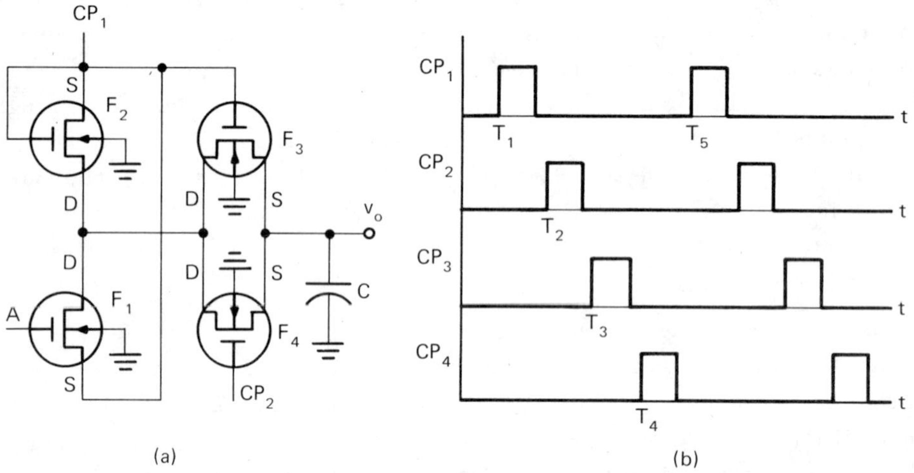

Figure 5-27. (a) A clocked dynamic four phase MOS inverter; (b) the four phase clock pulses.

Now let us consider the operation of this circuit. At T_1, CP_1 turns on F_2 and F_3. Then, the capacitor C will be charged through the series interconnection of the channels of F_2 and F_3. Thus, independent of the input at A, v_o becomes a 1 during CP_1. Now, at T_2, CP_2 is applied. (Note that now CP_1 is a 0.) If A represents a 1, then F_1 will be turned on. Similarly, CP_2 turns on F_4. Thus, C will be discharged through the series interconnection of the channels of F_1 and F_4. Note that CP_1 is now a 0 and the source of F_2 is now at zero potential. Thus, a 1 input produces a 0 output starting at T_2. Now consider that the input signal is a 0. Then, at T_2, F_1 will be cut off. Thus, C will not have a discharge path to zero potential. Thus, v_o remains a 1.

Note that if the input is a 1, then v_o will be a 0 between T_2 and T_5. Similarly, if the input is a 0, then v_o will be a 1 between T_2 and T_5. At T_5, CP_1 becomes a 1 and the process is repeated(i.e., v_o is set equal to a 1, etc). Thus, the charge in C can be refreshed.

Thus, this circuit functions as a NOT circuit for those times after the pulse of CP_2 and before the next pulse of CP_1(e.g.,between T_2 and T_5). During the pulses of CP_1(e.g.,between T_1 and T_2)the output is always a 1. The clock pulses CP_3 and CP_4 are used to control the remainder of the circuit. For instance, they are needed so that the output of a 1 during the pulse of CP_1 does not result in erroneous operation.

This is called *four phase* operation since there are four different clock pulses and, thus, there can be "four phases" of operation. Such dynamic MOS circuits are becoming increasingly important.

5.15 SOME GENERAL COMMENTS ABOUT LOGIC FAMILIES

We have considered the basic logic gates of various families. In LSI many interconnected gates are fabricated on one chip. The user must then connect the proper power supply voltage(s) and provide for the appropriate input and output signals. Care should be taken to see that the signals are of the proper polarities and are at the correct levels. The impedance of the input and output circuits should also be compatable. Appropriate values can usually be obtained from the manufacturer's specifications.

When small scale integration is used, gates of one logic family are usually interconnected by the user to obtain a switching function. Again, the correct power supply voltages must be obtained. If all the gates are from the same family and are produced by the same manufacturer, then there will probably be no incompatability of levels. However, care should be exercised. (Note that gates other than the basic ones are produced.) The previous comments about input and output circuits apply. In addition, the rated fan in and fan out should not be exceeded.

Comparison of Logic Families

We have indicated the various advantages and disadvantages of logic families. Let us now summerize the most important points.

At the present time, TTL is probably the most widely used logic family. Its switching speed is the fastest of any saturated logic circuit, about 5-12 ns(1 ns = 10^{-9}sec). The fan out is relatively high(10-20)and the power dissipated per gate is relatively low.

ECL is about 2 to 3 times as fast and has about twice the fan out of TTL. However, the power dissipated is 2 to 4 times as high and the noise immunity is much poorer than in TTL.

DTL is comparable to TTL except that its switching speed is slower and its fan out slightly less. The noise immunity is somewhat less than with TTL.

RTL has a switching speed slightly slower than TTL. However, its fan out and noise immunity are poorer.

COSMOS has a very low power dissipation, about 100 times less than TTL after transients have died away. However, its switching speed is about 10 times as long as that of TTL. COSMOS logic has the highest fan out, exceeding 50, of any commonly used logic and its noise immunity is as good as TTL. The comparitively low switching speed prevents the very wide adoption of COSMOS logic. However, it is used in small special purpose computers.

5.16 INTEGRATED CIRCUIT GATE CHARACTERISTICS

The gates that we have discussed can all be implemented as integrated circuits. Let us consider some of the details here. For instance, consider Figs. 5-24 and 5-25 which are COSMOS positive logic NOR and NAND gates. Each of these individual gates could be implemented as an integrated circuit on a single chip. In all likelihood, however, such an integrated circuit would contain several logic gates. For instance, a typical integrated circuit chip might contain 10 NOR gates, each with three inputs. Note that there are much more complicated chips containing hundreds of gates. We shall discuss some of these subsequently. Here we are considering some simple logic circuit chips.

Integrated circuit manuals give pertinent data concerning these chips. Let us consider some of the more commonly used terms.

Power Supply Voltage

This is the recommended supply voltage(e.g., V_{dd} of Fig. 5-24). A range of values is often given. For instance, 5-15 volts is typical. The device will perform satisfactorally if the supply voltage is kept within this range. Note that the operation of the device is usually a function of power supply voltage and, therefore, the data given in the manuals usually list several values corresponding to different power supply voltages. In addition, since characteristics vary from device to device, maximum, minimum, and typical values for the data are also given.

Output Device Current

This indicates the amount of current which is available to drive succeeding devices.

Output Voltage

This gives the range of output voltage levels for a 0 and for a 1.

Quiescent Device Current

This is the current drawn by the device when it is not changing its state. In COSMOS devices, this current will be extremely small except during the transient condition when the states are changing. In other devices(e.g.,TTL)the level may vary greatly from saturation to cut off values.

Quiescent Device Dissipation

This represents the power dissipation of all the gates under quiescent conditions. Note that the power dissipation during the transient will be much greater than this. This is true for all saturating logic. It is especially true for COSMOS logic.

Input Drive Requirements

This rating gives the voltage and current levels required to switch the state of the gates. The values which guarantee a 0 or a 1 state are given. Also, the value that will cause transition from one state to the other is given. Note that these "values"are usually specified as a range of values. For COSMOS circuits, the currents will be very low except under transient conditions.

Noise Data

The noise voltage levels needed to cause switching are given here. The noise immunity of the circuit can be calculated from this.

Dynamic Data Characteristics

Various quantities, such as the time required for the gate to switch from a 0 to a 1 or vice versa are given here.

Interconnection Data

Applicable quantities such as fan in and fan out are given here.

Note that the terminology used and the data given by various manufacturers will vary. However, basically, the same type of data is given.

These characteristics allow the circuit designer to incorporate the integrated circuit into his overall design. For instance, the required power supply voltage is given. The required driving signals are specified. These characteristics are used when the circuits driving the one in question are specified. Similarly, the specified output characteristics of the gates are used to determine if they can suitably drive the next stages of the device. In general, gates from one logic family of a given manufacturer can drive other gates of the same family. However, difficulties can arise if large fan in or fan out is required.

BIBLIOGRAPHY

Belove,C, Schachter, H., Schilling,D.L., "Digital and Analog Systems, Circuits and Devices", Chapter 5, McGraw-Hill Book Co., New York, 1973

Chirlian, P.M., "Electronic Circuits:Physical Principles, Analysis and Design", Chapters 3,4, and 15, McGraw-Hill Book Co., New York, 1971

Gray, P.E., and Searle,C.L., "Electronic Principle Physics, Models and Circuits", Chapters 4-10, 22, John Wiley and Sons,Inc., Englewood Cliffs, New York, 1969

Kohonen, T., "Digital Circuits and Devices", Chapters 6,7, Prentice Hall, Inc., Englewood Cliffs, N.J. 1972

Millman,J., and Halkias,C.C.,"Integrated Electronics:Analog and Digital Circuits and Systems", Chapters 3-7, 10, McGraw-Hill Book Co.,Inc., New York 1972

O'Malley, J.,"Introduction to the Digital Computer", Chapter 3, Holt, Rinehart and Winston, New York, 1972

PROBLEMS

5-1. Quantitatively describe the operation of the gate circuit of Fig. 5-6c. (Use as numerical parameter values those which were used in the text to quantitatively describe the operation of Fig. 5-6b.)

5-2. Determine the operation and noise levels that will cause erroneous readings for the circuit of Fig.5-6b if the diodes D_1 and D_2 are replaced by three diodes in series. Use all parameter values given in the text.

5-3. Repeat Prob. 5-2 for the gate circuit of Fig. 5-6c.

5-4. Determine the fan out of the gate circuit of Fig. 5-6c.

5-5. Determine the fan out of the circuit of Prob. 5-2. Compare this with the circuit of Fig. 5-6b.

5-6. Compute the power dissipation in the output transistor of the circuit of Fig. 5-6b, as a function of the fan out.

5-7. The parameters of the circuit of Fig. 5-9 are the same as those used for Fig. 5-6b except that

$R_{D1} = 1000$ ohms $\qquad$ $R_{D2} = 4000$ ohms

Compute the fan out of this circuit. Compare the results with those for Fig. 5-6b.

5-8. Calculate the noise voltage levels that will cause trouble for the circuit of Prob. 5-7.

5-9. Why is the model of Fig. 5-10b oversimplified?

5-10. Compute the fan out of the circuit of Fig. 5-11. Use all the parameter values given in the text.

5-11. Compute the noise levels that will cause trouble for the gate circuit of Fig. 5-11. Use all the parameter values given in the text.

5-12. If the diodes of Fig. 5-11b are the same as those used in Fig. 5-6, what will be the maximum negative voltage that can appear between the input and ground? Assume that reasonable negative voltages are set up on the transmission lines.

5-13. Discuss how increased noise immunity is obtained using HTL.

5-14. Quantitatively describe the operation of a TTL circuit using the active pull-up of Fig. 5-15. Use the parameter values given in the text.

5-15. Quantitatively compare power dissipation in passive and active pull-up circuits. Use the parameter values given in the text. Assume that the cut off transistor can be represented by a 15,000 ohm resistor between collector and base while the saturated one can be represented by a 100 ohm resistor.

5-16. Quantitatively describe the operation of the gate circuit of Fig. 5-16, where R_B = 450 ohms and R_C = 700 ohms. Assume that V_{CC} = 3.5 volts and the transistors are such that V_γ = 0.6 volt, V_{CE} = 0.3 volt and $V_{BE,sat}$ = 0.7 volt.

5-17. Repeat Prob. 5-16 for the DCTL circuit which is obtained from Fig. 5-16 by setting R_B = 0.

5-18. Repeat Prob. 5-16 but now assume that the circuit of Fig. 5-17 is driven. Use the parameters of Prob. 5-16.

5-19. Repeat Prob. 5-18 but now assume that R_B=0 in Fig. 5-17.

5-20. Repeat Prob. 5-19 but now assume that one of the driven transistors has a $V_{BE,sat}$ = 0.6 volt and V_γ = 0.5 volt.

5-21. Describe why ECL can be faster than TTL.

5-22. What are the advantages of TTL over ECL?

5-23. In the analysis of Fig. 5-20, we assumed that v_{EC} = 0.8 volt during conduction. If v_{EC} = 1.0 volt, would the circuit function properly?

5-24. Discuss the advantages of the active pull-up of the MOS inverter of Fig. 5-21.

5-25. Compare the operation of the NOT circuits of Figs. 5-21 and 5-23. What are the advantages of the circuit of Fig. 5-23?

5-26. Describe the operation of the circuit of Fig. 5-25.

5-27. Compare the various logic gates.

5-28. If speed is the most important factor, which logic gate family would be used?

5-29. If power dissipation is the most important factor, which logic gate family would be used?

5-30. If noise immunity is the most important factor, which logic gate family would be used?

5-31. Which gate family would be used at the present time for most digital computer applications? Is it one of those picked in Probs. 5-28 to 5-30? Why?

6

Sequential Circuits

We have thus far considered combinational logic circuits where the output signals, at any one time, depended only upon the input signals, at that time. Of course, there is a small time delay between the application of the input signal and the establishment of the output signal. Also, the output signals were *not* functions of any signals which occurred at past times. In other words, a combinational circuit does not have *memory* so past signals do not affect it. Very often it does not feed back signals from the output to the input. A combinational circuit is illustrated in block diagram form in Fig. 6-1a.

We shall now consider circuits where these restrictions are removed. That is, they can have memory so that output signals can be functions of past values of signals. Note that output signals can be fed back so that, at any one time, the output signals may be a function of past values of any output signals. The block diagram for such a circuit is given in Fig. 6-1b.

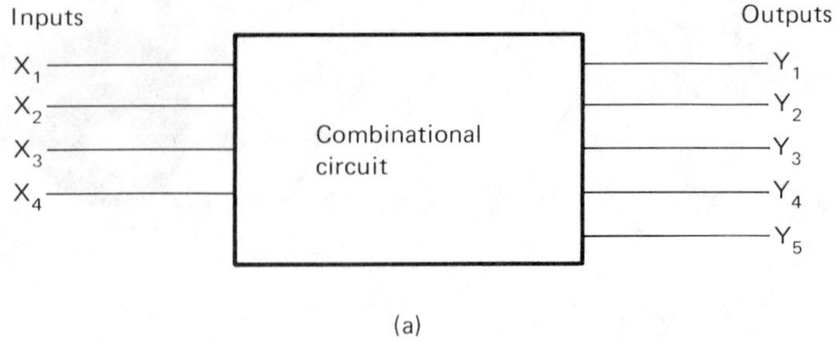

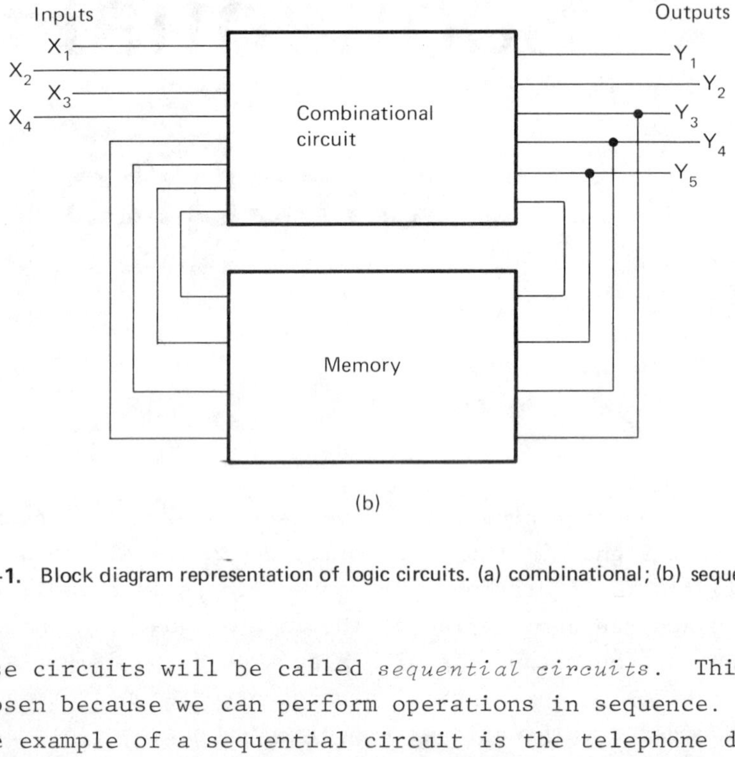

Figure 6-1. Block diagram representation of logic circuits. (a) combinational; (b) sequential.

These circuits will be called *sequential circuits*. This name is chosen because we can perform operations in sequence. A very simple example of a sequential circuit is the telephone dialing circuit. Here there is one dial, or set of push buttons. The first number is dialed and stored, then the same dial is used for the second number, etc. Note that this results in a great saving. If there were no memory, then a separate dial would have to be used for each digit of the telephone number.

Another example of a sequential circuit is a sequential adder. In Sec. 3-4 we discussed adder circuits. If combinational circuits are used, then an adder must be used for each digit. When sequential circuits are used, only one adder need be constructed.

It is used to add each digit in turn and the results are stored in the register(memory). We shall discuss this circuit in Chapter 10. Thus, sequential circuits can result in a great saving of components. More important, the calculations performed by the modern digital computer would be impractical or impossible to perform without using memory.

In general, there are two types of sequential circuits, *synchronous* and *asynchronous*. In a synchronous sequential circuit, the system responds to signals only at discrete intervals of time. A device called a *master clock* generates pulses and the circuits can respond to signals only when these clock pulses allow them to. In a computer, different circuits can respond at different rates. The clock pulses are slow enough so that every circuit has time to respond to its input signals. Of course, the slowest circuit now limits the speed of all circuits. (We shall modify this statement subsequently, see Fig. 11-7.)

In asynchronous circuits, each circuit responds at it own rate. Thus, it is faster than the synchronous case, but as we shall see, there are considerable design difficulties introduced in asynchronous sequential circuits. Therefore, at the present time, many computers utilize synchronous sequential circuits, however, there are those that utilize asynchronous ones or combinations of both synchronous and asynchronous ones.

In this chapter we shall consider the various types of sequential circuits and the techniques used in their analysis and design.

6.1 FLIP-FLOPS

A basic sequential circuit is called the *flip-flop* or *bistable multivibrator*. This is a unit whose output remains either a 0 or a 1 until it is switched from one to the other by one or more input signals. This accounts for the name flip-flop. The output of the flip-flop(either a 0 or a 1)is called the *state* of the flip-flop. A block diagram representation of a flip-flop is shown in Fig. 6-2. The output Y(the terminal marked with a 1)represents the state of the flip-flop. Usually, there is another output $\bar{Y}$ which is the complement of Y. A single flip-flop can be in one of two states. A memory, or register, can be made up of many flip-flops. The contents of the register or memory is determined by the state of all its flip-flops. Thus, if a memory has n flip-flops, it can have 2^n possible states. The variable Y is called the *state variable* of the flip-flop.

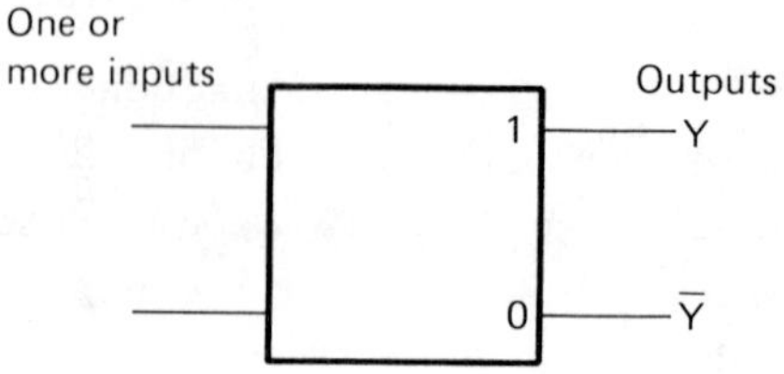

Figure 6-2. A general representation of a flip-flop.

Now we shall consider some specific flip-flops. We shall do this because they are important in their own right and also because it will lay the groundwork for a general discussion of sequential circuits. Many flip-flops will use combinational logic circuits in conjunction with the feedback of signals.

The R-S Flip-flop

A fundamental form of flip-flop is called the R-S flip-flop. Its block diagram is shown in Fig. 6-3*a*. The input leads are called R and S, which stand for reset and set, respectively. Let us call the value of Y(the output marked 1)the state of the flip-flop. We assume that the inputs are in the form of signals which are at levels corresponding to 0 or 1. If Y=0, and a 1 input pulse arrives at S, then the flip-flop will change its state so that Y=1. If Y=1, then a 1 pulse at S will make no change in Y. We assume that R=0 in both of the cases. Note that the S input sets the state of the flip-flop to 1.

The R input is the reset. That is, it resets the state of the flip-flop to zero. In all of the following we assume that S=0. If R=0 and Y=1, then the state does not change. If R=1 and Y=1 then the state resets to Y=0. If R=1 and Y=0, then the state remains at zero, and if R=0 and Y=0, then the state remains at 0.

In the previous discussion we did not consider the case R=1, S=1. This would attempt to simultaneously set and reset the

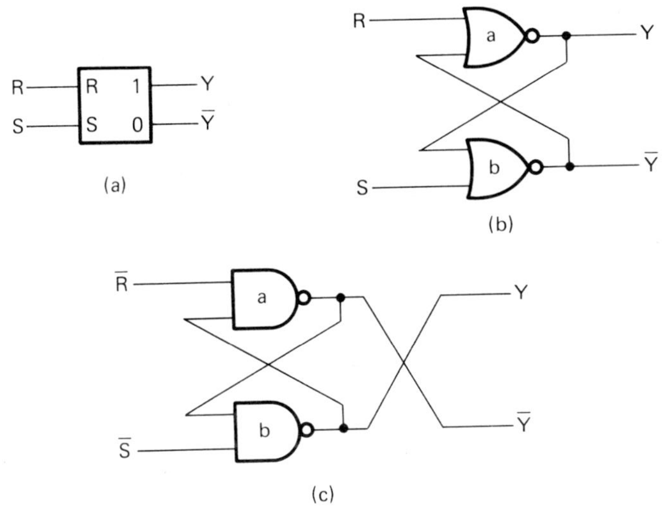

Figure 6-3. R-S flip-flops. (a) block diagram; (b) NOR gate realization; (c) NAND gate realization. The function of the 0's and 1's is interchanged in gates (b) and (c).

flip-flop. Thus, this is not allowed and the circuitry which supplies the input signals should be such that it does not occur.

All of the preceding information can be summarized in a table called a transition table, see Table 6-1. The first column indicates the state prior to the application of the input pulse. The second column indicates the input pulses and the last column indicates the state after response to the input signals. Note that we assume that it takes τ seconds for the flip-flop to change state.

TABLE 6-1
TRANSITION TABLE FOR FLIP-FLOP

Present state Y(t)	R	S	Next state Y(t+τ)
0	0	0	0
0	0	1	1
0	1	0	0
0	1	1	Indeterminate(not allowed)
1	0	0	1
1	0	1	1
1	1	0	0
1	1	1	Indeterminate(not allowed)

Another table, called the *excitation table*, which is used to characterize these circuits, gives the signal required to change the state in a specified way. This is

TABLE 6-2
EXCITATION TABLE FOR R S FLIP-FLOP

State change		Required input	
from	to	R	S
0	0	d	0
0	1	0	1
1	0	1	0
1	1	0	d

where the d's represent "don't cares". For instance, R=1, S=0 and R=0, S=0 will result in a state of 0 remaining unchanged.

We have now characterized the R-S flip-flop. Now let us discuss a circuit which implements it. The circuit of Fig.6-3b uses two NOR circuits in conjunction with feedback. Suppose that Y=0 and $\bar{Y}$=1. Now if we apply the signal which results in S=0, R=0, then Y, the output of NOR a, will remain 0 since $\bar{Y}$=1. Then, both inputs to NOR b will be 0 and $\bar{Y}$ remains a 1.

Now consider that the S input shifts to a 1 while R remains a 0. One input of NOR b is now a 1. Thus, its output changes to a 0. Note that this does not take place instantaneously, but requires T_1 seconds. Hence, after T_1 seconds, $\bar{Y}$=0 and both inputs to NOR a become 0's. Therefore, after another T_1 seconds, Y=1. Note that it requires $2T_1$ seconds for the circuits to change its state. We shall call

$$\tau = 2T_1 \tag{6-1}$$

the required time for the change of state.

To clarify these ideas let us plot the various signals as functions of time. To simplify the diagrams we shall omit the exponential rise of the pulses and just assume that the gates delay the signals. Here we assume that Y=0 and $\bar{Y}$=1. At $t=t_a$ we apply

S=1, keeping R=0. The resulting wave forms, which are explained in the previous paragraph, are shown in Fig. 6-4.

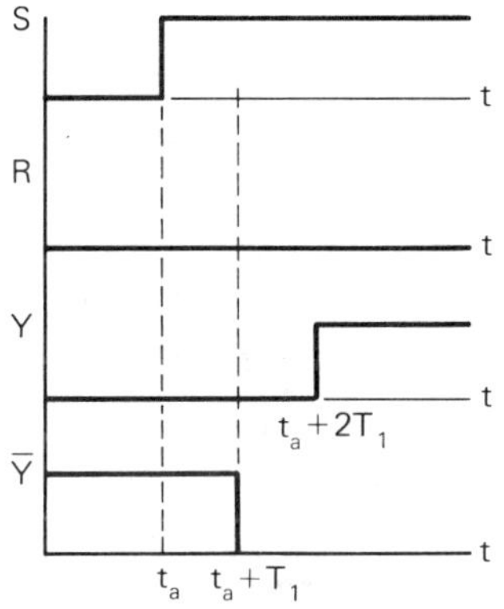

Figure 6-4. Some time responses in an R-S flip-flop which illustrate delays.

Now assume that the input shifts to the case S=0, R=0(the output prior to application of the signals is Y=1,$\bar{Y}$=0). Both inputs to NOR a will be unchanged. Thus, Y remains at 1. This maintains one input of NOR b at 1. Hence, $\bar{Y}$ remains at 0. Therefore, an S=1, R=0 input followed by an S=0,R=0 input changes the state only once(following the application of S=1,R=0). Note that the S=1, R=0 input must be maintained for at least τ seconds, to allow the state to change.

Now assume that we apply the input S=0, R=1 with the state Y=1, $\bar{Y}$=0. Since one input of NOR a is a 1, it will switch Y to 0 after T_1 seconds. Now both inputs of NOR b are 0. Hence, after another T_1 seconds, $\bar{Y}$=1.

If we now apply R=1, S=0, the input to NOR a will have two inputs of 1. Thus, Y remains at 0 and both inputs of NOR b remain at 0 and $\bar{Y}$ remains at 1.

Suppose that we apply the forbidden inputs S=1, R=1(with Y=0, $\bar{Y}$=1). Then, one input to each NOR circuit becomes 1. Then, the output is Y=0, $\bar{Y}$=0. This violates our condition that $\bar{Y}$ be the complement of Y, which is required in many logic circuits. Thus, such an input is not allowed. There are other problems that can result. These will be discussed in Sec. 6-6.

An R-S flip-flop using NAND gates is shown in Fig. 6-3c. In this flip-flop, 0's activate the circuit(i.e.,level changes from 1 to 0). (We shall see a circuit subsequently that rectifies this condition.) That is, a 0 on the set input causes Y to become a 1, while a 0 on the reset input causes Y to become a 0. For this reason, the inputs are labeled as complements, i.e., as $\bar{S}$ and $\bar{R}$. (Note that the actual inputs are $\bar{S}$ and $\bar{R}$.) For instance, when there is a 1 on the lower input of NAND b, then $\bar{S}=1$. We can say that S=0 since this will conform with the R-S flip-flop of Fig. 6-3b. However, the actual input is a 1.

If Y=0, $\bar{Y}=1$, and we apply $\bar{S}=0$, $\bar{R}=1$(equivalent to S=1, R=0)then one input to NAND b is a 0. Then, Y becomes a 1. Now all inputs to NAND a are 1's and $\bar{Y}$ becomes a 0.

Next let us apply $\bar{R}=0$, $\bar{S}=1$ with Y=1, $\bar{Y}=0$. Hence, one input to NAND a is a 0 so that $\bar{Y}$ becomes a 1. Now both inputs to NAND b are 1's so that Y becomes 0. We have considered the response of the flip-flop for a pair of inputs. Proceeding similarly, we can determine the state for all combinations of inputs and states.

Note that, in this case, the disallowed input is $\bar{R}=0$, $\bar{S}=0$.

Let us now illustrate the use of the R-S flip-flop of Fig.6-3b with a simple example. Suppose that we have a decimal number between 0 and 9, which is in binary form. Consider that this is the last digit of a number and we desire to use this to control round off. We want the output of the circuit which uses an R-S

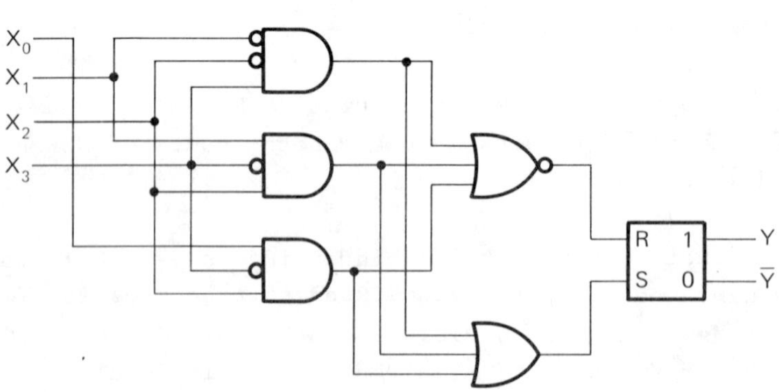

Figure 6-5. The round off control memory circuit.

flip-flop to be a 1 if the digit is 5 through 9 and the output to be a 0 otherwise. Thus, we want the input to the R-S flip-flop to be R=0, S=1 if the number if 5-9 and R=1, S=0 if the number is 0-4. Let us assume that the number is given by the binary digits $X_3X_2X_1X_0$. Then, the truth table for the input to the flip-flop is

Decimal Number	X_3	X_2	X_1	X_0	S	R
0	0	0	0	0	0	1
1	0	0	0	1	0	1
2	0	0	1	0	0	1
3	0	0	1	1	0	1
4	0	1	0	0	0	1
5	0	1	0	1	1	0
6	0	1	1	0	1	0
7	0	1	1	1	1	0
8	1	0	0	0	1	0
9	1	0	0	1	1	0

If we apply any of the minimizations discussed in Chapter 4, we obtain

$$S = \bar{X}_1\bar{X}_2X_3 + X_1X_2\bar{X}_3 + X_0X_2\bar{X}_3 \tag{6-3a}$$

$$R = \bar{S} \tag{6-3b}$$

A combinational logic gate circuit is used to realize the R and S signals which drive the R-S flip-flop. This circuit realizes the desired memory which is shown in Fig. 6-5.

Clocked R-S Flip-flop

The flip-flop that we considered responded to changes in its inputs whenever they arrived. We now want to obtain an R-S flip-flop which can be controlled by clock pulses. At the start, let us assume that we want a flip-flop that will only respond when pulses from a master clock are present. A simple form of clocked R-S flip-flop is shown in Fig. 6-6. The basic part of the circuit is an R-S flip-flop. However, the R and S inputs are applied through separate AND gates. One of the AND gate inputs is the clock pulse(i.e., that input is a 1 when CP=1). Consider

AND c. If the clock pulse is present, then its output is R. That is, if R=1, then the outout is 1, when R=0, the output is 0. When the clock pulse is 0, the output of gates c and d are 0. AND gate d serves the same function with the S signal. Thus, the input to the unclocked R-S flip-flop is 0 unless the clock pulse is present. Then, the flip-flop can only change its state when the clock pulse is present. Remember that an input R=0,S=0 leaves the state of the flip-flop unchanged. The time of the clock pulse τ_p must be long enough to allow the state of the flip-flop to change. In the next section we shall consider this in much greater detail. A block diagram representation of the R-S flip-flop is shown in Fig. 6-6c.

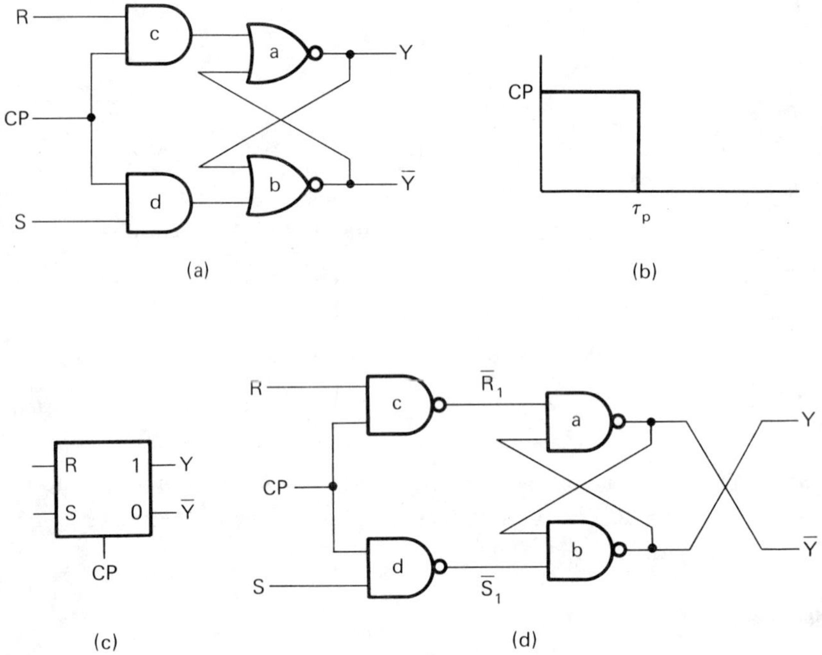

Figure 6-6. (a) A clocked R-S flip-flop using NOR gates; (b) a single clock pulse; (c) the block diagram for a clocked R-S flip-flop; (d) a clocked R-S flip-flop using NAND gates.

The type of synchronization just discussed is simple and results in simple circuits. However, we shall see that, at times, problems arise and more complex circuits must be used. These will be discussed in the next section.

Now let us consider a clocked R-S flip-flop which uses NAND gates, see Fig. 6-6d. Remember that the output of a NAND gate is 1 unless *all* its inputs are 1's. Thus, when the clock pulses are absent, the output of NAND c and d are 1's, i.e., $\bar{R}_1=1$, $\bar{S}_1=1$. This input causes no change in the values of Y and $\bar{Y}$. (Note that, in this case, $\bar{R}_1=0$, $\bar{S}_1=0$ is the disallowed case.)

When a clock pulse is present, then $\bar{R}_1=0$ if R=1. On the other hand, $\bar{R}_1=1$ if $R_1=0$. Thus, during a clock pulse, the output of NAND c is the complement of R, i.e., $\bar{R}$. Similarly, when a clock pulse is present, the output of NAND d is the complement of S, i.e., $\bar{S}$. Thus, the input of the nonclocked portion of this R-S flip-flop is the complement of the actual input. Hence, the clocked NAND gate R-S flip-flop responds to 1's. (The unclocked flip-flop responds to 0's.) An alternative way of stating this is that the flip-flops of Fig. 6-6a and d perform identically as far as the polarity of the input pulses is concerned.

Let us consider some other forms of flip-flops that are commonly used.

The D Flip-flop

We can modify the R-S flip-flop so that it has only one input called D. A block diagram for such a flip-flop is shown in Fig. 6-7a. If D=1, then Y=1 and $\bar{Y}=0$; if D=0, then Y=0 and $\bar{Y}=1$. The transition table is

TABLE 6-3
TRANSITION TABLE FOR D FLIP-FLOP

Present state Y(t)	D	Next state $Y(t+\tau)$
0	0	0
0	1	1
1	0	0
1	1	1

The D flip-flop can be simply implemented from an R-S flip-flop and a NOT gate, see Fig. 6-7b. If D=0 then the input circuit including the NOT gate sets R=1 and S=0. This will always reset the R-S flip-flop so that Y=0 and $\bar{Y}=1$. If D=1 then R=0, S=1. This will always set the R-S flip-flop so that Y=1, $\bar{Y}=0$.

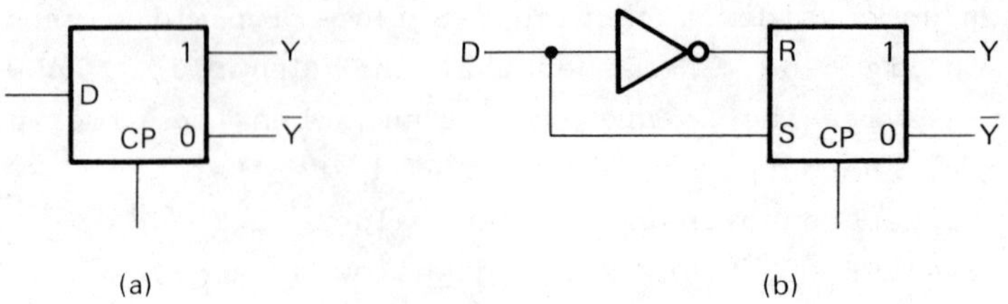

Figure 6-7. The D flip-flop. (a) block diagram; (b) an implementation.

The J-K Flip-flop

At times, it is desirable to have a flip-flop which responds to its two inputs in the same way as the R-S flip-flop. However we would like to allow the R=1, S=1 input. A flip-flop called the J-K flip-flop does this. The K(clear)input replaces the R input and the J input replaces the S input. The J-K flip-flop functions as does the R-S flip-flop except that now, if the J=1, K=1 input occurs, the output state changes. (i.e., each output is replaced by its complement) A block diagram for the J-K flip-flop is shown in Fig. 6-8a. This circuit is different from the previous ones in that it must be a clocked circuit. The transition and excitation tables for the J-K flip-flop are

TABLE 6-4
TRANSITION TABLE FOR THE J-K FLIP-FLOP

Present state Y(t)	J	K	New state $Y(t+\tau)$
0	0	0	0
0	0	1	0
0	1	0	1
0	1	1	1
1	0	0	1
1	0	1	0
1	1	0	1
1	1	1	0

where τ is the time between clock pulses.

TABLE 6-5
EXCITATION TABLE FOR THE J-K FLIP-FLOP

State change		Required input	
From	To	J	K
0	0	0	d
0	1	1	d
1	0	d	1
1	1	d	0

Note that there are many "don't care" cases in this table. This often simplifies the logic design.

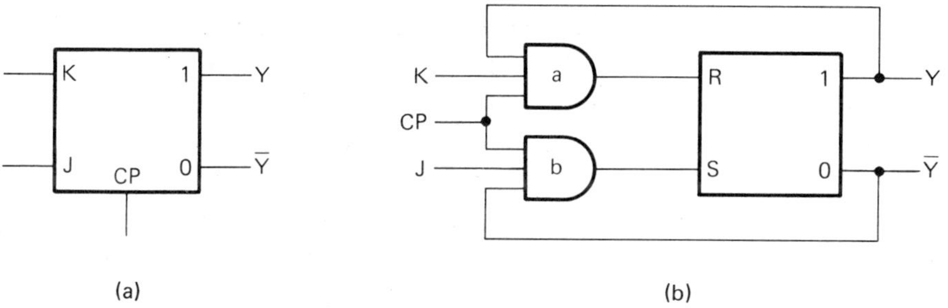

Figure 6-8. The clocked J-K flip-flop. (a) block diagram; (b) circuit implementation.

An implementation of the J-K flip-flop using AND gates and an R-S flip-flop is shown in Fig. 6-8b. Note that when the clock pulse is absent, the output of the AND gates are 0 and no input is applied to the R-S flip-flop.

Now suppose that Y=0 and K=0 and J=0 and a clock pulse is applied. Then, R=0 and S=0 and the outputs are unchanged. Suppose now that the J=0 and K=1 and that a clock pulse is applied. Since Y=0, then R=0 and S=0 since J=0. Thus, again the outputs are unchanged.

Now suppose that with Y=0, we have J=1 and K=0 during a clock pulse. All inputs of AND b will be 1. Thus, S=1. Also R=0 since the K and Y inputs of AND a are 0. Then, the R-S flip-flop changes its state to Y=1 and $\bar{Y}$=0. Note that this changes one input of AND b to 0. Then, S=0. The output of AND a is 0 since K=0. Thus, the output remains Y=1, $\bar{Y}$=0.

With the input Y=1, suppose that, during a clock pulse, J=1, K=0. Then, S=0 and R=0 and the output does not change.

Now with Y=1, assume that, during a clock pulse, we apply J=0, K=1. Then, the output of AND a is 1 while that of AND b is 0. Thus, R=1 and S=0. The R-S flip-flop switches back to Y=0 and $\bar{Y}$=1. Again, both AND gates have at least one input 0 and the output remains at Y=0, $\bar{Y}$=1.

Now let us consider the input K=1, J=1. Suppose that Y=0 and $\bar{Y}$=1 and this input is applied during a clock pulse. In this case, the output of AND a is 0 while that of AND b is 1. Thus, the feedback and the input and gates cause R=0 and S=1, and the R-S flip-flop changes its state to Y=1 and $\bar{Y}$=0. Now the output of AND a is a 1 and that of AND b is a 0, therefore, R=1, S=0. Thus, the R-S flip-flop would reset itself to Y=0, $\bar{Y}$=1. So we can see that if J=1, K=1 is applied, the flip-flop will *continuously* change its state as long as a clock pulse is applied. This is an *unstable operation* or *oscillation*. One way of preventing it is to make the length τ of the clock pulse exactly equal to the switching time of the R-S flip-flop. However, this requires precise adjustment. In addition, each output may not change at the same time, which can cause further problems. In the next section we shall consider better procedures for eliminating such unstable operation. For the time being, assume that the clock pulses are such that instability does not occur.

The T Flip-flop

We shall consider one more basic form of flip-flop called the T or toggle flip-flop. This has only one input and is such that the application of an input 1 signal causes it to change its state. That is, a 0 input leaves the output state unchanged while a 1 input causes it to change. The block diagram for this flip-flop is given in Fig. 6-9a. The transition table is

TABLE 6-6
TRANSITION TABLE FOR THE T FLIP-FLOP

Present state Y(t)	T	Next state Y(t+τ)
0	0	0
0	1	1
1	0	1
1	1	0

A circuit implementation using a J-K flip-flop is shown in Fig. 6-9b. Here we connect the two inputs of the J-K flip-flop together. If T=0, then J=0 and K=0 and the output is unchanged. If T=1 we have the case J=1, K=1 which causes the output to change. Of course, changes only can occur when the clock pulses are present.

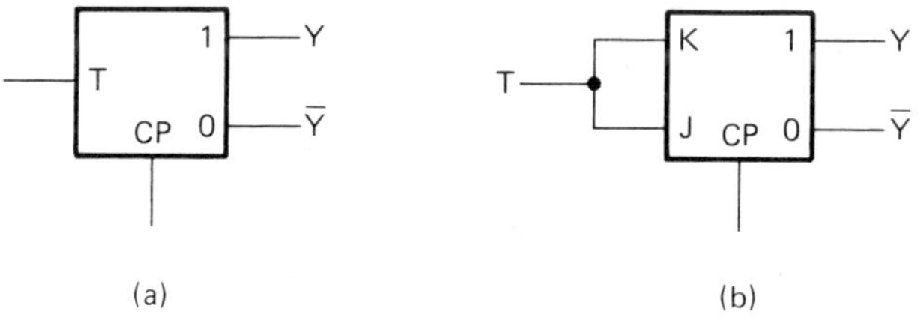

Figure 6-9. The clocked T flip-flop. (a) block diagram; (b) circuit implementation using a J-K flip-flop.

6.2 SYNCHRONIZING AND TRIGGERING OF FLIP-FLOPS

In the last section we considered the basic ideas of some clocked flip-flop circuits. Now we shall consider the timing in greater detail and discuss some additional circuits. Also, techniques for eliminating some problems that were discussed in Sec. 6-1 shall be covered.

In Sec. 6-1 we assumed that the clock pulses were on "long enough" for a change in state to take place. Let us see what is involved here. We shall work with the clocked R-S flip-flop of Fig. 6-6a. Normally, the clock pulse is a periodic signal whose duration is short in comparison to the period, see Fig. 6-10a.

The period of the clock pulse is τ_p and its duration is τ_c where τ_p is considerably greater than τ_c.

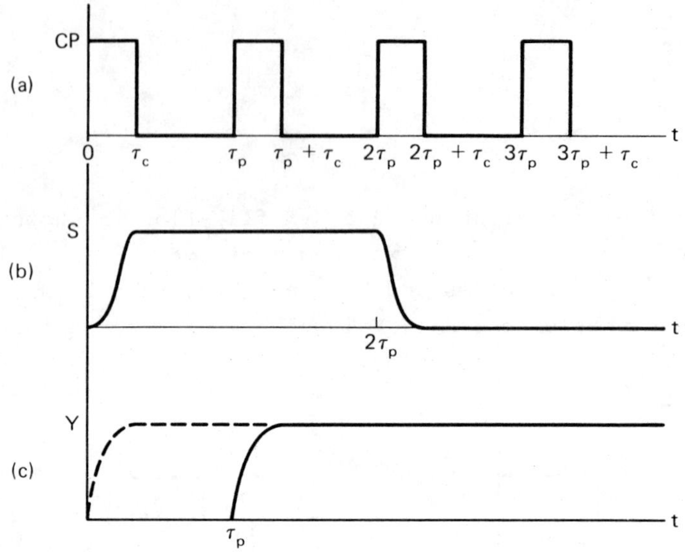

Figure 6-10. A timing diagram for an R-S flip-flop which is triggered on the leading edge of the clock pulse. (a) clock pulses; (b) S input; (c) X output, note the ambiguity in X.

Now consider that an S input is applied at the same time as the clock pulse (with X=0). This is shown in Fig. 6-10. The AND gates of Fig. 6-6a allow signals to be applied between 0 and τ_c. However, S is just building up during that time and it may not be large enough to cause the flip-flop to change its state. Thus, action may not take place until time τ_p when the next clock pulse allows the inputs to activate the R-S flip-flop. Now, Y changes as shown by the solid curve of Fig. 6-10c. However, it is possible that the circuit might have responded in the first timing pulse interval. Then, Y would be of the dashed form. Thus, Y is potentially ambiguous. Such ambiguities occur if the input signal is changing when the clock pulse is on. Figure 6-10 illustrates what is called *leading edge triggering*, since the leading edge of the pulse (i.e., the edge that occurs first in time) determines when the input signals are applied to the flip-flop and when switching starts.

One way of avoiding the ambiguity of simple leading edge triggering is to delay the input signals so that they do not change until the clock pulse has passed, see Fig. 6-11. We shall consider the implementation of these circuits subsequently. For the time being, let us just consider that the delay is available. The input signal must be delayed by at least τ_c seconds. That is, the input signal must not start until the *trailing edge* of the clock pulse has passed. The possible ambiguity has been removed, since the S input is 0 until after the first clock pulse. At $t=\tau_p$, the second clock pulse arrives. Then, S=1 and the flip-flop changes state so that Y=1.

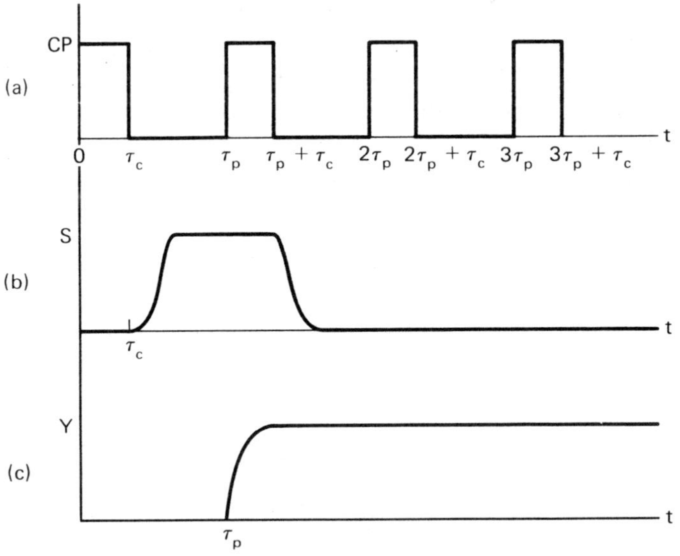

Figure 6-11. A timing diagram for an R-S flip-flop which introduces delay to avoid ambiguous response. (a) clock pulses; (b) S imput delayed to trailing edge of clock pulse; (c) X output.

One way of implementing the delayed output is to put a delay at the output of each flip-flop or other logic element. This is illustrated in Fig. 6-12. The delay network could consist of a length of transmission line. This is awkward since, if the delay is relatively long, very long lengths of transmission line are needed. The delay could be established by driving a cascade of NOT circuits. Since each one takes a finite time to respond, the output would be delayed. In either of these cases, any shift in the timing of the clock pulses might cause the delay to be too short, reinstating ambiguity. Thus, flip-flops with built in delay could only be used with specific clock pulses.

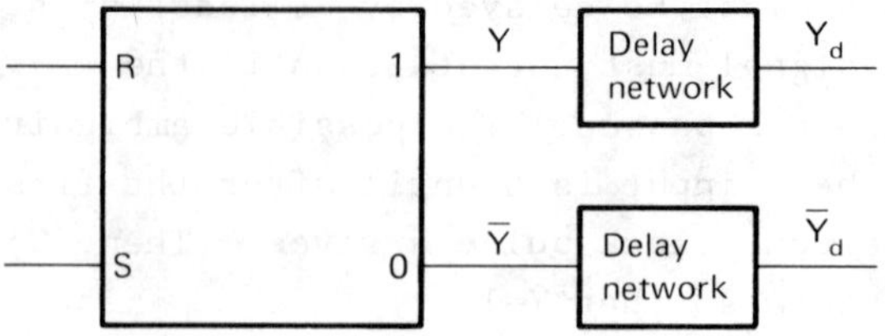

Figure 6-12. An R-S flip-flop with delay networks.

Usually, the best type of delay is derived from the clock pulses. Suppose that we had a circuit which would start to respond with the leading edge of the clock pulse but whose output does not change until the trailing edge of the clock pulse had arrived. Assume that all circuits are of this type. Then, the inputs to any gates(which come from the outputs of other gates)would not change until after the clock pulse. This is the desired situation. It can be implemented relatively easily. Consider the circuit of Fig. 6-13, which consists of one R-S flip-flop driven by another one. For the time being, let us ignore the clock pulse. The Y output of the first flip-flop is connected to the S input of the second. Similarly, the $\bar{Y}$ output of the first is connected to the R input of the second.

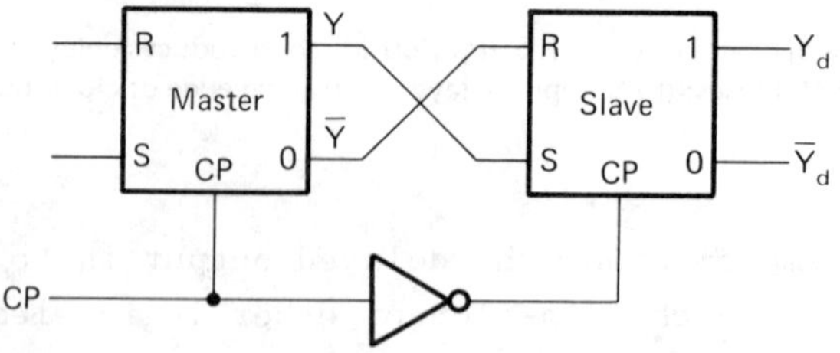

Figure 6-13. An R-S master-slave flip-flop.

Now, if the first flip-flop has an output Y=1, $\bar{Y}$=0, then the input to the second is S=1, R=0. The output will be Y_d=1, $\bar{Y}_d$=0.

That is, the second flip-flop will be set by its S=1, R=0 input. If the first flip-flop has an output Y=0, $\bar{Y}$=1, then the second has an input R=1, S=0. This will reset the second flip-flop so that $Y_d=0$, $\bar{Y}_d=1$. Thus, the output of the second flip-flop will always be the *same* as the first. For this reason, the second flip-flop is called the *slave* while the first is called the *master*.

Now let us consider the effect of clock pulses on the flip-flop. The clock pulses are directly applied to the clock terminal of the master but they are passed through a NOT circuit before they are applied to the slave. Thus, there will be a 1 applied to the CP terminal of the slave only when a zero is applied to the CP terminal of the master. Assume that an input is applied to the master. When the clock pulse arrives, it will change its state. During the clock pulse Y and $\bar{Y}$ achieve their final values. However, the slave *cannot* change because there is a 0 input at its clock pulse terminal. After the clock pulse is over, a 1 will appear at the slave's clock pulse terminal. Thus, it can now change its state. Note that the slave does not change its state until after the clock pulse is past. Then the master switches on the leading edge of the clock pulse while the slave switches on the trailing edge. This is the operation we desired. That is, if all the flip-flops are master-slave types, then the outputs of any flip-flop cannot change during the clock pulse. If these outputs are inputs to other flip-flops, then these inputs cannot change during the clock pulse.

A timing diagram is shown in Fig. 6-14. Assume that the S input is from another master-slave flip-flop, so that it does not start until the trailing edge of the clock pulse. The master flip-flop does not start to change its state until time $t=2\tau_p$. (Note that at $t=\tau_p$, the S input was zero.) The output of the slave Y_d does not start to change until $t=2\tau_p+\tau_C$, at which point, Y has become established. Thus, the master-slave flip-flop eliminates the problems associated with flip-flops being triggered by changing signals. The time delay of the slave is exactly fixed at τ_C. Thus, if there is some shift in the timing signals, then the slave delay also shifts.

We have illustrated master-slave flip-flops using the R-S flip-flop. Actually, any one can be put into a master-slave configuration. Let us consider the J-K master-slave flip-flop, see Fig. 6-15. The basic circuit of Fig. 6-8b is used, except that the R-S flip-flop is replaced by a master-slave R-S flip-flop and the internal AND gates are combined with the input AND gates of the clocked R-S flip-flop.

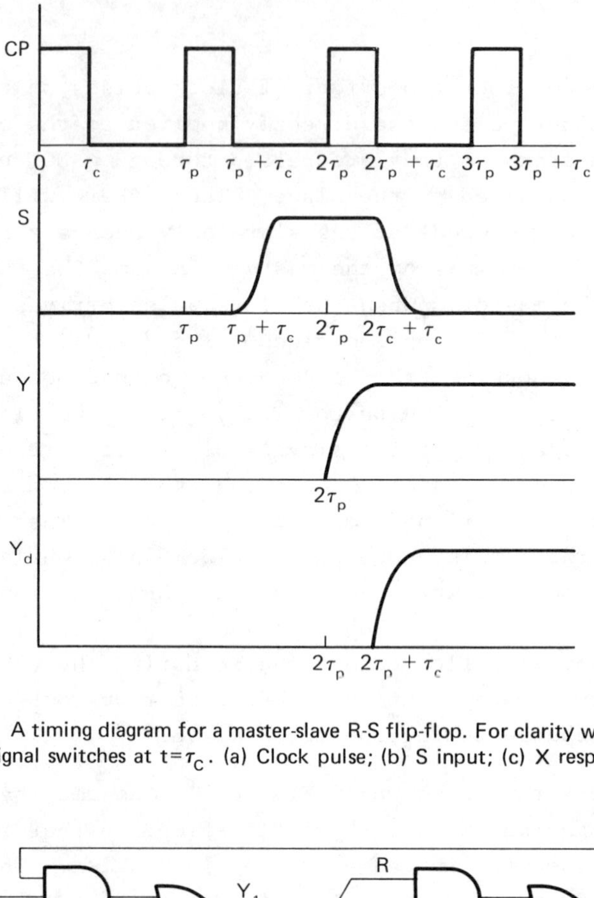

Figure 6-14. A timing diagram for a master-slave R-S flip-flop. For clarity we assume that the driving signal switches at $t = \tau_C$. (a) Clock pulse; (b) S input; (c) X response; (d) X_d output.

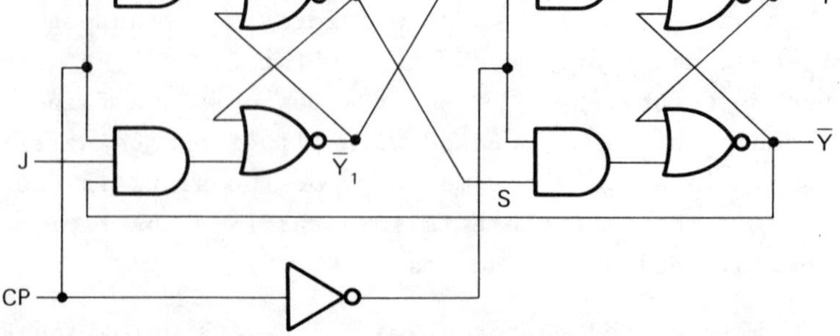

Figure 6-15. A J-K master-slave flip-flop.

This circuit functions in a manner similar to the ordinary J-K flip-flop except that the slave does not change state until after the clock pulse is passed and the master can only change its state during a clock pulse.

Now consider the operation. Suppose that Y=0, $\bar{Y}$=1, and we apply J=1, K=0. Then, during the clock pulse, the output of the master becomes Y_1=0, $\bar{Y}_1$=1. Once the clock pulse is over, the slave can change its state. Since the R input is 0 and the S input is 1, the output becomes Y=1, $\bar{Y}$=0. The operation follows that for the J-K flip-flop with the exception of the time delay between master and slave. Thus, all the other possible inputs will not be considered.

One case we will consider is J=1, K=1. Suppose that Y=0, $\bar{Y}$=1. Then, during the clock pulse, $\bar{Y}_1$ becomes a 0 and Y_1 a 1. However, the slave does not change its state. After the clock pulse, the slave is able to change its state and $\bar{Y}$ becomes a 0 and Y becomes a 1. In this case, instability does not occur because now the master cannot change its state until the next clock pulse occurs. Thus, the output remains constant. By the next clock pulse it is assumed that the inputs will be changed since it is assumed that they have also come from a master-slave flip-flop. For instance, in Fig. 6-14b, the input(S in this case)falls off after the clock pulse at $2\tau_p$ is over and is 0 by the time that the next clock pulse arrives. Of course, if the logic maintains the K=1, J=1 input, then the output state will again change after the next clock pulse.

A J-K master-slave flip-flop using NAND gates is shown in Fig. 6-16. This uses the R-S flip-flop of Fig. 6-6d as the master and slave portions of the R-S flip-flop. Otherwise, the basic ideas are essentially the same as those of Fig. 6-15. Again, since the input of the master R-S flip-flop is a NAND gate, we do not have to use a separate AND gate to combine the output feedback signal with the input signal.

We have illustrated master-slave J-K and R-S flip-flops. Actually, any of the flip-flops discussed could be implemented in a master-slave configuration. For instance, in Figs.6-15 or 6-16, if we connect the J and K inputs together, a T master-slave flip-flop is obtained.

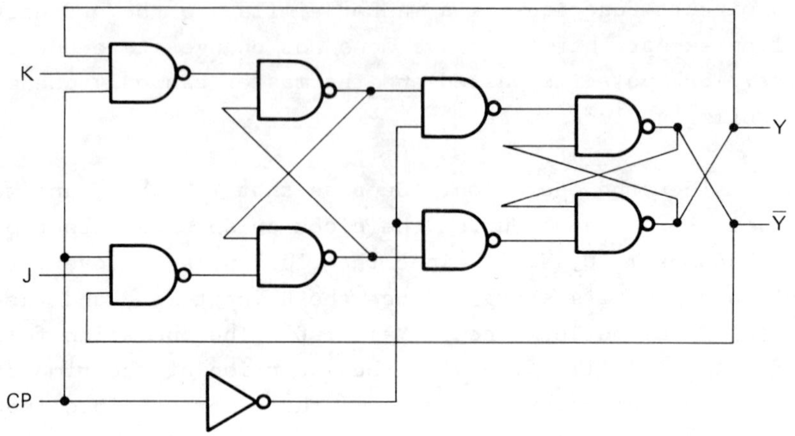

Figure 6-16. A J-K master-slave flip-flop using NAND gates.

In the remainder of the book, unless we state otherwise, it will be assumed that master-slave flip-flops are used when we work with synchronous sequential circuits.

6.3 SOME BASIC ANALYSIS TOOLS FOR SYNCHRONOUS SEQUENTIAL CIRCUITS

In general, sequential circuits consist of very many flip-flops and logic elements. In this section we shall consider an efficient procedure that can be used to analyze synchronous sequential circuits.

Transition Tables

We can characterize the operation of a sequential circuit by a table. Suppose that we have the clocked sequential circuit of Fig. 6-17. Note that, for simplification, we often do not show the clock pulse connection. However, all CP terminals are connected to the clock circuit. The blocks represent master-slave flip-flops. We use D and T flip-flops here. Remember that the state of the T flip-flop changes whenever a 1 is applied and the state of a D flip-flop is 1 if the input is a 1 and is 0 if the input is 0.

There are three classes of quantities that we are concerned with; the input variables, in this case, just one, $x(t)$; the output variables, in this case, just one $z(t)$; and the internal states of the flip-flops which, in this case, are $y_1(t)$ and $y_2(t)$.

When an input is applied, it will, in general, cause the state to change. In characterizing the network, we list the present values of the input and the current states. Then, the next state is listed. The *current* value of the output is also listed. Note that the output does not change until after the next clock pulse. In Table 6-7 we characterize the circuit of Fig. 6-17.

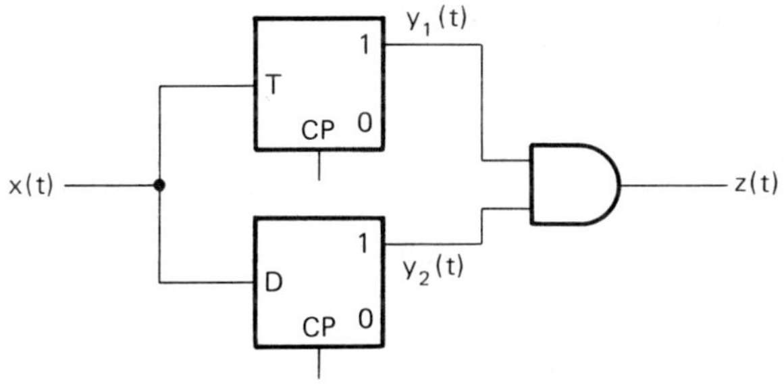

Figure 6-17. A simple sequential network.

TABLE 6-7
A TABULAR CHARACTERIZATION OF THE SEQUENTIAL NETWORK OF FIG. 6.17

Present Input	Present State		Next State		Present Output
$x(t)$	$y_1(t)$	$y_2(t)$	$y_1(t+\tau_p)$	$y_2(t+\tau_p)$	$z(t)$
0	0	0	0	0	0
0	0	1	0	0	0
0	1	0	1	0	0
0	1	1	1	0	1
1	0	0	1	1	0
1	0	1	1	1	0
1	1	0	0	1	0
1	1	1	0	1	1

This table provides the pertinant information. However, since all possible combinations of inputs and states are listed, a table of this type will be very long when complex circuits are encountered. Thus, a short hand notation is used to condense the table. The next state and present output are combined into a single term, which is illustrated in Fig. 6-18. Note that the slash separates the *next* state form the *present* output. Table 6-8 is such a condensed table.

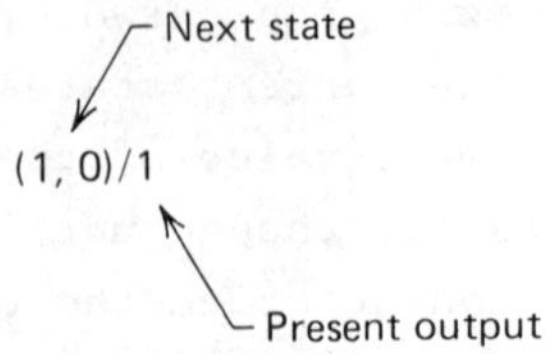

Figure 6-18. An illustration of the notation used in the transition table.

TABLE 6-8
TRANSITION TABLE FOR THE NETWORK OF FIG. 6-17

Current State y_1y_2	Current Input x: 0	Current Input x: 1	
(0,0)	(0,0)/0	(1,1)/0	Next state and current output
(0,1)	(0,0)/0	(1,1)/0	
(1,0)	(1,0)/0	(0,1)/0	
(1,1)	(1,0)/1	(0,1)/1	

Table 6-8 provides all the information of Table 6-7 in a much more condensed form. At times, the current state is often represented by a single subscripted variable. For instance, (0,0) could be written as q_0; (0,1) as q_1, (1,0) as q_2; and (1,1) as q_3. This is especially helpful in design procedures when we may not have a specific binary representation of the state. In this case, we can rewrite Table 6-8 as

TABLE 6-9
STATE TABLE FOR THE NETWORK OF FIG. 6-17

Current state	Current Input x: 0	Current Input x: 1
q_0	$q_0/0$	$q_3/1$
q_1	$q_0/0$	$q_1/1$
q_2	$q_2/0$	$q_1/0$
q_3	$q_2/1$	$q_1/1$

It often clarifies the state or transition table if we represent it by a drawing. Such drawings, called *state transition diagrams*, are shown in Fig. 6-19. The circles represent the states. The lines represent transition from one state to another. The arrow heads indicate the direction of the transition. The numbers separated by a slash represent the current input and current output, in the following manner:

current input/current output

Note that this is different notation from the representation in the table. Now suppose that we start in state (0,1) and we apply the input signals in the following sequence(each input coincides with a clock pulse)1, 1, 0, 1, 0, 1, that is

$$\begin{aligned} x(0) &= 1 \\ x(\tau_p) &= 1 \\ x(2\tau_p) &= 0 \\ x(3\tau_p) &= 1 \\ x(4\tau_p) &= 0 \\ x(5\tau_p) &= 1 \end{aligned} \tag{6-4}$$

Now we want to determine the output signal $z(0)$, $z(\tau_p)$, $z(2\tau_p)$, ... The first transition is to state (1,1) and the current output is 0. The next transition is to state (0,1) and the "current" output then is 1. Proceeding in this way, we obtain

$$\begin{aligned} z(0) &= 0 \\ z(\tau_p) &= 1 \\ z(2\tau_p) &= 0 \\ z(3\tau_p) &= 0 \\ z(4\tau_p) &= 1 \end{aligned} \tag{6-5}$$

$z(5\tau_p) = 0$

The state sequence is

$$q_1 \rightarrow q_3 \rightarrow q_1 \rightarrow q_0 \rightarrow q_3 \rightarrow q_2 \rightarrow q_1 \qquad (6\text{-}6)$$

Note that the sequence depends upon the assumed starting state. That is, if we started with state q_3, a different sequence of states and outputs would be obtained.

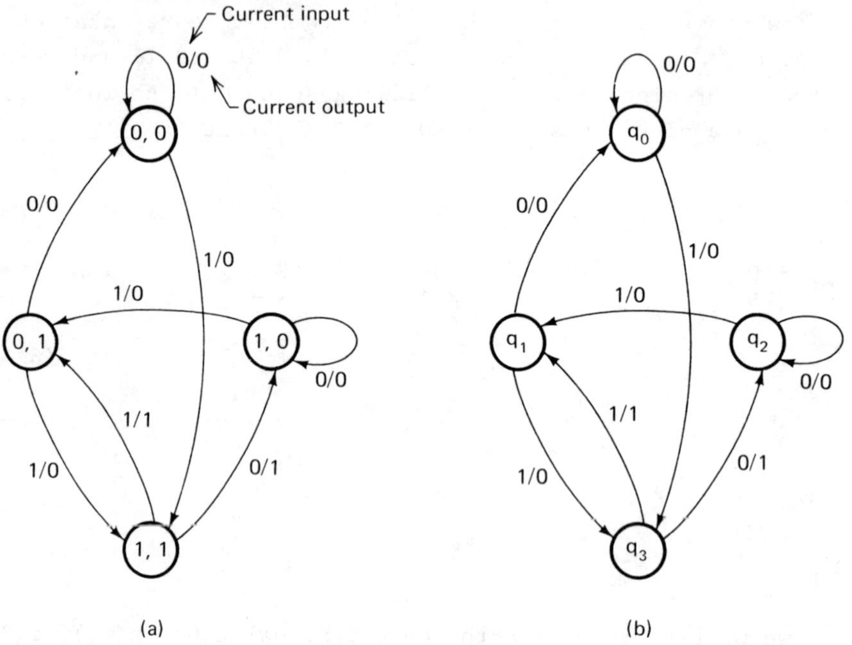

Figure 6-19. State transition diagrams for the network of Fig. 6-17. (a) from the transition table; (b) from the state table.

We have considered a network with a single input and a single output. This need not be the case. For instance, a more complex case is shown in Fig. 6-20. In order to analyze this network, we must organize our equations. Let us write general logical expressions for the output of the commonly used flip-flops. Then, we can obtain the logical expressions for the network.

D Flip-flop

The state of a D flip-flop is 1 if the input is 1 and 0 other-

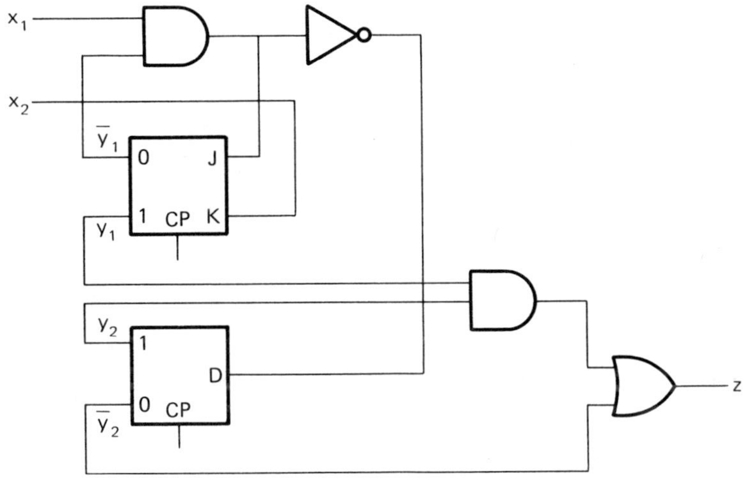

Figure 6-20. A more complex sequential network.

wise. Thus, we can write

$$y(t+\tau_p) = D(t) \qquad (6\text{-}7)$$

where y and D are identified in Fig. 6-7a. Thus, if D(t) = 1, then y(t+1) = 1, etc.

J-K Flip-flop

The logic equation is

$$y(t+\tau_p) = \bar{y}(t)J(t) + y(t)\bar{K}(t) \qquad (6\text{-}8)$$

see Fig. 6-8a. Note that if y=0 and we apply J=1, K=0, then $\bar{y}J=1$. Thus, $y(t+\tau_p) = 1$ as it should. If we check all possible inputs and outputs, the validity of Eq.(6-8) will be demonstrated. We leave this as an exercise for the reader.

R-S Flip-flop

In this case we have

$$y(t+\tau_p) = S(t) + y(t)\bar{R}(t) \qquad R(t)S(t) = 1 \text{ not allowed} \qquad (6\text{-}9)$$

Again, this can be verified by checking all possible combinations.

T Flip-flop

In this case, the state changes if T=1. Thus, a logical expression that represents this is

$$y(t+\tau_p) = \bar{y}(t)T(t) + y(t)\bar{T}(t) \tag{6-10}$$

Note that this can be obtained from Eq. (6-8) by setting

$$J = K = T$$

Now let us analyze Fig. 6-20. Using the circuit, we obtain

$$J(t) = x_1(t)\bar{y}_1(t) \tag{6-11a}$$

$$K(t) = x_2(t) \tag{6-11b}$$

$$D(t) = \bar{J}(t) \tag{6-11c}$$

Then, substitution in Eqs. (6-7) and (6-8) yields

$$y_1(t+\tau_p) = \bar{y}_1(t)x_1(t)\bar{y}_1(t) + y_1(t)\bar{x}_2(t) \tag{6-12a}$$

$$y_2(t+\tau_p) = \bar{x}_1(t) + y_1(t) \tag{6-12b}$$

From Fig. 6-20, we have

$$z(t) = y_1(t)y_2(t) + \bar{y}_2(t) \tag{6-12c}$$

Simplifying Eqs. (6-12), we obtain

$$y_1(t+\tau_p) = x_1(t)\bar{y}_1(t) + y_1(t)\bar{x}_2(t) \tag{6-13a}$$

$$y_2(t+\tau_p) = \bar{x}_1(t) + y_1(t) \tag{6-13b}$$

$$z(t) = y_1(t) + \bar{y}_2(t) \tag{6-13c}$$

Then, substituting in Eqs. (6-13) we can determine the transition table.

TABLE 6-10
TRANSITION TABLE FOR THE NETWORK OF FIGURE 6-20

Current state (y_1, y_2)	Current Input(x_1, x_2) (0,0)	(0,1)	(1,0)	(1,1)
(0,0)	(0,1)/1	(0,1)/1	(1,0)/1	(1,0)/1
(0,1)	(0,1)/0	(0,1)/0	(1,0)/0	(1,0)/0
(1,0)	(1,1)/1	(0,1)/1	(1,1)/1	(0,1)/1
(1,1)	(1,1)/1	(0,1)/1	(1,1)/1	(0,1)/1

We can use the table to plot the state transition diagram for this network, see Fig. 6-21. The notation used is

x_1x_2/z

If there is more than one set of inputs that will cause the same transition between the same two states, they are enclosed in parentheses and only one arrow is drawn. For instance, both x_1= 0 and x_2=1 and x_1=0, x_2=0 will cause the transition from y_1=0, y_2=0 to y_1=0,y_2=1. The output sequence for a sequence of inputs is determined for this diagram just as it was for Fig. 6-20.

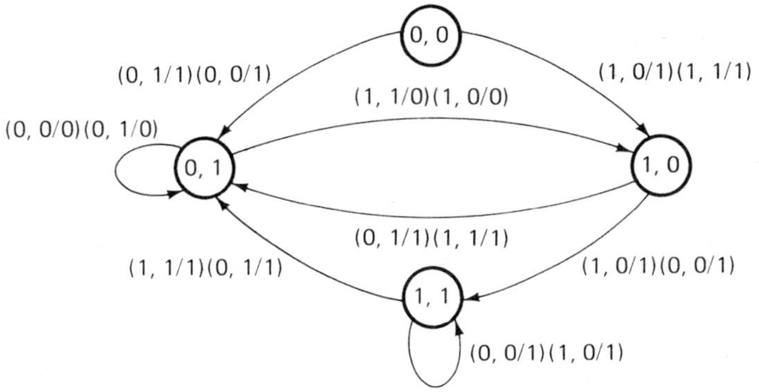

Figure 6-21. A state transition diagram for the network of Fig. 6-20.

Equations 6-12 were relatively short. Thus, Eqs.(6-13) could be obtained by inspection. In general, this may not be the case, and all the techniques of Chapter 4 can be used to simplify the expressions. Note that the right hand side of the equation which gives the state or output, will just be simple combinational logic expressions. Thus, these simplification techniques can be applied to them. (We shall illustrate this in Sec. 6-5.)

Next we shall consider some practical circuits and their transition tables and diagrams. Subsequently, we shall discuss general design.

6.4 REGISTERS

Registers are used to store information in computers. In this section we shall discuss such registers, not only to consider them, but also to introduce some simple design ideas. These will be expanded in the next section.

Information is stored in a register as a sequence of 1's and 0's. This can be supplied to the register in one of two ways. One binary digit can be supplied at each clock pulse. For instance, suppose that the register stores the output of a sequential adder, see the introduction to this chapter. In this case, the addition circuit supplies only one digit at a time and these would be supplied to the register at successive clock pulses. This type of operation is called *series or serial operation* of the register. That is, data is inputted to the register in *series*. Similarly, data can be outputted from the register in series, that is one digit for each clock pulse. Again, this type of operation would be useful if the information were to be supplied to a sequential adder. We have ignored the details of addition in this simple discussion. We shall consider them again in Chapter 10.

There is another type of register operation called *parallel*. In this case, all data is inputted or outputted during a single clock pulse period. This is much faster than series operation. However, many more gates and flip-flops are required with such operation.

Shift Register

A shift register is one where the information is transmitted down the register with each clock pulse. Often, the data is entered one binary digit per clock pulse. This is illustrated in

Fig. 6-22. At the start, the register has the contents 0000. On the next clock pulse, a 1 is entered. On the next pulse, a 0 is entered. The 1 that is already in the register shifts one space to the right, see Fig. 6-22c. On the next pulse a 0 is again entered and the 1 is shifted to the right again as shown in Fig. 6-22d. On the next clock pulse a 1 is entered. Thus, it appears in the leftmost position and the original 1 is again shifted to the fourth(rightmost)position. On the next clock pulse a 0 is again entered and the contents of the registers are shifted to the right again. Note that the previous rightmost term is now lost, see Fig. 6-22f. Of course, it could have been used by some other logic circuit before this was done.

(a) Input 0	0	0	0	0
(b) Input 1	1	0	0	0
(c) Input 0	0	1	0	0
(d) Input 0	0	0	1	0
(e) Input 1	1	0	0	1
(f) Input 0	0	1	0	0
(g) Input 1	1	0	1	0
(h) Input 1	1	1	0	1
(j) Input 1	1	1	1	0

Figure 6-22. Illustration of the information stored in a shift register on successive clock pulses.

Now let us see how we can implement a simple shift register. In Fig. 6-23a, we have a three stage shift register using J-K flip-flops. Remember that these are master-slave flip-flops. (Note that we have drawn the J on top for convenience.) Suppose that all the states are 0, i.e., $y_1=y_2=y_3=0$. If the input X is 0, then the states will remain 0. Now suppose that X=1. Then, $J_1=1$, $K_1=0$. Then *after* the next clock pulse, $y_1=1$. However, the state of the second flip-flop cannot change until the next clock pulse occurs. Thus, it remains $y_2=0$. Now suppose that the input is 0. Then, $J_1=0$, $K_1=1$. Thus, *after* the(next)clock pulse, $y_1=0$. However, *during* the clock pulse, y_1 has not as yet changed. Then, $J_2=1$, $K_2=0$, during that pulse. This actuates the second flip-flop. Hence, after the clock pulse, $y_2=1$. Thus, the 1 has shifted to the right, and the shift register will operate as desired. Since the input J=1, K=1 never occurs for any of the flip-flops, we can replace the J-K flip-flops by R-S flip-flops here, without changing the operation.

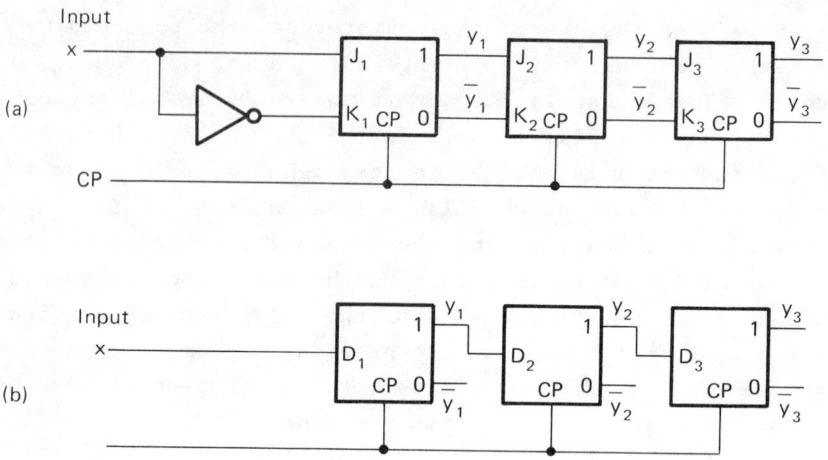

Figure 6-23. Shift registers. (a) using J-K flip-flops; (b) using D flip-flops. Note R-S flip-flops can replace the J-K flip-flops of Fig. 6-23a.

In Fig. 6-23b we have implemented a shift register using D flip-flops. The basic ideas are the same and the details of operation will be left to the reader.

A state transition diagram for these flip-flops is shown in Fig. 6-24. This diagram should be compared with Fig. 6-23 to gain familiarity with the operation. (Note that y_1, y_2 and y_3 are the outputs.)

We have constructed a shift register which can store three binary digits. To store more digits, we need only add more flip-flops to the right in Fig. 6-23.

In an actual computer, the shift registers are more versatile than those of Fig. 6-23. For instance, at times, we do not want to shift when a clock pulse occurs. Sometimes we want to enter data on the left side and shift to the right. Other times, we want to enter data on the right and shift to the left. Finally, there are times when we want to clear the register, that is, to set all its states to 0. Control signals are present which indicate which type of operation is desired. Let us now design such a register.

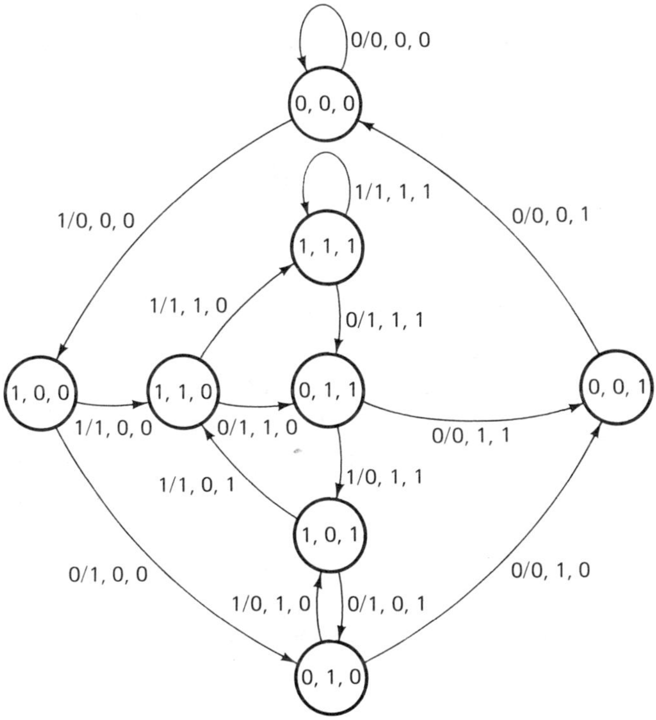

Figure 6-24. State transition diagram for the shift registers of Fig. 6-23.

Let us assume that there are three control leads, c_0, c_1, and c_2. The operation is to be such that if $c_0=1$, $c_1=0$ and $c_2=0$, then the input is at the left and the register shifts right; if $c_0=0$, $c_1=1$, and $c_2=0$, then the input is at the right and the register shifts left; if $c_0=0$, $c_1=0$ and $c_2=1$, then the register clears, i.e., all states are set to 0; and if $c_0=0$, $c_1=0$ and $c_2=0$, there is no input or shift.

Let us realize the shift register using D flip-flops. In order to obtain the logical expression to design the circuit, we need a truth table type of formulation. In general sequential circuit design, which we shall discuss in the next section, this is obtained from the state transition table. However, in this case, we can obtain the table directly. With D flip-flops, the output state is the same as the input. Thus, the following table is that part of the truth table for which D=1. (Remember that we need only those terms for which D=1 to obtain the logical expression.) Thus, for the first flip-flop, we have

c_0	c_1	c_2	x_L	y_1	y_2	D_1
1	d	d	1	d	d	1
d	1	d	d	d	1	1
0	0	0	d	1	d	1

The don't cares in the table result because we assume that no two c's will ever be 1 at the same time. Note that D_1 will be 1 if $c_0=1$, and the left input $x_L=1$. That is, if $c_0=1$, the input is from the right. Thus, $D_1=1$ if $y_2=1$. Note that flip-flop 2 is to the right of flip-flop 1. Thus, if we are shifting to the left, the input to flip-flop 1 is y_2, the state of flip-flop 2. The only other time that the input to the flip-flop is a 1 is if its present output is a 1 and if $c_1=c_2=c_3=0$, that is, when there are to be no shifts. Using the table we can write the logical expression

$$D_1 = c_0x_L + c_1y_2 + \bar{c}_0\bar{c}_1\bar{c}_2\, y_1 \tag{6-14a}$$

Note that if the don't cares in the table were not present, then considerably more complex expressions would result.

The equation for the other two flip-flops can be obtained from Eq. (6-14a) if we replace x_L by the left input to the flip-flop, y_2 by the right input, and y_1 by the state of the flip-flop in question. Thus,

$$D_2 = c_0y_1 + c_1y_3 + \bar{c}_0\bar{c}_1\bar{c}_2\, y_2 \tag{6-14b}$$

$$D_3 = c_0y_2 + c_1x_R + \bar{c}_0\bar{c}_1\bar{c}_2\, y_3 \tag{6-14c}$$

where x_R is the input signal when the register is to be shifted to the left, i.e., the right input.

The complete circuit is shown in Fig. 6-25. Note that each flip-flop has three AND gates and an OR gate associated with it. These correspond to each of the terms in Eqs.(6-14). The NOR gate is used to realize $\bar{c}_0\bar{c}_1\bar{c}_2$.

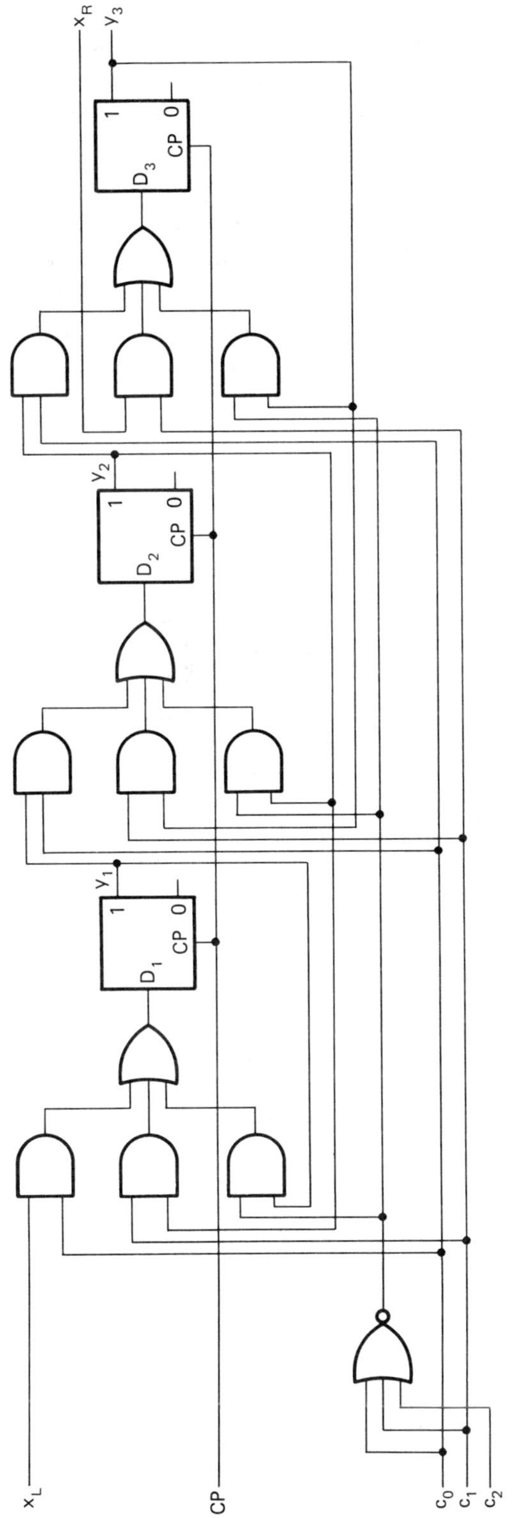

Figure 6-25. The controlled shift register.

The last term in each Eq. (6-14) is used when shifting does not occur. An alternative way of doing this, which reduces the number of gates required, is to prevent the clock pulses from reaching the flip-flop when we do not require shifting. If we add the logic which will do this, then the last terms can be omitted from Eq. (6-14). A circuit which accomplishes this is shown in Fig. 6-26. Instead of redrawing the entire circuit, only one stage is drawn. AND a will apply the left input signal if $c_0=1$. Similarly, AND b will apply the right input signal if $c_1=1$. If $c_2=1$, then c_0 and $c_1=0$ and the input to the flip-flop will be 0. This will set y=0. This clears the register. If c_0, c_1 and c_2 are all 0's, then the output of AND c will be 0. In this case, clock pulses will not be applied to the (any) flip-flop and states will not change at the clock pulse time.

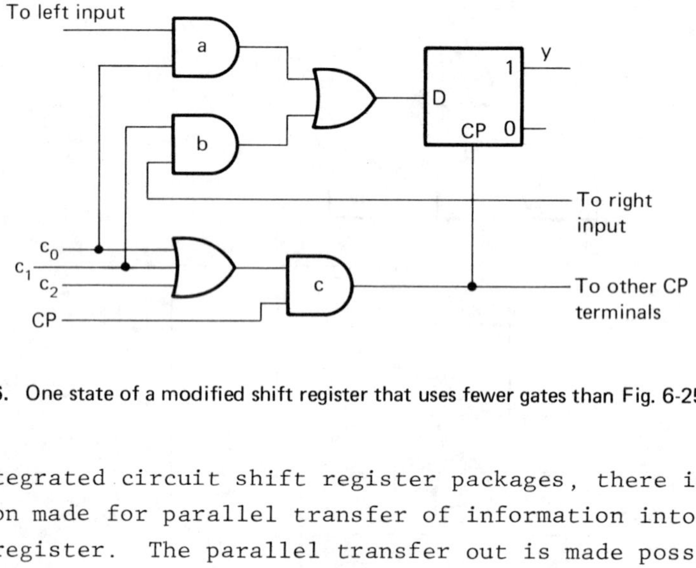

Figure 6-26. One state of a modified shift register that uses fewer gates than Fig. 6-25.

In integrated circuit shift register packages, there is often provision made for parallel transfer of information into and out of the register. The parallel transfer out is made possible by having access to the y_1, y_2 and y_3 leads. To obtain parallel as well as series input, the circuit of Fig. 6-27 can be used. Again, to avoid duplication, we only show one stage. An extra control signal c_3 has been added. c_3 is 0 unless parallel input is desired. If $c_3=0$, then no signals pass through AND p and the operation is essentially as in the circuit of Fig. 6-26. If $c_3=1$, with all other control signals 0, then

$$D_j = x_j$$

where there is a separate x_j input for each flip-flop(i.e., x_1 is applied to D_1, x_2 to D_2, etc). Thus, parallel input is obtained. If $c_3=0$, then series input is used and the circuit operates as an ordinary shift register.

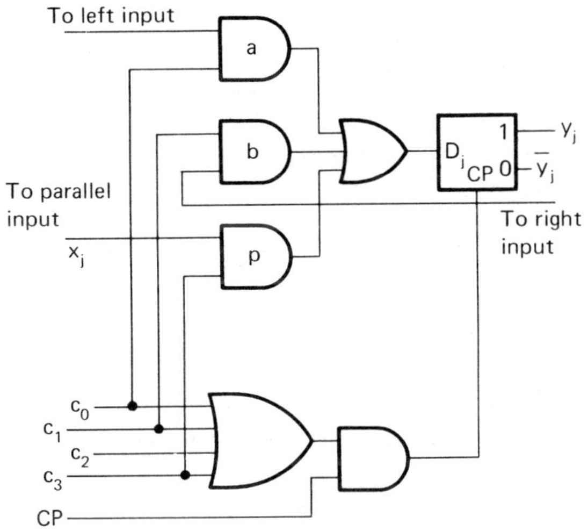

Figure 6-27. One stage of a shift register that allows parallel as well as series input.

Note that this circuit can also be used as both a *serial-to-parallel converter* and a *parallel-to-serial converter*. For instance, the data can be inputted serially. This data can then be used in a parallel fashion.

6.5 DESIGN OF SYNCHRONOUS SEQUENTIAL CIRCUITS

In this section we shall discuss a procedure used to design synchronous sequential circuits. There are several distinct steps and each shall be considered in turn. We will start by obtaining the network from a specified transition table or diagram. Then, we shall consider how the transition table is obtained from general specifications. The simplification of the table will also be discussed.

Determination of Network from Transition Table

We shall start by considering that the transition table is available and that the network is to be determined. Let us illustrate this with Table 6-8. (That was obtained from the network of Fig. 6-17. However, we shall assume that this is not known.) Then, repeating this table we have

TABLE 6-11
A TRANSITION TABLE

Current state	Current input x	
(y_1y_2)	0	1
(0,0)	(0,0)/0	(1,1)/0
(0,1)	(0,0)/0	(1,1)/0
(1,0)	(1,0)/0	(0,1)/0
(1,1)	(1,0)/1	(0,1)/1

In order to synthesize the network, we shall use a truth table type of formulation. This should include inputs to the flip-flops. In general, the designer can choose any type of flip-flop to realize the network. However, the choice is not arbitrary since the proper type can result in a much simpler network. Unfortunately, it is usually not obvious which flip-flops are to be used. The design can be repeated using different flip-flops and then all the networks can be studied to see which is optimum. This is a tedious procedure. Nevertheless, it often should be done.

Let us assume that y_1 is the state of a T flip-flop and y_2 is the state of a D flip-flop. This is an arbitrary choice which was done to conform with the original network. Now let us obtain the truth table. This will contain the inputs, present states, next states, present inputs to the flip-flops and the present outputs. That table is

TABLE 6-12
THE TRUTH TABLE FOR THE NETWORK WHOSE TRANSITION TABLE IS GIVEN IN TABLE 6-11

Input	Current State		Next State		Flip-flop input		Current output
$x(t)$	$y_1(t)$	$y_2(t)$	$y_1(t+\tau_p)$	$y_2(t+\tau_p)$	$T_1(t)$	$D_2(t)$	$z(t)$
0	0	0	0	0	0	0	0
0	0	1	0	0	0	0	0

0	1	0	1	0	0	0	0
0	1	1	1	0	0	0	1
1	0	0	1	1	1	1	0
1	0	1	1	1	1	1	0
1	1	0	0	1	1	1	0
1	1	1	0	1	1	1	1

The columns for $y_1(t+\tau_p)$, $y_2(t+\tau_p)$ and $z(t)$ are obtained directly from the transition table. The columns for $T_1(t)$ and $D_2(t)$ are obtained from a knowledge of the excitation tables of the T and D flip-flops. For instance, a T flip-flop changes its state when T=1 and the output of a D flip-flop is equal to D. Now we use the truth table to obtain logical expressions for all the flip-flop inputs and the output. The procedures of Chapter 4 are used here. Then, considering the 1 terms in the T_1 column, we have

$$T_1(t) = x(t)\bar{y}_1(t)\bar{y}_2(t) + x(t)\bar{y}_1(t)y_2(t) + x(t)y_1(t)\bar{y}_2(t) + x(t)y_1(t)y_2(t) \qquad (6\text{-}15)$$

Since these all occur at the same time, it is convenient to omit time specifications. Thus, we write

$$T_1 = x\,\bar{y}_1\bar{y}_2 + x\,\bar{y}_1y_2 + x\,y_1\bar{y}_2 + x\,y_1y_2 \qquad (6\text{-}16a)$$

Similarly,

$$D_2 = x\,\bar{y}_1\bar{y}_2 + x\,\bar{y}_1y_2 + x\,y_1\bar{y}_2 + x\,y_1y_2 \qquad (6\text{-}16b)$$

$$z = \bar{x}\,y_1y_2 + x\,y_1y_2 \qquad (6\text{-}16c)$$

The first step is to simplify these functions. They represent a multiple output network so that the procedures discussed in Sec. 4-4 can be used. Actually the network is so simple that we can consider the functions separately. Let us use Karnaugh maps for the simplification. Then, from Eq. (6-16a) we obtain the Karnaugh map of Fig. 6-28a. (Note that to avoid confusion, we placed the x on the left hand side and y_1 and y_2 across the top.

The conventional notation would have x and y_1 across the top and y_2 on the left hand side.) From these Karnaugh maps, we obtain

$$T_1 = x \tag{6-17a}$$

and

$$D_2 = x \tag{6-17b}$$

The Karnaugh map of Fig. 6-28b is used to minimize Eq.(6-16c). This yields

$$z = y_1 y_2 \tag{6-17c}$$

Equations (6-17) allow us to construct the network. It is shown in Fig. 6-17.

As a further example, now let us realize the network using all D flip-flops. Table 6-12 now must be modified by changing T_1 to D_1. In this case, we have

TABLE 6-13
TRUTH TABLE FOR THE NETWORK WHOSE TRANSITION TABLE IS GIVEN IN TABLE 6-11 WHEN D FLIP-FLOPS ARE USED

Input t=t	Current state t=t		Next state t=t+τ		Flip-flop input t=t		Present output t=t
x	y_1	y_2	y_1	y_2	D_1	D_2	z
0	0	0	0	0	0	0	0
0	0	1	0	0	0	0	0
0	1	0	1	0	1	0	0
0	1	1	1	0	1	0	1
1	0	0	1	1	1	1	0
1	0	1	1	1	1	1	0
1	1	0	0	1	0	1	0
1	1	1	0	1	0	1	1

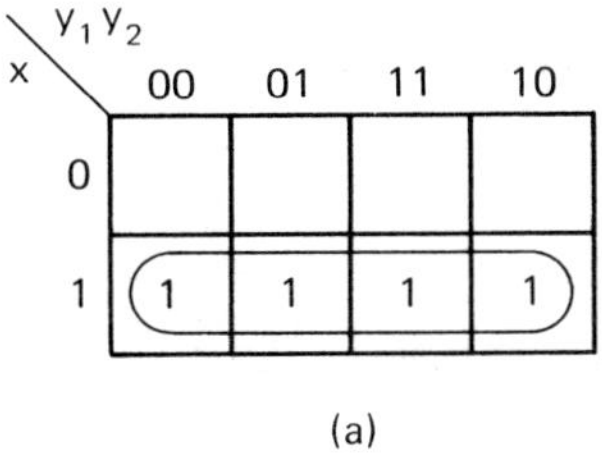

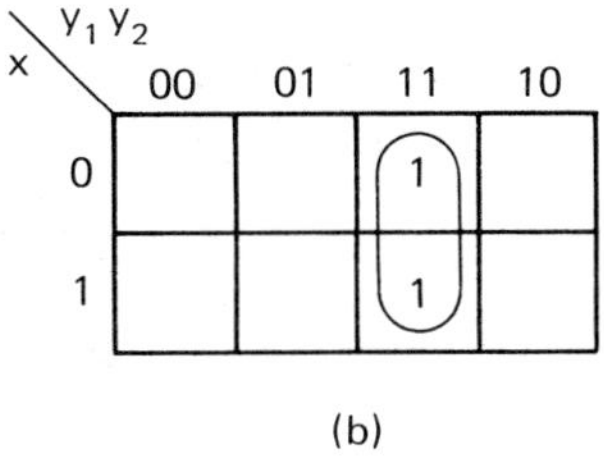

Figure 6-28. Karnaugh maps. (a) used to minimize Eqs. (6-16a) and (1-16b); (b) used to minimize Eq. (6-16c). Note that the labeling is not conventional.

All the columns are the same as before except that for D_1. Now we have

$$D_1 = \bar{x}\, y_1\bar{y}_2 + \bar{x}\, y_1 y_2 + x\, \bar{y}_1\bar{y}_2 + x\, \bar{y}_1 y_2 \tag{6-18}$$

Minimizing this we obtain

$$D_1 = \bar{x}y_1 + x\bar{y}_1 = x \oplus y_1 \tag{6-19}$$

This is the exclusive OR, EXOR operation, see Eq. (3-15). Thus, using the previous results, we have

$$D_2 = x$$

$$z = y_1 y_2$$

Thus, the network has the form shown in Fig. 6-29. If the EXOR gate were not used, a NOT gate, 2 AND gates and an OR gate would have to be used. Thus, the complexity of the network can vary greatly with the type of flip-flop used.

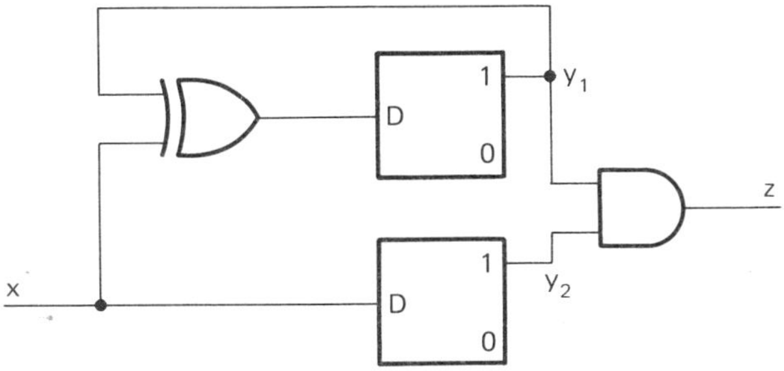

Figure 6-29. A realization, using D flip-flops, of the network shose transition table is given in Table 6-11.

As a final example illustrating the different choices of gate types, let us realize this network using J-K flip-flops. Now the truth table becomes

TABLE 6-14
TRUTH TABLE FOR THE NETWORK WHOSE TRANSITION TABLE IS GIVEN IN TABLE 6-11 WHEN J-K FLIP-FLOPS ARE USED

Input	Current state		Next state		Flip-flop input		Current output
x	y_1	y_2	y_1	y_2	J_1K_1	J_2K_2	z
0	0	0	0	0	0 d	0 d	0
0	0	1	0	0	0 d	d 1	0
0	1	0	1	0	d 0	0 d	0
0	1	1	1	0	d 0	d 1	1
1	0	0	1	1	1 d	1 d	0
1	0	1	1	1	1 d	d 0	0
1	1	0	0	1	d 1	1 d	0
1	1	1	0	1	d 1	d 0	1

We can write the logical expressions. However, these steps can be skipped and the Karnaugh maps can be obtained directly from the truth table. For instance, if we obtain the map for J_1, then, corresponding to each $J_1=1$, we mark a 1 directly on the Karnaugh map. The maps for the various functions are shown in Fig. 6-30. Thus, we have

$$J_1 = x \tag{6-20a}$$

$$K_1 = x \tag{6-20b}$$

$$J_2 = x \tag{6-20c}$$

$$K_2 = \bar{x} \tag{6-20d}$$

The expression for z is the same as before.

$$z = y_1y_2$$

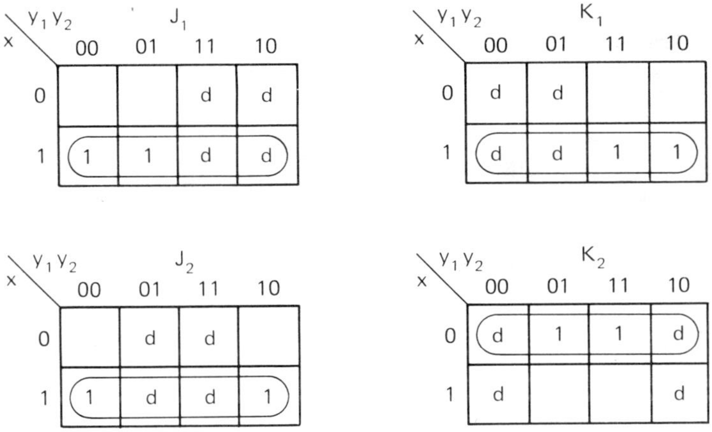

Figure 6-30. Karnaugh map used to minimize the J and K function in Table 6-14. Note that the labeling is not conventional.

The realization for the network is given in Fig. 6-31.

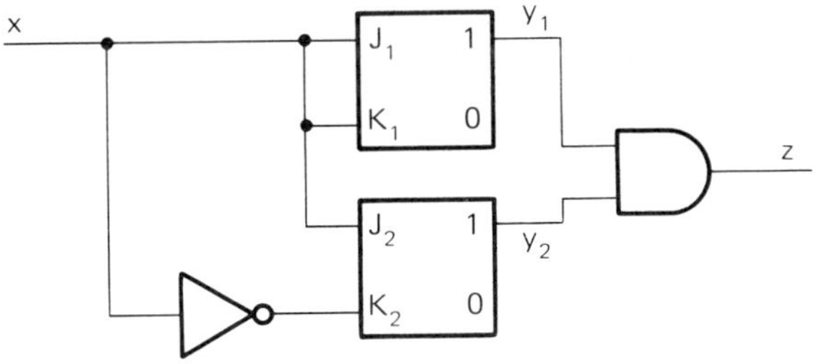

Figure 6-31. A realization, using J-K flip-flops, of the network whose transition table is given in Table 6-11.

Assignments of Binary Values to States

The networks that we have discussed all realize the transition diagram of Fig. 6-19a or 6-19b. In Fig. 6-19b, the states do not have binary values assigned to them. Note that we do not have to know the binary values assigned to the states to determine the output sequence for a given input sequence. That is, the sequence is independent of the binary values assigned to the states. Thus,we can use different states of binary variables and still obtain the same input and output responses. Let us illustrate this. Using Fig. 6-19b we can obtain the following transition table.

TABLE 6-15
A TRANSITION TABLE

Current states (y_1y_2)	Current input x: 0	1
q_0	$q_0/0$	$q_3/0$
q_1	$q_0/0$	$q_3/0$
q_2	$q_2/0$	$q_1/0$
q_3	$q_2/1$	$q_1/1$

Although the assigned binary values of the states do not affect the input and output sequence, they will affect the complexity of the network. However, there is no way of telling beforehand which assignment to use. Just as in the case of the choice of flip-flops, we can try all possible assignments of binary values to the states. However, in a complex network, there will be very many choices and this will become an extremely tedious procedure. If there are n states, then we need at least K bits, where K is the smallest integer satisfying

$$K \geq \log_2 n$$

If there are many states, there can be many bits and, thus, very many choices. (At times, more bits than the minimum are used. Such realizations are sometimes superior to the ones using the minimum number of bits. However, there is no algorithm for generating these realizations.)

Let us now consider an example where we assign states which are different from those used before. In particular, let

$$q_0 = 1,1 \tag{6-21a}$$

$$q_1 = 0,1 \tag{6-21b}$$

$$q_2 = 0,0 \tag{6-21c}$$

$$q_3 = 1,0 \tag{6-21d}$$

Then, the transition table becomes

TABLE 6-16

Current state (y_1y_2)	Current input x: 0	1
(1,1)	(1,1)/0	(1,0)/0
(0,1)	(1,1)/0	(1,0)/0
(0,0)	(0,0)/0	(0,1)/0
(1,0)	(0,0)/1	(0,1)/1

Let us realize this using D flip-flops. The truth table is

TABLE 6-17
TRUTH TABLE FOR TABLE 6-16 USING D-FLIP-FLOPS

Input	Current state		Next state		Current flip-flop input		Current output
x	y_1	y_2	y_1	y_2	J_1K_1	J_2K_2	z
0	0	0	0	0	0 d	0 d	0
0	0	1	1	1	1 d	d 0	0
0	1	0	0	0	d 1	0 d	1
0	1	1	1	1	d 0	d 0	0
1	0	0	0	1	0 d	1 d	0
1	0	1	1	0	1 d	d 1	0
1	1	0	0	1	d 1	1 d	1
1	1	1	1	0	d 0	d 1	0

In Fig. 6-32 we have the various Karnaugh maps which are used to minimize the switching functions for J_1, K_1, J_2, K_2 and z. Thus, we have

$$J_1 = y_2 \tag{6-22a}$$

$$K_1 = \bar{y}_2 \tag{6-22b}$$

$$J_2 = x \tag{6-22c}$$

$$K_2 = x \tag{6-22d}$$

$$z = y_1\bar{y}_2 \tag{6-22e}$$

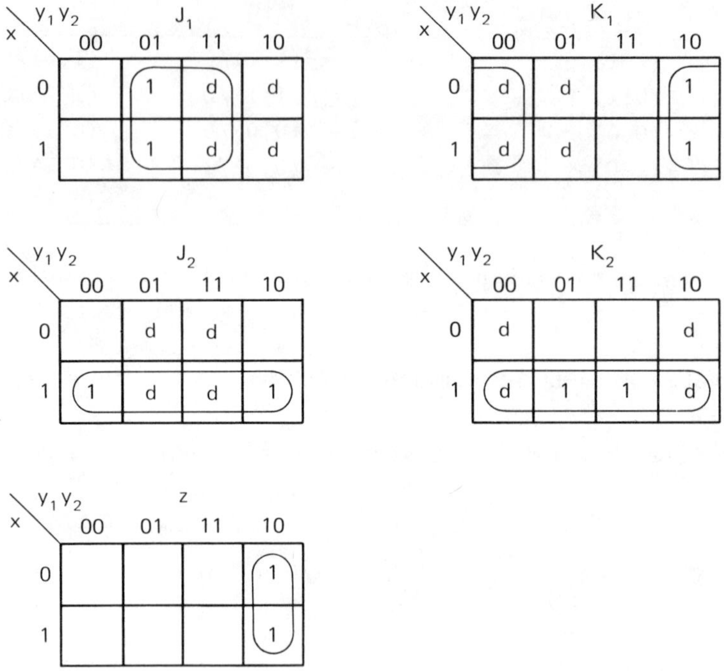

Figure 6-32. Karnaugh maps used to minimize the J, K and z functions of Table 6-17. Note that the labeling is not conventional.

The realization of the network is shown in Fig. 6-33. It is somewhat simpler than that of Fig. 6-31. In large complex networks, it is possible that differences in assignment of states can result in considerable simplification.

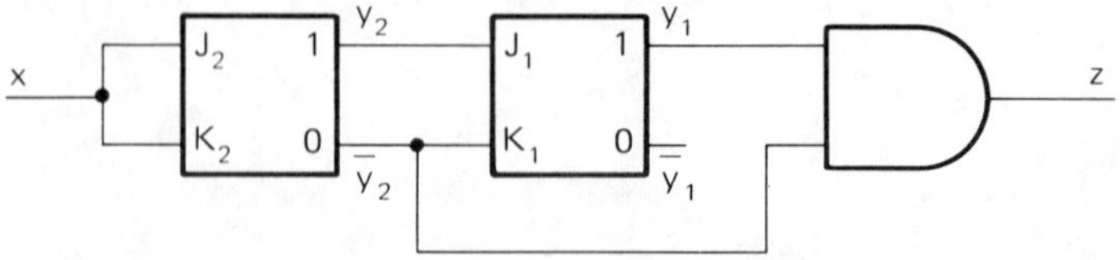

Figure 6-33. A realization, using J-K flip-flops of the network whose transition table is given in Table 6-15. The assignments of Eqs. (6-21) are used.

Reduction of the Transition Table and Diagram

At times, two rows in the transition table are identical. For instance, in Table 6-15 the rows for q_0 and q_1 are such. Note that the same inputs produce *both* the *same transitions* and the *same outputs*. When this happens, we can eliminate one state. For instance, if we are in state q_1 and the current input is a 0, then the next state will be q_0 with a current output 0. The same statement could be made for state q_0. An equivalent set of statements could be made for x=1. Thus, there would be no change in the results if we replaced q_1 by q_0 wherever it occurred and eliminated the row for q_1. (In general, we try to eliminate more than one row by repeating this procedure, if possible.) Note that, in this case, this procedure does not reduce the number of flip-flops required because now there are three states. However, two flip-flops are still required. If we could reduce the number of states by one more, then one less flip-flop could be used. In a practical design, it is unlikely that this will happen since the amount of storage required can usually be determined by the statement of the problem. Note that in Table 6-15, we cannot combine q_2 and q_3 since the output terms are different. In Sec. 6-7 we shall further discuss some related ideas.

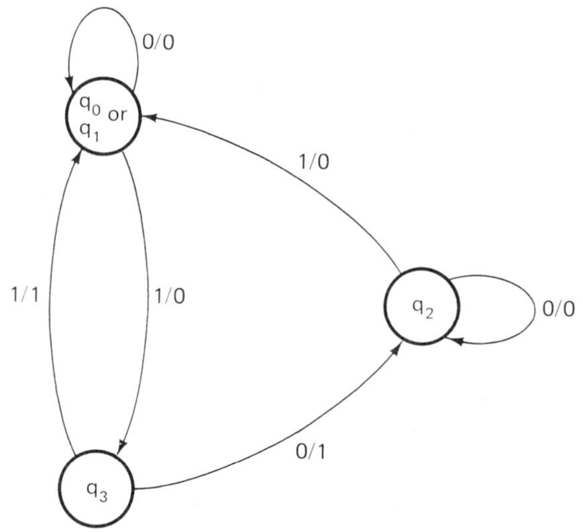

Figure 6-34. A reduced form of Figure 6-19b.

Eliminating a state can reduce the number of logic gates since additional "don't cares" can appear. This will be illustrated later in this section when Fig. 6-39 is realized. The details of the present example proceed as before. A reduced state transition diagram that is equivalent to Fig.6-19b is given in Fig. 6-34.

Determination of State Transition Tables and Diagrams

We shall now consider some more complex examples where we shall determine the state transition tables from some "verbal" specifications. That is, we shall know the task the circuit is to perform and then shall determine the state transition table and, finally, the network.

Before we consider some specific examples, let us consider some definitions. Certain sequential circuits' outputs are only functions of the present state and not of the present input. For instance, in the examples just considered, $z=y_1y_2$ which is a function only of the states and not of the input. Of course, the states are functions of the input. Such a device is called a *Moore machine* or *Moore network*. If the output is a direct function of the input and the states(e.g., $z = xy_1y_2$)then the device is called a *Mealy machine* or *Mealy network.* G.H.Mealy and E.F.Moore are the names of the early investigators of these devices.

If a Moore circuit has no input, then it is called an *autonomous network*. (Note that a clock pulse is not considered to be an input.) It might seem as though such a network would have no use. However, they are actually very important. We shall consider several of them here. One use is as a counter. Suppose that we have a synchronous sequential network which changes its state with each clock pulse. By determining the state of the network, we can determine how many clock pulses have elapsed. Such a circuit can have many uses. It can be used to keep track of the number of steps performed by a computer. Also, suppose that the clock pulses come from impulses received from an electric eye on a production line. The counter can then determine the number of items produced. Similarly, the number of dimes passing a point on a change sorter can be counted. Another use of a counter is to very accurately determine the frequency of an oscillator. The output of the oscillator is first shaped to be a pulse, see Sec. 12-2, and applied to the clock pulse terminal through an AND gate. The other input to the AND gate is a single pulse whose time duration is known very accurately. Thus, the oscillator is effectively connected to the clock pulse terminal for a standard length of time. The counter is then used to determine the number of cycles of the oscillator and the frequency can, thus, be accurately determined.

Let us now design a counter that will count from 0 to 8 and then reset itself. This is a modulo 8 counter. Actual counters may count to much higher values. However, this just involves a duplication of circuitry. We will keep the number small to avoid obscuring the design with unnecessary details.

The state transition diagram for a molulo 8 counter is shown in Fig. 6-35. No input values are given since there are no inputs. The output values are the same as the past state. There are 8 states and we shall use the following state assignment.

TABLE 6-18
TRANSITION TABLE FOR A MODULE 8 COUNTER

Current state (y_1, y_2, y_3)	Current input none
(0,0,0)	(0,0,1)/(0,0,0)
(0,0,1)	(0,1,0)/(0,0,1)
(0,1,0)	(0,1,1)/(0,1,0)
(0,1,1)	(1,0,0)/(0,1,1)
(1,0,0)	(1,0,1)/(1,0,0)
(1,0,1)	(1,1,0)/(1,0,1)
(1,1,0)	(1,1,1)/(1,1,0)
(1,1,1)	(0,0,0)/(1,1,1)

Note that three states are needed. In general, if there are n flip-flops, then

$$2^n$$

values can be stored. Since 8 values are to be stored, we must use n=3.

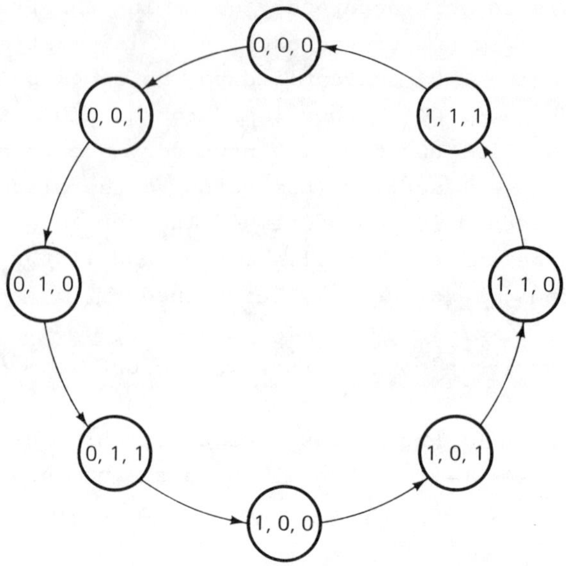

Figure 6-35. State transition diagram for a modulo 8 counter.

Now let us obtain the truth table for this device. We shall omit the output since it is the same as the current state. Thus, the y_1, y_2 and y_3 leads can be used as the output leads. We shall build the counter using J-K flip-flops.

TABLE 6-19
TRUTH TABLE FOR THE MODULO 8 COUNTER

Current state			Next state			Current flip-flop inputs		
y_1	y_2	y_3	y_1	y_2	y_3	J_1K_1	J_2K_2	J_3K_3
0	0	0	0	0	1	0 d	0 d	1 d
0	0	1	0	1	0	0 d	1 d	d 1
0	1	0	0	1	1	0 d	d 0	1 d
0	1	1	1	0	0	1 d	d 1	d 1
1	0	0	1	0	1	d 0	0 d	1 d
1	0	1	1	1	0	d 0	1 d	d 1
1	1	0	1	1	1	d 0	d 0	1 d
1	1	1	0	0	0	d 1	d 1	d 1

The Karnaugh maps corresponding to this truth table are given in Fig. 6-36. Thus,

$$J_1 = y_2y_3 \qquad (6\text{-}23a)$$

$$K_1 = y_2y_3 \qquad (6\text{-}23b)$$

$$J_2 = y_3 \qquad (6\text{-}23c)$$

$$K_2 = y_3 \qquad (6\text{-}23d)$$

$$J_3 = 1 \qquad (6\text{-}23e)$$

$$K_3 = 1 \qquad (6\text{-}23f)$$

The network that realizes this is shown in Fig. 6-37. Note that the J-K flip-flops are actually all connected as T flip-flops.

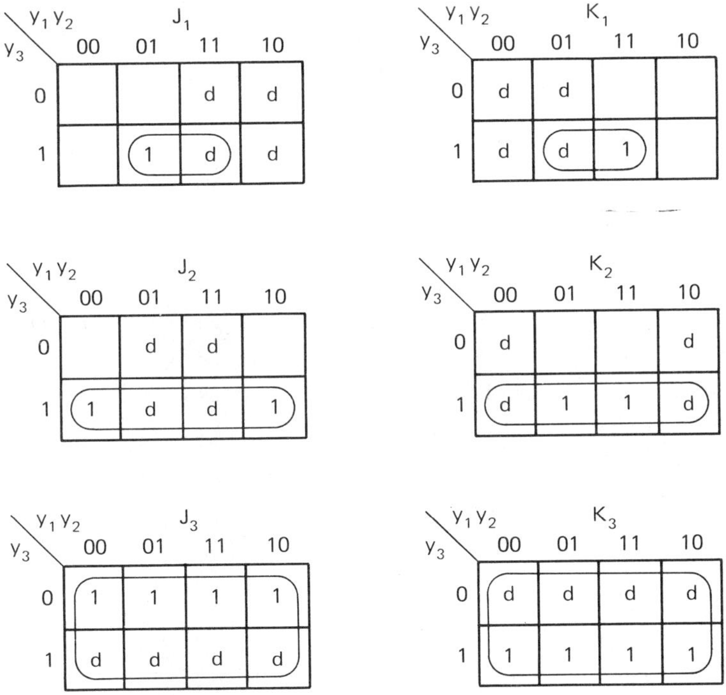

Figure 6-36. Karnaugh maps used in the design of the modulo 8 counter.

The inputs J_3 and K_3 are always 1's. They are connected to the clock pulse terminal so that whenever a clock pulse occurs, $J_3=1$ and $K_3=1$.

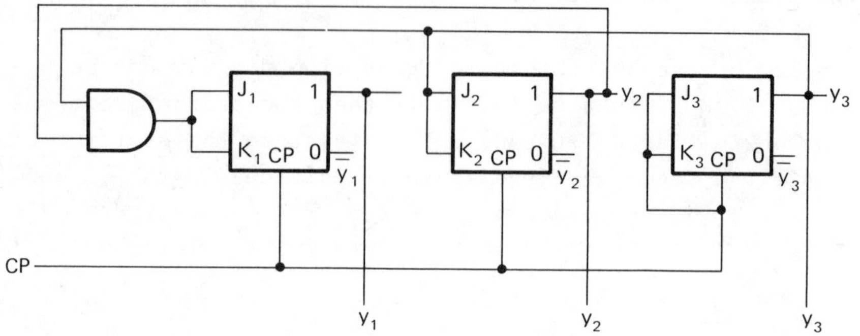

Figure 6-37. A modulo 8 counter realized with J-K flip-flops.

We designed the counter so that the binary numbers corresponding to 1-8 were generated in sequence. Actually, we need not do this. For instance, the sequence could be 000, 101, 001, 110, 010, 111, 011 and 100 or any other sequence. This is equivalent to changing the assignment of the binary values of the states which was discussed earlier in the section. It is not necessary that the counter have 8 states. For instance, using 3 flip-flops, we could count the sequence 000, 001, 010, 011, 100, 101 and then repeat. The details of generating such networks will be left to the reader. Such sequence generators are used to control various types of operations. For instance, in the previous section, we considered a shift register whose mode of operation was controlled by a set of control variables, see Fig. 6-25. Such a sequence generator could be used to control the operation of the shift register, i.e., so that it varied its operation in accordance with a desired sequence of commands.

We have considered circuits that generate a sequence of numbers. Now let us design a circuit that will identify when a specified sequence occurs. For instance, suppose that we want the output of the network to be a 1 each time that the sequence

010 occurs in the input. For instance, the sequence

01101010

would produce two 1's.

Now let us see how the state transition diagram can be obtained. If a 1 input follows a 1, then the desired sequence cannot be initiated. Let us call this state the starting state q_0. A first segment of the state transition diagram is shown in Fig. 6-38a. If there are any successive 1 inputs, the circuit should remain in state q_0 with no output, see Fig. 6-38b. Any 0 input is a possible first term in the desired 010 sequence. This should initiate a new state, q_1. Of course, the output should remain a 0. This next step is shown in Fig. 6-38c. A 0 input should leave the state unchanged. This is also shown in Fig. 6-38c. (Note that *any* sequence of 0's followed by a 10 produces the desired 010 sequence.)

If a 1 input *follows* a 0 input, then a *possible* 010 sequence may occur. Thus, this should lead to a new state q_2, see Fig. 6-38d. Finally, if the state is q_2 and the next input is a 0, then the sequence 010 has been received. Thus, a 1 should be outputted. This should also result in the next state's being q_1, since a 0 was inputted, see Fig. 6-38e. On the other hand, if we are in state q_2 and the next input is a 1, then we must return to state q_0, with 0 output, since the 010 sequence cannot be initiated until the next 0 is inputted. This transition is also shown in Fig. 6-38e. Figure 6-38e is the complete state transition diagram since all possible input sequences have been considered. The state transition table is

TABLE 6-20
TRANSITION TABLE FOR FIG. 6-38

Current state	Current input x	
	0	1
q_0	$q_1/0$	$q_0/0$
q_1	$q_1/0$	$q_2/0$
q_2	$q_1/1$	$q_0/0$

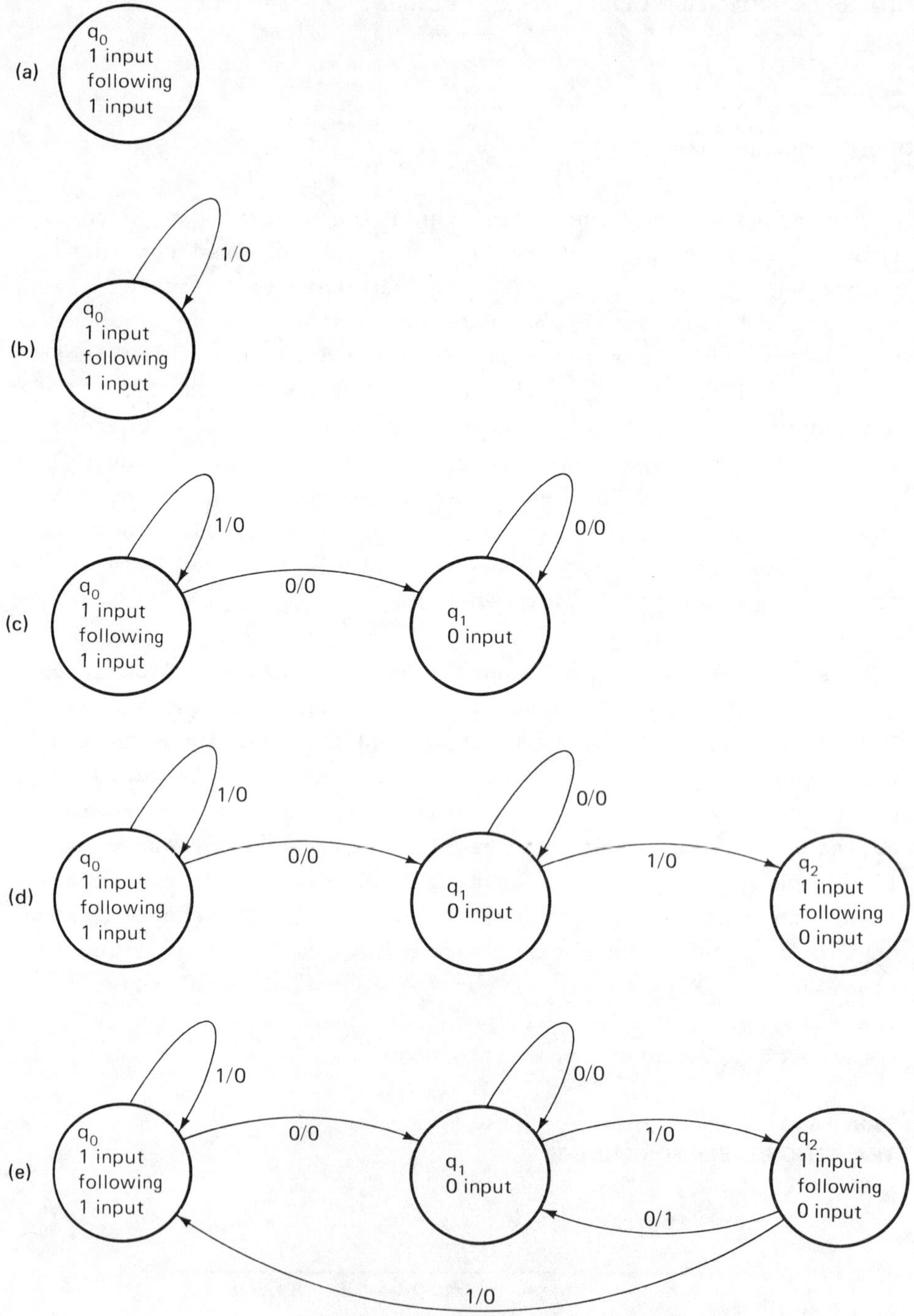

Figure 6-38. State transition diagram for the 010 sequence detector. (a) first step; (b) second step; (c) third step; (d) fourth step; (e) final diagram.

Let us use the following binary assignments (for purposes of example, we do not make the assignments "agree" with the subscripts).

$q_0 = 1,0$

$q_1 = 0,0$

$q_2 = 0,1$

The truth table is then

TABLE 6-21
TRUTH TABLE CORRESPONDING TO TABLE 6-20

Current Input	Current State		Next state		Current flip-flop input		Current output
x	y_1	y_2	y_1	y_2	D_1	D_2	z
0	0	0	0	0	0	0	0
0	0	1	0	0	0	0	1
0	1	0	0	0	0	0	0
0	1	1	---------		d	d	d
1	0	0	0	1	0	1	0
1	0	1	1	0	1	0	0
1	1	0	1	0	1	0	0
1	1	1	---------		d	d	d

Note that the state 1,1 will never occur. Thus, we "don't care" about the flip-flop inputs or the z outputs in these cases. These "don't cares" in the truth table result in a simplification of the logic functions. Proceeding as before, we obtain

$$D_1 = x\ y_2 + x\ y_1 = x(y_1+y_2) \tag{6-24a}$$

$$D_2 = x\ \bar{y}_1\bar{y}_2 \tag{6-24b}$$

$$z = \bar{x}\ y_2$$

The network which realizes the 010 sequence detector is shown in Fig. 6-39a. A typical timing diagram is shown in Fig. 6-39b.

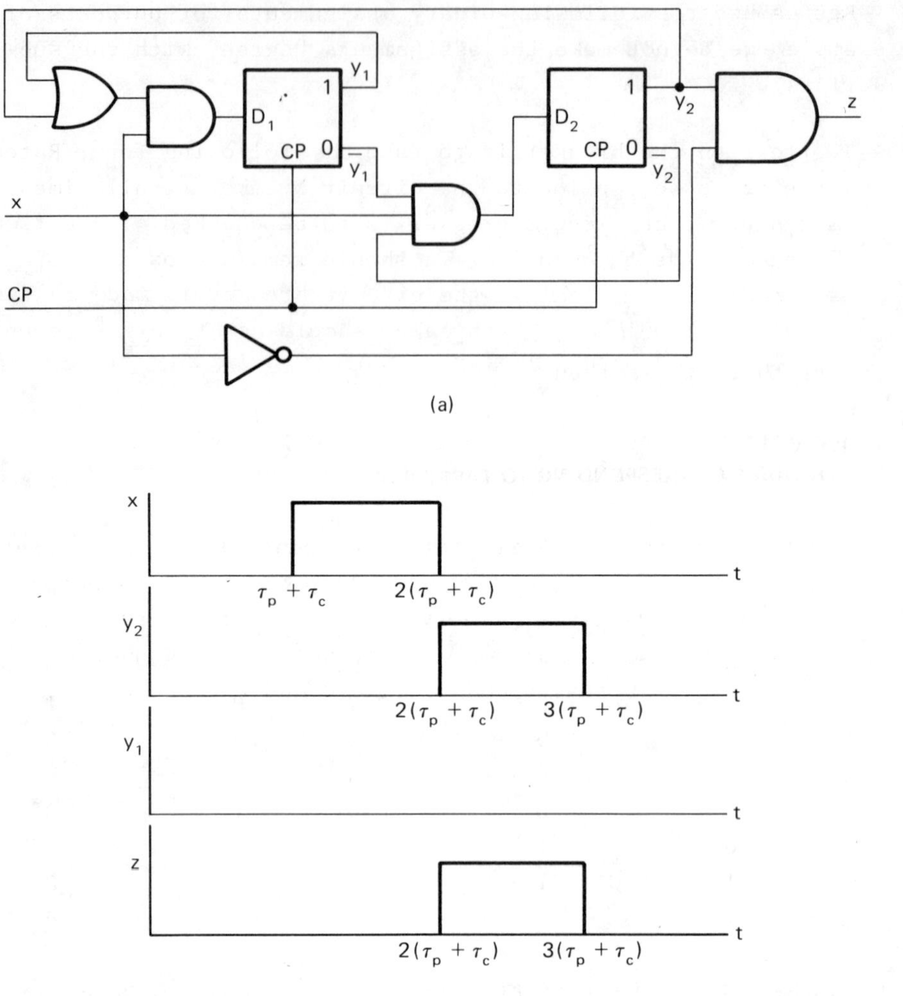

Figure 6-39. (a) A realization of the 010 sequence detector using D flip-flops; (b) a timing diagram for the sequence detector.

We have considered circuits that can generate a specified sequence of numbers and others that can determine when a specified sequence has occurred. We shall discuss the computer applications of such circuits in Chapter 11. Now let us consider a simple noncomputer application. Suppose that we want to obtain a simple system that automatically tests electrical circuit breakers. The circuit breaker is a switch which functions in the following manner. Once the switch is closed, it remains closed unless the current through it exceeds a predetermined value, say 100 amps. Then, the circuit breaker switch opens and remains open until it is manually reset.

The test on the breaker is to take the following form: Rated voltage is to be applied to the circuit breaker at all times. At the start, the circuit breaker is to be checked with currents of 0 and 50 amps through it. It should remain closed under these conditions. Finally, the circuit breaker is made to carry a current of 150 amps. The breaker should now trip(i.e.,open). We have used an oversimplified test to keep the example simple.

A circuit which can be used to perform the test is shown in Fig.6-40a. The resistors R_1 and R_2 are loads. Their values are such that when sw_1 is closed, the current through the circuit breaker is 50 amps and when both sw_1 and sw_2 are closed, the current is 150 amps. The test procedure is the following: The circuit breaker is connected into the test set up, turned on, i.e.,manually reset, and then a start button is pushed. The circuit breaker is tested under no load and then under 50 and 150 amp loads. The output voltage is applied to a voltage modifying circuit, which converts the ac to dc and reduces the level of voltage to that of a logic 1. Note that if the circuit breaker is open, then the output of the voltage modifying circuit is a 0; when the circuit breaker is closed, the output is a 1.

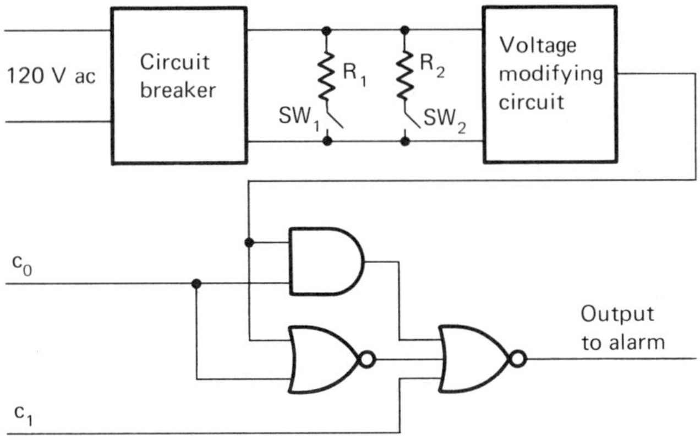

Figure 6-40. A simple circuit breaker test apparatus.

Now consider the circuit. A sequence generator is used to generate a four bit sequence ($c_0c_1c_2c_3$). The c_0 and c_1 are used in the logic, c_2 controls a relay that closes sw_1, and c_3 controls a relay that controls sw_2. For instance, when $c_2=1$, then sw_1 is closed. If $c_2=0$, then sw_1 is open. The sequence generator produces the following sequence after the start button is pushed.

c_0	c_1	c_2	c_3
1	0	0	0
1	0	1	0
0	1	1	1
0	0	1	1

Now consider the logic circuit. When c_0 is 1 and c_1 is 0, the output will be a 0 only if the circuit breaker is closed. When c_0 is 0 and c_1 is 0, the logic circuit output will be 0 only if the circuit breaker is open. If $c_1=1$, then the output will always be 0. The logic circuit output is used to sound an alarm which indicates that the circuit breaker should be rejected. The c_1 signal causes the logic circuit output to be 0 independent of the condition of the circuit breaker. This is included to allow time for the circuit breaker to open after sw_2 closes. If c_1 were not included, then there would be a short alarm given after the 0011 signal were given since the circuit breaker cannot open instantaneously after sw_2 closes.

In this circuit we did not use a sequence detector. However, it could be used to replace the combinational logic circuit to detect the output sequence. For instance, if the sequence 110 did not occur at the output of the voltage modifying circuit, an alarm would sound. (It is assumed that the start button which initiates the test generates signals that will reset all the flip-flops to 0 state in both the sequence generator and in the sequence detector and also that the momentary transient that occurs when sw_2 is closed is suppressed.)

We have realized a number of synchronous sequential networks in this section. In subsequent chapters, we shall consider the realization of additional sequential circuits.

6.6 ASYNCHRONOUS SEQUENTIAL CIRCUITS-RACES-HAZARDS

Now let us consider unclocked or asynchronous sequential circuits. Actually, we have already discussed these. For instance, a flip-flop is an asynchronous circuit. Even if it is a clocked flip-flop, its internal behavior is asynchronous. For instance, consider Fig. 6-6a. Suppose that $Y=1$ and $\bar{Y}=0$, and we apply the input $S=0$, $R=1$. During a clock pulse, Y becomes 0 and after that $\bar{Y}$ becomes 1. These changes are asynchronous. That is, even though they take place during a clock pulse, the rate at which Y changes and then $\bar{Y}$ changes is not controlled by the clock pulse. Of course, the rate of change has to be fast enough if proper operation is to result.

Portions of sequential circuits operate asynchronously. For instance, in Fig. 6-39, the logic gates respond to their inputs without regard to the clock pulses whereas the flip-flops are synchronized. Of course, there are many sequential circuits which operate independently of any clock pulse.

Let us illustrate the analysis of an asynchronous circuit by considering one which has the following switching expressions.

$$y_1 = \bar{x}_1 \bar{y}_2 \tag{6-25a}$$

$$y_2 = (\bar{x}_1\bar{x}_2 + x_1x_2)\bar{y}_1 \tag{6-25b}$$

where y_1 and y_2 are the states of the system and x_1 and x_2 are the independent inputs. Now let us construct a transition table for *this* system. In it we list the present state and present input and the succeeding (next) state, or *successor* state is given for every possible combination of inputs. The values in the table are found by substitution in Eqs. (6-25). For instance, if $y_1=0$, $y_2=0$, $x_1=0$, $x_2=0$, then $y_1=1$ and $y_2=1$.

TABLE 6-22
TRANSITION TABLE FOR EQS. (6-25)

Present state y_1y_2	Present input x_1x_2: 0 0	0 1	1 1	1 0	
(0,0)	(1,1)	(1,0)	(0,1)	(0,0)	successor states
(0,1)	(0,1)	(0,0)	(0,1)	(0,0)	
(1,1)	(0,0)	(0,0)	(0,0)	(0,0)	
(1,0)	(1,0)	(1,0)	(0,0)	(0,0)	

Note that, in the table, the ordering of y_1, y_2 and x_1, x_2 is such that there is only a single change of variable beteeen adjacent terms.

The details of the operation are considerably different from the synchronous case. For instance, suppose that the present state is (0,1), i.e., $y_1=0$, $y_2=1$, and the present input is $x_1=0$, $x_2=1$ which does not change, then the next state will be

$$y_1 = 0 \tag{6-26a}$$

$$y_2 = 0 \tag{6-26b}$$

The network will not remain in this state. If Eqs.(6-26) are substituted in Eqs.(6-25) with $x_1=0$, $x_2=1$, then the state becomes $y_1=1$, $y_2=0$, (1,0). Note that we do not have to substitute in Eqs.(6-25) since this can be determined from Table 6-22. Now, if this is the state, then the next state will be the same, (1,0). Thus, the system remains in this state. Thus, an asynchronous network may go through a sequence of states before it comes to its final state. This is called a *cycle*.

Let us illustrate this be rewriting the excitation table. We shall leave off the parentheses for clarity.

TABLE 6-23
EXCITATION TABLE ILLUSTRATING RACES AND CYCLES

Present state y_1y_2	Present input x_1x_2: 0 0	0 1	1 1	1 0
0 0	1 1	1 0	0 1	0 0
0 1	0 1	0 0	0 1	0 0
1 1	0 0	0 0	0 0	0 0
1 0	1 0	1 0	0 0	0 0

In the $x_1=0, x_2=1$ column, we have marked the transition that takes place from a present state of (0,1) to the final state of (1,0). The (1,0) is encircled in the table to indicate that it is a final state.

Now suppose that, with an input which remains fixed at $x_1=0$, $x_2=0$, we start from the state (0,0). The state will change to (1,1). This will change to (0,0) which will change to (1,1). This will continuously repeat. Thus, an *oscillation* will be set up. This *unstable condition* is indicated by the dashed lines in the first column. Such operation should always be avoided.

Another effect can occur. Assume that the state is (0,0) and that the inputs are $x_1=1, x_2=0$. Now the inputs are switched so that the new inputs are $x_1=0, x_2=0$. The next state should be (1,1). However, y_1 and y_2 *may not change simultaneously*. Suppose that y_1 changes before y_2. The state will shift to (1,0). Now, if the state is (1,0) and the input is $x_1=0, x_2=0$, then the next state should be (1,0). That is, the state remains at (1,0) which is stable operation. This is indicated by the left hand solid arrow in the first column. On the other hand, suppose that y_2 changes faster than y_1. Then, the next state will be (0,1). If the state is (0,1), and the input is $x_1=0$, $x_2=0$, then the next state will be (0,1). This, then, is a stable state and the circuit remains in this state. We see that the transition from (0,0) can take one of three paths depending upon the speed of the various transitions. This is called a *race*. Note that races result when *more* than one of the binary variables must change simultaneously since these variables may not all change at the same time. This type of race is called a *critical race* because the final state can be more than one value and, thus, erroneous results can occur.

Let us consider another case. Suppose that the state is (1,0) and we apply the input $x_1=1$, $x_2=1$. Now the state cycles through (0,0) and then (0,1) and it remains at (0,1). Thus, this is the final stable state. This is a normal allowable cycle.

We have not marked all the possible transitions in Table 6-23. If we did, the table would become cluttered. To avoid this, the table is often rewritten omitting the numerical values for the successor states. Then, the various transitions are marked. Such a table is shown in Table 6-24.

TABLE 6-24
TRANSITION TABLE FOR EQS. (6-25)

Present state	Present input x_1x_2			
y_1y_2	0 0	0 1	1 1	1 0
0 0				
0 1				
1 1				
1 0				

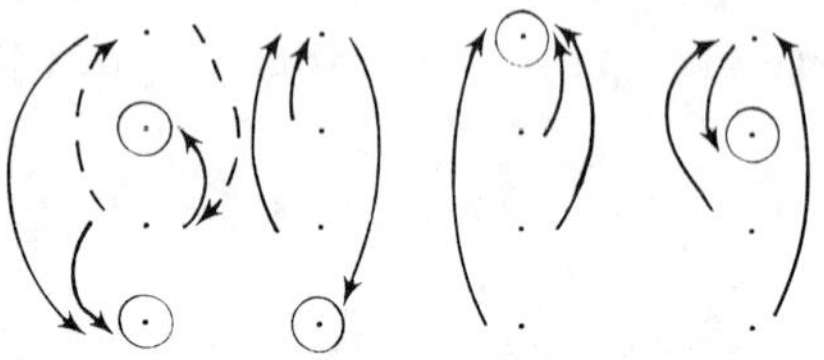

The intermediate states are indicated by dots. The stable final states are circled.

Let us consider the circuit which realizes Eq.(6-25), see Fig. 6-41. (Note that $\bar{x}_1\bar{x}_2 + x_1x_2 = \overline{x_1 \oplus x_2}$. That is, it is the complement of the EXOR operation.) Now suppose that we switch from an input $x_1=1, x_2=0$ to $x_1=0, x_2=0$. In this case, the three race conditions indicated in the $x_1=0, x_2=0$ column result. If we appropriately add delay at the input of one of the gates, then the "winner of the race" will be determined and the ambiguity will be removed. For instance, suppose that we add an additional state y_3 such that

$$y_3 = x_1$$

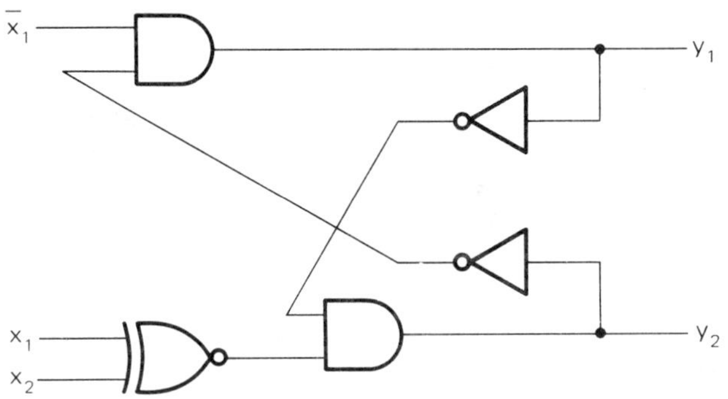

Figure 6-41. A network which realizes Eqs. (6.25).

and then we modify Eqs. (6-25) to read

$$y_1 = \bar{y}_3\bar{y}_2 \qquad (6\text{-}27a)$$

$$y_2 = (\bar{x}_1\bar{x}_2 + x_1x_2)\bar{y}_1 \qquad (6\text{-}27b)$$

$$y_3 = x_1 \qquad (6\text{-}27c)$$

These are equivalent to Eqs. (6-26) in that, upon substitution of x_1 for y_3, we obtain the same y_1 and y_2. However, now the realization is as in Fig. 6-42. Suppose we are in the state (0,0) and the input changes from $x_1=1, x_2=0$ to $x_1=0, x_2=0$; then the delay introduced by the two added NOT gates will cause y_2 to change before y_1 and the output will be $y_1=0, y_2=1$. Thus, the race has been eliminated. Note that if $x_1=1, x_2=0$, then the present state is $y_1=0, y_2=0, y_3=1$. If the input becomes $x_1=0, x_2=0$, then, substituting in Eqs. (6-27), we obtain the next state as $y_1=0, y_2=1$ and $y_3=0$. Thus, the addition of additional states can eliminate races. That is equivalent to saying that the addition of delay can eliminate races. The delay added by the additional gates must be sufficiently long. For instance, in Fig. 6-42, we assume that the two added NOT gates add enough delay so that y_2 changes before the upper input to the upper AND gate changes.

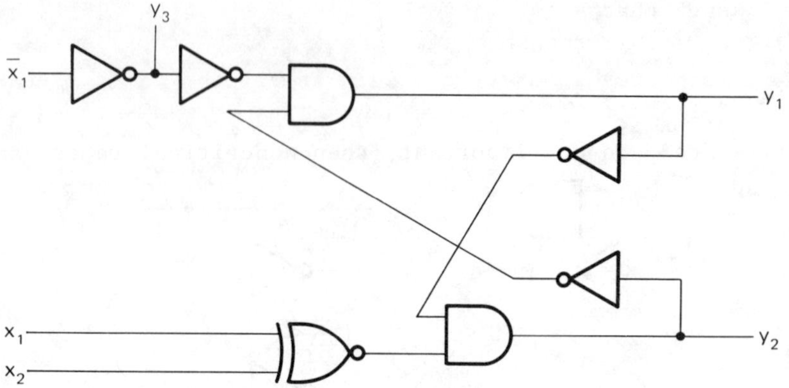

Figure 6-42. A modification of Fig. 6-41 which realizes Eqs. (6-26). Note that the addition of delay has eliminated the race condition.

We have eliminated the race by the addition of delay. At times, the race can be eliminated by assigning different binary variables to the states, see Sec. 6-5.

We have thus far considered that races are to be avoided. Actually, there is one type of race that can be tolerated. Consider the simple excitation table which follows.

TABLE 6-25
A SIMPLE TRANSITION TABLE ILLUSTRATING A NON-CRITICAL RACE

Present state y_1y_2	Present input x = 0	Present input x = 1
0 0	1 1	1 1
0 1	(0 1)	1 1
1 1	0 1	1 0
1 0	0 1	1 0

Now suppose that we have an initial state (0,0) with a constant input x=0. Then, the next state is (1,1) and then (0,1), which is a stable state. The transition from (0,0) to (1,1) involves a change of more than one variable. Thus, a race exists. Now suppose that the transition from (0,0) is to (0,1). This is the same stable state and will remain there. If the transition from (0,0) is to (1,0), then the next state will be (0,1), again the same stable state. Thus, the result of all the paths is the same stable state. This is called a *non-critical race*. If the intermediate states are not important, then noncritical races can be tolerated.

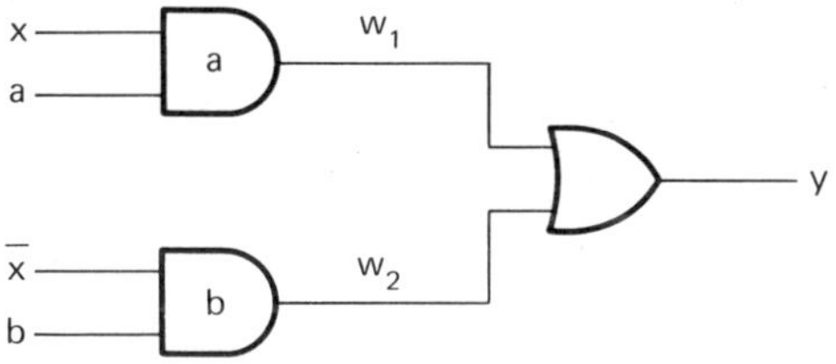

Figure 6-43. A simple gate network.

Hazards

Another problem can arise in asynchronous circuits or in the asynchronous portion of a synchronous one or in combinational circuits. Consider the circuit of Fig. 6-43. An expression for the output y is

$$y = x\,a + \bar{x}\,b \tag{6-28}$$

Now suppose that x switches from a 1 to a 0 while a=b=1. The output of AND a becomes a 0 while that of AND b becomes a 1. Thus, y should remain constant at a 1. However, it is possible that the switching of the AND gates may not occur at exactly the same time. This is illustrated in the timing diagram of Fig. 6-44. Suppose, as illustrated, that w_1 switches at T_1 while w_2 switches at T_2 where $T_2 > T_1$. Then, there will be a short period of time $T_2 - T_1$ when both inputs to the OR gate will be 0. In this interval y will be 0. Thus, we have the following situation. A change in x should produce no change in y(with a=1,b=1). However, because of unequal time delays, y becomes 0 for a short time. Such short changes in the variable can produce spurious responses and must be avoided. If the change in a *single* variable causes a momentary change in other variables, which should not occur, then a *hazard* is said to exist. This is actually called a *static hazard*.

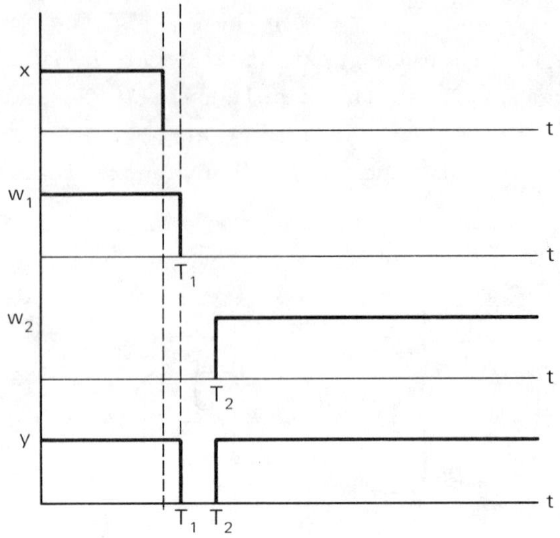

Figure 6-44. A timing diagram for the network of Fig. 6-43.

Now let us see if we can eliminate the hazard. First we shall review some terminology. Two products of input variables are said to be *adjacent* if they differ by just one variable which is complemented in one and is uncomplemented in the other. For instance, xab and $\bar{x}ab$ are adjacent. Now we can determine if a hazard exists by studying the Karnaugh map. If there are two adjacent input products and they do not lie in the same subcube, then a hazard exists. When the network is realized, each subcube results in an AND gate. Thus, if two adjacent input products are in different subcubes, then a situation such as that in Fig. 6-43 will exist and a hazard will result. Let us illustrate this. If we write Eq. (6-28) in the unminimized canonical sum of products form, it becomes

$$y = xa(b+\bar{b}) + \bar{x}b(a+\bar{a}) = xab + xa\bar{b} + \bar{x}ab + \bar{x}\bar{a}b \qquad (6\text{-}29)$$

The Karnaugh map which corresponds to this is shown in Figs. 6-45. In Fig. 6-45a we have chosen subcubes which yield the minimal realization of Eq. (6-28) as Fig. 6-43. This contains a hazard. Note that the arrow shows the two adjacent product terms which lie in different subcubes.

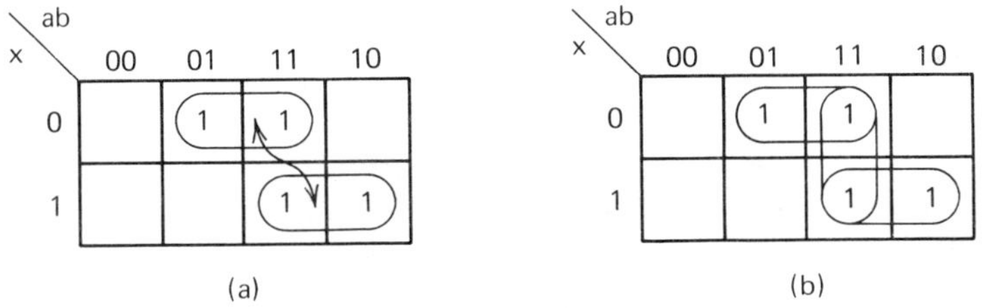

Figure 6-45. Karnaugh maps for Eq. (6-28). (a) Minimal realization; (b) hazard-free realization.

To eliminate the hazard, we can add a third subcube as shown in Fig. 6-45b. Now the adjacent product terms lie in the same subcube. Thus, we have

$$y = xa + \bar{x}b + ab \qquad (6\text{-}30)$$

This realization is shown in Fig. 6-46. If a=1 and b=1, then the output of AND c will constantly be a 1 independent of x. Thus, we have eliminated the hazard.

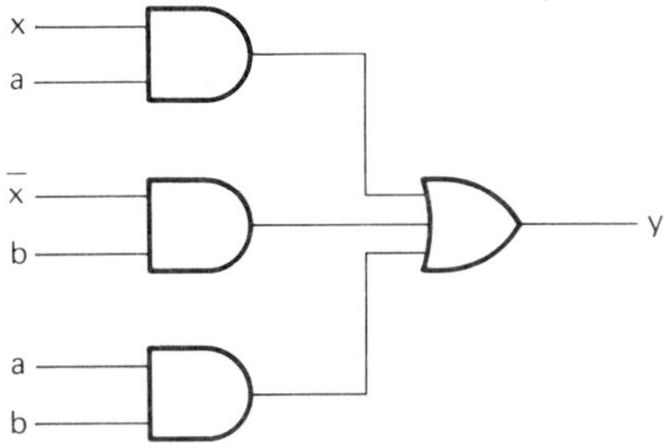

Figure 6-46. A hazard-free realization of Eq. (6-27).

Note that Eq. (6-30) is a redundant expression. That is, it can be reduced. However, it is not reduced and, to eliminate the hazard, the circuit is realized exactly according to Eq. (6-30). This increases the network complexity. In general, the price paid for the elimination of hazards is increased network complexity. Thus, the elimination of hazards requires that a compromise be made.

We have considered hazards based on only one input variable since hazards are usually eliminated only for single input variations.

Another form of hazard can exist. Suppose that, upon switching the input variable, the output variable is to switch from 0 to 1. There may be a brief period when it shifts back and forth, e.g., the output sequence is 0101. The intermediate 0 and 1 persist only for a very short time. This is called a *dynamic hazard*. Circuits which have been made free of static hazards are also free of dynamic hazards.

A third type of hazard occurs when a change in input variable affects one part of a circuit before another. Thus, the circuit can take on one stable state rather than another. This is analogous to a race condition, and is called an *essential hazard*. Such hazards can always be eliminated by incorporating the appropriate delay.

6.7 DESIGN OF ASYNCHRONOUS SEQUENTIAL CIRCUITS

We shall now consider the design of asynchronous sequential circuits. Although there are similarities with the design of synchronous circuits, there are important differences. We shall start by formalizing some definitions. In a synchronous circuit, there are pulses applied to the inputs. These, in conjunction with the clock pulses, cause changes in state. In an asynchronous circuit, we can speak of the inputs as *levels* which are either 0 or 1. These levels can persist for, or change at, arbitrary times. To make the design tractable, and to eliminate race type operation, the operation is usually restricted to one of several modes. We shall consider them here.

Fundamental Mode Operation

A basic sequential circuit is shown in Fig. 6-1b. We shall redraw it here in a somewhat more specific way. This circuit is shown in Fig. 6-47. Here we have a combinational circuit with feedback through delay elements. We discussed, in the last section, that such a circuit could cycle through a sequence of states before reaching a stable state. When a stable state has been reached, the system stops changing. Thus, the inputs and outputs of each delay will then become equal.

$$y_{1d} = y_1 \tag{6-31a}$$

$$y_{2d} = y_2 \tag{6-31b}$$

$$y_{3d} = y_3 \tag{6-31c}$$

.
.
.

Then, when all the inputs and outputs of the delays are equal, the system is said to be in its stable state. The set of variables $x_1, x_2, \ldots$ is called the *input state* while the $y_1, y_2, \ldots$ are called the *internal state*.

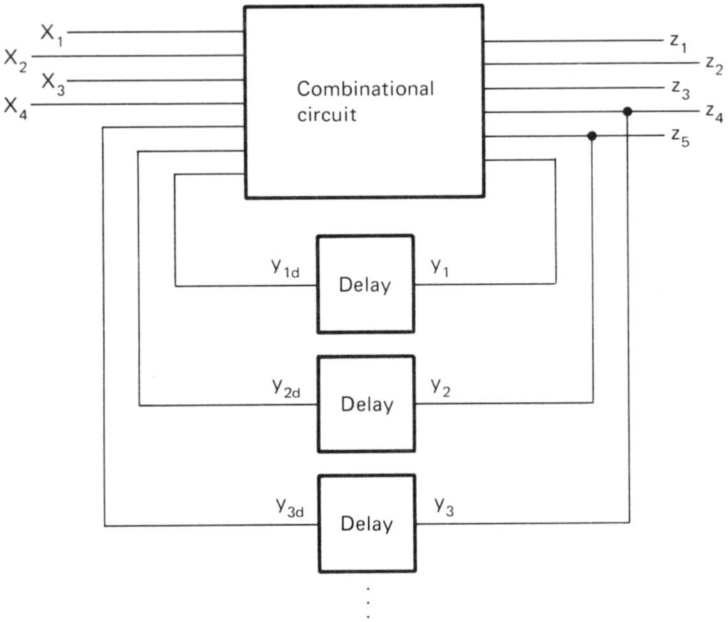

Figure 6-47. The representation of a simple asynchronous sequential circuit.

If a system is operating in a stable state, then it can only change its state if the input state changes. We shall restrict our operation in the following ways: (1)Only one input variable can change at a time; (2)no other change in any input variable can occur until the system has settled down to a stable state. When the operation satisfies conditions (1)and (2), it is said to be *fundamental mode operation.*

Now let us consider the design of a fundamental mode sequential circuit. The specifications for sequential circuit are often done by specifying the output signal for a given sequence of input signals. For instance, suppose that we have two input signals x_1 and x_2 and two output signals y_1 and y_2. Suppose that we want a logic circuit to perform in the following way:

$$y_1=1 \text{ if and only if } x_1=1 \text{ and } x_2=1, \text{ following } x_1=0, x_2=1 \qquad (6\text{-}32a)$$

$$y_2=1 \text{ if and only if } x_1=1 \text{ and } x_2=1, \text{ following } x_1=1, x_2=0 \qquad (6\text{-}32b)$$

There are five possible states of input variable combinations; ($x_1=0,x_2=0$), ($x_1=1,x_2=0$), ($x_1=0,x_2=1$), ($x_1=1,x_2=1$, following $x_1=0,x_2=1$) and ($x_1=1,x_2=1$, following $x_1=1,x_2=0$). Note that we only consider different states for different input *sequences* for the states immediately preceding the input state $x_1=1,x_2=1$, since any other sequence has no effect on the output. A timing diagram for a typical sequence is shown in Fig. 6-48. We assume that the initial state is $x_1=0,x_2=0$. The five states are designated by number which is shown in Fig. 6-48. We list them here

1	$x_1=0$	$x_2=0$	
2	$x_1=0$	$x_2=1$	
3	$x_1=1$	$x_2=1$	following $x_1=0,x_2=1$
4	$x_1=1$	$x_2=0$	
5	$x_1=1$	$x_2=1$	following $x_1=1,x_2=0$

The next step in the design is to set up some tables, called *primitive flow tables*. These show both the transition between states as well as the stable states. Note that, in the design, we are primarily concerned with the input states. These will be utilized to obtain the desired output. We will start by illustrating just a partial table. (Remember that this relates the input states. Thus, it differs from tables that we have previously considered.)

TABLE 6-26
PARTIAL PRIMITIVE FLOW TABLE

"Present State"	Input x_1x_2 0 0	0 1	1 1	1 0	
1	(1)/0,0 →	2 ↓	-		successor state/output y_1y_2
2		(2)/0,0 →	3 ↓		
3			(3)/1,0		

In this table we show the "present state", then the "next state", and the resulting state. The output is also shown. However, we only list the output for stable states. We shall discuss this subsequently.

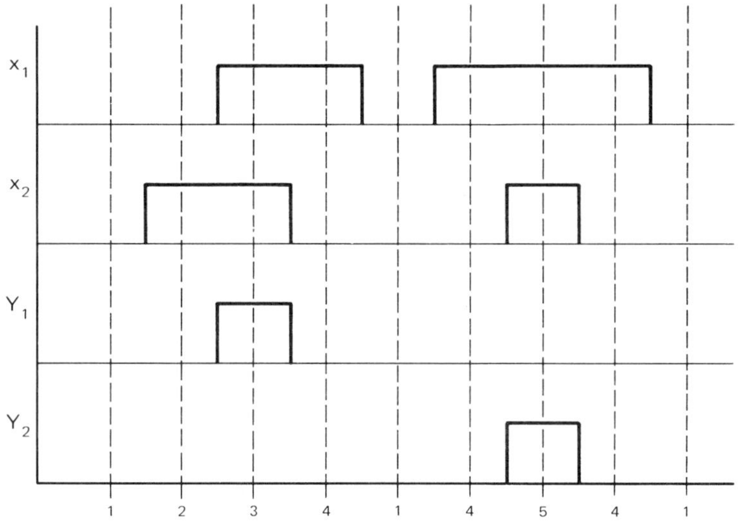

Figure 6-48. A typical timing diagram showing the assigned states for the network specified in (6-32).

Now let us consider Table 6-26. Let us start with a present input state of 1 (i.e., $x_1=x_2=0$). Now suppose that the input is switched to $x_1=0, x_2=1$. This will lead to state 2 . We note this by drawing a horizontal move into the $x_1=0,x_2=1$ column. This is labeled 2. However, it is not a stable state since this corresponds to a "present input" of $x_1=0,x_2=0$. Thus, the next move is a vertical one to the "present state" 2 line. Now we encircle the state since the present input and states are the same. Thus, this is a stable state. The output is also shown for the stable state. Next we assume that the input changes to $x_1=1,x_2=1$. Again, a similar set of transitions is shown.

Note the dash - in the 1 1 portion of the first row. We cannot make the change from $x_1=0,x_2=0$ to $x_1=1,x_2=1$ since only one input at a time is allowed to change. Thus, we put a dash in that position. Note that this is a type of "don't care" since this input condition can never occur. We can put down any state we wish there.

Now let us form the complete primitive flow table. We do not show arrows in this table to avoid cluttering it.

TABLE 6-27
PRIMITIVE FLOW TABLE

"Present input state"	Input x_1x_2			
	0 0	0 1	1 1	1 0
(1)	(1)/0,0	2	-	4
(2)	1	(2)/0,0	3	-
(3)	-	2	(3)/1,0	4
(4)	1	-	5	(4)/0,0
(5)	-	2	(5)/0,1	4

This flow table can be used in the synthesis of the network. However, before this is done, we shall reduce the table since this will result in a simplified network.

Reduction of the Primitive Flow Table

Suppose that two rows of the flow table are identical. This would indicate that the same new inputs would result in the same successor states. Thus, it does not matter which state we were in originally. In this case, we can combine rows. Then there will be fewer states and the table will be simpler. This is similar to the discussion in Sec. 6-5. There we indicated that the merging of rows could only take place if the rows, including the output, were identical. Essentially the same holds true here. However, except for stable states, the output is not specified. Note that there will be one and only one stable state in each row. We will not be able to merge rows which have different stable states in the same column. However, for the unstable state, we do not have to consider the outputs. Actually, there are times when we want to specify the outputs during the transition times. We shall consider this subsequently.

Since Table 6-27 is short, the merging of rows can be done by inspection. It is desirable to systematize the procedure, since this is helpful when large tables are considered.

We draw a diagram called a *merger diagram*. In this, we put down all the states. Then, each row is compared with all the others and any compatible(i.e., combinable)pair of states is connected by a straight line, see Fig. 6-49. Any two states can be combined if they are connected by a straight line. Similarly, if there are three states, *each* one of which is connected to(all) the other two, then these three rows can be merged. In addition, if there are four states, *each* of which is connected to all of the other three states, then these can be merged. In general, if there are n states, *each* of which are connected to *all other* n-1 states, then these states can be merged.

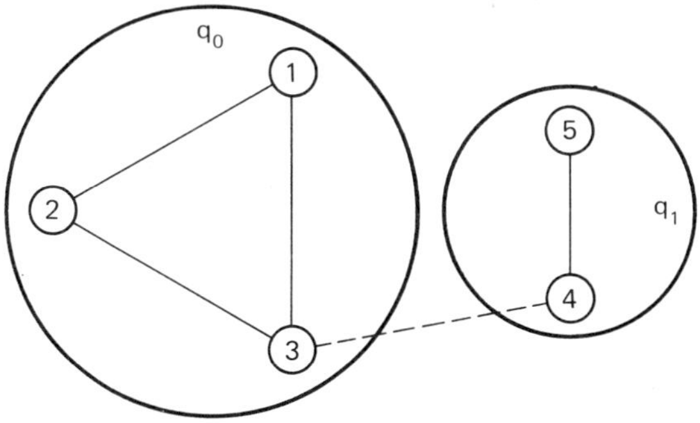

Figure 6-49. Merger diagram for the primitive flow table, 6-27. The dashed line should be ignored.

The merger indicated in Fig. 6-49 is the obvious one to use. However, suppose that the extra line shown dashed was added to the diagram. Then, we could divide these into two states q_0 and q_1 as shown, or divide into three states, 1 and 2, 3 and 4, and 5. Probably, the first division (indicated on the figure) is preferable since only two states are needed. However, in complex cases, several trial designs may have to be made. We shall merge as shown in Fig. 6-49. Now we merge the rows as indicated. This yields

TABLE 6-28
MERGED FLOW TABLE

Present state	Present input x_1x_2 0 0	0 1	1 1	1 0	
q_0	(1)/0,0	(2)/0,0	(3)/1,0	4	successor states of x_1 and x_2 and outputs
q_1	1	2	(5)/0,1	(4)/0,0	

Remember that these tables give the input states plus the output. Now let us consider q_0 and q_1 which represent the internal state of the system. That is, each row in the merged flow table represents a different mode of response of the system. Hence, it must be a different internal state. A transition diagram representing this is shown in Fig. 6-50. Just as in the case of synchronous sequential circuits, we must now assign binary variables to the states. To avoid having races, we must attempt to do this in such a way that no two internal state variables can change simultaneously. Of course, since we have only two internal state variables, we cannot have a race. Let us assign

$$q_0 = 0 \tag{6-33a}$$

$$q_1 = 1 \tag{6-33b}$$

We call the internal state y. To synthesize the network, we shall use Table 6-28 to set up the following truth table

TABLE 6-29
TRUTH TABLE USED TO REALIZE THE NETWORK WHOSE MERGED FLOW TABLE IS GIVEN IN TABLE 6-28

Input		Current state	Next state
x_1	x_2	y	y
0	0	0	0
0	1	0	0
1	0	0	1
1	1	0	0
0	0	1	0
0	1	1	0
1	0	1	1
1	1	1	1

Figure 6-50. State transition diagram for the states of Table 6-28.

The Karnaugh map for this table is given in Fig. 6-51. Thus,

$$y = x_1\bar{x}_2 + x_1 y \tag{6-34}$$

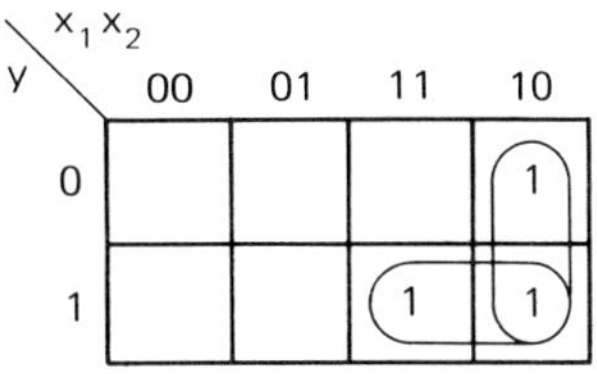

Figure 6-51. The Karnaugh map for Table 6-29.

Note that, when the Karnaugh map minimization procedure is used, it should be done in such a way that hazards are avoided, see

Sec. 6-6. Now let us consider the output. Compare Tables 6-27 and 6-28. The z_1 output will only be 1 when $y=0, x_1=1, x_2=1$. Similarly, z_2 will be 1 only when $y=1$ and $x_1=1, x_2=1$. Thus, we can write

$$z_1 = x_1 x_2 \bar{y} \tag{6-35a}$$

$$z_2 = x_1 x_2 y \tag{6-35b}$$

The network realization then takes the form shown in Fig. 6-52.

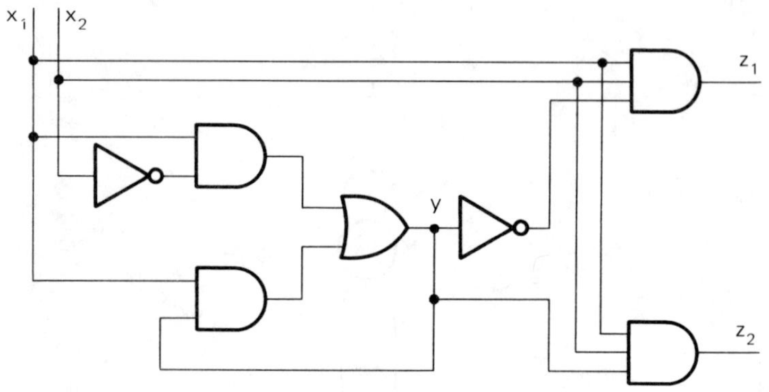

Figure 6-52. A network that realizes the requirements of Eqs. (6-32).

To provide a more complex example, let us not combine the last two rows of Table 6-27. Thus, we have

TABLE 6-30
MERGED FLOW TABLE

Present state	Present inputs x_1x_2 0 0	0 1	1 1	1 0
q_0	(1)/0,0	(2)/0,0	(3)/1,0	4
q_1	1	-	5	(4)/0,0
q_2	-	2	(5)/0,1	4

Then, the state transition diagram is shown in Fig. 6-53. There are three states. Thus, we need two binary variables. We did not assign the state (1,1) to avoid having a race condition. In general, the state assignment should be such that there are no races. If this cannot be done, then extra delay can be incorporated, see Sec. 6-6. The states are

$$q_0 = (0,0) \tag{6-36a}$$

$$q_1 = (0,1) \tag{6-36b}$$

$$q_2 = (1,0) \tag{6-36c}$$

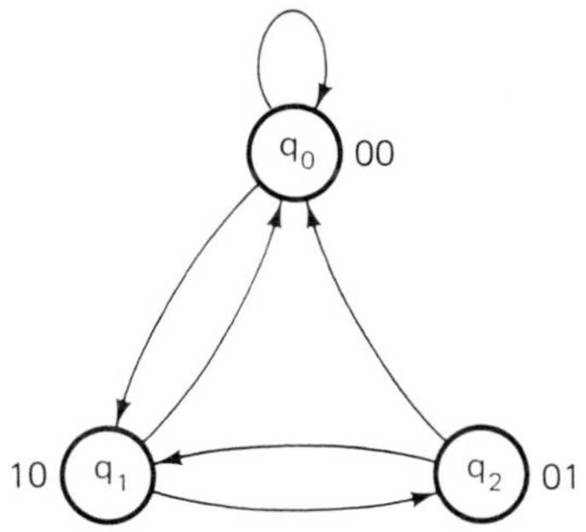

Figure 6-53. A state transition diagram for the states of Table 6-30.

Then, the truth table is

TABLE 6-31
TRUTH TABLE USED TO REALIZE THE NETWORK WHOSE MERGED FLOW TABLE IS GIVEN IN TABLE 6-30

Input		Present state		Next state	
x_1	x_2	y_1	y_2	y_1	y_2
0	0	0	0	0	0
0	1	0	0	0	0
1	0	0	0	0	1
1	1	0	0	0	0
0	0	0	1	0	0
0	1	0	1	-	-
1	0	0	1	0	1
1	1	0	1	1	0
0	0	1	0	-	-
0	1	1	0	0	0
1	0	1	0	0	1
1	1	1	0	1	0
0	0	1	1	-	-
0	1	1	1	-	-
1	0	1	1	-	-
1	1	1	1	-	-

The Karnaugh maps for y_1 and y_2(which are obtained from the lines where y_1=1 and y_2=1, respectively)is given in Fig. 6-54. Thus, we have

$$y_1 = x_2y_2 \tag{6-37a}$$

$$y_2 = x_1\bar{x}_2 \tag{6-37b}$$

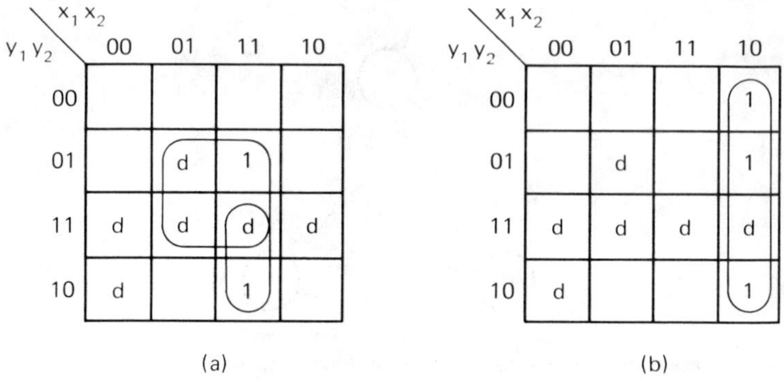

Figure 6-54. Karnaugh maps for Table 6-31. (a) for y_1; (b) for y_2.

Then, from Table 6-30, we obtain

$$z_1 = \bar{y}_1\bar{y}_2x_1x_2 \tag{6-38a}$$

$$z_2 = y_1\bar{y}_2 \tag{6-38b}$$

The gate circuit realization is shown in Fig. 6-55. Note that this realization uses the *same* number of gates as does Fig.6-52. In general, the reduction of the primitive flow table may not result in the reduction of the number of gates. All that can be stated is that reduction of the primitive flow table reduces the number of internal states. If these states are represented by flip-flops, then the reduction of the primitive flow table can result in a minimum number of gates.

At times, we want to specify the output during the intermediate unstable states. This can eliminate considerable output transients during the transition between states. In this case, separate states must be defined for these intermediate values. In general, fewer rows of the primitive flow table can be merged in this case.

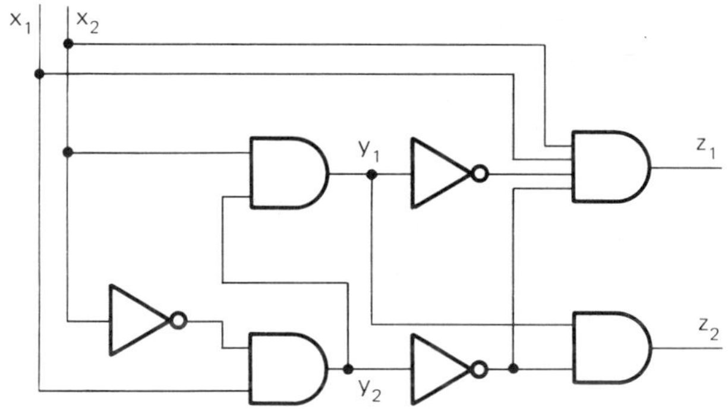

Figure 6-55. A network that realizes the requirements of Eqs. (6-32).

Pulse Mode Operation

We have discussed fundamental mode operation. There is another mode of operation that is used. In this case, the input signals consist of short pulses, which must satisfy two requirements: (1) they must be long enough so that the states can change; (2) they must be short enough so that all inputs are 0 after the memory elements have changed their states. This is called *pulse mode operation*. In this case, a pulse is applied at the input. The network changes its state. By the time that the circuit has settled down, the pulse must be removed. The network's state then remains unchanged until the next input. Design with circuits of this type is difficult because of the precise timing of the pulses that is required.

6.8 INTEGRATED CIRCUIT IMPLEMENTATION OF SEQUENTIAL CIRCUITS

In this chapter we have discussed the fundamental ideas behind the analysis and design of sequential circuits. Many complete sequential circuit can be purchased from manufacturers in the form of integrated circuits. In this section we will discuss some such commercially available integrated circuit sequential circuits. Since they are in integrated circuit form, the complete circuit is fabricated on a single chip. There are many such circuits available. We shall only consider a representative sample.

In Secs. 6-1 and 6-2 we discussed flip-flops. Figure 6-56 illustrates a commercially available type CD4027A COSMOS, J-K flip-flop chip. Actually, there are two flip-flops on a single chip. Figure 6-56a shows the basic pin connections. That is, each integrated circuit is mounted and leads are brought out. This diagram specifies the lead locations. In Fig. 6-56b, a logic circuit for one of the flip-flops is shown, and in Fig. 6-56c, the actual circuit is shown. Note that the dual number, e.g., 11/5, at the terminals of Figs. 6-56b and c refers to terminal connections (i.e., terminal assignments). (There are two sets of numbers since there are two flip-flops on the chip.)

The boxes marked TG on Fig. 6-56b are toggles. That is, they function as do T flip-flops. This J-K flip-flop also incorporates set and reset leads which can perform the set and reset functions independent of the clock pulses. Of course, the J and K inputs initiate action in accordance with the clock pulses.

Now let us consider an integrated circuit shift register. In Fig. 6-57a we illustrate a type SN54165 eight bit TTL shift register. Note that this shift register can receive its data either serially or in parallel. However, provision is only made for

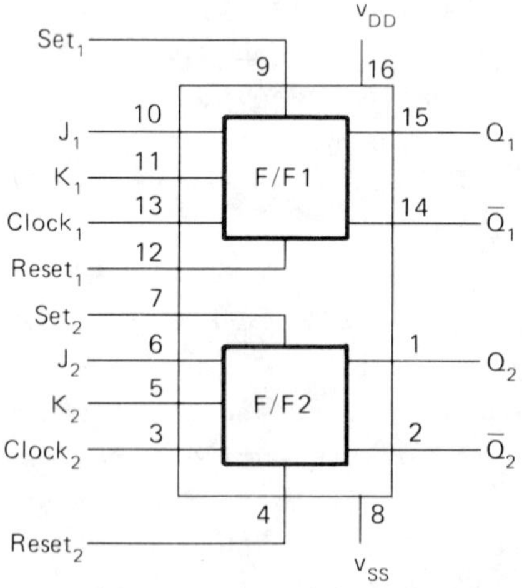

(a)

Figure 6-56. A COSMOS dual J-K flip-flop Type CD4027A. (a) Pin diagram; (b) logic circuit; (c) schematic diagram (Courtesy RCA, Inc.).

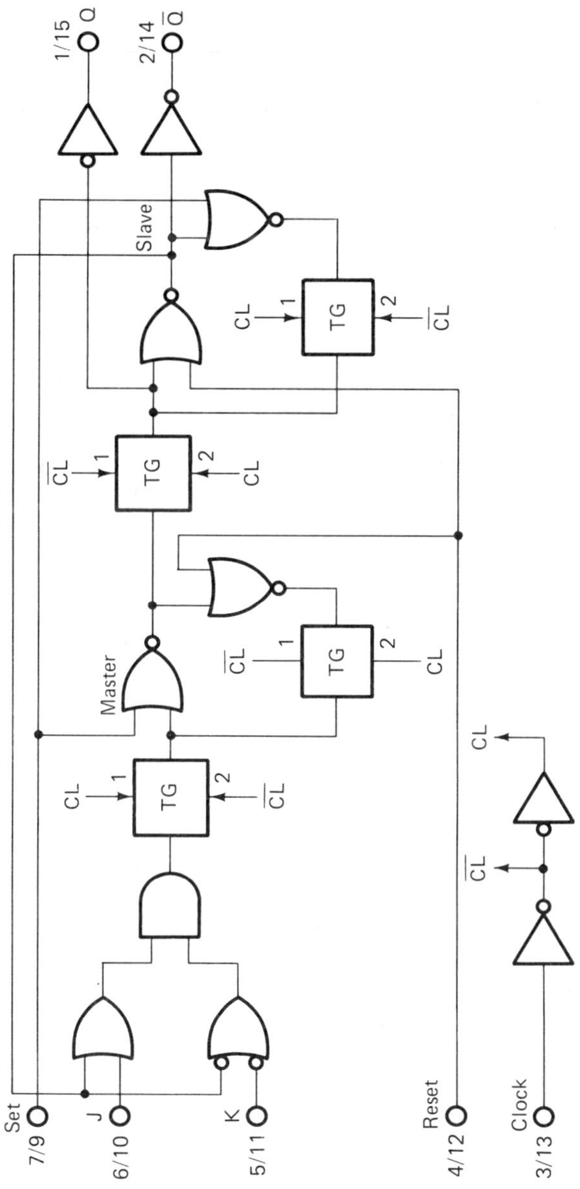

Figure 6-56. (Continued) part (b).

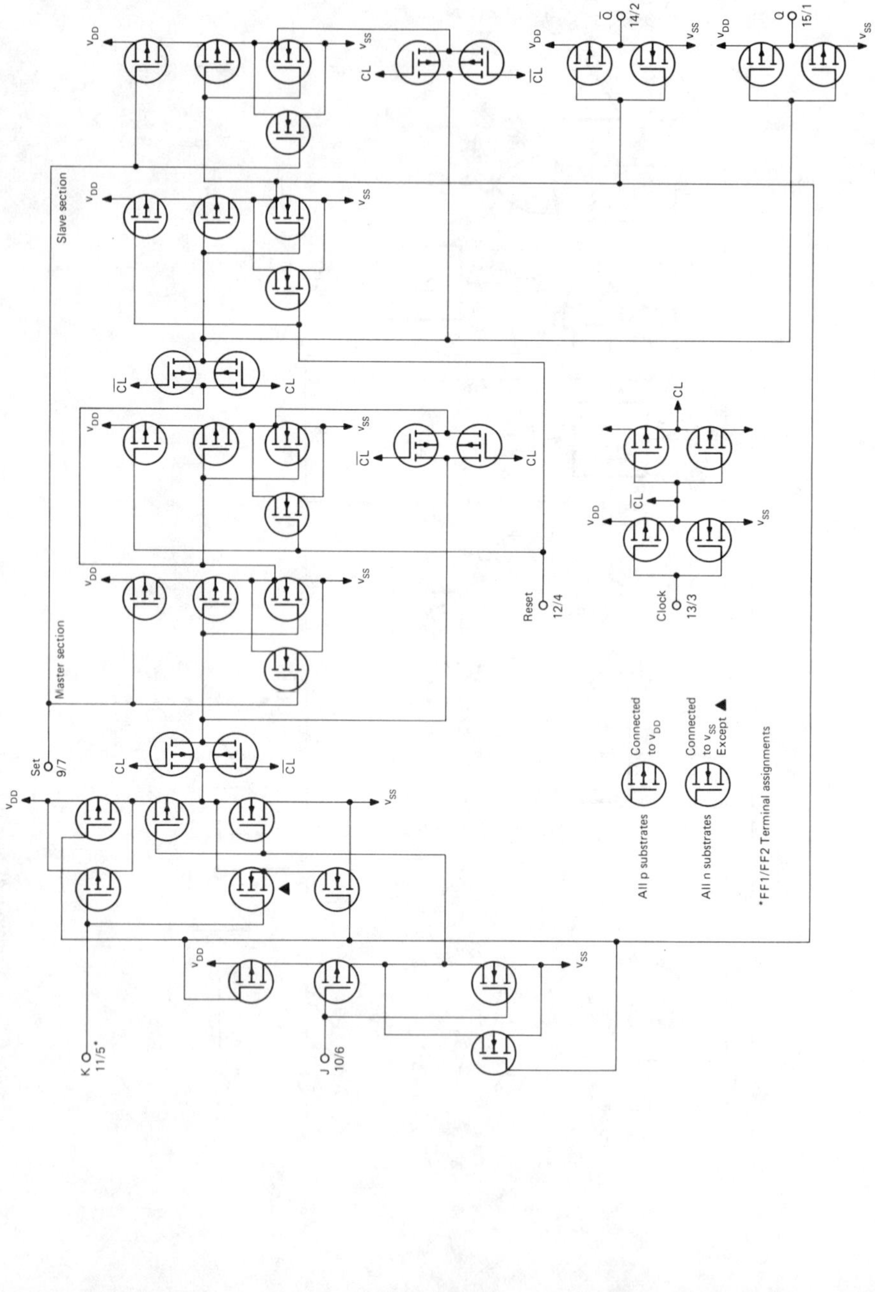

Figure 6-56. (Continued) part (c).

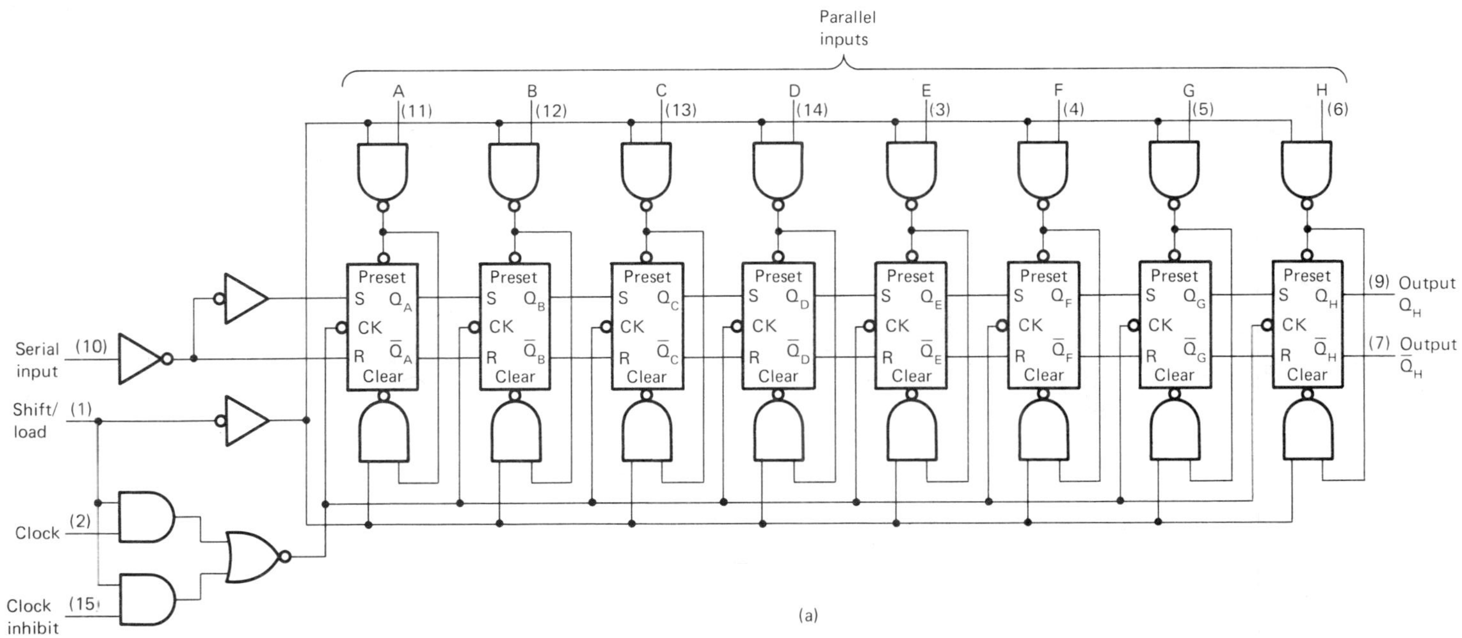

Figure 6-57. A TTL eight bit shift register Type SN54165. (a) the logic diagram; (b) typical input circuits; (c) typical output circuit; (d) function table (Courtesy Texas Instruments, Inc.).

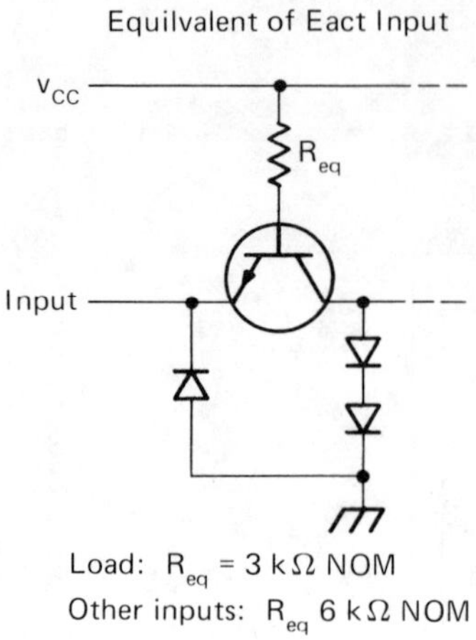

(b)

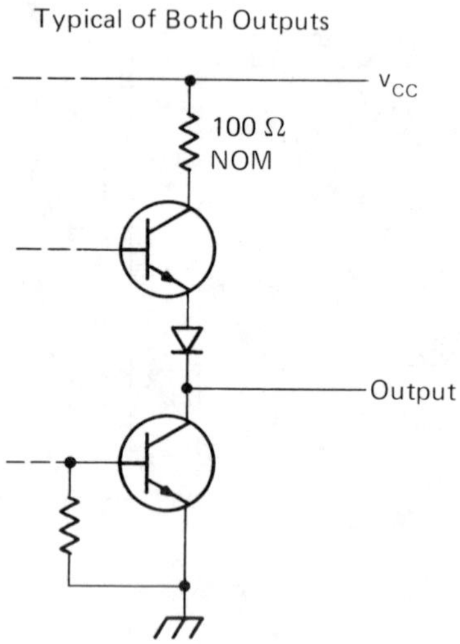

(c)

Function Table

Inputs					Internal Outputs		Output
Shift/ load	Clock inhibit	Clock	Serial	Parallel A . . . H	Q_A	Q_B	Q_H
L	X	X	X	a . . . h	a	b	h
H	L	L	X	X	Q_{A0}	Q_{B0}	Q_{H0}
H	L	↑	H	X	H	Q_{An}	Q_{Gn}
H	L	↑	L	X	L	Q_{An}	Q_{Gn}
H	H	↑	X	X	Q_{A0}	Q_{B0}	Q_{H0}

(d)

Figure 6-57. (Continued) part (b), (c) and (d).

serial output. (Parallel output would only require access to the points marked $Q_A, Q_B, \ldots, Q_G$ and their complements.) In Figs. 6-57b and c, typical input and output circuits are given. In addition, data such as that discussed at the end of Sec. 5-16 is also provided. Thus, the interaction of this shift register with the other elements in the circuit can be determined.

An abbreviated truth table is also given in the specifications. This is shown in Fig. 6-57d. The column entitled "internal outputs" lists the condition of the internal flip-flop's outputs. However, for this particular integrated circuit, these (parallel) outputs are not available. In the function table, L stands for low and H for high, see Sec. 5-2. If positive logic is used, the H corresponds to a 1 and L to a 0. With negative logic, the reverse is true. The X's in the table are "don't cares". The symbols Q_{An} and Q_{Gn} are equal to the levels of Q_A and Q_G before the most recent clock pulse. The symbols $Q_{A0}, \ldots Q_{G0}$ are equal to the values of $Q_A, \ldots Q_G$ *before* the indicated steady state conditions were applied. The upward pointing arrow indicates a transition from a low to a high level. Note that this only occurs in the clock column and indicates that the transition is initiated by the leading edge of the clock pulse as it switches from low to high.

We have illustrated two integrated circuits in this section. Of course, very many other integrated circuits are available. For instance, there are counters, parallel and series input and output shift registers, etc. The data given by manufacturers may differ in detail. However, similar methods are used to present the information.

BIBLIOGRAPHY

Booth, T. L., "Digital Networks and Computer Systems", Chapter 7 and 8, John Wiley and Sons, Inc., New York 1971

Givone, D. D., "Introduction to Switching Circuit Theory", Chapters 8-10, 12, McGraw-Hill Book Co., New York, 1970

Klir, G. J., "Introduction to the Methodology of Switching Circuits" Chapters 7-9, D. VanNostrand Co., New York, 1972

Mano,M.M., "Computer Logic Design", Chapters 6-7, Prentice Hall, Inc., Englewood Cliffs,N.J.,1972

Rhyne,V.T., "Fundamentals of Digital System Design", Chapters 5-6, Prentice Hall, Inc., Englewood Cliffs, N.J., 1973

PROBLEMS

6-1. Discuss the difference between combination and sequential circuits.

6-2. Discuss the difference between synchronous and asynchronous sequential circuits.

6-3. Determine the behavior of the R-S flip-flop for all possible inputs and all possible states

6-4. Determine a logic circuit which will produce an output of 1 if a four digit binary number is equal to or greater than 7_{10}. Use an R-S flip-flop. (Actually, this could be realized by a combinational circuit but it is done in this way as an exercise.)

6-5. Repeat Prob.6-4 but now obtain an output of 1 if the number is more than 6 or less than 3.

6-6. Repeat Prob. 6-4 using a D flip-flop.

6-7. Repeat Prob. 6-5 using a D flip-flop.

6-8. Repeat Prob. 6-4 using a J-K flip-flop.

6-9. Repeat Prob. 6-5 using a J-K flip-flop.

6-10. Repeat Prob. 6-4 using a T flip-flop.

6-11. Repeat Prob. 6-5 using a T flip-flop.

6-12. Deisgn a logic circuit which functions as a J-K flip-flop. That is, it has two inputs J and K, etc., but which uses a D flip-flop in its implementation.

6-13. Repeat Prob.6-12 now using a T flip-flop.

6-14. Discuss the advantage of master-slave type flip-flops.

6-15. Draw a timing diagram for all inputs and states of the R-S master-slave flip-flops.

6-16. Repeat Prob. 6-15 for the J-K master-slave flip-flop of Fig.6-15.

6-17. Repeat Prob. 6-16 for the J-K master-slave flip-flop of Fig. 6-16.

6-18. Discuss the differences in the type of gate circuits(e.g., TTL)that could be used to realize the J-K flip-flop of Figs. 6-15 and 6-16.

6-19. Obtain the transition table and transition diagram for the network of Fig. 6-58

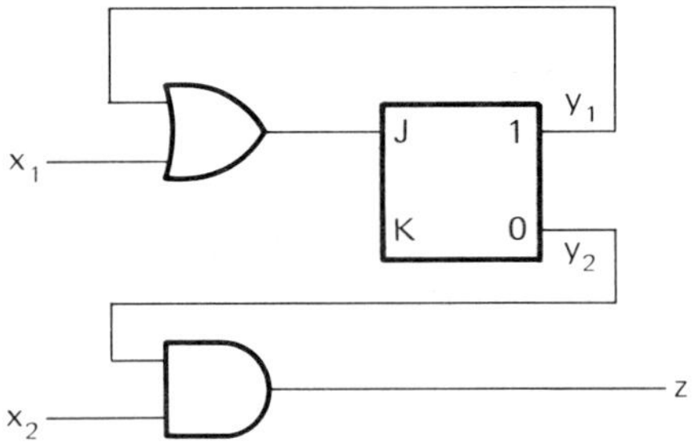

Figure 6-58.

6-20. Repeat Prob. 6-19 for the network of Fig. 6-59.

6-21. Discuss the difference between series and parallel operation of registers.

6-22. Modify the shift register of Fig. 6-25 to use J-K flip-flops.

6-23. Repeat Prob. 6-22 using T flip-flops.

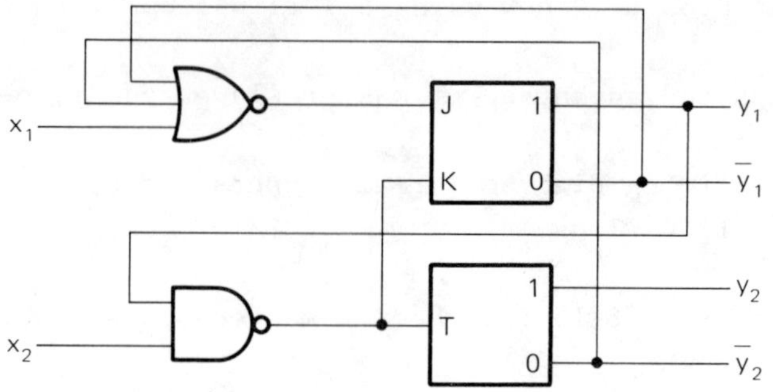

Figure 6-59.

6-24. Design a three bit register which is controlled by four input signals, c_0, c_1, c_2, c_3, in the following way:

c_0 = 1, enter right , shift left
c_1 = 1, enter left , shift right
c_2 = 1, set register to 101
c_3 = 1, enter sequence 000(clear)
all c = 0 no shift

Only one c will be nonzero at a time.

6-25. Modify the shift register of Prob. 6-24 so that parallel transfer can be used. An additional control signal will be required.

6-26. Realize a network using a D flip-flop that realizes the transition table of Prob. 6-19.

6-27. Repeat Prob. 6-26 for the transition table of Prob. 6-20.

6-28. Repeat Prob. 6-27 but now use T flip-flops.

6-29. Design a network using D flip-flops, which realizes the transition diagram of Fig. 6-60. Use the following assignments: q_0=0,0; q_1=0,1; q_2=1,0; and q_3=1,1.

6-30. Repeat Prob. 6-29 but now use J-K flip-flops.

6-31. Repeat Prob. 6-29 but now use T flip-flops.

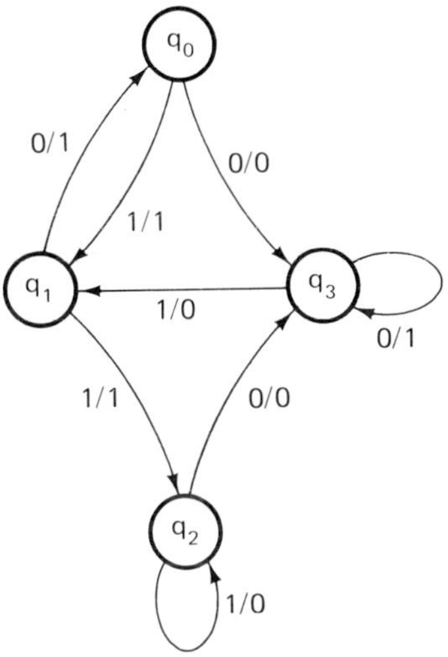

Figure 6-60.

6-32. Repeat Prob. 6-29 but now use the state assignments q_0=1,1; q_1=0,1; q_2 = 0,0; and q_3 = 1,0.

6-33. Repeat Prob. 6-30 but now use the state assignments of Prob. 6-32.

6-34. Repeat Prob. 6-31 but now use the state assignments of Prob. 6-32.

6-35. Design a network using J-K flip-flops, which realizes the transition diagram of Fig. 6-61. Use the following state assignments:

q_0=0,0,0; q_1=0,0,1; q_2=0,1,0; q_3=0,1,1;
q_4=1,0,0; q_5=1,0,1; q_6=1,1,0; q_7=1,1,1

6-36. Repeat Prob. 6-35 but now use D flip-flops.

6-37. Repeat Prob. 6-35 but now use T flip-flops.

6-38. Repeat Prob. 6-35 but now use the following assignment of states:

q_0=0,0,1; q_1=0,0,0; q_2=0,1,0; q_3=0,1,1;
q_4=1,0,1; q_5=1,0,0; q_6=1,1,1; q_7=1,1,0

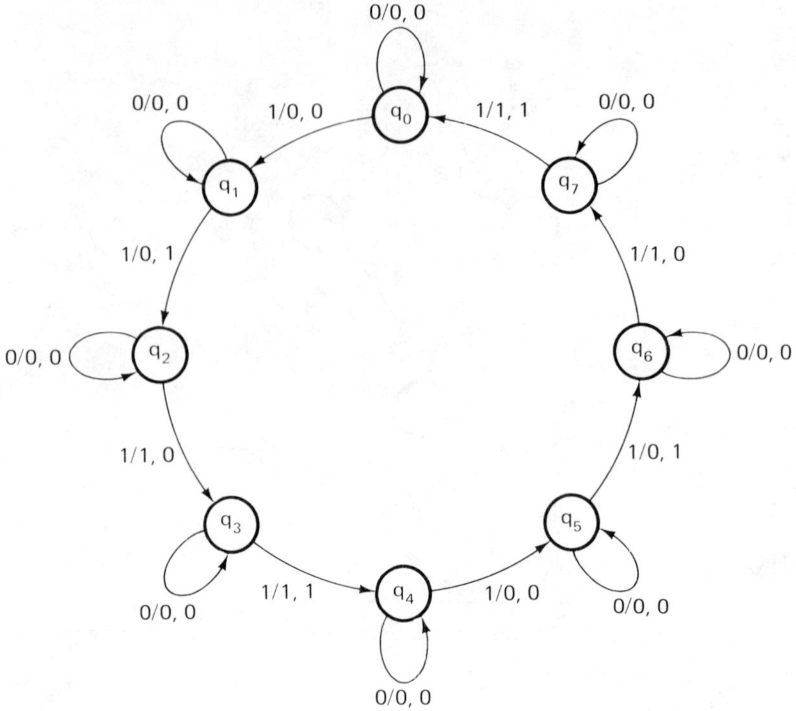

Figure 6-61.

6-39. Repeat Prob. 6-36 but now use the state assignments of Prob. 6-38.

6-40. Repeat Prob. 6-37 but now use the state assignments of Prob. 6-38.

6-41. Design a counter which produces the following sequence:

001, 000, 010, 011, 100, 111, 110, 101

and then repeats. Use J-K flip-flops.

6-42. Repeat Prob. 6-41 but now use D flip-flops.

6-43. Repeat Prob. 6-41 for the sequence

001, 010, 111, 110, 011

6-44. Repeat Prob. 6-43 but now use T flip-flops.

6-45. Design a circuit which will produce an output of 1 whenever the sequence 0110 occurs at its input. Use J-K flip-flops.

6-46. Repeat Prob. 6-45 for the sequence 0111.

6-47. Repeat Prob. 6-45 but now use D flip-flops.

6-48. Repeat Prob. 6-46 but now use T flip-flops.

6-49. Design a circuit which will receive an input of five pulses, one at each clock pulse, and produces an output of 1 if a majority of them are 1's. Assume that the input stops after 5 pulses.

6-50. Repeat Prob. 6-49 but have the circuit reset itself after each five clock pulses. The circuit then produces a second output after the next five input pulses, etc.

6-51. Determine an excitation table for a network characterized by the switching functions

$$y_1 = x_1 + x_2\bar{y}_2$$

$$y_2 = \bar{y}_2 + x_2$$

Examine it for races and oscillations.

6-52. Repeat Prob. 6-51 for

$$y_1 = x_1 + \bar{x}_2 + \bar{y}_2 + \bar{y}_1$$

$$y_2 = x_1 + x_2\bar{y}_1 + \bar{y}_2 y_1$$

6-53. The Karnaugh map of Fig. 6-62 is to be used to realize a function y. Obtain a minimal gate realization. Then obtain one which is hazard free.

x_3 \ x_1x_2	00	01	11	10
0	1			
1	1	1	1	

Figure 6-62.

6-54. Repeat Prob. 6-52 for the Karnaugh map of Fig. 6-63.

x_3x_4 \ x_1x_2	00	01	11	10
00		1	1	
01			1	
11	1	1	1	
10				

Figure 6-63.

6-55. Discuss the compromises made in obtaining a hazard free network.

6-56. Describe the differences between levels and pulses.

6-57. Design a fundamental mode asynchronous sequential circuit with two inputs x_1 and x_2 and two outputs z_1 and z_2. The z_1 output is to be a 1 if and only if $x_1=1$ and $x_2=0$ which follows $x_1=1$, $x_2=1$, and the z_2 output is to be a 1 if and only if $x_1=1, x_2=0$ following $x_1=0, x_2=1$.

6-58. Repeat the design of Prob. 6-57 with the following specifications on the output. z_1 is to be a 1 if and only if $x_1=1, x_2=1$ and this state follows the sequence $x_1=0, x_2=1, x_1=0, x_2=0$. $z_2=1$ if and only if $x_1=1$, $x_2=1$ which follows $x_1=1, x_2=0$.

7

Memories

An important part of any digital computer is the *memory* in which data can be stored. In this chapter, we shall consider the various types of memories that are used.

The register discussed in the last chapter can be used as a form of memory. However, it is common usage to associate the word memory with a device which can store many thousands of bits. Since a register stores only a small number of bits, it is usually not called a memory.

Memories are made up of elements, each of which can store a single bit. The flip-flop is such an element. Special magnetic elements are used also. We shall consider both types of memories in this chapter. The semiconductor flip-flop memory has the advantage that it is faster and can be fabricated using integrated circuit techniques. Thus, it is also small. The magnetic memories, at the present time, are cheaper to construct and they will not lose their stored information in the event of a power failure. For this reason, these memories are said to be *nonvolatile*, while the semiconductor memories are called *volatile* memories.

In the future, as LSI techniques are improved, we shall see a large increase in the use of semiconductor memories. At the present time, magnetic core memories are widely used. However, semiconductor memories are used to some degree in most computers. In fact, relatively large memories can be obtained using LSI techniques.

The memories that we have discussed make up what is called the *main memory* or *inner memory*. This stores data and instructions that are used during the actual running of a computer program. There are other memories, called *auxiliary memories*, that are used to store information such as data to be used at a future time. When the computer program using this information is to be executed, the information is put into the main memory. Auxiliary memories usually consist of storage on magnetic drums, magnetic tape or magnetic discs. Other storage techniques such as those using magnetic bubbles show promise in this area.

At times, errors result when words are written on, or read from, magnetic memories. The frequency of occurrance of these errors is very low. However, such reading and writing is done so often that error detecting and/or correcting codes are almost always used in conjunction with these memories. Such codes will be discussed in Sec. 8-4. At least one bit must be used for a parity check bit. However, there are times when many more check bits are used. For instance, 8 bits have been used in a Hamming error correcting code in a memory. The use of so many bits for error correcting may seem wasteful. However, it may be necessary to insure reliable operation.

7.1 MAGNETIC CORE MEMORIES

The basic element in a magnetic core memory is a small doughnut shaped piece of ferromagnetic material called a *core*, see Fig. 7-1. These cores are made very small since this allows faster switching speed and reduces the size of the memory. The outer diameter is typically 0.5 mm or less.

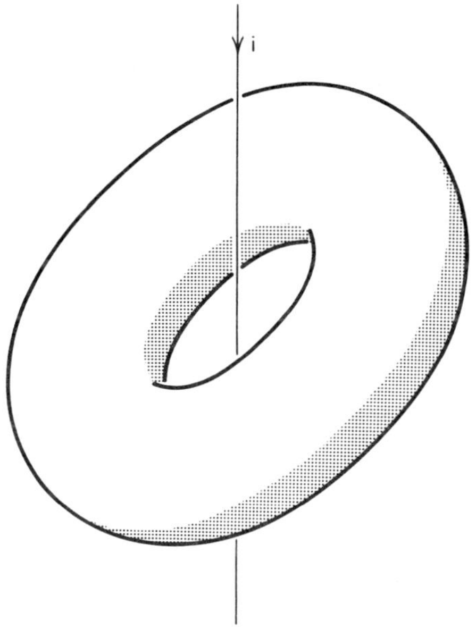

Figure 7-1. A magnetic core.

In an actual memory, several wires are passed through the core. At present, let us assume that there is only one wire. When a current i is in the wire, a magnetic flux will be set up in the core. The operation of ferromagnetic cores is very nonlinear. A typical plot of magnetic flux versus current is shown in Fig. 7-2. This is called a *hysteresis loop*. If we start out with an unmagnetized core, the flux will build up along the dashed line. However, if we now vary the current, the solid curve will be traversed. Thus, once current has been passed through the core for the first time, we shall always be on the solid curve. If i is greater than a critical value, which we call I_c, the flux will essentially be at ϕ_1, which we call the 1 level (i.e., the level corresponding to a 1). If i is less than $-I_c$, then the flux will be $-\phi_1$ which corresponds to a 0. Hence, ϕ can be "flipped" between ϕ_1 and $-\phi_1$. Therefore, the core can act as a flip-flop element and store information just as an ordinary flip-flop does.

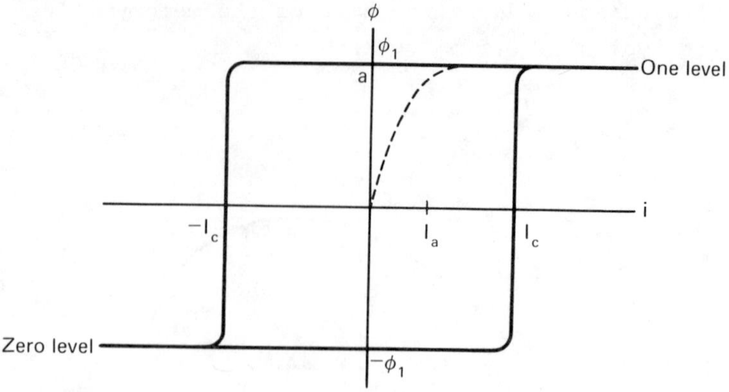

Figure 7-2. The hysteresis loop of a magnetic core.

Let us study the hysteresis loop in detail. Suppose that $i=I_c$ and then we reduce i to 0. We will move to point a. Note that ϕ remains essentially constant at ϕ_1. If we start at $i=I_c$ and then change the current to any I_1 where

$$I_1 > -I_c \tag{7-1}$$

then the flux will remain essentially constant at ϕ_1.

If we make i more negative than $-I_c$, then the flux will switch to $-\phi_1$, the 0 level. It will essentially remain at this level until we make the current more positive than I_c, at which point the flux will switch to $+\phi_1$. Note that turning off the current does not change the flux. Thus, even if the power supply is interrupted, the information stored in the memory is not lost. This is why magnetic cores are nonvolatile memories. Then, in order to switch the core from its 0 level to its 1 level, we must apply a current greater than I_c. That is,

$$i > I_c \tag{7-2a}$$

To switch the core from its 1 level to its 0 level, we must supply a current less than $-I_c$.

$$i < -I_c \tag{7-2b}$$

The current is in the form of pulses. If the pulse height satisfies Eqs.(7-2) and is long enough, then the core will be switched.

We must have a way of determining if the core is at a 0 level or at a 1 level. Suppose that a second wire, called a *sense wire*, is passed through the core, see Fig. 7-3. This wire carries no current. If the flux changes, then a voltage will be induced in the sense wire. This voltage is proportional to the rate of change of the flux

$$v_{ab} = k\frac{d\phi}{dt} \tag{7-3}$$

where k is a constant.

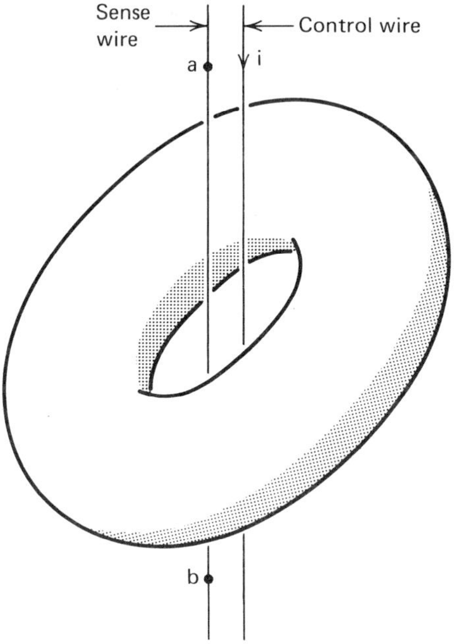

Figure 7-3. A magnetic core with a sense wire.

Now suppose that the core is at the 1 level with i=0. Now, to determine the state of the core, we pass a current $i < -I_c$ through the control wire. The flux will change from ϕ_1 to $-\phi_1$. This change in flux will induce a voltage v_{ab}. If the core had been at the 0 level, then the current $i < -I_c$ would have produced almost no change in flux and v_{ab} would almost be 0. Thus, by passing a current $i < -I_c$ through the control wire and observing the resultant voltage across the sense wire, we can determine the state of the core. Actually, v_{ab} is relatively small even when the core is switched from a 0 to a 1. The voltage v_{ab} is amplified by a *sense amplifier* to obtain a pulse of reasonable voltage. Of course, if the state of the core is 0, then the amplified v_{ab} will still be essentially 0.

Whenever we read the state of the core, we set its state to 0. This is called a *destructive readout*. (In flip-flops, the readout is *nondestructive*.) If we want to save the data stored in a core for further use, we must make provision to read the value back in. For instance, if the readout indicated that the core were at the 1 level, then, after readout, we must write a 1 back. That is, we must apply a pulse of current $i > I_c$ into the control line to reset the state to a 1. We shall consider the details of this subsequently.

In an actual memory, there are many cores. To facilitate the reading and writing, two control lines are used. (The reason for this will be explained subsequently.) Such a core is shown in Fig. 7-4a. A symbolic representation is shown in Fig. 7-4b. The actual spatial arrangement of the control wires is such that, if i_1 and i_2 are both positive, the core acts as though there were a current $i_1 + i_2$ through a single control wire.

Now let us define a current I_a where

$$I_c/2 < I_a < I_c \tag{7-4}$$

If both i_1 and $i_2 = I_a$, then the effective magnitizing current will be

$$i_1 + i_2 = 2I_a > I_c \tag{7-5}$$

In this case, the core will be switched. On the other hand, if I_1=0 and i_2=I_a, then the effective current will be I_a which is less than I_c and the core will not switch.

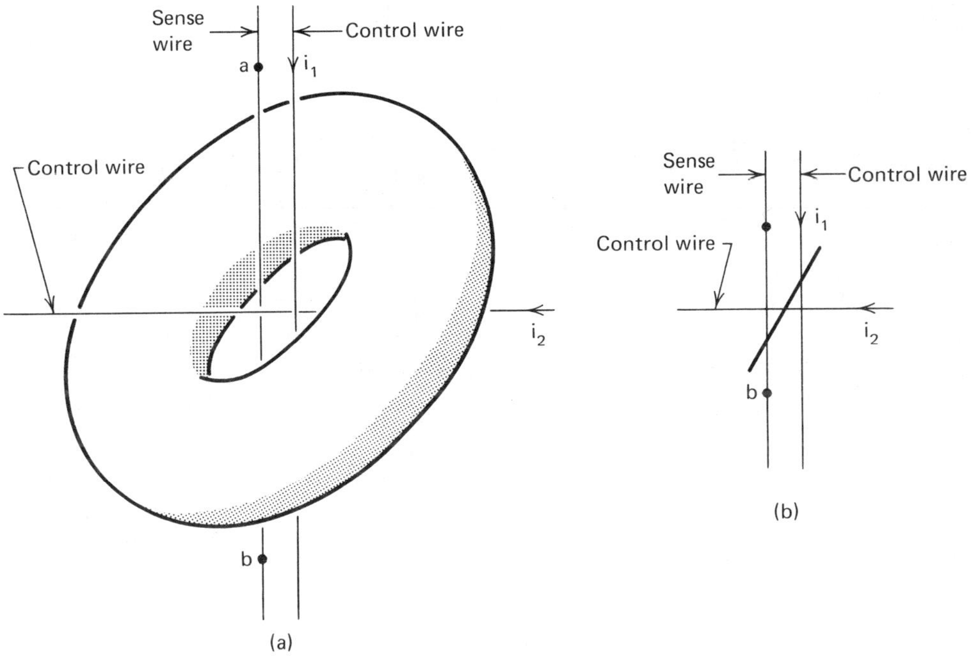

Figure 7-4. (a) A magnetic core with two control wires; (b) a diagrammatic representation of this.

Now let us consider a simple memory, that stores 16 bits. It is called a 2D memory since it is in a single plane. The memory is illustrated in Fig. 7-5. The x and y wires are control lines used to set the core to a 0 or to a 1 for reading and writing. If a pulse of I_a is put on both x_j and y_k, then the state of the core jk will be set equal to 1. For instance, if the x_0 and y_2 wires each carry a current of I_a, then the state of core 0,2 will be set to a 1. If both the x and y wires carry a current of I_a, the core is said to be *selected*. Now consider core 0,3. It only carries a current in one of its control wires. It is said to be *half selected*. Of course, this does not switch the core. Now, by sending current through the appropriate control wires, we can set the state of any desired core to 1. If the polarity of the control current is reversed, we can set the state of any desired core to 0.

In general, it can be shown that the most economical use of space and wires is made if the number of cores is an integral power of 2. Thus, almost all memories are made in accordance with this.

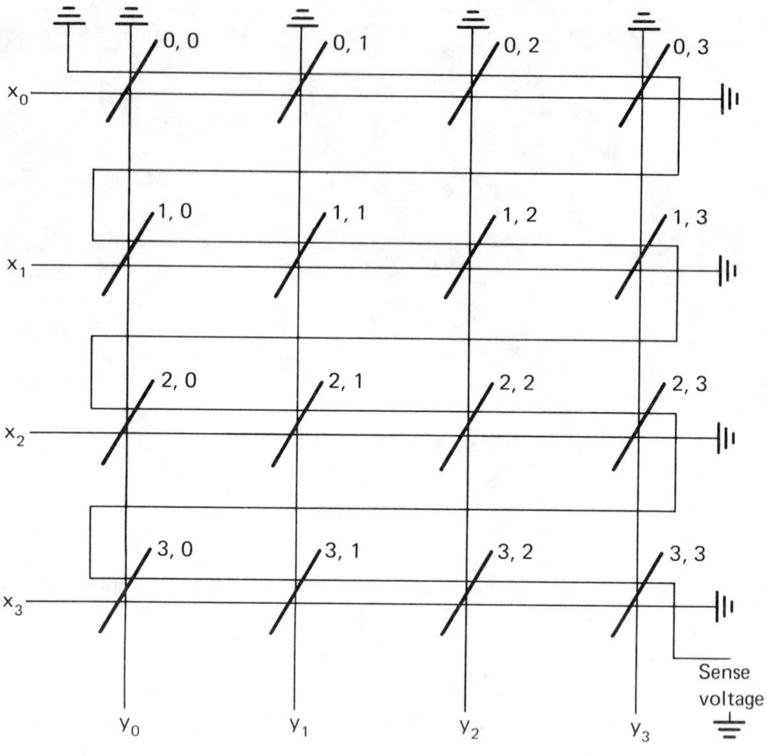

Figure 7-5. A basic 2D memory plane.

The sense wire passes through all the cores. Thus, when we want to read the value of a particular core, e.g., core 3,1, we send a current of $-I_a$ through both x_3 and y_1. The voltage appearing at the sense voltage output then gives the value stored in core 3,1. After reading, the state of core 3,1 is set to 0. Actually, the sense voltage is amplified and is then stored in a register called the *memory buffer register*. If this value is a 1, then instructions are generated to write a 1 on core 3,1. That is, a pulse of I_a is simultaneously sent down the x_3 and y_1 lines.

The location of a core in the memory is called its *address*. In the simple memory we have discussed, the x address would be controlled by two binary digits, as would the y. In Fig. 7-6, we illustrate a simple circuit that will select the proper control line in accordance with a two bit address. It will then apply a pulse of the proper polarity to that line. The input a_1a_0 is the binary representation of the subscript of x. The circuit is such that only one AND gate will have an output of 1. For instance, if $a_0=1$ and $a_1=1$, then only the output of AND 3 will be 1. The details of the operation will be left to the reader. Since the currents from the AND gates are usually not of sufficient amplitude to drive the control wires, amplification is used. The amplifiers, which are shown in Fig. 7-6, are such that the polarity of their output currents can be switched. This is done by the read-write control. Then, if we want to write a 1, the output will be I_a. If we want to read, or clear, then the amplified output will be $-I_a$. At times, other electronic configurations are used. However, Fig. 7-6 illustrates the basic ideas.

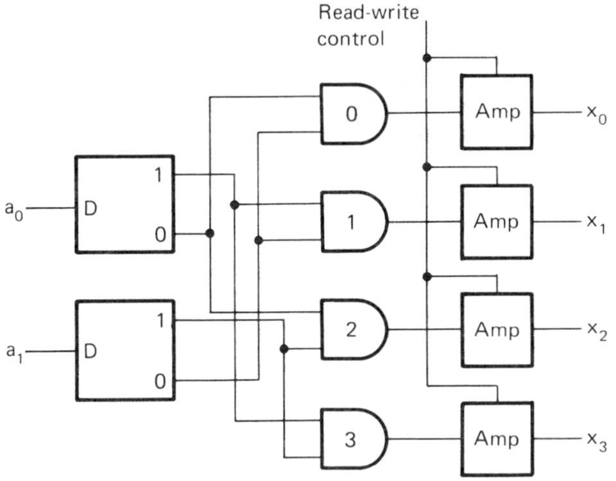

Figure 7-6. A simple control line driver.

When very many cores are to be used, as they are in a typical memory, then the 2D arrangement becomes cumbersome. Let us modify the simple 2D memory slightly and then see how it is used in a larger memory. Consider Fig. 7-5, but now assume that the sense wire is used to carry a current. Suppose that the current in x_j is I_a, the current in y_k is I_a, and the current in the sense wire is $-I_a$. The net magnetizing current through core j,k will be

$$I_a + I_a - I_a = I_a \tag{7-6}$$

Thus, the core will not be switched. The sense lead is now called an *inhibit lead*. Note that if there is current in the inhibit lead, then switching will not take place. If we make the current in the inhibit wire zero, then Eq.(7-6) becomes $I_a+I_a+0=2I_a$ and switching does take place. Adding the inhibit lead may seem to just add complexity. However, when we consider large memories, we shall see that it is useful. Note that we use the inhibit lead only when we do not want to write a 1. Therefore, it is not used during reading. Hence, the same lead could be used for the inhibit and sense operations. However, this makes the electronics complicated. Therefore, when inhibit leads are used, a separate sense wire is also strung through the cores.

More complex memories have many planes. Each plane is similar to Fig. 7-4. In Fig. 7-7 we show the arrangement of a typical plane. We do not show one side of the control wires grounded because, in the complete memory, they will be interconnected to other such planes. The inhibit wire is also shown.

Note the winding path taken by the sense wire. When one core is sensed, several other cores will be half selected. In this case, the cores do not switch. However, the flux does change slightly and this induces small voltages in the sense line. Each of these is very much less than the actual voltage produced by a sensed core with a state of 1, and, thus, does not cause trouble. However, the sense wire passes through many half selected cores. The sum of the small voltages could cause trouble. The sense voltage lead is routed so that, on the average, half of the "half selected" sense voltages will be of one polarity and the other half will be of the other polarity. Thus, the small voltages will not add up to a large enough value to cause trouble. We have considered a memory plane with 16 cores. Actual memory planes have many more. For this reason, this half selected sensing problem can be important.

The 3D Memory

In order to form a complete memory, many planes of the type shown in Fig. 7-7 are interconnected in a three dimensional array. An array of three such planes is shown in Fig. 7-8. All of the x_1 leads of each plane are connected in series. Similarly, all of the x_2 leads are connected in series, etc. That is, each con-

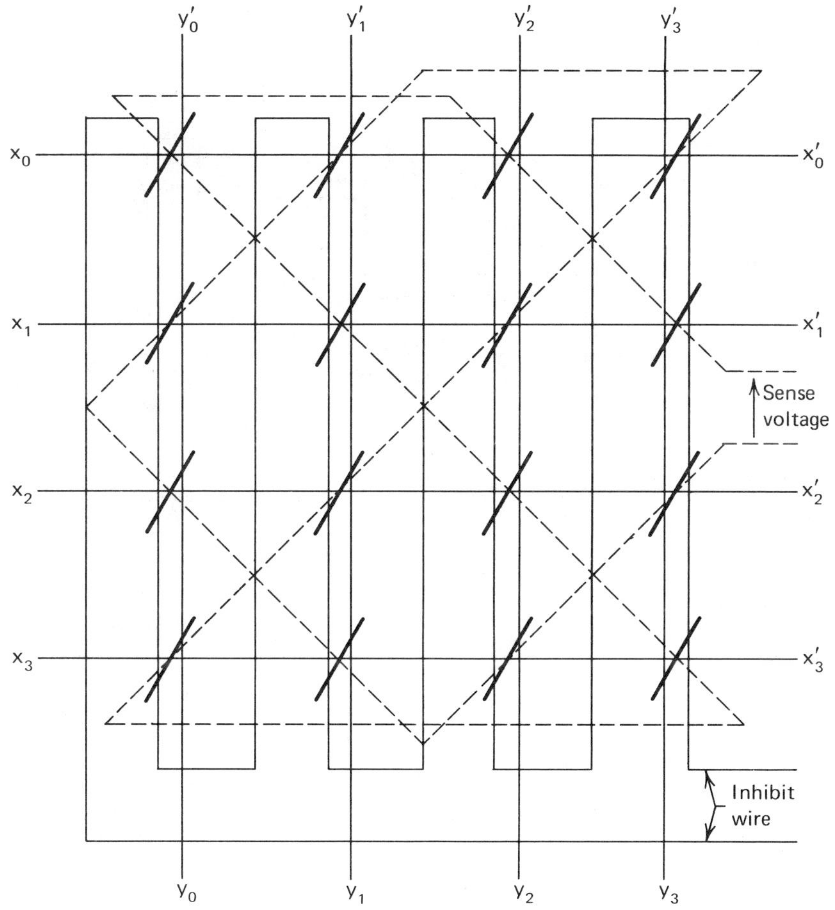

Figure 7-7. One plane of a 3D memory system.

trol lead(x and y)in each plane is connected in series. Thus, if we place a current of I_a in the x_3 and y_2 leads, with no current in any inhibit lead, then core 3,2 in each plane will have a 1 written on it.

Each plane has separate inhibit leads. These are not interconnected. Similarly, each plane has separate noninterconnected sense leads.

All of the cores in the same position on each plane make up a word. For instance, all of the 2,3 cores constitute word 2,3 In Fig. 7-8 each word will have three bits, corresponding to the three planes. In actual memories, there are 8, 16, 32 or 64

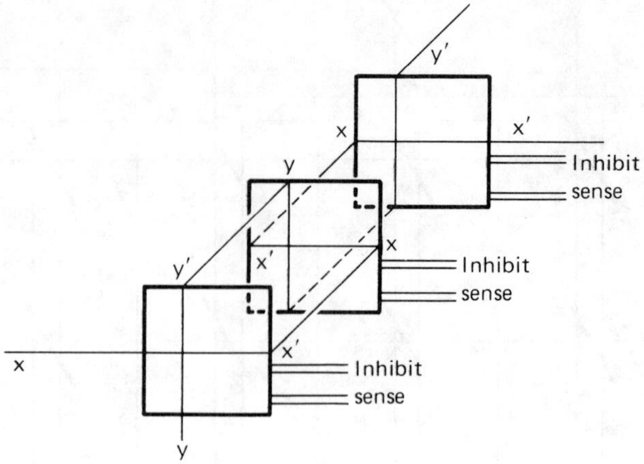

Figure 7-8. Arrangement of planes in a 3D memory system. Only one x and one y wire is shown.

or more bits to a word. That is, we build the memory such that the number of planes is equal to the desired number of bits in the word.

Now let us consider the procedure for writing a word. First it is stored in a flip-flop register, called the memory buffer register. The memory location is put into the word position storage register which is called the *memory address register.* Two circuits of the type shown in Fig. 7-6 are used for this purpose. Suppose that we want to write the word in the 2,3 position. All of the 2,3 cores are cleared. Let us use the following notation. i_{xj} is the current in the x_j lead, i_{yj} is the current in the y_j lead and i_{ik} is the current in the inhibit lead of the kth plane. All the cores in the 2,3 position are set equal to 0. This is done by passing the following currents

$$i_{x2} = -I_a \tag{7-7a}$$

$$i_{y3} = -I_a \tag{7-7b}$$

$$i_{i1} = i_{i2} = i_{i3} = 0 \tag{7-7c}$$

Now we write the desired word. In this case, we set

$$i_{x2} = I_a \tag{7-8a}$$

$$i_{y3} = I_a \tag{7-8b}$$

Suppose that we want to have 1's in the first and third bit positions of the word and a 0 in the second bit position. Then, the inhibit currents are set equal to zero in the first and third planes. That is

$$i_{i1} = 0 \tag{7-8c}$$

$$i_{i2} = -I_a \tag{7-8d}$$

$$i_{i3} = 0 \tag{7-8e}$$

Thus, no 1 will be written in the second plane. The inhibit current is obtained by an amplifier whose output is controlled by the memory buffer register which stored the number to be written. Note that this procedure causes the entire word to be written at one time. That is, the word is written in parallel.

Now let us consider a read cycle. Each sense wire is connected to one bit of a memory buffer register. This information is supplied to the register in parallel, see Sec. 6-4. Now suppose we want to read the word in the 2,3 position. We set

$$i_{x2} = -I_a \tag{7-9a}$$

$$i_{y3} = -I_a \tag{7-9b}$$

$$i_{i1} = i_{i2} = i_{i3} = 0 \tag{7-9c}$$

This sets all the 2,3 cores to 0 and each plane's sense lead supplies a voltage to its sense amplifier. The voltage at each sense lead will be either a 0 or a 1 depending upon the value stored in the core. For instance, if the word 101 were stored, the sense wire from plane one would have a pulse induced in it, the sense lead from plane two would not, etc. Thus, each bit of the word is simultaneously supplied to, and stored in, the memory buffer register. Thus, only one read cycle is needed to supply all the bits of the word. The state of a flip-flop register can be read nondestructively. Thus, this value can be used by the rest of the computer. In addition, a write cycle is now initiated. The word is written from the memory buffer register back into the 2,3 word position. The memory cycle is then completed. Note that reading and writing are done at separate times. Thus, one memory buffer register can be used for both reading and writing. Note that the memory buffer register must store the number of bits in a word. The memory may store thousands of words.

The 2 1/2 D Memory

There is another memory organization that is used. This requires that only three wires be threaded through the cores. Thus, it is more economical to wire. However, it requires more electronics. This is called the 2 1/2 D memory. No inhibit wire is used. Instead, the y control leads of each memory plane are driven separately. For instance, if we want to write on core 2,3 in the second plane, then we would set $i_{x2}=I_a$. This would half select all the 2,j cores, j=1,2,.... However, only i_{y2} in the *second plane* would be set equal to I_a. Thus, only the 2,3 core in the second plane would be written on.

Note that if there are 4 y leads and 3 planes, then 12 y amplifiers and associated circuits would be needed in the 2 1/2 memory. In contrast, the 3D memories would only need 4 such amplifiers for the y leads and 3 amplifiers, etc. for the inhibit leads.

In the 2 1/2 D memory or in the 2 D memory, there is no inhibit lead. Thus, cores are selected by only two coordinates, x and y. (In the 3 D case, a core is selected using the x and y coordinates plus the plane.) In the 2 1/2 D or 2 D cases, it becomes convenient to have the x coordinates define the words and the y coordinates define the bits of the word.

7.2 MAGNETIC FILM MEMORIES

Another form of magnetic memory uses thin films of magnetic material which are deposited on a glass substrate or on a wire. These memories are somewhat faster than core memories. However, at present, in the larger sized memories, the core memories are cheaper and occupy much less space.

The magnetic films are fabricated to be anisotropic. That is, their magnetic properties vary with direction. They are characterized as having a hard axis and an easy axis. Along the easy axis, the magnetic characteristics exhibit the hysteresis loop of Fig. 7-2. Along the hard axis, there is almost no hysteresis loop(i.e., if the current is reduced to zero, the flux becomes zero)and considerably more current is needed to change the flux.

There are several configurations that are used. One is shown in Fig. 7-9. Here the magnetic film is plated on a wire that is used both as an x lead and as a sense lead. We shall see that $i_{xj}=0$ during sensing and, thus, this can be done. The y leads are in the form of relatively broad wires, or straps(see Fig.7-9). This is so that current in them magnetizes a sufficiently large area. In a sixteen bit plane, there would be four y leads and four plated wires. In Fig. 7-9 we show two such elements.

The current along the y strap sets up a magnetic flux along the hard axis direction, see Fig. 7-9. The x current is through the wire and sets up flux in the easy axis direction. 0's and 1's are stored by flux which lies along the easy axis. The direction of the flux determines if a 0 or a 1 is represented.

Current in the y strap tends to align the flux 90° to the easy axis(i.e., along the hard axis). Current in the x wire is adjusted so that it is not enough to flip the direction of flux from one easy direction to the other, that is, to flip the direction 180°. However, if current is in the y strap, then the flux will be rotated almost 90°. Then the current in the x wire need only flip the flux a little more than 90° and this is enough to flip the flux. Thus, the i_x is adjusted so that, in order to write, we must have current in both the x and y leads. Note that the direction of the x current is reversed depending upon whether a 0 or a 1 is to be written. However, the y strap current's only function is to shift the direction by 90°. The direction is unimportant. Thus, the direction of i_y need never be changed. This reduces the electronics needed.

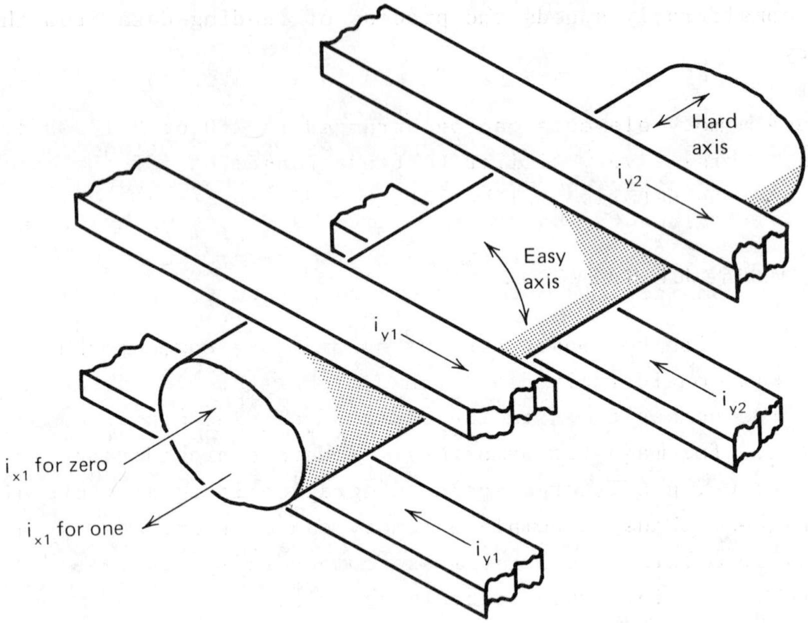

Figure 7-9. Two magnetic thin cylindrical film memory elements.

Now let us consider reading. In this case, we only supply current to the y strap. Since there is no current in the x lead, it can now be used for sensing. The desired memory location is picked by pulsing current through the appropriate y strap and sensing the voltage across the appropriate x line. Now let us consider the actual details. A pulse of current is passed through the y strap. This causes the flux to rotate almost 90°. The direction of rotation will depend upon whether a 0 or a 1 is stored. This rotation of flux will induce a voltage in the sense line. The direction of flux will determine the polarity of the voltage. Thus, the polarity of the sense voltage determines if a 0 or a 1 is stored. Note that both 0 and 1 produce an output in this case.

When a sense current is put through the y strap, the flux direction rotates from the easy axis towards the hard axis. However, the rotation will be less than 90°. When the sense current is removed, the magnetic properties of the material are such that the flux rotates *back* to the easy axis. Thus, the readout is *nondestructive*. This nondestructive readout is one of the primary advantages of this type of memory. Additional time does not have to be spent rewriting destroyed data back into the memory. This considerably speeds the process of reading data from the memory.

These memory elements can be arranged in 2 D or 2 1/2 D arrangements. This is very similar to the arrangement for the core memory, and the details will be left to the reader.

7.3 SEMICONDUCTOR MEMORIES

Memories can be constructed using only semiconductor devices. These are considerably faster than magnetic device memories(e.g., they have access times in the order of nanoseconds in comparison to 1μsec. for magnetic memories). Modern semiconductors are constructed using LSI(large scale integration)integrated circuit techniques. Thus, a complete memory can be fabricated on a single silicon chip. Such chips are commercially available. Thus, semiconductor integrated circuit memories are very attractive. Magnetic memories are, at times, cheaper. However, as LSI techniques develop, the cost advantage should shift to semiconductor devices. The one advantage that magnetic memories possess is that they are nonvolatile. That is, in the event of a power failure, they retain their information. This is an advantage but it may be offset by others. In addition, the contents of the memory can periodically(e.g., every two minutes)be stored on a magnetic disc or drum. Thus, a record would be preserved in the case of a power failure. Disc and drum storage are discussed in the next section.

In the memories that we have considered, each bit is stored in an element which can be written on or read from. For instance, a magnetic core was such an element which we could call a *magnetic memory element*. Figure 7-10a illustrates a *semiconductor memory element*. It uses an R-S flip-flop to store the information. If the write signal is 0, then the input to the flip-flop is R=0, S=0. Hence, the state of y can only change during a write pulse. If the write signal is a 1, and the input signal is a 0, then R=1,S=0. In this case the state of the flip-flop will be reset to y=0. If, during a write pulse, the input signal is a 1, then R=0, S=1, and the state will be set to y=1. Then, the state of the flip-flop will always be equal to the input signal that is present when the write control lead has a 1 on it. (Note that this will probably be a master-slave flip-flop which is operated in accordance with the discussion of Sec. 6-2. We have not shown the clock lead connections.) We say that the state of the memory element is equal to the state of its flip-flop.

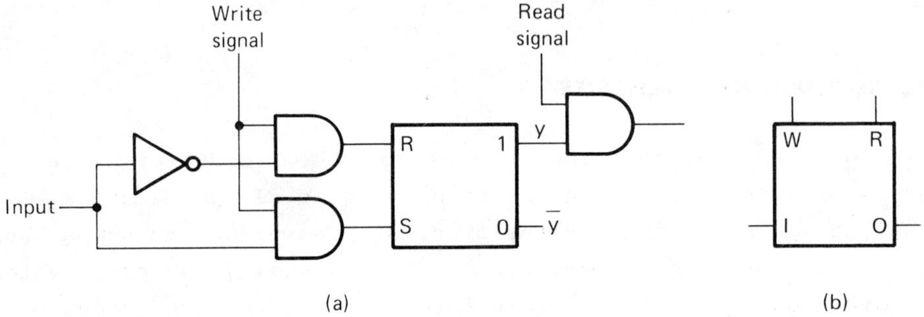

Figure 7-10. (a) A semiconductor memory element; (b) its block diagram.

The output of the memory element will be 0 unless a 1 is applied along the read signal lead. In this case, the output is the state y. Thus, we can read and write on the semiconductor memory element. A block diagram representation of this memory element is shown in Fig. 7-10b.

The gates and flip-flop of the semiconductor memory can be fabricated using either transistors or MOSFETs. If transistors are used, the operation will be faster. However, the integrated circuit fabrication is simpler, and more elements can be placed on a single chip if we use MOS devices. Often a compromise is made in that the memory elements themselves are MOS devices while the associated circuitry used to drive the memory and extract information from it uses (integrated) transistors. In the usual memory, there are very many memory elements and relatively few elements in the associated circuitry. (e.g., there is a single memory buffer register.) However, the associated circuitry may be used several times, each time that a single memory element is written on or read from. Thus, a compromise such as the one above becomes a reasonable one.

In Sec. 5-14 we discussed dynamic MOS gates. Here the information is retained by charge stored in the gate to channel capacitance of the MOSFET. This charge will slowly "leak off" and, to prevent the information from becoming lost, it is periodically refreshed. Such MOS gates are also used in memory devices.

Now let us consider a complete small memory unit. We shall use a 16 bit memory. The memories are arranged in a 2 D or a 2 1/2 D arrangement, since there is no inhibit wire. Thus, it is desirable that the x variables define words and the y variables define bits in the word. This is essentially the same as the case of the 2 D or 2 1/2 D magnetic core memory. The memory is shown in Fig. 7-11. Note that we have numbered each memory element with a double number(e.g.,2,3). The first number indicates the word and the second the bit. The x's select the desired word. For instance, suppose that we want to write on the x_3 word position. Then, a 1 is placed on the x_3 write line. Then, in the fourth row, the memory element 3,0 will set itself to the value of d_0. That is, the state of the flip-flop of the memory unit will be a 1 if d_0 is a 1, and will be a 0 if d_0 is a 0. Similarly, the memory element 3,1 will set itself to the state of input d_1 and memory element 3,2 will set itself to the state of d_2 and memory element 3,3 will set itself to the state of d_3.

The memory element in all the other rows will be unaffected since their write leads will have 0's on them.

If we want to read a word, then we place a 1 on the appropriate read lead. For instance, if we want to read the word corresponding to x_2, then x_2's read lead will have a 1 placed on it. The state of all the other read and write leads will be 0. Then, all the outputs except those in the x_2 row will be 0. In addition, b_0 will equal the state of memory element 2,0, b_1 will equal the state of memory element 2,1, b_2 will equal the state of memory element 2,2 and b_3 will equal the state of memory element 2,3. Thus, we have obtained the desired output. These outputs will only be present while there is a pulse on the read line. Thus, the outputs are supplied to a register which stores them until they are needed. This is the memory buffer register. Note that the readout is nondestructive.

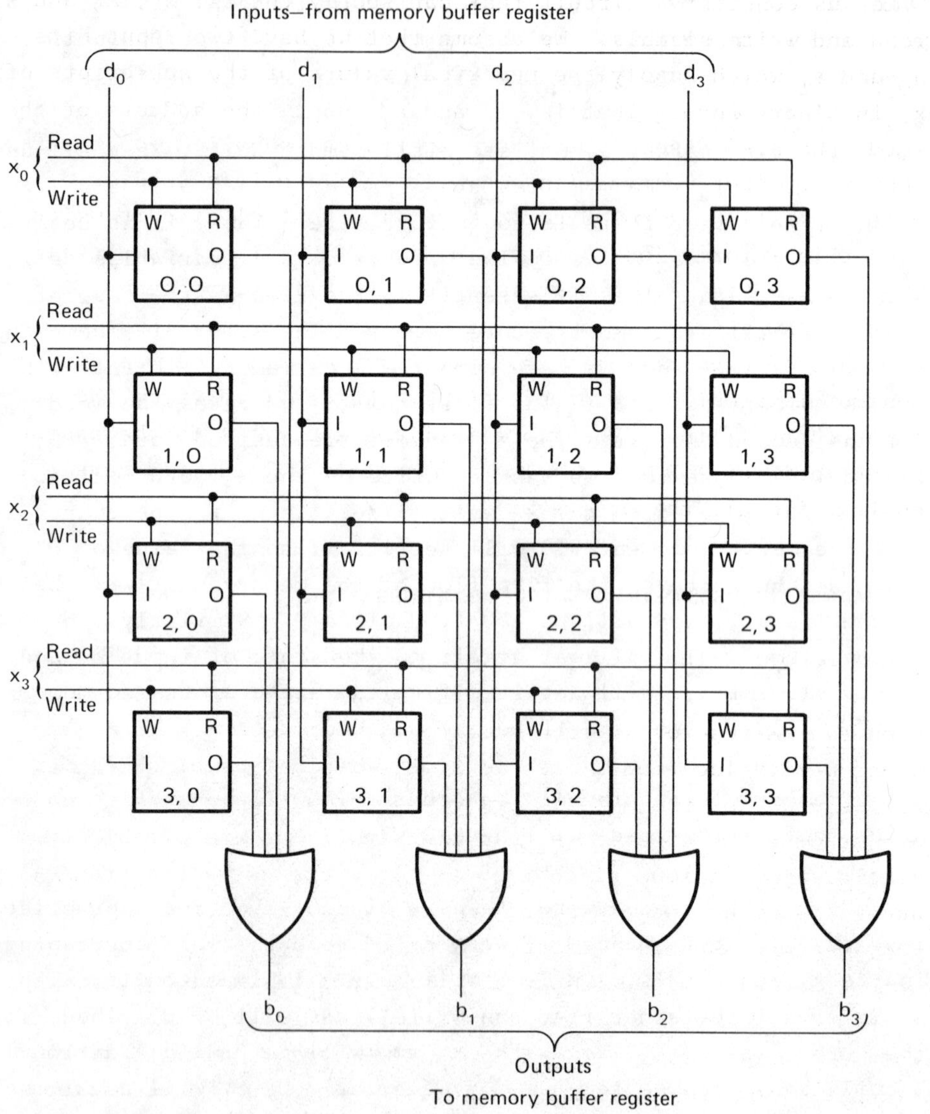

Figure 7-11. A semiconductor memory plane.

The inputs d_0, d_1, d_2 and d_3 are usually stored in a register before they are transferred to the memory. This may be the same memory buffer register used to store the output. Note that since we do not read and write at the same time, we can use the same register.

Let us consider a circuit that can supply the x_0, x_1, x_2 and x_3 read and write signals. We assume that we have two input bits, a_0 and a_1 which supply the numerical values of the subscripts of x, in binary form. That is, a_0 and a_1 supply the address of the word. We also assume that there are pulses supplied, on the read and write input control lines, which indicate if a read or a write operation is to be performed. The control circuit which uses this information is shown in Fig. 7-12. It is essentially the same as Fig. 7-6 except that now the flip-flops each drive a pair of AND gates. Each AND gate has an additional input. One AND gate of each pair has an input which is from the read input control line while the other AND gate has an input from the write input control line. The two flip-flops apply 1's to the appropriate pair of AND gates. Thus, the desired lead will have a 1 on it while all the others will be 0.

7.4 MAGNETIC DISCS-DRUM AND TAPE MEMORIES

The memories that we have thus far considered are called *random access memories* (RAM), since the contents of any memory location can be read or written on directly, without regard to any other memory location. This is a very rapid type of operation and is used for the main memory of almost all computers. There are other types of storage that are slower but are often used for special applications since they are less exepnsive than random access memories and can store extremely large amounts of data. For instance, suppose that a bank wants to process its depositors' accounts. All this data would be too much for most main memories to store. However, all the data can be stored on a magnetic tape. When a program using the records is to be run, only the appropriate data from the tape is read into the main random access memory. Then, the program is executed. Programs, as well as data, can also be stored on tape. When they are to be used, they are read into the RAM. These slower storage techniques, that we

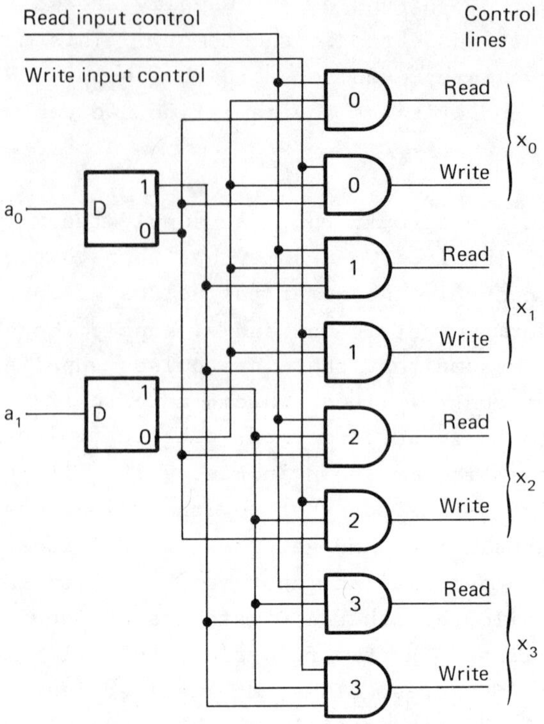

Figure 7-12. A control circuit for the memory of Fig. 7-11.

shall discuss in this section, store information until a program is to be executed. Then, it is placed in the RAM.

The techniques that we shall discuss store the information on ferromagnetic films. In contrast to the main memory, this storage is dynamic in that the film moves in relation to a pick up device. Basically, in all such storage, a thin layer of ferromagnetic material is mounted on some support mechanism, and moved under a magnetic *head*. This is diagrammatically represented in Fig. 7-13. The head is made of ferromagnetic material and has a coil around it. There is a small gap in the ferromagnetic core. The head is put in very close proximity to the magnetic film.

If current is passed through the coil, magnetic flux will be set up as indicated. This will magnetize a region of the film. The magnetic film is moved under the head. Thus, by applying pulses of current to the coil, we can magnetize pulse shaped

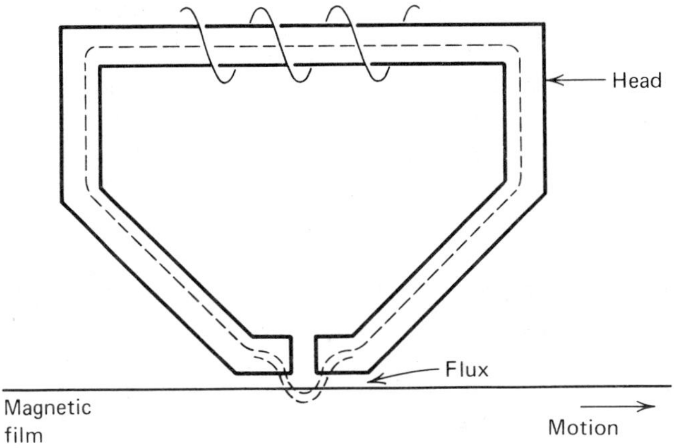

Figure 7-13. A magnetic recording head.

regions on the film, and write information on the film. Actually, the magnetic regions are rounded because of the recording properties. By reversing the direction of the current, we can reverse the polarity of the flux and the resultant magnetism.

Now suppose that there is no current in the coil and that the film is moved under the head. When a magnetized region passes under the head, flux will be induced in it. When the magnetized region passes, the flux will fall back to zero. Thus, this changing flux will induce a voltage in the coil. Hence, information can be read from the film. There are several procedures that are used to move the film in relation to the head. We shall consider them later in this section. First we shall consider the techniques used for recording 0's and 1's on the magnetic film.

Return to Zero (RTZ)

In this technique 1's are recorded as positive pulses and 0's as negative pulses. For example, the current pulses used to record the sequence 0101100 are shown in Fig. 7-14a. Positive currents of I_0, which saturate the magnetic material, are used to record 1's. Similarly, a current of $-I_0$, which saturates the material in the reverse direction, is used to write 0's. This procedure is straightforward. However, it suffers from two disadvantages. First, there are relatively large spaces between the pulses. This is done to prevent the pulses from interferring with each other. In a system such as this, there are mechanical tolerances so that the positions of the 0's and the 1's may be displaced from their ideal locations. Thus, the relatively wide spacing is used. The disadvantage of this is that the number of bits that can be stored per mm of magnetic material is reduced.

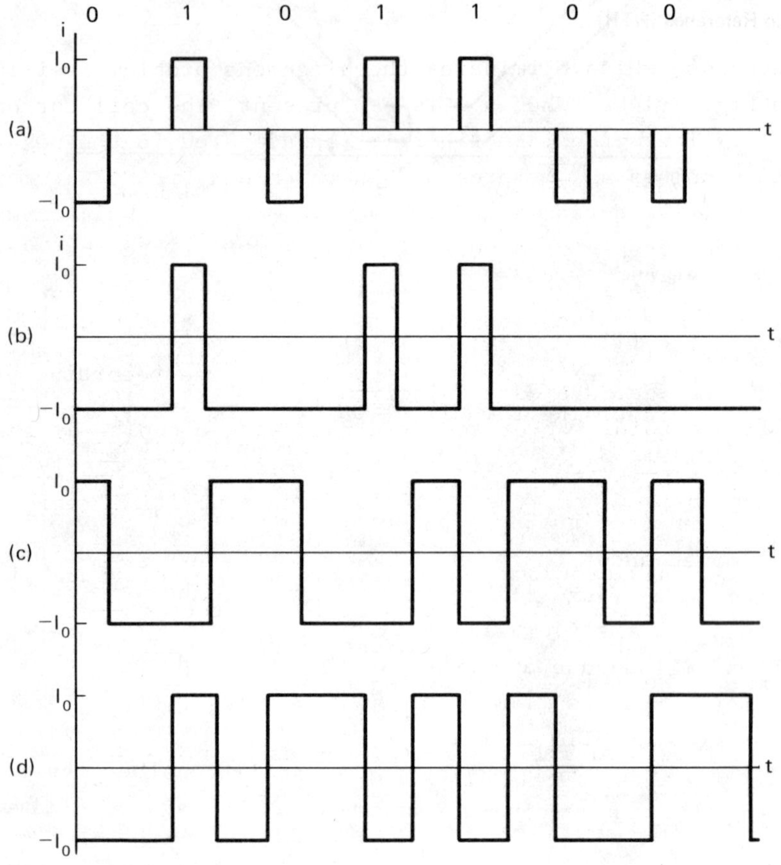

Figure 7-14. Magnetic surface recording techniques used to record the sequence 0101100. (a) return to zero; (b) return to reference; (c) phase modulation; (d) frequency modulation.

The second problem that results occurs when we want to rerecord. Suppose that we want to rerecord a 0 where a 1 had been recorded. A current of $-I_0$ would be put through the head coil. This will reverse the saturation in the magnetic film and change the 1 to a 0. However, suppose that the position of the previously recorded 1 is not exactly under the head when the 0 is recorded. Then, there will be a small region where the positively magnitized region of the 1 remains. This will be followed by the negatively magnitized region of the 0. This positive region will cause spurious responses and, thus, must be avoided. For these two reasons, the return to zero method is almost never used.

Return to Reference (RTR)

A technique which eliminates the alignment problem is illustrated in Fig. 7-14b. When a 1 is *not* present, the coil current is made $-I_0$. Thus, the magnetic film is saturated in the negative direction unless a 1 is present, in which case it is saturated in the positive direction. Now the presence of a pulse indicates a 1 and the absense of a pulse indicates a 0. Note that another advantage of this procedure is that we do not have to detect the polarity of the pulse, but only its presense. There must still be relatively large spaces between the pulses but the problem of writing over recorded data is eliminated. If a 0 is written, its area will certainly cover the original space occupied by the 1. Thus, there will be no residual positive flux. Similarly, if a 1 is written over a 0, there will only be a positive region surrounded by a negative region. Thus, there is no need for exact alignment. This procedure is called return to reference since, when a 1 is not present, the magnetizing current is returned to the reference value, $-I_0$. (We have not discussed how we time the pulses. We shall do so subsequently.)

Note that both of the previously discussed procedures suffer from the disadvantage that they make inefficient use of the magnetic material because of the large spacing. We shall now consider procedures that eliminate this problem.

Phase Modulation—Manchester Method

In this procedure, 0's and 1's are indicated by a change in flux or, equivalently, are recorded by a change in current in the magnetic head. The change always occurs in the middle of the time allocated for each bit. This is illustrated in Fig. 7-14c. Note that, in the *middle of the time reserved for a bit,* 1's always have positive going changes in level while 0's have negative going changes in level. The recording electronics must be more complex in this case. The current polarity at the start of each bit time must be sensed. If it is positive, and a 1 is to be recorded, then the current is switched to $-I_0$, a negative value. Thus, when the middle of the bit time is reached, the current can be switched to $+I_0$, which yields a positive change. Of course, a converse operation is performed with 0's.

We have shown this technique to the same time scale as the return to zero and return to reference techniques. Actually, there need not be spaces between pulses here, since there is no danger of positive and negative pulses overlapping. Thus, in a given length of magnetic material, we can store many more bits. For this reason, phase modulation and the techniques that we shall discuss next have replaced the return to zero and return to reference techniques.

Frequency Modulation

This method is illustrated in Fig. 7-14d. Here, the stored flux, or, equivalently, the head current, is always reversed in the middle of the bit time if a 1 is to be recorded. If a 0 is recorded, then there is no flux reversal in the middle of the bit time. The flux is always changed at the beginning of a bit time. In contrast to phase modulation, we do not have to determine if a change has to be made at the start of the bit time.

Drum Memories

The fastest dynamic magnetic film storage uses a drum coated with magnetic material, see Fig. 7-15. The diameter of this drum can vary from 5 cm. to about 80 cm. The length ranges from 0.5 cm. to about 100 cm. The drums are rotated at high speeds(500-100,000 rpm). Many fixed magnetic recording and reading heads are spaced along the drum. The region on the drum under the head is called a *track*, see Fig. 7-15. One track has timing pulses permanently recorded on it. These pulses occur at each bit interval. They serve as "clock pulses" to indicate where the recorded pulses should be.

There are several recording techniques used. Words can be recorded in *sequence* on a single track. In this case, a single word is read or written by a single head. The region of the

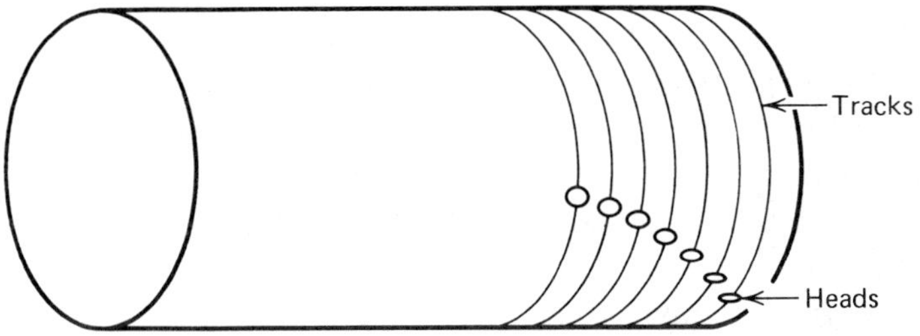

Figure 7-15. A magnetic drum memory.

track which contains a word is called a *sector*. There is usually one track which has pulses only at the beginning of each sector. Using these pulses and the "clock pulses", the appropriate sector can be found and the desired word written or read.

Another recording scheme uses several heads to write a word. For instance, if the word has 8 bits, then eight heads would be used to record it. In this case, the information can be read in or out in parallel. Some recording schemes use combination of series and parallel techniques(e.g., 4 heads are used to read an 8 bit word).

The primary disadvantage of these mechanical memories is that they are relatively slow. For instance, suppose that we want to read a word that has just passed under the heads. We must wait for a complete revolution of the drum before the word is in position to be read. On the average, we must wait the time required for one half revolution. This is called the *access time* which is typically 100×10^{-6} sec. or 100 microsec(μsec.). This is much longer than the access time of magnetic core or semiconductor memories which can be very much less than 1 μsec. The primary advantage of drums is their large capacity. A large drum can store 10^8 bits which is usually much larger than the main memory. Thus, drums are used as auxiliary memories. For instance, many programs and their data may be stored on the drum. When *one* program is to be run, it and its data are read from the drum into the main memory. This is done only once prior to execution. The main memory may be used thousands of times during the execution of the program. However, the drum is not used for this and excessive time is not consumed. In a drum memory, there is only one drum and it cannot be removed.

Disc Memories

In this form of recording, the magnetic film is put on a disc which resembles a phonograph record. Both upper and lower surfaces are coated. The tracks are concentric circles. The discs are rotated(just as a phonograph record, but at a higher speed). To change tracks, the head must be moved radially along the surface of the disc. This must be a relatively precise operation. There are separate heads and tracks for each side of the disc. The access time is longer for a disc than for a drum(e.g.,30 milisec. for a disc and 8 milisec. for a drum). This is because the correct position of the track must not only be rotated under the head, but the head must be moved into the proper position. The access time can be reduced by having several heads for each disc.

Many disc devices are such that the disc is permanently mounted. However, there are others where the disc can be removed and another disc, with different stored information, can be mounted and used.

A typical disc can store 10^6-10^7 bits. Often, groups of discs and their heads are arranged in a *disc pack*. In this case, the operation is the same except that there are many discs rotated simultaneously. Each has its own head(s). These may be mechanically coupled.

Tape Memories

Data can be stored on magnetic tape. There are several tracks recorded, some of which are used for error detection, timing, and locating information. Since long tapes can be used, the amount of storage on a tape can be extremely large. There can be from 80 to 400 bits stored on a mm. of tape. A tape can be a thousand meters long and there can be many tracks. If 6 of these tracks are used for data storage, then about 10^8 bits can be stored on a tape. Unlike drum storage, the tape can be demounted from the tape drive and another tape mounted. Tapes are relatively inexpensive. Thus, the amount of information stored on a tape is almost unlimited.

The primary disadvantage of tape memory is the long access time. An entire tape may have to be wound under the head before the desired data is obtained. Very high speed devices are built which will move the tape from 50 to 400 cm/sec. However, the access time is still very long in comparison with other storage systems.

Tape memories are called sequential access memories since data is stored sequentially and we must run through the entire tape between the present location under the head and the location of the desired data. In discs or drums, the access time is faster since we can switch tracks or position heads and do not have to run through all the intervening data. This is called *direct access storage*. Of course, the fastest storage is random access storage.

7.5 READ ONLY MEMORIES-DECODERS

At times, we work with memories which contain stored data which is not to be changed. For instance, suppose that data is in binary form and we want to change it to another form. Note that the binary representation of decimal numbers can be considered to be a code. Other binary codes are, at times, used to represent decimal numbers. We shall discuss the reasons for this in Chapter 8. Suppose that data is represented in binary form and we want to change it to a different code. A memory could be encoded with the necessary information to convert from one code to the other. There would never be any reason to change this memory. Thus, the memory could be manufactured with the required information stored in it, and no writing provision would be made. In this section, we shall discuss these *read only memories* (ROM).

We have only considered one use of ROMs. Special programs that are repeated over and over are stored in them. These may be, for instance, instruction decoders which are used in large computers or they may be used to control triginometric function generators which are used in a desk calculator.

Let us consider a read only memory. It is arranged in a rectangular array form. It has a set of input leads. When a 1 is supplied to one of these leads, it outputs a single word which is determined by the design of the ROM. The outputted word depends upon which input lead is supplying the 1. A simple ROM memory is shown in Fig. 7-16. The *input* to the ROM memory is supplied to one of the *word lines*, the *output word* occurs in the *bit lines*. Suppose that a 1 is put on word line 0. (We assume that a 1 is a positive pulse and a 0 is zero volts.) The diodes a and b will conduct and connect bit line 1 and 3 to the word line 0. Since this line has a 1 on it, the output word will be 1010. Now suppose that a 1 is put on word line 1. Then, the diodes c, d and e conduct and the output is 1101. Similarly, an input on word line 2 produces an output 0010, and an input on word line 3 produces 0110. We have just shown output lines. Usually, they supply values to a buffer memory register. The input may be supplied from a register.

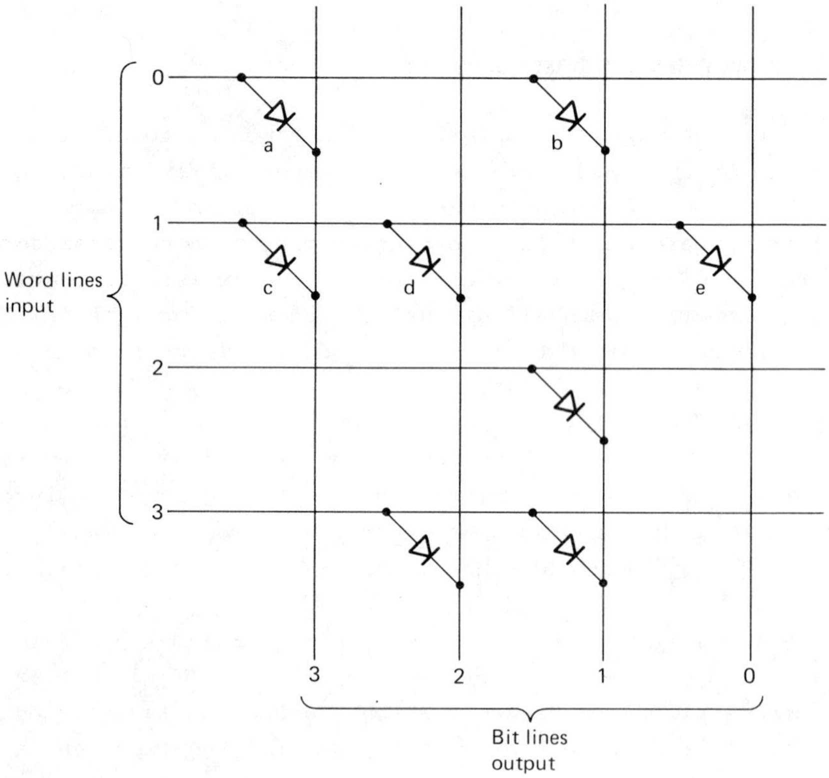

Figure 7-16. A simple diode ROM.

We have considered the ROM using diodes. Actually, many devices are used. For instance, in Fig. 7-17, we have a ROM that is equivalent to Fig. 7-16 which uses enhancement MOSFETs. All of the drain leads are connected to the positive voltage. If there is no signal on the word line, then the gates are open circuited and the FET acts as an open circuit. If there is a 1 on the word line, then there is a large positive gate voltage and the FET saturates. Then, the power supply is effectively connected to the corresponding bit line and it achieves a large positive potential. Thus, that line has an effective state of 1. Hence, the ROM of Fig. 7-17 will produce the same output as that of Fig. 7-16.

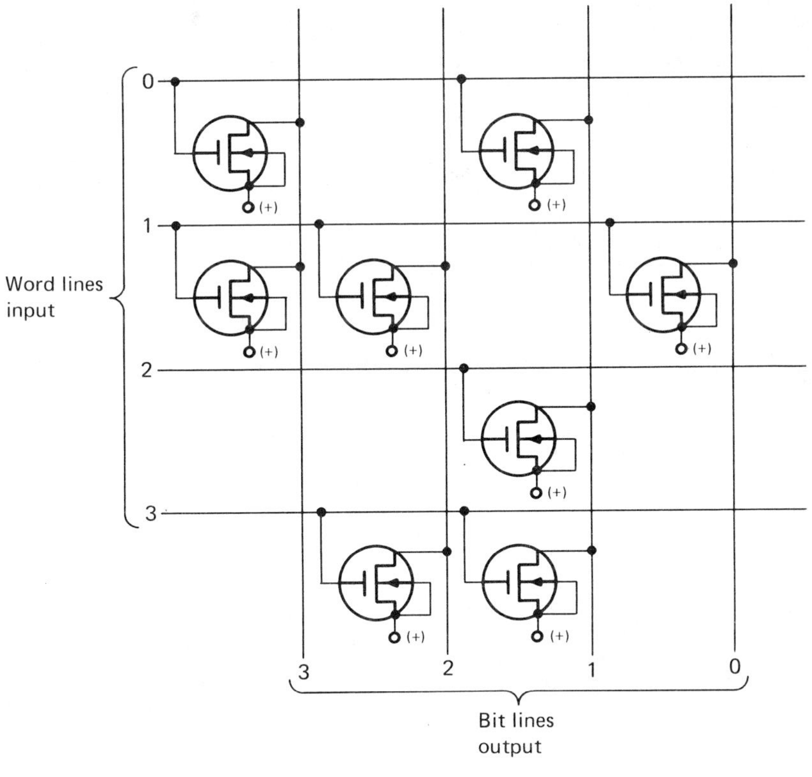

Figure 7-17. An enhancement MOSFET ROM that is equivalent to Fig. 7-16.

We have illustrated the ROM using MOSFETs. Junction transistors can also be used. The same basic ideas are used. Now, the word lines are used to supply the base current to turn the transistor on and off.

Resistors and capacitors can be used in place of the diodes to couple the word lines to the bit lines. The only problem with this procedure is that they are bilateral and undesirable coupling can result. For instance, consider Fig. 7-16 but assume that resistors have replaced the diodes. Now consider that a signal is on word line 2. Then there will be an output on bit line 1. Resistor b can couple this back onto word line 0. Thus, spurious signals can result. The magnitude of this spurious signal depends upon the magnitudes of the resistances, the output impedance of the amplifiers used to drive the word lines and the input impedance of the device driven by the bit lines.

As an additional example, let us obtain a ROM which will convert from a binary representation to a different code called a Gray code(In Sec. 8-2, we shall discuss the use of such codes.)

Binary Number	Gray Code
000	000
001	001
010	011
011	010
100	110
101	111
110	101
111	100

Each binary number represents an input line. That is, each binary causes a 1 to be placed on an appropriate word line. Later in this section we shall discuss a device which supplies signals to the word lines. For the time being, let us just assume that the signals are available. The ROM is shown in Fig. 7-18. As an example, if the input signal is 100, then word line 4 will have a 1 placed on it. Then, the output will be coupled to bit lines 2 and 1. Thus, the output is 110.

We must now consider a circuit that will drive the appropriate word lines. This is called a *decoder*. Actually, we have already considered such circuits. For instance, Fig. 7-6 illustrated a circuit which supplied a 1 to one of four amplifier inputs. The decoder circuit for Fig. 7-18 is shown in Fig. 7-19. The flip-flops make up a memory which stores the input signal. The remainder of the circuit constitutes the decoder. Note that all

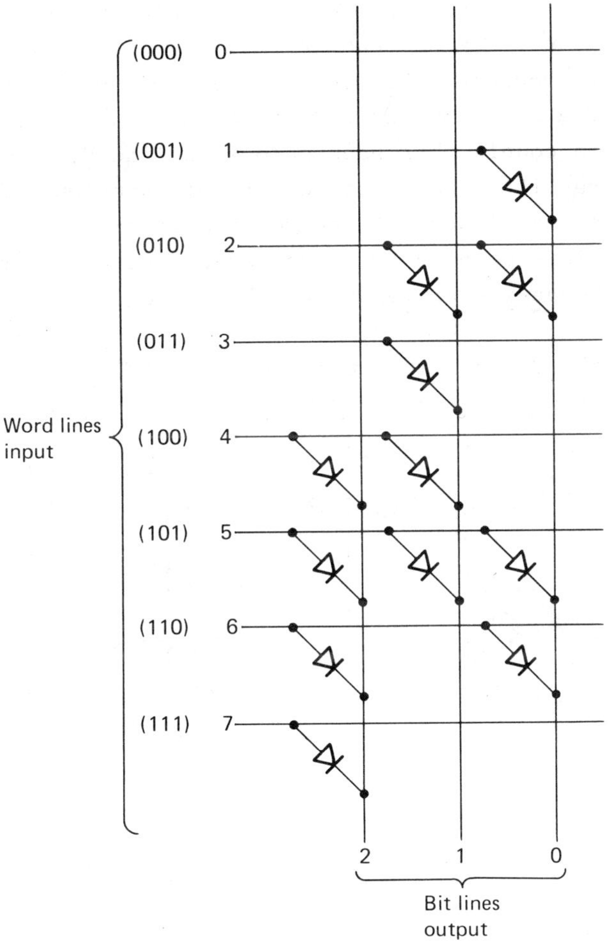

Figure 7-18. A ROM the converts three bit binary numbers into a Gray code.

three inputs to AND j, j=0,1,2,3,4,5,6,7, will only be 1 when the binary number $a_2a_1a_0$ is equal to j.

We have shown an output line and AND gate for an input of 000. Actually, this is not needed since, see Fig. 7-18, there is no coupling between the word zero line and any of the output lines. AND 0 is shown in Fig. 7-19 for completeness.

We can take a somewhat different viewpoint. Consider the 0 bit line of Fig. 7-18. That will have a 1 output if any of word lines 1,2, 5 or 6 has a 1 on it. Thus, bit line 9 can be considered to be a logic function of the input. We can formalize this so that we can design the output of a specific bit line as a given function of the input variables.

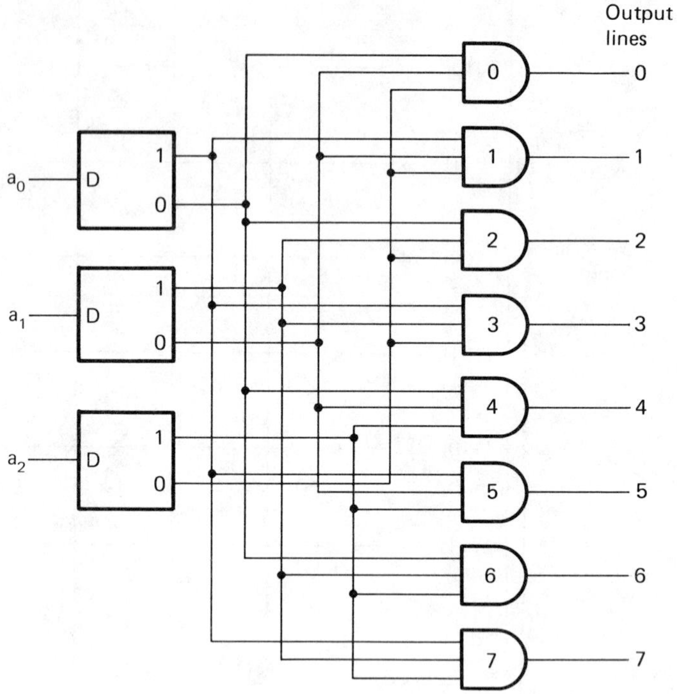

Figure 7-19. A driver circuit for the ROM of Fig. 7-16.

Let us realize some arbitrary functions. It is most convenient if these are in the nonminimized, canonical sum of products form. Suppose we have

$$y_0 = x_1\bar{x}_2\bar{x}_3 + x_1\bar{x}_2x_3 + \bar{x}_1x_2\bar{x}_3 + x_1x_2x_3 \tag{7-10a}$$

$$y_1 = \bar{x}_1x_2x_3 + \bar{x}_1x_2\bar{x}_3 + \bar{x}_1\bar{x}_2\bar{x}_3 \tag{7-10b}$$

$$y_2 = \bar{x}_1\bar{x}_2x_3 + x_1x_2\bar{x}_3 + x_1x_2x_3 + \bar{x}_1\bar{x}_2\bar{x}_3 + \bar{x}_1x_2\bar{x}_3 \tag{7-10c}$$

A ROM network which realizes this is shown in Fig. 7-20. Each input line represents one minterm. That is, the signal on the input line labeled $\bar{x}_1\bar{x}_2\bar{x}_3$ is 1 if x_1=0, x_2=0, x_3=0, etc. Then, there are diodes connecting each function to each of its required minterms. For instance, from the diodes connected to bit line y_1 we see that it will be 1 if $\bar{x}_1\bar{x}_2\bar{x}_3$ OR $\bar{x}_1x_2\bar{x}_3$ OR $\bar{x}_1x_2x_3$ are 1's. Thus, we have realized the desired function.

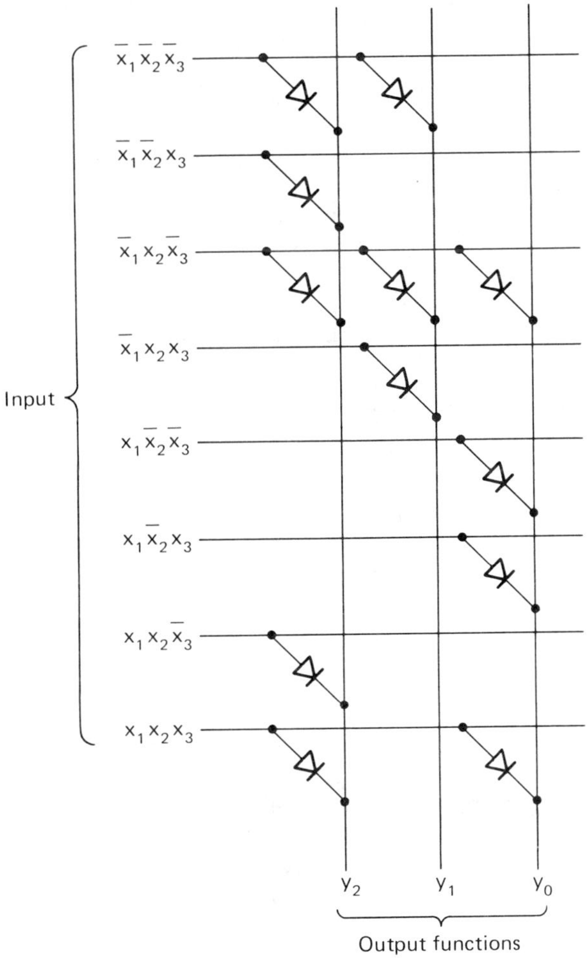

Figure 7-20. A ROM that realizes Eqs. (7-10).

Now let us see how we can realize the driving functions. For each variable we need both the variable and its complement. This can be realized from a flip-flop circuit. Such a circuit is shown in Fig. 7-21. This is then used to drive the decoder used in Fig. 7-22. That is, the x_1 and $\bar{x}_1$ outputs of Fig. 7-21 are connected to the x_1 and $\bar{x}_1$ inputs of Fig. 7-22, etc. The outputs are all the functions we desire. This acts in essentially the same way as the decoder of Fig. 7-19. Thus, the details of the interconnection of all the circuits will not be shown.

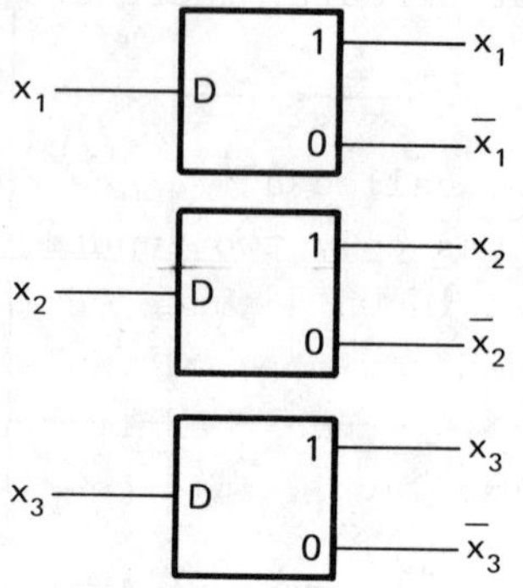

Figure 7-21. Flip-flops used to drive a three variable decoder.

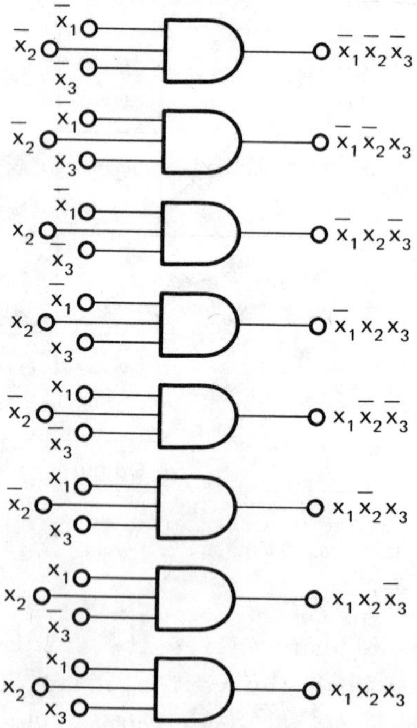

Figure 7-22. A simple one level decoder that can drive the ROM of Fig. 7-18.

This decoder is a simple one which uses only one level of logic (i.e., gates do not drive other gates). Thus, it is fast. In general, if there are n variables, a maximum of 2^n gates will be needed. Each gate has n inputs. Note, see Chapter 5, that increasing the number of inputs increases the complexity of the gate. For this type of circuit, there are, at most, $n \times 2^n$ inputs.

A multilevel decoder, called a *tree decoder*, is shown in Fig. 7-23. Every AND gate has only two inputs. Each level introduces another input. We have shown a three variable decoder. However, the basic layout is used for more variables. For instance, if there were four variables, then the first two levels would be as in Fig. 7-23. The third level would consist of 16 gates, in eight pairs. Each pair would have two inputs. One of the inputs of each gate in the pair would come from one of the outputs of Fig. 7-23. In addition, one gate in each pair would have an input from x_4, the other gate would have an input from $\bar{x}_4$. The number of AND gates required is

$$G = \sum_{k=2}^{N} 2^k \qquad (7\text{-}11)$$

where N is the number of variables. The total number of inputs required is 2G.

If N is a large number, then the total number of inputs will be less than with the single level decoder. The single level decoder will have fewer gates and will be faster. However, for a large number of variables, it becomes complex electronically.

Another decoder circuit that is used is the *matrix decoder*. This is also a multilevel decoder with two inputs per gate. The lines are arranged matrix fashion. In Fig. 7-24, we illustrate this for a three variable decoder. There are 6 lines, with inputs x_1x_2, $x_1\bar{x}_2$, $\bar{x}_1x_2$, $\bar{x}_1\bar{x}_2$, x_3 and $\bar{x}_3$. Then, AND gates are connected between every possible pair of three lines. Note that this uses a total of 12 AND gates and 24 inputs. The corresponding tree decoder also used 12 gates and 24 inputs. The single level decoder used 8 gates with 24 inputs. It appears as though the single level decoder is the simplest. However, when the number of gates increases, the matrix decoder will have many fewer inputs and gates than the tree decoder, and will have many fewer inputs than the one level decoder.

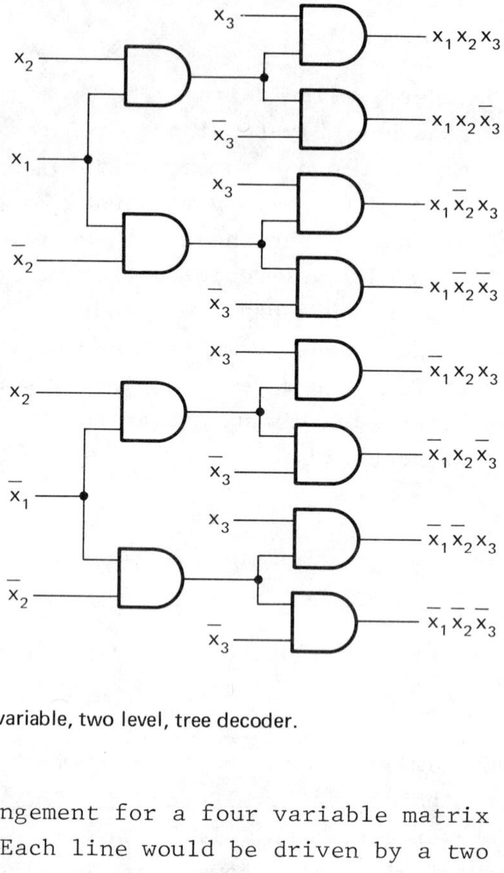

Figure 7-23. A three variable, two level, tree decoder.

The line arrangement for a four variable matrix decode is shown in Fig. 7-25. Each line would be driven by a two input AND gate. Two input AND gates would be connected between each pair of lines. The outputs of these second level AND gates would be the output of the decoder.

If more inputs are used, these ideas are extended. For instance, if there were 8 input variables, then these will be divided into sets of 4 each. Each pair would then be used as the inputs to a four variable matrix decoder. All possible pairs of outputs would then be combined using AND gates.

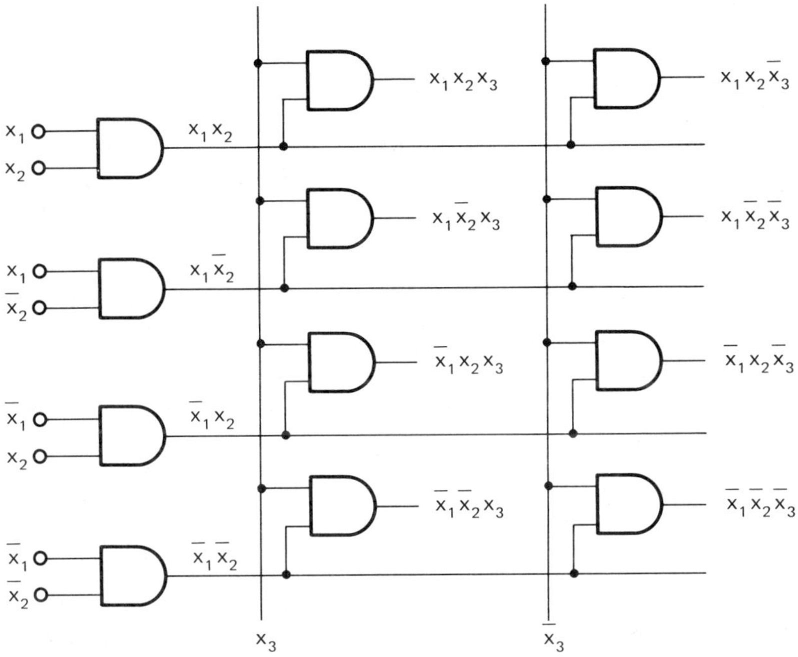

Figure 7-24. A three variable, two level matrix decoder.

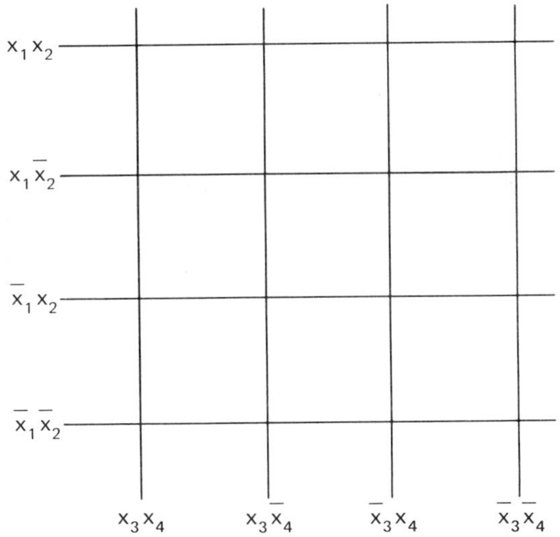

Figure 7-25. Line arrangements in a four variable matrix decoder.

BIBLIOGRAPHY

Booth, T.L., "Digital Networks and Computer Systems", Chapter 7, John Wiley and Sons, Inc., New York, 1971

Kohonen, T., "Digital Circuits and Devices", Chapter 9, Prentice Hall, Inc., Englewood Cliffs, N.J. 1972

Mano, M.M., "Computer Logic Design", Chapter 8, Prentice Hall, Inc. Englewood Cliffs, N.J., 1972

O'Malley, J., "Introduction to the Digital Computer", Chapters 4 and 9, Holt, Rinehart and Winston, Inc., New York, 1972

PROBLEMS

7-1. Discuss the basic storage properties of a magnetic core.

7-2. Describe and compare the organization of 2 D, 3D and 2 1/2 D memories.

7-3. A 3 D memory is to be constructed which will store 16,384 16 bit words. How many cores are needed? How many cores lie in each plane?

7-4. Modify the control line driver of Fig. 7-6 so that it uses T flip-flops.

7-5. Why is the magnetic memory a nonvolatile memory?

7-6. Draw a 64 bit 2 D magnetic film memory.

7-7. Describe the complete process of reading and writing on magnetic memories.

7-8. Obtain a semiconductor memory element which uses a D flip-flop.

7-9. Repeat Prob. 7-8 for a T flip-flop.

7-10. Modify Fig. 7-11 so that the inputs are taken from, and the outputs are supplied to, a single memory buffer register.

7-11. Describe the complete process of reading and writing for a semiconductor memory.

7-12. Describe the process of dynamic magnetic film recording.

7-13. Represent the dynamic magnetic film storage for sequence 011000110101 using a return to zero technique.

7-14. Repeat Prob. 7-13 using a return to reference technique.

7-15. Repeat Prob. 7-13 using a phase modulation technique.

7-16. Repeat Prob. 7-13 using a frequency modulation technique.

7-17. Why are frequency and phase modulation superior to return to zero and return to reference recording techniques?

7-18. Describe and contrast drum, disc and tape storage.

7-19. Modify the ROM of Fig. 7-16 so that it uses transistors in place of the diodes.

7-20. Repeat Prob. 7-19 but now use resistors. Quantitatively consider the problem of spurious signals.

7-21. Obtain a ROM which converts 4 bit numbers into ordinary binary in accordance with the following table.

Given numbers	New Numbers
0000	0000
0001	0001
0011	0010
0010	0011
0110	0100
0111	0101
0101	0110
0100	0111
1100	1000
1101	1001
1111	1010
1110	1011
1010	1100
1011	1101
1001	1110
1000	1111

7-22. Obtain a ROM which realizes the following functions

$$y_1 = x_1x_2x_3 + x_1\bar{x}_2x_3 + \bar{x}_1\bar{x}_2x_3$$

$$y_2 = x_1x_2\bar{x}_3 + \bar{x}_1\bar{x}_2\bar{x}_3 + \bar{x}_1x_2x_3 + x_1\bar{x}_2x_3$$

$$y_3 = x_1x_2x_3 + \bar{x}_1x_2\bar{x}_3 + x_1x_2\bar{x}_3 + \bar{x}_1\bar{x}_2\bar{x}_3$$

7-23. Repeat Prob. 7-22 for the functions

$$y_1 = x_1x_2 + \bar{x}_1x_3 + \bar{x}_2x_3$$

$$y_2 = x_1 + x_2\bar{x}_3$$

$$y_3 = x_1 + x_2$$

$$y_4 = x_1 + \bar{x}_2 + x_1\bar{x}_3$$

7-24. Design a one level decoder for four input variables.

7-25. Repeat Prob. 7-24 for a tree decoder.

7-26. Repeat Prob. 7-24 for a matrix decoder. Show all gates.

8

Binary Codes

We have thus far discussed the components of digital systems. Before turning our attention to the complete digital computer, we must consider several other topics. In this chapter, we shall discuss procedures for supplying data in a form which the computer can process. Since digital systems handle all data in binary form, all information supplied ot the computer must ultimately be converted into a binary code. These binary codes will be the subject of this chapter.

We shall start by considering numerical codes. Subsequently, we shall discuss how alphabetic data is handled.

8.1 BINARY CODED DECIMAL CODE (BCD CODE)

In digital computers, all signals are in the form of 0's and 1's. These lend themselves to binary arithmetic. Almost all large computers perform binary arithmetic. However, computer *users* work in the decimal system. In most computers, there are circuits which perform a conversion from decimal to binary when a number is supplied to the computer, and vice versa. However, in some computers, the numbers are retained in decimal form within the computer. The arithmetic operations are not as simple in this case but if they are relatively few in number, then the elimination of the conversion of radix results in an overall saving. There must be some provision for representing decimal numbers using 0's and 1's. What is done is that each decimal digit is represented by a sequence of binary digits. That is, there is a *binary code* which represents each of 0,1,2,3,4,5,6,7,8 and 9. Such codes are called *binary coded decimal* (BCD) codes. BCD type codes are often used in the input and output of data for larger computers.

The BCD codes are special examples of *weighted* codes. In such codes, each bit is assigned a weight. A weight just means a numerical value here. For instance, a very common code is the 8421 code. Here four bits are used to represent a decimal digit. If the first(rightmost)bit is not 0, then it represents a weight of 1. If the second bit is not 0, then it represents a weight of 2, etc. Thus, the number 1010 is equivalent to

$$8 + 0 + 2 + 0 = 10$$

Similarly, 1001 is equivalent to

$$8 + 0 + 0 + 1 = 9$$

Thus, a table of the 8421 BCD code is

TABLE 8-1
8421 BCD CODE REPRESENTATION

Decimal digit	(8 4 2 1)
0	0 0 0 0
1	0 0 0 1
2	0 0 1 0
3	0 0 1 1
4	0 1 0 0
5	0 1 0 1
6	0 1 1 0
7	0 1 1 1
8	1 0 0 0
9	1 0 0 1

Note that in the 8421 BCD code, each decimal digit is represented simply by its binary number.

As an example, we shall write 8624 using the 8421 BCD code. From table 8-1, we have

8 → 1000
6 → 0110
2 → 0010
4 → 0100

Thus, the 8421 BCD representation of 8624 is

1000 0110 0010 0100 = 1000011000100100

The weighted code that we have considered appears logical since each decimal digit is represented by its binary equivalent. However, it is not the only one that could be used. For instance, another weighted BCD code that is sometimes used is the 2421 code. Here the weights are 2, 4, 2 and 1. For instance, 1101 is equivalent to $2 + 4 + 0 + 1 = 7_{10}$. Negative numbers can also be used as the weights. For instance, at times, the weights 8, 4, -2 and -1 are used. Here 1011 is equivalent to $8 + 0 - 2 - 1 = 5_{10}$. Table 8-2 represents the decimal representations for these two codes.

TABLE 8-2
TWO DIFFERENT BCD CODES

Decimal digit	2 4 2 1 representation	8 4 -2 -1 representation
0	0 0 0 0	0 0 0 0
1	0 0 0 1	0 1 1 1
2	0 0 1 0	0 1 1 0
3	0 0 1 1	0 1 0 1
4	0 1 0 0	0 1 0 0
5	1 0 1 1	1 0 1 1
6	1 1 0 0	1 0 1 0
7	1 1 0 1	1 0 0 1
8	1 1 1 0	1 0 0 0
9	1 1 1 1	1 1 1 1

Then, 8624 in the 2421 representation is

1110 1100 0010 0100

Note that there is some freedom of choice here. For instance, the 2421 representation of 4 could be either 0100 or 1010. The choices made in Table 8-2 are standard ones which actually simplify arithmetic in those computers which perform arithmetic directly.

We have considered three weighted codes here. Actually, there are very many other weighted BCD codes.

There is another BCD code that has been used at times. The following scheme is used. 3 is added to each digit. The resulting base 10 number for each digit is then represented by its ordinary 8421 BCD representation. This is called an *excess* 3 code. For instance,

8476 → 11 7 10 9 → 1011 0111 1010 1001

The excess 3 code has the simple arithmetic feature discussed in conjunction with the 2421 and the 8421 codes.

8.2 REFLECTED CODES-GRAY CODES

Many times, digital systems are used in conjunction with analog ones. For instance, a digital computer may be used in a control system that varies the position of the rudder of a ship. Or a digital system may be used in the guidance system of a space probe. The position of the shaft of the rudder or of the inertial guidance platforms of the space probe must be supplied to the computer. This is done by an analog to digital converter. In such mechanical systems, the conversion is often a commutator connected to a rotating shaft. As an example, the commutator may have many rings. Each ring corresponds to a binary digit. Brushes make contact with the rings. Each ring is divided into segments which are insulated from each other, so that parts of the ring correspond to a 0 and the remaining parts correspond to a 1. Suppose that there are four rings. Then, the position of the shaft can be encoded into a four bit binary number. Suppose that the 8421 BCD code is used here. Then, the shaft position can be divided into 10 parts. (If 8 rings are used, then the position can be divided into 100 parts, etc.) If ordinary binary representation is used, then 4 rings can represent 16 positions and 8 rings can represent 256 positions, etc.

The shaft position is then given the appropriate numerical value. However, a problem can arise here when the position of the shaft is such that one or more brushes are at the junction(s) between segments of the rings simultaneously. In this case, large errors can arise. Let us illustrate this using an 8421 BCD code. Suppose that the position changes from 3, represented by 0011 to 4, represented by 0100. Three bits must change *simultaneously*. However, because of mechanical tolerances, this simultaneous change may not occur. In this case, errors can result. For instance, suppose that the second bit changes before the others. Then, the number 0000 results. This represents 0, which is not between 3 and 4 and is a serious error. This type of error can be avoided if adjacent numbers change by only one bit. Then, there is no shaft position where the contact to more than one segment must change simultaneously. A new BCD code must be used. (Note that it is not necessary that BCD codes be used here.) A code which does this is called a *reflected* code. An example of a reflected code is given in Table 8-3.

Note that each successive number changes by only one binary digit. Thus, as the position of the shaft is changed, only one commutator segment will ever be crossed, at a time. Thus, we do not have the previous problem of several bits having to change simultaneously. We have illustrated the use of a reflected code with a rotary shaft. However, such codes are often used with many types of analog to digital conversion.

TABLE 8-3
A REFLECTED BINARY CODE

Decimal number	Reflected binary number
0	0000
1	0001
2	0011
3	0010
4	0110
5	0111
6	0101
7	0100
8	1100
9	1101
10	1111
11	1110
12	1010
13	1011
14	1001
15	1000

Note the dashed line in Table 8-3. Except for the leftmost bit, there is a mirror symmetry(i.e.,reflection)about this line. This is why this is called a reflected code. This type of reflected code is also called a *Gray code*.

Let us now discuss a general procedure for obtaining a Gray code. We assume that we start with the ordinary binary representation of a number. First, place a 0 to the left of the number. Then, starting from the left, use each adjacent pair of bits and apply the exclusive OR operation (EXOR) to them. Each result of this operation yields a digit of the Gray code. Let us illustrate it with an example. Let us obtain the Gray code for 7_{10} = 111. Add a 0 to the left. This yields 0111. Then, the EXOR operations are

$0 \oplus 1 = 1$

$1 \oplus 1 = 0$

$1 \oplus 1 = 0$

Thus, the Gray code is 100 or, equivalently, 0100. Similarly, the Gray code for $65_{10} = 1000001_2$ is 1100001.

To convert from a Gray code to an ordinary binary number, we can use the following procedure. Examine the number. Each digit in the binary code is either the same as the corresponding digit of the Gray code or its complement. To determine which, count the number of 1's to the *left* of the digit in question. If it is an even number, then the digit is unchanged. If the number of 1's to the left is odd, then take the complement. For example, if we have 1100001 in Gray code, then the first digit of the regular binary number remains a 1 since there are an even number (zero) of ones to the left. The second digit must be changed since there is an odd number of ones to the left in the original (Gray)number. Thus, we can change the 1 to a 0. Proceeding in this way, we obtain $1000001_2 = 65_{10}$, as we should.

8.3 ALPHANUMERIC CODES

We have thus far concentrated on numbers. However, often computers process data which contains letters or other symbols. For instance, lists of names can be alphabetized. Student's grades can be stored and printed by a computer. Also, computer programs are written in terms of words and symbols. Thus, the computer must be able to process all of these. All information in a digital computer is stored or processed in binary form(i.e., as 0's and 1's). Thus, all data, letters, symbols, etc. must have a binary representation or binary code. Various codes are used for this purpose.

Before discussing the codes, let us discuss some background material. Information is not only processed within the computer. It must also be transmitted to it by the user and the answers must be transmitted from the computer to the user. At times, this information must be transmitted over great distances. For instance, the user may be transmitting his information, using a teletype and a telephone coupler, over telephone lines. The coupler generates an appropriate digital code which can be transmitted over telephone lines. Even when the user is close to the computer, information must be transmitted, by means of a suitable code, to and from such input-output devices as teletypes, card readers, etc. These are external to the central processor of the computer. We shall consider some codes that are in use. In Chapter 6 we considered the electronic circuits that are associated with coding and decoding(i.e., sequence generating and detecting circuits). These codes provide for transmitting 26 letters, 10 digits, and various symbols. They are called *alphanumeric* codes since numbers and letters are transmitted. Two representative codes are given in Table 8-4. One is called the EBCDIC(extended BCD interchange code). This code uses eight bits. The other is called the ASCII(American standard code for information interchange.) Note that there are several versions of this code. A recent modification of this code is called ANSCII(American national standard code for information interchange).

In both of these codes, the numerical values, in binary, of the alphabet characters are in increasing order. This is done to allow such things as alphabetizing to be done easily.

TABLE 8-4
SOME ALPHANUMERIC CODES

	ASCII code	EBCDIC code
A	100 0001	1100 0001
B	100 0010	1100 0010
C	100 0011	1100 0011
D	100 0100	1100 0100
E	100 0101	1100 0101
F	100 0110	1100 0110
G	100 0111	1100 0111
H	100 1000	1100 1000
I	100 1001	1100 1001
J	100 1010	1101 0001
K	100 1011	1101 0010
L	100 1100	1101 0011
M	100 1101	1101 0100
N	100 1110	1101 0101
O	100 1111	1101 0110
P	101 0000	1101 0111
Q	101 0001	1101 1000
R	101 0010	1101 1001

S	101 0011	1110 0010
T	101 0100	1110 0011
U	101 0101	1110 0100
V	101 0110	1110 0101
W	101 0111	1110 0110
X	101 1000	1110 0111
Y	101 1001	1110 1000
Z	101 1010	1110 1001
0	011 0000	1111 0000
1	011 0001	1111 0001
2	011 0010	1111 0010
3	011 0011	1111 0011
4	011 0100	1111 0100
5	011 0101	1111 0101
6	011 0110	1111 0110
7	011 0111	1111 0111
8	011 1000	1111 1000
9	011 1001	1111 1001
blank	000 0000	0100 0000
.	010 1110	0100 1011
(	010 1000	0100 1101
+	010 1011	0100 1110
)	101 1101	0100 1111
(	101 1011	0101 1010
$	010 0100	0101 1011
*	010 1010	0101 1100
)	010 1001	0101 1101
;	011 1011	0101 1110
,	010 1100	0110 1011
-	010 1101	0110 1101
=	011 1101	0111 1110

We have shown some symbols such as +, -, *, etc. here. Actually, provision for many more symbols is made. We have just given a representative list here. Note that the codes may vary from computer to computer. The ones given here are fairly representative.

These codes appear to be unduly long. For instance, the reader may ask why use seven or eight bits to transmit a decimal digit when BCD codes require only four bits. There are several reasons for this. The codes transmit much more information than just the ten digits. For purposes of detection, it is desirable to have each symbol the same length. Another reason for having the *code word*(i.e., the number of bits for each symbol)longer than necessary is to detect or eliminate errors which occur during transmission. We shall discuss this next.

8.4 ERROR DETECTING AND CORRECTING CODES

When data is transmitted from a remote location to a computer via telephone, or other communication channels, there is a reasonable probability that some errors will occur. For instance, noise introduced on the line may make a 0 appear as a 1 or vice versa. Thus, it is desirable to have some way of knowing if an error has occurred, and, if possible, to correct it.

The usual transmission channels are not very noisy and the probability of an error in a single bit is usually low. Thus, the probability of an error in more than one bit in a code word is very low. Let us assume that this is so and discuss a simple procedure that will allow *single* errors to be detected. In this way an error message can be generated and the word can be retransmitted.

We shall illustrate this with the 8421 BCD code given in Table 8-1. An extra bit called a *parity bit* is added to the code. This bit value is such that the number of 1's is always even(or odd). This is called an *even(odd) parity check.* The resulting code is called a parity checking code. For instance, an even parity checking code is illustrated in Table 8-5.

Note that, in each case, the parity bit is such that the number of 1's is even. (Note that zero is considered to be an even number.)

If we are to have error detection, then the code must be inefficient in the sense that more bits than necessary are used simply to transmit the information. For instance, the change of a single bit must not produce another valid code word if an error is to be detected. If the change of a single bit produced a

TABLE 8-5
8421 BCD CODE WITH EVEN PARITY CHECK

Decimal digit	(8 4 2 1 p)
0	0 0 0 0 0
1	0 0 0 1 1
2	0 0 1 0 1
3	0 0 1 1 0
4	0 1 0 0 1
5	0 1 0 1 0
6	0 1 1 0 0
7	0 1 1 1 1
8	1 0 0 0 1
9	1 0 0 1 0

valid code word, then there would be no way of knowing that an error had occurred, since the changed word could have been the one transmitted. That is, to detect an error, it must result in a "word" that is not in the table. To be able to describe things such as this numerically, we define the *Hamming distance* between two code words as the number of bits which must be changed to change one word into the other. The *minimum distance* of a code is the smallest distance between *any* two words of the code. For instance, in Table 8-1, the minimum distance is one, while in Table 8-5, the minimum distance is two.

If a code is to be such that a single error can be detected, then the *minimum distance must be at least two*. Again, if the minimum distance is one, then a single error can produce another valid code word and an error cannot be detected. If the minimum distance is three, then two errors can be detected, etc.

If the minimum distance is three, or more, then single errors can be not only detected, but they can also be corrected. That is, there is sufficient information present so that we not only know that an error has occurred, but we can also determine in what digit the error has occurred. Once this is known, the errors can be corrected by replacing the erroneous digit by its complement, that is by changing a 0 to a 1 or vice versa. We shall now illustrate a code which is error correcting if single errors occur. It is called a *Hamming code*. In these codes, several parity checking bits are added. Parity is then checked for groups of bits. Consider a code which has four bits. The 8421 code is one such. Let us add three parity bits so that a code word is of the form

$$p_1 \quad p_2 \quad w_3 \quad p_4 \quad w_5 \quad w_6 \quad w_7 \tag{8-1}$$

The subscripts indicate the position of the bit in the word. Note that w_3, w_5, w_6 and w_7 would be the word if no parity checking were used. Then, the parity bits are adjusted so that there are an even number of 1's in the following groups of positions labeled c_0, c_1 and c_2.

$$c_2 \text{ positions } 4\ 5\ 6\ 7 \tag{8-2a}$$

$$c_1 \text{ positions } 2\ 3\ 6\ 7 \tag{8-2b}$$

$$c_0 \text{ positions } 1\ 3\ 5\ 7 \tag{8-2c}$$

Note that only one parity bit is in each checking position. Thus, the parity bits do not interact with each other.

As an example, let us use the 8421 BCD code. The Hamming code is

TABLE 8-6
8421 BCD CODE WITH HAMMING ERROR CORRECTION

Digit	Position	1	2	3	4	5	6	7
		p_1	p_2	8	p_4	4	2	1
0		0	0	0	0	0	0	0
1		1	1	0	1	0	0	1
2		0	1	0	1	0	1	0
3		1	0	0	0	0	1	1
4		1	0	0	1	1	0	0
5		0	1	0	0	1	0	1
6		1	1	0	0	1	1	0
7		0	0	0	1	1	1	1
8		1	1	1	0	0	0	0
9		0	0	1	1	0	0	1

When the word is received, the three parity checks are made and any error is detected and corrected.

For instance, suppose that 7 is incorrectly transmitted as 0001011, then c_2 will fail, c_1 will not and c_0 will fail. Thus, the error must be in position 5. Note that c_0 and c_2 have errors in position 5 and 7 in common. Thus, if c_0 and c_2 fail, the error could be either in position 5 or in position 7. However, c_1 did not fail. Therefore, the error is not in position 7. Thus, the error is in position 5. Complementing the fifth bit, we obtain the correct value 0001111. Comparing Eqs,(8-2), we obtain the following table for the position of failure. Here T indicates no failure and F indicates a failure.

TABLE 8-7
POSITION OF FAILURE FOR HAMMING CODE OF TABLE 8-6

Test			Digit of failure
c_2	c_1	c_0	
T	T	T	None
T	T	F	1
T	F	T	2
T	F	F	3
F	T	T	4
F	T	F	5
F	F	T	6
F	F	F	7

In the above table, if we consider the trues(no error) as 0's and the falses(error) as 1's, then the binary number resulting from the parity check indicates the position of the error. Note that transmission errors in the parity bits are also corrected.

The Hamming detecting procedure can be applied to codes with more than four(original) bits. In general, the parity checking bits are placed in position 1, 2, 4, 8, 16, 32, 64,..., 2^n. The bits which are checked can be determined using the following scheme. p_1 checks all bits whose position(subscript) in binary has a 1 in the 2^0 position. For instance, 001, 101, 111 all have

1's in that position. Then, p_2 is a parity check for all bits whose binary representation has a 1 in the 2^1 position(i.e., 010, 111, 011)etc. Using this scheme, Hamming codes can be generated for words of any length. Note that no parity bit checks any other parity bit. Thus, the parity bits are all independent. We have illustrated some error correcting codes. However, there are very many such codes that are used.

BIBLIOGRAPHY

Booth, T.L., "Digital Networks and Computer Systems", Chapter 2, John Wiley and Sons, Inc., New York, 1971

Flores, I., "Computer Design", Chapter 9, Prentice Hall, Inc., Englewood Cliffs, N.J., 1967

Kohavi,Z., "Switching and Finite Automatic Theory", Chapter 1, McGraw-Hill Book Company, N.Y. 1970

Mano,M.M., "Computer Logic Design", Chapter 1, Prentice Hall, Inc., Englewood Cliffs, N.J. 1972

O'Malley, J., "Introduction to the Digital Computer", Chapter 5 Holt, Rinehart and Winston, Inc., New York, 1972

Steinhart, R.F. and Pollack, S.V., "Programming the IBM System/360", Chapter 2, Holt, Rinehart and Winston, Inc., New York, 1970

PROBLEMS

8-1. Write the following numbers using an 8421 BCD code.

1796

437892

1264.36

8-2. Repeat Prob. 8-1 using a 2421 BCD code.

8-3. Repeat Prob. 8-1 using an 84-2-1 BCD code.

8-4. Write a table for a 74-2-1 BCD code.

8-5. Obtain a BCD code which uses different weights than those used previously.

8-6. Obtain a Gray code for numbers between 0 and 32.

8-7. Obtain the Gray code representation for 1785_{10} and 1236_{10}.

8-8. Obtain the ordinary binary representations of the following which are given in Gray code.

1010110

111011011

8-9. Write the ASCII representation of the word BOOK. If this is to be stored in a register which stores 4n bits(n=1,2,...) what is the smallest value that n can have?

8-10. Repeat Prob. 8-8 using an EBCDIC representation.

8-11. Rewrite Table 8-5 but now use odd parity checking.

8-12. Modify Table 8-2 so that an even parity check can be made.

8-13. The following data are sent with 8421 BCD code with even parity checking. Which are in error? Assume that there is, at most, one error.

01011

10001

01111

8-14. The following are data sent with an 8421 BCD code with Hamming error correction. Idnetify the data sent correcting any errors if necessary. Assume that there is, at most, one error.

1101001
1101010
1001101
0000001
1000000
1001111

8-15. Obtain the Hamming error correction code for the 2421 BCD code.

8-16. Discuss how Hamming error correction would be applied to an 8 bit(uncorrected) code.

9

Basic Computer Arithmetic

In this chapter we shall consider the details of the arithmetic that is used in a digital computer. In Chapter 2 we discussed the basic ideas of addition using binary numbers. Here we shall consider all pertinant aspects of binary arithmetic and relate them to digital computers.

9.1 ACCUMULATORS-MODULAR ARITHMETIC

In Sec. 2-2 we discussed addition in the binary system. Let us now see how this is implemented. Addition is performed in a device called an *accumulator*. Actually, an accumulator can be considered to be a register which adds. For instance, suppose that we have an 8 bit accumulator. Then, as a register, it can store an 8 bit binary number. However, if a binary number is added to the accumulator, then the new stored(binary) number will be the sum of the original stored number plus the added number. In Chapter 10 we shall actually consider details of hardware that make up an accumulator. In this chapter we shall consider the arithmetic details of the accumulator's operation.

In the simplest case, the number stored in the accumulator is, as discussed, the sum of two numbers which are added. However, problems can arise if the numbers added are sufficiently large. In this case, an overflow can result, see Sec. 2-2. For instance, an overflow will occur if we are working with an 8 bit accumulator(i.e., it can store 8 binary digits)and if the sum of the two numbers which are added results in a 9 bit binary number. Since such overflows result in a loss of the most significant figure, see Sec. 2-2, they must be avoided. Provision is usually made to indicate that an overflow has occurred, so that the computer user is warned. Computation may stop when an overflow occurs. In the next section we shall see that, at times, this process can be used to advantage. However, for the time being, let us assume that overflow is undesirable.

Since the number of bits in a register or accumulator is limited, arithmetic using them assumes a special form, which we shall now discuss. We shall start with decimal numbers and then use binary numbers. Consider an accumulator which can store three decimal numbers. An example of such a "decimal accumulator" is an automobile odometer. Let us assume that it is set at zero and numbers are added to it, one at a time. The sequence of its readings will be

000
001
002
003
.
.
.
009
010
011
012 (9-1)
.
.
.

996
997
998
999
000
001
002
.
.
.

The accumulator can store numbers from 000 to 999. If the sum becomes greater than 999, then the *accumulator cycles again*. This is a characteristic which is common to almost all accumulators including those found in computers. Thus, if the accumulator reads 041, this could actually represent any of the numbers

$$\begin{array}{r} 41 \\ 1041 \\ 2041 \\ \vdots \end{array} \tag{9-2}$$

(There could also be negative representations. However, we shall defer a discussion of these.) We shall now discuss an arithmetic which can be used to not only characterize registers and accumulators, but also to study arithmetic processes used with accumulators. Using this we shall place restrictions on the input to accumulators so that they do not overflow. That is, these restrictions will insure that the uncertainty of sequence (9-2) does not occur. We shall also use this arithmetic to find ways of using overflow for useful purposes.

Now let us define a mathematical process called *modular arithmetic* which is of fundamental importance in registers. Two integers M and N are said to be *congruent, modulo k* if M/k and N/k have the same remainder. In this case, we write

$$M \equiv_k N \tag{9-3}$$

k is called the *modulus*, and is an integer. Let us consider an example

$$17 \equiv_3 29 \tag{9-4}$$

Note that 17/3 = 5 + 2/3, and 29/3 = 9 + 2/3. In each case, the remainder is 2. Thus, 17 and 29 are congruent, modulo 3. Other terminology and symbols are used. For instance, in (9-3), M and N are also said to be *equivalent, modulo k*. Another symbolic representation for (9-3) is

$$M \equiv N(\text{mod } k) \tag{9-5}$$

If Eq.(9-3) is satisfied, then we can write

$$\frac{M}{k} - \frac{N}{k} = I \tag{9-6}$$

where I is an integer. Note that if M/k and N/k have the same remainder, then their difference will be an integer. Thus, Eq. (9-6) is a criterion which can be used to test if two numbers are congruent, modulo k. Now let us see how modular arithmetic applies to registers. Consider the register of (9-1) and the example (9-2). For the register, we demonstrated that the numbers, 41_{10}, 1041_{10}, 2041_{10}... would all produce the same reading. We can now express this using modular arithmetic.

$$41 \equiv_{1000} 1041 \equiv_{1000} 2041 \equiv_{1000} \cdots \tag{9-7}$$

Note that 41/1000, 1041/1000, 2041/1000,... all have the remainder 41. Thus, we can use modular arithmetic to characterize the reading of the register.

Let us now discuss binary accumulators or registers. As an example, we consider a three bit binary accumulator. (This contains fewer bits than those usually encountered. This has been done to keep the example simple.) Assume that the register is set at 0 and numbers are added to it one at a time.

$$
\begin{array}{l}
000 \\
001 \\
010 \\
011 \\
100 \\
101 \\
110 \\
111 \\
000 \\
001 \\
010 \\
\vdots
\end{array}
\tag{9-8}
$$

These correspond to the decimal numbers 0,1,2,3,4,5,6,7,.... Now consider a specific number. A reading of 101 could correspond to 5, 13, 21, 29.....(We shall again restrict outselves to positive integers.) In terms of modular arithmetic, we have

$$5 \equiv_8 13 \equiv_8 21 \equiv_8 29 \equiv_8 \cdots \tag{9-9}$$

Note that for each of these N/8 yields a remainder of 5.

Let us generalize these results. Consider a register whose highest number is R. For instance, in the register of (9-8), the register ranges between 0 and 7. Thus, R=7. We either want the register to store a number N, or the accumulator to compute a sum which is equal to N without overflow. That is, the value stored is to be the *true value* of N. Then,

$$N \leq R \tag{9-10}$$

where we consider that N and R are positive integers. If we are to avoid overflow, then relation (9-10) must be satisfied. Let us now consider this in terms of a binary register. If the register has b bits, then $R = 2^b-1$. Thus, relation (9-10) becomes

$$N < 2^b - 1 \qquad (9\text{-}11)$$

Relation (9-11) is valid for a binary radix accumulator or register. At times, but usually not in computers, accumulators are constructed using other radices. For instance, an automobile odometer is a base 10 register. In general, if there are d digits in the accumulator, then $R = r^d-1$. Thus, we can generalize relation (9-11) to be

$$N < r^d - 1 \qquad (9\text{-}12)$$

Thus far we have only considered positive numbers. Of course, almost all computers work with negative numbers as well. Thus, some provision must be made to store the sign of the number. The leftmost digit of the register is usually used for this purpose. In general, *a 0 or a 1 in the leftmost digit does not simply replace the sign*. For instance, in one system, 0111 would represent 7 while 1001 would represent -7. (note $001_2 \neq 7_{10}$) We shall consider such systems in the next section.

Now, if we have a K bit register, only K-1 of the bits are available for the magnitude of the number. Thus, relation (9-11) becomes

$$|N| \leq 2^{K-1} - 1 \qquad (9\text{-}13)$$

if there is to be no overflow. We shall modify this somewhat in the next section.

We have defined congruence. Now let us consider arithmetic (e.g., addition and subtraction) on a modular basis, since this is the basis on which registers and accumulators function.

Addition

Let us assume that we are doing addition on a binary basis and that the accumulators or registers have k binary bits. Also assume that k_1 of these are available for storing the magnitude of the number. For instance, if one bit is always 0 if the number is positive, then

$$k_1 = k - 1 \tag{9-14}$$

We shall again restrict ourselves to positive integers. The binary addition process follows that discussed earlier in this section. Since there are k bits, the modulus of our modular arithmetic system is 2^k. Now suppose that we add two integers N_1 and N_2 and that after adding them, the resulting total in the accumulator is T. There may be overflow. However, whether there is or not, the total stored in the accumulator is

$$T \equiv_{2^{k_1}} N_1 + N_2 \tag{9-15}$$

T is the smallest positive integer which satisfies Eq.(9-15). Note that the accumulator reading cannot exceed $2^k - 1$. For instance, if $k_1 = 4$ and $N_1 = 6$ and $N_2 = 4$, then

$$T \equiv_{16} 6 + 4$$

In this case, the value read on the register will equal the true sum 10. Of course, these numbers will be in binary form. We write them in base 10 for human convenience. Now suppose that N_1=12 and N_2=8. Then, the smallest integer value of T which satisfies this is 4. This will be the value indicated by the accumulator. Note that an overflow has occurred in this case. This also could have been determined from relation (9-13).

Subtraction

We shall now consider subtraction. At the start we shall discuss the basic ideas and then apply these to computer subtraction. We will briefly consider radix 10 and then turn our attention to the binary case. Suppose that we subtract

$$\begin{array}{rl} 2965_{10} & \text{minuend} \\ -\ 1432_{10} & \text{subtrahend} \\ \hline 1533_{10} & \text{difference} \end{array}$$

Here each digit of the subtrahend is subtracted from the corresponding digit of the minuend to obtain the digits of the difference. Now suppose that a digit of the subtrahend is greater than the corresponding digit of the minuend. In this case, we "borrow" from the digit to the left. Let us illustrate this with the subtraction 6345 - 3726. We write this as

```
 -1 -1
  6345     →      5 13  3 15
- 3726          - 3  7  2  6
  2619            2  6  1  9
```

The first digit of the subtrahend, 6, is greater than the first digit of the minuend 5. Thus, we borrow 10 from the next column. Column 2 now represents 30 while the first column is now 15. Then, 15-6 yields 9. In the second column, we now have 30-20 which is represented by 1 in the second digit of the answer. Now consider the third column. 300 is less than 700 and thus we borrow 1000 from the fourth column and the subtraction proceeds. Let us consider an additional example.

```
 -1 -1 -1
  6345     →     5 12 13 15
- 1346           1  3  4  6
  4999           4  9  9  9
```

Notice how the borrowing in the first column causes borrowing to take place in the remaining columns.

We have discussed the subtraction of positive integers. It is assumed that the reader is familiar with the rules of arithmetic which relate to subtraction of negative numbers. Similarly, it is assumed that the reader is familiar with the arithmetic operation that results if the subtrahend is larger than the minuend [i.e., a-b = - (b-a)].

Now let us work with subtraction on a digital basis. For instance,

```
 -1
 ↷
 1011   →    0 10 1 1
-0101      - 0  1 0 1
-----      ----------
 0110        0  1 1 0
```

In the first two columns, the digits of the subtrahend are not greater than those of the minuend and "simple subtraction" takes place. In the third column, we must borrow. That is, 4_{10} is greater than 0. Thus, 8_{10} is borrowed from the fourth column.

As a second example, let us consider the subtraction

```
 -1 -1 -1
 1000   →    0 1 1 10
-0011       -0 0 1  1
-----       ---------
 0101        0 1 0  1
```

Since 1 is greater than 0, we borrow $2_{10} = 10_2$ from the second column. However, the second column of the minuend was 0. Thus, we must borrow $4_{10} = 100_2$ from the third column. This is also 0. Thus, we must borrow $8_{10} = 1000_2$ from the fourth column. The final representation is 0 1 1(10). Observe the digit 10 representing 2_{10}. Note that in the second column we have borrowed four from the third column, but we have also supplied 2 to the first column. Thus, 10_2 is left. This is written as 1_2 since it is in the second column. When several zeros are adjacent to each other, a borrow propagates through all of them. This can be awkward. In the next section we shall investigate procedures which remove some of this awkwardness.

For illustrative purposes, let us consider one final example where the minuend is less than the subtrahend. In this case, we shall subtract the minuend *from* the subtrahend and write the number as a negative one. For instance,

```
  1010          {  1111 }
- 1111        - { -1010 }
------           ------
- 0101           -0101
```

Now let us consider subtraction on the basis of registers and accumulators. Again, for clarity, we shall start the discussion by considering base 10. Consider a three digit accumulator. As it is cycled(i.e., 1 is added successively)we have

$$\begin{array}{l} \vdots \\ 996 \\ 997 \\ 998 \\ 999 \\ 000 \\ 001 \\ 002 \\ 003 \\ 004 \\ 005 \\ \vdots \\ 998 \\ 999 \\ 000 \\ 001 \\ \vdots \end{array} \tag{9-16}$$

Suppose we have 5 entered in the accumulator and we subtract 2. The accumulator is cycled backward two steps to 3. Now suppose that we have 5-7. The accumulator is stepped back 7 steps to 998. The answer should be -2. Thus, an underflow(i.e., the equivalent of an overflow when the number becomes too small,) has resulted. Let us consider this on the basis of modular arithmetic.

$$\frac{-2}{1000} = 0 - \frac{-2}{1000}$$

$$\frac{998}{1000} = 1 - \frac{2}{1000}$$

Each of these have the same remainder(i.e., -2). Thus, we can say

$$-2 \equiv_{1000} 998 \qquad (9\text{-}17)$$

Thus, on a modular basis, we have the correct answer. Actually, this is not a real overflow since we can write 5-7 = -(7-5) = -(2). Thus, if the minuend and the subtrahend had been interchanged *before* subtraction, then a number within the range of the accumulator would have been obtained. This assumes that there is some provision for obtaining the correct sign. We shall consider such procedures in the next section.

An overflow can occur if the magnitude of the answer is too large. For instance, 999 - (-20) = 1019 which is a true overflow. Let us consider another aspect of the previous example. 998/1000 = 0 + 998/1000. In this case, the remainder is not -2 but 998. Consider Eq. (9-15). This states that two integers are equivalent, modulo k if their difference, divided by k is an integer. Then,

$$\frac{999}{1000} - \frac{(-2)}{1000} = 1 \qquad (9\text{-}18)$$

Thus, the definition is satisfied. Now let us consider base 2. We shall again illustrate the ideas with a three bit register. As it is cycled, the readings are

⋮
100
101
110
111
000
001
010
011
100
101
110
111
000
001
010
⋮

Now let us perform 5-7 or 101-111. Cycling the register backward from 101 we obtain 110. On a modular basis, we have

$$6 \equiv_8 -2 \tag{9-19}$$

Note that 6/8 = 1 - 2/8. Thus, we have achieved the correct answer on a modular arithmetic basis. Again, an overflow appears to have resulted. However, this can be avoided if we write 101-111 = -(111-101) = -010. Thus, as long as we can keep track of the minus signs, we will not have an overflow unless Eq. (9-13) is violated. That is, an overflow will only result if the magnitude of the answer is too large(in this case, greater than 7). Actually, we have oversimplified here. In the next section we we shall consider subtraction and negative numbers in greater detail.

We have considered subtraction by cycling the accumulator backwards. This procedure is convenient if the accumulator is an automobile odometer. However, in a digital computer, such backward cycling of an accumulator is awkward. Circuits can be built which can add and, similarly, other circuits can be built which can subtract, but these are different circuits. Circuits can be built which can both add and subtract but these are more complex than those which only add(or subtract). We shall next consider a subtraction procedure which essentially uses only an adding circuit and, thus, reduces the complexity of the computer.

9.2 COMPLEMENT ARITHMETIC

To reduce the complexity of a digital computer, it would be desirable to utilize the same circuit to perform both operations of addition and subtraction. In this section, we shall demonstrate how an adding accumulator can also be used to perform subtraction by taking account of its "overflow properties"

Let us illustrate the procedure that we shall use. Suppose we want to compute the difference of two numbers A and B. Now let us assume that R-1 is the largest number stored in the accumulator. The accumulator operations are thus defined modulo R. The difference that we want is

$$D = A - B \tag{9-20}$$

Now suppose that we form the number N in the following way

$$N = R - B \tag{9-21}$$

Note that we have performed a subtraction here. However, it is a special subtraction in that the minuend is equal to R. We shall subsequently see that this *particular* subtraction is especially simple and does not require special subtractor circuits in the digital computer. Now suppose we take the *sum* of A and N.

$$A + N = A + R - B$$

Since R/R = 1, an integer, we have

$$D \equiv_R A-B \equiv_R A+N \tag{9-22}$$

Thus, the sum A+N and the difference D are congruent, modulo R. Hence, each produces the same result in the accumulator. Let us illustrate this using base 10. Suppose that A=80 and B=6 and R=1000. That is, there are three decimal digits. Then,

$$A-B = 80 - 6 = 74$$

Now

$$N = R - B = 994$$

$$A + N = 80 + 994 = 1074$$

$$1074 \equiv_{1000} 74$$

Thus, the accumulator will store the correct value of 74. Let us now consider an example in binary. We shall assume a seven bit accumulator. Then,

$$A = 80_{10} = 1010000_2$$

$$B = 6_{10} = 0000110_2$$

Since there are 7 bits, R will be an 8 bit number.

$$R = 128_{10} = 10000000_2$$

Then,

$$N = R - B = 10000000 - 0000110$$

Performing the subtraction, we obtain

$$N = 1111010$$

Hence,

$$A + N = 1010000 + 1111010 = 11001010$$

$$11001010_2 \equiv_{128} 1001010_2$$

The above statement is equivalent to

$$202_{10} \equiv_{128} 74_{10}$$

Thus, we obtain the desired result.

The process of subtracting R - B is awkward in digital systems since R will equal 1 followed by a string of zeros. Thus, extensive carrying will have to be performed. However, we can simplify this. We shall illustrate this simplification with an example. Suppose that

$$R = 10000000$$

We can write this as

$$R = 1111111 + 0000001$$

When we take R-B we can subtract B from 1111111 and then add 0000001 to the result. No carrying will be required here since each digit of 1111111 will always be greater than the corresponding digit of B. Note that R will have one more digit than B. Remember that R is a number one greater than the maximum number stored by the accumulator. Let us consider the subtraction of a number from 111...111, where each has the same number of bits.

For example

```
  1111111
- 1011010
  0100101
```

Note that the answer is equal to the subtrahend with each digit reversed. That is, 0's are replaced by 1's and vice versa. Note that this will always happen independent of the number of binary digits. The subtrahend and minuend must have the same number of digits. Thus, if needed, we must fill in 0's to the left. For instance, if we have $11111_2 - 101_2$ we write

```
  11111
- 00101
  11010
```

Thus, to obtain R-B, we reverse the digits of B, after filling in 0's if necessary, and then add 00...01.

The process of reversing the binary digits is not only simple for humans, it is also simple to perform in the computer. In fact, as we have seen, many computer circuits are such that both a binary number and a number with reversed binary digits are simultaneously available. Thus, no circuitry need be added to reverse the bits. Hence, subtraction can be performed using adding accumulators, without any special subtracting accumulator's being necessary.

We shall also demonstrate that the ideas just discussed will enable us to easily represent negative numbers. Let us now formalize some of the previous discussion by defining some mathematical terms.

The r's Complement

We define the r's complement of a number M in radix r using the following relation

$$C_r = r^k - M \qquad (9\text{-}23)$$

If we are working with a register, then in general, the maximum number stored is r^k-1. Thus, if we have a k digit binary register, i.e., k is the total number of bits, then the 2's complement is

$$C_2 = 2^k - M \tag{9-24}$$

For instance, in the previous discussion, R-B represents the 2's complement of B. (The definition of a complement can be broadened on a modular arithmetic basis. However, this definition will suffice for our purposes.)

Thus, if we want to subtract two numbers, i.e., A-B, we can accomplish this by adding A and the 2's complement of B. This assumes that we are using modular arithmetic and binary digits and that the exponent k in Eq. (9-24) is equal to the number of bits of the register.

Let us summarize the previous results. We use the following procedure to obtain the 2's complement of a binary number. (1) Write the number with the proper number of digits. (2) Change the bits to their opposite values(i.e., replace 0's by 1's and vice versa); and (3) add one. For example, if k=7, let us obtain the 2's complement of 101_2. Then we have $101_2 = 0000101_2$. Changing the bits and adding 1, we have

$$\begin{array}{r} 1111010 \\ +\ 0000001 \\ \hline 1111011 \end{array}$$

Thus, 1111011 is the 2's complement we desire.

We can generalize the above result to the r's complement. For instance, suppose we want to obtain the 10's complement of a number. From (9-23) we have

$$C_{10} = 10^k - M = 10000\ldots0 - M \tag{9-25}$$

This can be written as

$$C_{10} = 999\ldots999 - M + 1 \tag{9-26}$$

Use of 2's Complement to Represent Negative Numbers

The 2's complement is very useful if we want to represent a negative number. We have mentioned that the leftmost bit can carry sign information. Let us now formalize this. If a number B is the negative of another number A, then

$$A + B = 0 \tag{9-27}$$

Thus, we can express B uniquely in terms of A. In the case of computers, we must work with modular arithmetic. Thus, if we have a k bit register

$$A + B \equiv_{2^k} 0 \tag{9-28}$$

Now there is *not* a unique relationship between A and B since there is an ambiguity of 2^k. For purposes of example, assume that A is negative and B is positive. Now assume that the register has k bits. If the magnitude of the number is not to produce an overflow, then it must be less than $2^{k-1} - 1$, see Eq.(9-13). Let us assume that this is so. If B represents the number and

$$|B| \leq 2^{k-1} - 1 \tag{9-29}$$

then the 2's complement of B will be

$$C_2 = 2^k - B \tag{9-30}$$

This will always be a positive number. From Eq.(9-27) we have

$$B = -A \tag{9-31}$$

Then, substituting in Eq.(9-30), we obtain

$$C_2 = 2^k + A \tag{9-32}$$

In terms of modular arithmetic, we have

$$C_2 \equiv_{2^k} A \equiv_{2^k} - B \tag{9-33}$$

Thus, in all operations involving modular arithmetic(e.g., using accumulators)we can replace a negative number by the 2's complement of its magnitude. Let us illustrate this with an example. Suppose that we have a four bit binary register. For positive numbers we used only three of the bits. That is, for positive numbers, the leftmost bit is 0. Now we shall use all four bits since negative numbers will be considered. Consider -1. We represent it by the 2's complement of +1. Thus, we use

$$C_2 = 1110 + 0001 = 1111$$

Similarly, for -2, we use the 2's complement of +2. Hence,

$$C_2 = 1101 + 0001 = 1110$$

etc.

In Table 9-1 we represent the 2's complement of the negative numbers for -1 to -7.

TABLE 9-1
FOUR BIT BINARY REPRESENTATION OF NUMBERS FROM –7 TO +7. THE 2'S COMPLEMENT IS USED FOR THE NEGATIVE NUMBERS

Decimal number	Binary number
7	0111
6	0110
5	0101
4	0100
3	0011
2	0010
1	0001
0	0000
-1	1111
-2	1110
-3	1101
-4	1100
-5	1011
-6	1010
-7	1001

Note that if the first digit is a 0, then the number is positive.

while if the first digit is a 1, the number is negative. *Note that here we do not make a number negative by simply replacing the leading 0 by a 1.* Again note that if we have an accumulator or register using k bits, we can store an integer whose magnitude is equal to or less than $2^{k-1}-1$. Finally, consider the negative numbers. If the leftmost digit is ignored, then their decimal equivalents are 1 to 7 in *reverse* order.

Note that there is a four digit binary number that we have *not* used. It is 1000. This is the 2's complement of 8_{10}. Thus, using 4 binary digits we can represent numbers for -8 to +7 using 2's complement. In general, when 2's complement is used with k bits, we can represent a number N in the range

$$2^{k-1} - 1 \geq N \geq -2^{k-1} \tag{9-34}$$

Note that there is an "extra" negative number.

9.3 2'S COMPLEMENT SUBTRACTION

We can now formalize our rules of subtraction. At the start of the last section, we demonstrated that subtraction could be accomplished by replacing the subtrahend by its 2's complement and adding. The previous results which demonstrate that negative numbers can be represented by their 2's complement substantiate this. In the previous example, we did not allow for the sign bit since, for simplicity, we only worked with positive differences. When we work in general, we must include the sign bit to account for negative integers. Note that all the results are still valid. For instance,

$$11101 \underset{2^3}{\equiv} 1101 \underset{2^3}{\equiv} 101$$

The follows from the fundamental definition of binary numbers.

$$a_n a_{n-1} \ldots a_2 a_1 a_0 = a_n 2^n + a_{n-1} 2^{n-1} + \ldots + a_2 2^2 + a_1 2^1 + a_0 2^0$$

where a_j=0 or 1. If we represent this modulo 2^k, then all coefficients whose subscripts are k or higher can be ignored since they represent integral multiples of 2^k. Thus, if we have a four bit register and we want to subtract A-B where

$A = 0011_2 = 3_{10}$

$B = 0100_2 = 4_{10}$

we obtain the 2's complement of B

$$\begin{array}{r} 1011 \\ +\ 0001 \\ \hline 1100 \end{array}$$

Then, adding this to A, we obtain

$A - B = 1111$

This represents -1_{10} as it should, see Table 9-1.

Now let us consider another example. Suppose we want to subtract A - B where

$A = 1111_2 = -1_{10}$

$B = 0100_2 = 4_{10}$

From the previous example, the 2's complement of B is 1100. Then adding this to A, we obtain

$$\begin{array}{r} 1111 \\ 1100 \\ \hline 11011 \end{array}$$

Note that there is an apparent overflow. However, from the previous discussion we have

$11011_2 \equiv_{2^3} 1011_2$

That is, we can *ignore* the fifth digit, which has been caused by a carry. Hence,

$A - B = 1011_2 = -5_{10}$

The answer is correct. Note that the above was *not* an example of an overflow. The only time that a *true overflow* occurs is when the magnitude of the resulting number is too large, see Eq.(9-34). In this case, the allowable range is between 2^3-1 and -2^3.

Actually, we can study the leftmost digit to determine if a true overflow has occurred. The only way that a true overflow can possibly occur is if we add two numbers of the same sign or subtract two numbers of opposite sign. Any other operation will result in an answer which is smaller in magnitude than the magnitude of at least one of the original numbers. Addition is always performed. When we want to subtract, we take the 2's complement and add. If we are actually adding two positive numbers, then the leftmost digit of each will be a 0. If their sum causes a 1 to be carried into the leftmost digit then this is actually attempting to represent the magnitude and thus, will result in a true overflow. On the other hand, if we add a positive to a negative number, then a true overflow cannot occur. In this case, if a 1 is carried into the leftmost digit, it will be added to a 1(representing a negative number)which is already there. This will be carried out(i.e., there will be a 1 carried to the left of this "leftmost" column). In this case, there is not a true overflow and this carried digit is ignored. Now consider adding two negative numbers. In the 2's complement, the large magnitudes will have "smaller" representations(e.g., $-1_{10} = 1111_2$ while $-7_{10} = 1001_2$). An overflow can occur if we add two negative numbers and their magnitudes are large enough. Assume that we have one additional column to the left. If, as a result of addition, a number is carried into this extra column(i.e., an overflow appears to result)then we are actually adding two small magnitude negative numbers. (Note that they appear as large numbers if they are not considered on a 2's complement basis.) Thus, a true overflow does *not* result. On the other hand, when adding two negative numbers, if a carry into the leftmost column occurs without a carry out of this column, then a true overflow will occur. Then, summarizing the previous discussion, the only time that a true overflow occurs is when a carry into the leftmost column occurs *without* a carry out. Circuits can be built which sense when *both* of these "occur"(i.e., a carry in without a carry out). Thus, the user can be warned of an overflow and/or computation can be terminated.

We have thus far considered integers. Now let us discuss numbers with fractional parts. Actually, we shall not discuss the problem completely here. In Sec. 9-7 we shall consider the representation of such numbers in more detail. However, the discussion here will provide some of the mathematical details. We start by assuming that the numbers all have an equal number of digits preceding the binary point and also that they have an equal number of digits following the binary point. If this is not true, then we can always add right or left 0's to accomplish it. For instance, suppose we want to perform the subtraction A - B where $A = 1011.11_2$ and $B = 101.101_2$. We rewrite these as

$$A = 1011.110_2 \quad \text{and} \quad B = 0101.101_2$$

Now consider the binary number

$$A = a_n a_{n-1} \ldots a_2 a_1 a_0 . a_{-1} a_{-2} \ldots a_{-k} = a_n 2^n + a_{n-1} 2^{n-1} + \ldots + a_2 2^2$$
$$+ a_1 2 + a_0 2^0 + a_{-1} 2^{-1} + a_{-2} 2^{-2} + \ldots + a_{-k} 2^{-k} \qquad (9\text{-}35a)$$

We can write this as

$$A = 2^{-k}(a_n 2^{n+k} + a_{n-1} 2^{n-1+k} + \ldots + a_2 2^{2+k} + a_1 2^{1+k} + a_0 2^k + a_{-1} 2^{k-1}$$
$$+ a_{-k} 2^{k-2} + \ldots + 2_{-k} 2^0) \qquad (9\text{-}35b)$$

Thus, we can represent A as a binary integer multiplied by 2^{-k}. We represent B in a similar way. If we add or subtract A and B, we can perform the operation on the integers and then multiply the result by 2^{-k}. Let us illustrate this with some examples. Suppose

$$A = 0111.110_2 \quad \text{and} \quad B = 0101.101_2$$

We then have

$$2^3 A = 0111110_2$$

$$2^3 B = 0101101_2$$

Now suppose we want A - B. Take the 2's complement of 2^3B. This yields 1010011_2. Adding this to 2^3A, we obtain

$$2^3(A-B) = 10010001$$

Note that we have carried an extra digit. Then, on the basis of modular arithmetic, we have

$$2^3(A-B) = 0010001$$

Dividing by 2^3 we have

$$A - B = 0010.001$$

Note that multiplying and dividing by 2^3 just involved shifting the binary point first three digits to the left and then three digits to the right. After some practice is obtained, this step can be omitted. Note that, if this is done, in the previous example, then when we obtain the 2's complement we reverse the digits and then add 0000.001_2.

Most computers use the 2's complement system for handling subtraction or, equivalently, negative numbers. However, there are other systems that can and are, at times, used. We shall consider them briefly.

9.4 I's COMPLEMENT REPRESENTATION

When we take the 2's complement of a negative number, we perform two operations. First. all the digits are reversed(i.e., 1's become 0's and vice versa). Next we add 1. We can form a different complement by omitting the second step. That is, we reverse the digits but do not add 1. Now, instead of forming 2^k+N we form 2^k-1+N, where N is a negative number. This is called taking the 1's complement. The 1's complement is simpler to obtain but calculations with it are somewhat more difficult. We shall illustrate this subsequently. Let us consider how numbers would be represented in 1's complement. To do this, we shall write a table which corresponds to Table 9-1.

TABLE 9-2
FOUR BIT BINARY REPRESENTATION OF NUMBERS FROM –7 to +7. THE 1'S COMPLEMENT IS USED FOR THE NEGATIVE NUMBERS

Decimal number	Binary number
7	0111
6	0110
5	0101
4	0100
3	0011
2	0010
1	0001
0	0000
-0	1111
-1	1110
-2	1101
-3	1100
-4	1011
-5	1010
-6	1001
-7	1000

One complexity can immediately be seen. There are two 0's. To designate this, one is labled -0. The other is commonly called +0. They yield equivalent results when used in arithmetic. Because of the extra zero, Eq.(9-34) becomes

$$|N| \leq 2^{k-1} - 1$$

We now consider addition and subtraction. Note that subtraction is performed by adding the complement of the subtrahend. Thus, we need only speak of addition. If two positive numbers are added, the result will be correct. However, if a positive and a negative number are added, the result will be in error by 1 if the result is positive. Arithmetic performed modulo 2^k will be correct if 2's complement is used. However, in this case, the 2's complement of the negative numbers was obtained "incorrectly". Note that, in 1's complement, negative numbers are represented by a "code", see Table 9-2, which is *not* correct modulo 2^k. If a positive and a negative number are added and the result is negative, then no correction is necessary since the code corrects for it. For instance, when the number is printed it would be decoded properly. That is, in both 2's complement and 1's complement representation of negative numbers, decoding is used before the numbers are printed out to be read by humans. Compare Tables 9-1 and 9-2. The representation for negative numbers differs by 1.

When two negative numbers are added, then 1 must be added to the sum. Note that each number will be "in error" by 1. Thus, the sum will be "in error" by two. The "code" corrects for 1. Thus, we see that, at times, a 1 must be added while at other times, this is not necessary. If the cases when 1 must be added are studied, it can be seen that this occurs if, and only if, a 1 is carried beyond the "leftmost" column. If this occurs, then 1 is added. Let us consider this. Add 5 and -2. Then,

```
  0101
+ 1101
 10010
```

There is a 1 carried "beyond" the leftmost column. Then, add 1. Thus, the result is

$$0011_2 = 3_{10}$$

which is correct. The leftmost carried 1 is omitted since we work on the basis of modular arithmetic.

It is sometimes considered that the discarded digit is carried around and added to the number. This is called an *end around carry*. Let us illustrate this with the previous example.

```
  0101
+ 1101                                    (9-36)
1 0010
└──→1   end around carry
  0011
```

Note that the end around carry is a symbolic representation of the two steps (discarding one digit and then adding 1). Actually, the end around carry is not difficult to implement in a computer. Note that true overflow must be guarded against here just as in the case of 2's complement.

Let us compare 1's complement with 2's complement. The 1's complement is easier to obtain since we only have to reverse the digits. However, 1's complement has disadvantages. The end around carry must be performed. There are also two 0's in the 1's complement which require us to keep track of more data.

9.5 SIGNED MAGNITUDE REPRESENTATION

A representation which is easy for prople to use is one where the leftmost digit represents the sign and the remaining digits represent the magnitude. Thus, 0001 would equal 1_{10} while 1001 would equal -1_{10}. Similarly, $0011_2 = 3_{10}$ and $1011_2 = -3_{10}$. Thus, humans need only remember the positive integers. However, computation with these numbers is more difficult. For instance, if a positive and a negative number are added, the result will not be correct, modulo 2^K, nor can it be made correct simply by adding 1. Thus, this system introduces complexities and is not often used.

9.6 THE EXCESS 2^{K-1} REPRESENTATION

Let us consider another representation that is used at times. Assume that we have a register or accumulator with k digits and that the largest magnitude we encounter is 2^{k-1}. Hence, if N is a number within the range of the accumulator, then 2^{k-1}+N will always be a nonnegative number. Consider a four bit register where all the positive numbers are represented by three bit numbers. That is, the first bit is 0. Adding 2^4 will just change the leading 0 to a 1. For instance,

$$0110_2 = 6_{10} \tag{9-37a}$$

$$1110_2 = 6_{10} + 2_{10}^{4} \tag{9-37b}$$

Now consider a negative number. To obtain the 2's complement, we take 2^k+N, see Eq.(9-32). However, now we take 2^{k-1}+N. In binary, the difference will be in the leftmost digit. Since it is 1 in the 2's complement case, it will be 0 in the excess 2^{k-1} representation. For instance, see Table 9-2, the four bit 2's complement representation of -5 is 1011; in excess 2^3 representation it is 0011.

In summary, the excess 2^{k-1} representation is closely related to the 2's complement representation. In fact, they are the same except that the leftmost digit is reversed(e.g., a leading 0 represents a positive number in 2's complement and a negative number in excess 2^{k-1} representation).

9.7 FLOATING POINT NUMBERS

Often, especially in scientific or engineering work, we encounter numbers with a very large or very small magnitude. These are usually represented using a multiplier which is an integral power of ten. For instance, we write 1,630,000 as 1.63×10^6. Also 0.214×10^{-16} represents .00000000000000000214. This type of notation is called *floating point* notation, since the position of the decimal point is allowed to vary, or float. That is, changing the exponent effectively changes the location of the decimal point. Floating point notation can be contrasted with *fixed point notation*. Here the location of the decimal point is fixed. For instance, numbers with no fractional part are fixed point since they have a fixed number of digits(zero)after the binary point.

Let us consider some terminology. If we have the number 0.23×10^{14} then 0.23 is called the *fractional part* while 14 is called the *exponent*. (At times, the fractional part is also called the *mantissa*.)

Floating point numbers are also used in digital computers. Of course, there, binary rather than decimal representation is used. Let us illustrate this.

Instead of using powers of 10_{10}, the binary system uses powers of $10_2 = 2_{10}$. (We shall consider exceptions to this subsequently.) For instance, we can write the number $64_{10} = 2_{10}{}^7$ as

$$2^7_{10} = 0.1_2 \times 2^8 \qquad (9\text{-}38)$$

The computer binary representation would be

$$0.1_2 \times 10_2^{1000_2}$$

Often, to avoid confusion for humans, the exponent is expressed in decimal form. The $10_2 = 2_{10}$ is also often expressed in decimal terms. Let us consider some additional examples. Here we shall express the exponents in binary form.

$$0.10110 \times 2^{100} = 1011.0 \tag{9-39a}$$

$$0.1011011 \times 2^{10000} = 0.1011011 \times 2^{16} = 1011011000000000. \tag{9-39b}$$

$$0.1011 \times 2^{-100} = 0.1011 \times 2^{-4} = 0.00001011 \tag{9-39c}$$

The same terminology is used for base 2 as for base 10. For instance, in Eq.(9-39b), 0.1011011_2 is the fractional part while 10000_2 is the exponent. We have written the number with a 0 preceding the binary point. In computers this is not done since it would be wasteful of storage space to store a known 0. Thus, for example, it is conventional to write Eq.(9-39b) as

$$.1011011 \times 2^{10000} \tag{9-40}$$

Now consider the leftmost digit of the fractional part. It is a 1. It is conventional to make this a 1. Note that we can make the first digit *either* a zero *or* a 1, by properly changing the exponent. Let us demonstrate this. A binary number representation of a fraction is

$$.a_{-1}a_{-2}\ldots a_{-k} = a_{-1}2^{-1}+a_{-2}2^{-2}+\ldots+a_{-k}2^{-k} \tag{9-41}$$

Note that the a_j's are either 1 or 0. Consider that starting from the binary point, many of the digits are 0. Assume that a_{-j} is the first nonzero digit(i.e., $a_{-j}=1$). In that case, Eq. (9-41) can be rewritten as

$$.a_{-1}a_{-2}\ldots a_{-k} = \left(a_{-j}2^{-1}+a_{-(j+1)}2^{-2}+\ldots+a_{-k}2^{-(k-j+1)}\right)2^{-(j-1)} \tag{9-42}$$

Hence, by appropriately adjusting the exponent, we can always make the leading digit a 1. (The one exception to this is if all the digits are 0, that is, the number represented is a zero.) Let us illustrate this with an example.

$$.00101 = .101 \times 2^{-2}$$

The number $.101 \times 2^{-2}$ is said to be *scaled*. Thus, the fractional part is written as a binary fraction which lies between 0.5_{10} and 1_{10}. (i.e., $0.5 \leq N < 1$) We shall subsequently see that scaling allows us to store the greatest number of significant figures in a given amount of storage.

The fractional part and the exponent can each be either positive or negative. Thus, in order to store a floating point number we must store two numbers and their signs. All the representations used in the previous section can be used here. To understand something of the storage of numbers, let us digress and consider some elementary definitions of storage.

A register stores a given number of binary digits, or bits. Groups of these are called *bytes*. Typically, a group of eight bits or, at times, four bits is called a byte. Often, many of the registers in a computer are of the same length. This is termed a *word*. For instance, a typical word could consist of two(eight bit)bytes or 16 bits.

In many computers, the word length is fixed at an integral number of bytes. Such a word is illustrated in Fig. 9-1 where each byte is four bits. There, the word is 16 bits. Some computers are organized so that the word length is fixed. For instance, we can speak of a computer's having 32 bit words. Such fixed word lengths are usually used when scientific calculations are performed. Many computers, and data processing devices are designed so that they can transfer information among themselves. In such cases, information storage is usually designated on the basis of four bit bytes. On the other hand, there are computers where the word length is variable. That is, the memory is organized in such a way that the user can specify the number of bytes in a word and these all need not be of the same length.

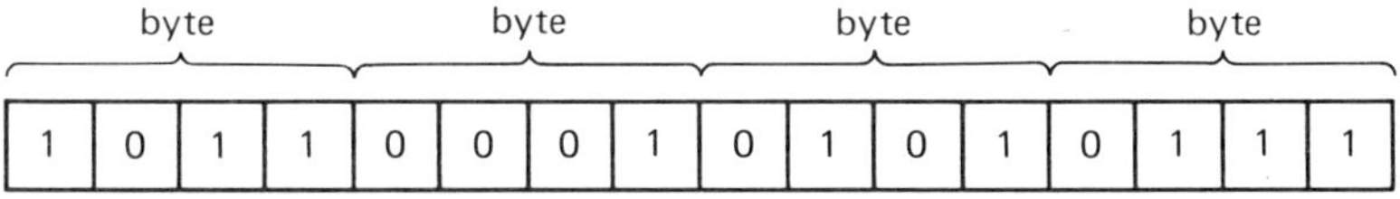

Figure 9-1. A 16 bit register. Four bit bytes are illustrated.

For purposes of the discussion of floating point numbers, let us assume that all words are the same length. When we considered integers, the word length determined the maximum magnitude of the number, see Eq.(9-34). In the case of floating point numbers, the magnitude is determined in large part by the exponent. In this case, the problem of determining the largest magnitude is somewhat more complex. Let us consider an example to illustrate this. We shall assume that the word length is 32 bits. We must store both the exponent and the fractional part. Either of these may be positive or negative. Therefore, there are two signs which must be stored. Whatever system is used to store negative numbers(e.g., 2's complement, etc.), one bit is used to store the sign. Thus, two of the 32 bits are used to store the signs. Typically, six bits will be used to store the "magnitude" of the exponent and the remaining 24 bits are used to store the "magnitude" of the fractional part. Let us consider the implications of this. Considering that the maximum positive number that can be stored for the exponent is

$$2^6 - 1 = 63_{10}$$

then,

$$2^{63} = 9.223 \times 10^{18}$$

This represents a large number. For instance, if this multiplies a fractional number between .5 and 1, it can result in very large numbers. Conversely if the exponent is negative, multiplication by 2^{-63} can result in extremely small numbers. (Note that in 2's complement or excess 2^{n-1}, there can be one more negative number represented. Thus, the preceding figure for negative numbers would be 2^{-64}.) Hence, the range of allowed numbers, when floating point(i.e., scientific)notation is used, is very large.

In some computers, the floating point notation is not written to base 2 but to base 8 or base 16. The numbers are, of course, stored in binary. Let us illustrate this. Suppose a hexidecimal base is used. If the exponent stored is positive with a binary representation $000011_2 = 3_{10}$, then the "exponent component" of the number would be

$$16^3 = 4096_{10}$$

If six bits are used for the exponent, then the largest multiplier stored for the exponent is, as before, $2^6 - 1 = 63$. Thus, the largest numerical value of the exponent part is

$$16^{63} = 7.23 \times 10^{75}$$

Therefore, by using the hexidecimal system, the largest positive number becomes much greater than if a binary system is used. Remember that the actual stored numbers are in binary form. Scaling is somewhat more complex in this hexidecimal system.

Now let us consider the fractional part. In the previous example, we stored this in 24 bits. This represents a fraction. That is, the binary point is at the left of the digits. If we consider the 24 bit binary number, we must shift the binary point 24 digits to the left. This is equivalent to dividing by 2^{24}. Thus, the largest value that this fraction can take on is $(2^{24}-1)/2^{24} = .99999994$. The fractional part will always be scaled by adjusting the exponent, so that its leading binary digit is 1. Since the largest number of 9's in the decimal number is 7, then 24 bits lead to no more than seven significant figures in the fractional part. Remember that the scaling keeps the number N in the range $0.5 \leq N < 1$. The problem of significant figures can also be studied by considering the following. The smallest increment in the binary fraction is $1/2^{24} = .596 \times 10^{-7} = .0000000596$. There are seven 0's. Thus, again, we can only have seven significant figures. That is, the smallest decimal step we can take is $.596 \times 10^{-7}$. Hence, the fractional part in this example can have up to seven significant figures. For example, we could not distinguish between .12461281 and .12461282 because we could not take small enough steps.

At times, a hexidecimal(or octal)basis is ued for the fractional part. The details of such use are similar and will be left to the reader. Note that the digits of the register are always binary.

Let us illustrate these ideas. Assume that we use decimal registers since we are more familiar with numbers to base 10. Suppose that we can store two digits for the exponent and four for the fractional part. In addition, there are two signs stored. Then, the exponent value can range between 10^{-99} and 10^{99}. Now let us represent the number 1439. We can write this as 0.1439 x 10^4. Now let us represent 15963. Only four digits can be stored in the fractional part. The last digit will then be lost. Thus, we have 0.1596 x 10^5 as the stored number. This could represent 15960, 15961, ..., 15969. For this reason, we say that this register has a *precision* of four significant figures. Note that because of the exponent we can represent very large numbers but only to four significant figures. For instance, 1,596,421 and 1,596,287 would be represented identically as .1596 x 10^7. In computer registers, the same concepts of precision apply. Of course, the numbers are stored in binary. The precision of a computer is usually rated in decimal terms for the convenience of the human users. Typically, there may be 7 or 8 significant figures. At times, this is not enough. Most computers, and their associated software, are such that additional registers can be used to store additional digits of the fractional part. This usage is termed *double precision*, since the number of available significant figures is increased. Usually, it is (approximately) doubled.

Note that we have not indicated any storage location for the binary point. This is because the binary numbers are scaled so that the number immediately to the right of the binary point is a 1. (All numbers to the left of the binary point are 0's.) Thus, there is no reason to store the location of the binary point since its position is fixed at the left of the number. Scaling also optimizes precision. For instance, $.1011011_2 \times 2^8 = .001011011_2 \times 2^{10}$. However, if we can only store seven bits, then the second number would become $.0010110 \times 2^{10}$. This is a loss of significant figures.

Let us now consider how arithmetic is performed with numbers written in floating point notation.

Addition and Subtraction

When numbers are added or subtracted, their radix points are aligned. Equivalently, when floating point notation is used, we must adjust the exponents so that they are equal. We assume that the binary fractional parts are scaled(i.e., zero to the left of the binary point and the first digit to the right of the binary point a 1). Now let us consider addition. Suppose we want to add $.1011011 \times 2^3$ and $.1101111 \times 2^6$. Assume, for purposes of simplicity, that there are seven significant figures. Then, to align the binary points, we write the addition as

$$\begin{array}{r} .1101111 \times 2^6 \\ + \; .0001011011 \times 2^6 \\ \hline .1111010011 \times 2^6 \end{array}$$

There is a problem here. To accurately represent this sum, ten significant figures are needed. However, only seven can be stored. Thus, the above addition actually would become

$$\begin{array}{l} .1101111 \times 2^6 \\ .0001011 \times 2^6 \\ \hline .1111010 \times 2^6 \end{array}$$

Thus, the resultant answer is *inaccurate*. This loss of accuracy is a roundoff error(other roundoff errors were discussed in Sec. 2-1). When many calculations are performed, the accumulated roundoff error can become very large. At times, double precision calculations must be used to prevent large errors even though large accuracy is not needed in the final answer.

Multiplication and Division

When numbers in scientific notation are multiplied or divided, we follow the rules for exponential notation. For instance, if

$$A = M_a 2^{E_a} \tag{9-43a}$$

$$B = M_b 2^{E_b} \tag{9-43b}$$

where M_a and M_b are the fractional parts and E_a and E_b are the exponents, respectively, then

$$A \cdot B = M_a M_b 2^{E_a + E_b} \tag{9-44}$$

That is, when two numbers are multiplied, we multiply the fractional parts and add the exponents.

Similarly,

$$\frac{A}{B} = \frac{M_a}{M_b} 2^{E_a - E_b} \tag{9-45}$$

When two numbers are divided, we divide the fractional parts and subtract the exponents.

Note that multiplication and division do not require that the binary points be aligned. Thus, we do not have this complexity and the associated roundoff error. However, roundoff error can result. For instance, consider the multiplication A·B where

$A = .1011 \times 2^4$

$B = .1001 \times 2^9$

$A \cdot B = .011100011 \times 2^{13}$

This would be represented as

$A \cdot B = .11100011 \times 2^{12}$

Note that we had four significant figures in each of A and B. Assume that this is all that can be stored. Then, the above product actually would be stored as

$A \cdot B = .1110 \times 2^{12}$

Again, a roundoff error results. Similarly, roundoff errors occur with division. This will be discussed in greater detail in Chapter 10.

All this discussion of roundoff error does not imply that computer calculations are inaccurate. Actually, many significant figures can be stored and very accurate results can ususally be obtained. However, the possibility of roundoff error should always be considered.

There is another factor which should be discussed when roundoff error is considered. The data used in many scientific calculations is obtained from measurements. These contain only a limited number of significant figures. It is unreasonable to assume that calculations involving these figures should result in more significant figures than are supplied. Thus, to be realistic, if calculations tend to "introduce more significant figures", as in the case of the previous multiplication, then the excess significant figures should be rounded off.

Some computers have special floating point hardware(i.e., special circuits that perform floating point arithmetic automatically). For instance, when multiplication is performed, the exponents would automatically be added without having to provide special instructions. This hardware gives high speed operation. However, many computers do not have it. Software, i.e., programs, are written which direct each step of the floating point operation. The use of the software simplifies the computer hardware. However, it results in slower execution of floating point operations.

BIBLIOGRAPHY

Booth, T.L., "Digital Networks and Computer Systems", Chapter 2, John Wiley and Sons, Inc., New York, 1971

Hill,F.J. and Peterson,G.R., "Introduction to Switching Theory and Logical Design", Chapter 2, John Wiley and Sons,Inc., New York, 1974

Mano,M.M., "Computer Logic Design", Chapter 1, Prentice Hall, Inc., Englewood Cliffs, N.J., 1972

O'Malley, J., "Introduction to the Digital Computer", Chapter 8, Holt, Rinehart and Winston, Inc., New York, 1972

PROBLEMS

9-1. Determine 4 integers which are congruent, modulo 3, to 7.

9-2. Repeat Prob. 9-1 for modulo 2.

9-3. A binary register has 32 bits. What is the largest positive number that can be stored in this register?

9-4. Repeat Prob. 9-3 but now assume that one bit of the register is used as a sign bit.

9-5. We want to perform the addition $729_{10} + 1436_{10}$ in a binary accumulator. How many bits are required?

9-6. Repeat Prob. 2-18 but now subtract the numbers.

9-7. Repeat Prob. 2-19 but now subtract the numbers.

9-8. Repeat Prob. 2-20 but now subtract the numbers.

9-9. Perform the subtraction of Prob. 9-6 but now use the procedure of Sec. 9-2.

9-10. Obtain the 10's complement of the following numbers.

123_{10}, 146_{10}, 937.12_{10}

9-11. Obtain the 2's complement of

10101, 110101, 101.101

9-12. Use the 2's complement to perform the subtraction of Prob. 9-6.

9-13. Demonstrate that when using accumulators, a negative number can be represented by its 2's complement.

9-14. Extend Table 9-1 for numbers ranging from 15 to -15. Use five bits.

9-15. Add the following two numbers which are given in 2's complement

0110, 1011

9-16. Can -16 be included in the table of Prob. 9-14? If the answer is yes, include it.

9-17. Demonstrate how, in your own words, the leftmost digit can be used to determine if a "true overflow" has occured in a register.

9-18. Use the 2's complement to perform the following subtraction.

1011.0101
101.1101

9-19. Obtain the 1's complement of the numbers in Prob. 9-11.

9-20. Extend Table 9-2 for numbers ranging from 15 to -15. Use 5 bits.

9-21. Can -16 be included in the table of Prob. 9-20. Compare this answer with that of Prob. 9-16. What is the difference?

9-22. Add the following numbers which are given in 1's complement representation.

0110, 1011

9-23. Write a table which gives numbers ranging from 7 to -7 using signed magnitude representation.

9-24. Repeat Prob. 9-23 but now use the excess 2^3 representation.

9-25. Write the following numbers in floating point binary.

126, 1479, 138×10^5, 1.427×10^3

9-26. The following pairs of numbers represent the fractional part and the exponent(in base 2)of a number. Express the number in base 10. The exponent is given in 2's complement. Assume that the fractional part is positive.

.1011, 0110
.1101, 1011
.101, 111110
.1011, 0111110101

9-27. A register has 64 bits. 12 bits are used to store the exponent of a number and the remaining 52 bits are used for the fractional part. What is the range of numbers that can be stored? What is the precision?

9-28. Add the following two numbers. Assume that the register can store seven bits for the fractional part. What is the round-off error?

$.1101101 \times 2^6$

$.1011011 \times 2^5$

9-29. Repeat Prob. 9-28 for

$.1101101 \times 2^{16}$

$.1101101 \times 2^3$

9-30. Repeat Prob. 9-28 for

$.1011101 \times 2^{-8}$

$.1011101 \times 2^{-10}$

9-31. Perform the following multiplication. Assume that only seven significant figures can be stored for the fractional part.

$(.1011011 \times 2^9)(.1011111 \times 2^{16})$

9-32. Repeat Prob. 9-30 for

$(.10111 \times 2^{-9})(.1011101 \times 2^5)$

10

Computation Operations

In this chapter we shall discuss some of the computer operations as well as the circuits that are used when computations are performed. The basic arithmetic operations of addition and subtraction will be discussed. (It is assumed that the reader is familiar with all the introductory material discussed in Chapter 9. The operations of multiplication and division will then be developed. Other mathematical operations shall also be considered.

The material presented in this chapter lays the groundwork for the discussion of the complete digital computer which will be given in the next chapter. Actually, it will be helpful for the understanding of the material discussed in this chapter if some ideas of complete digital computers were known before the material of the present chapter were discussed. Thus, we shall briefly consider here some ideas of the elements of the digital computer which will be discussed in greater detail in Chapter 11.

Computers can be considered to consist of four basic units: the *central processing unit* (CPU), the *control unit* (CU), the *main memory unit* (MMU), and various input-output (IO) devices. The computations are actually performed in the central processing unit. These operations are directed by the control unit, which also directs the operation of the entire computer. The main memory unit is a random access memory, see Chapter 7. Its function is to store the data and results of computation. In addition, such things as the program which directs computation are also stored in the main memory unit. All data is inputted and/or outputted using input-output devices such as card readers and printers. In this chapter we shall consider important aspects of the central processor unit. In particular, we shall consider arithmetic operations and the hardware which can be used to perform them. The important accumulator register will be discussed here. We shall also discuss a notation which is used to describe the computer operations. In the next chapter, we shall discuss the complete computer.

10.1 ADDITION

The fundamental arithmetic operations in a digital computer are addition and subtraction. Almost all other operations are performed using addition and subtraction, and the circuits associated with them. Actually, subtraction can be considered to be a special case of addition. In this section we shall consider addition circuits. It is assumed that the reader is familiar with Sec. 3-4 where the full and half adders were discussed.

We shall start by reviewing the *full adder*. This is a circuit which adds two binary digits and a carry digit. This yields their sum plus a carry digit. From Eq. (3-23), we have

$$\begin{array}{r} c_{j-1} \\ a_j \\ \underline{b_j} \\ c_j s_j \end{array}$$

where a_j, b_j and c_{j-1} are the digits to be added, s_j is the sum and c_j is the carry digit. In Eq.(3-29) we expressed c_j and s_j as

$$s_j = a_j \oplus b_j \oplus c_{j-1} \tag{10-1a}$$

$$c_j = a_j b_j + b_j c_{j-1} + c_{j-1} a_j \tag{10-2b}$$

Remember that we shall be adding two N bit binary numbers. Here we have considered the addition of their j^{th} digit, plus the carry resulting from the addition of the $j-1^{st}$ digits. A half adder sums two binary digits but does not make provision for the input of a carry term from the previous digit. Realizations for half and full adders are given in Figs.3-17 through 3-19. A block diagram representation of a full adder is given in Fig.10-1.

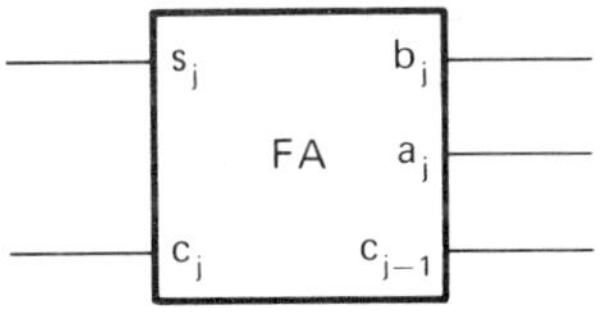

Figure 10-1. The block diagram of a full adder.

Now let us discuss a circuit that adds two three bit binary numbers, see Fig. 10-2. Actual adders add many more bits. However, all the fundamental principles can be described using three bits without cluttering the circuit with excessive detail.

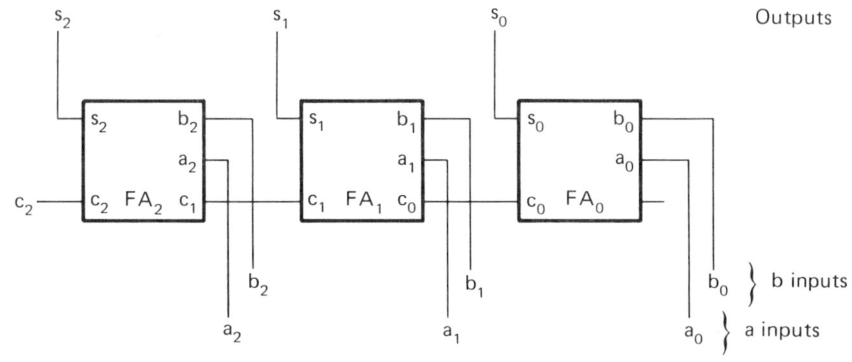

Figure 10-2. A three bit adder.

The circuit uses three full adders. We assume that the a_j and b_j inputs are stored in registers during the addition process. Note that the inputs to each adder are the ones called for in Eqs. (10-1). Thus, the sum digits will be the correct one. FA_0 adds the rightmost bits. Thus, there is no carry input for this adder. Hence, it could be a half adder. The carry digit from the leftmost adder is not used. (It could be used. We shall consider this subsequently.)

Carry Propagation

The diagram of Fig. 10-2 seems to imply that, if we apply the inputs $a_0, a_1, a_2, b_0, b_1, b_2$, then the output will be obtained after the time interval required by a full adder to produce its output. Actually, this is not the case. FA_1 does not have all its input signals until it has received the c_0 input from FA_0. Thus, if it takes Δt sec. for a full adder to respond to its input signal, FA_1 cannot complete its response until $2\Delta t$ sec. have elapsed. Thus, FA_2 will not receive its carry input for $2\Delta t$ sec. Thus, the final sum will not be achieved until $3\Delta t$ sec. have elapsed. If the register adds n bits, then the delay will be $n\Delta t$ sec. This delay is said to occur because of *carry propagation*.

At times, such delays are acceptable. However, it is often desirable to speed up the process. One way of doing this is called *carry look ahead* where the carry bits are calculated without waiting for the adder to do so. Let us illustrate this. From Eq. (10-1) we have

$$c_0 = a_0 b_0 \tag{10-2a}$$

$$c_1 = a_1 b_1 + b_1 c_0 + c_0 a_1 \tag{10-2b}$$

$$c_2 = a_2 b_2 + b_2 c_1 + c_1 a_2 \tag{10-2c}$$

We would like to express c_1 and c_2 in terms of the a's and b's only. Substituting Eq. (10-2a) into Eq. (10-2b), we obtain

$$c_1 = a_1 b_1 + b_1 b_0 a_0 + a_1 a_0 b_0 \tag{10-3a}$$

Substituting Eq.(10-3a) into Eq.(10-2c), we obtain

$$c_2 = a_2b_2+b_2a_1b_1+b_2b_1b_0a_0+b_2a_1a_0b_0+a_2a_1b_1+a_2b_1b_0a_0+a_2a_1a_0b_0 \quad (10\text{-}3b)$$

In Fig. 10-3 we have illustrated a three bit adder which utilizes Eq.(10-3a) to eliminate the effect of carry propagation to the second adder. To eliminate it in a four bit adder, we would also implement Eq.(10-3b) with 7 AND gates and one OR gate to provide the c_2 input. (Note that the adders can be simplified considerably since they do not have to output the carry bit.)

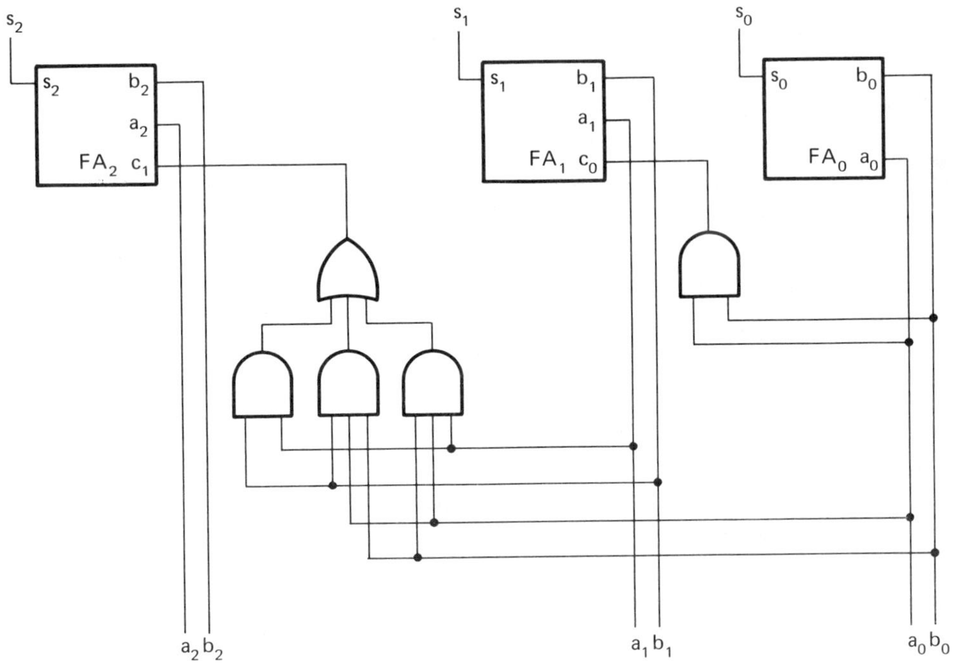

Figure 10-3. A three bit adder that does not suffer from carry propagation.

If this implementation is carried out, the carry propagation effect is eliminated. However, in large adders, the number of gates required, and the associated fan out becomes very large. There are other implementations which use fewer gates, and are not as fast. These are discussed in the Rhyne reference cited in the bibliography at the end of this chapter.

In the remainder of this book, we shall draw adder circuits as shown in Fig. 10-2. However, there may be gates which are not shown that are added to eliminate the effects of carry propagation.

A Simple Accumulator

In most computer circuits, we do not just have a simple adder, but an adder and register combined. Let us call the register A. The number in the register is added to another number which is supplied from an external circuit. This sum is placed in register A. Thus, the original number in the register is replaced by the sum. This device is called an *accumulator*. Actually, accumulators do much more than we have indicated here. Thus, the description of this section will be of a simple accumulator. In Secs. 10-2 and 10-4 we shall consider the accumulator in detail.

We want to implement a circuit which has a register storing a binary number and which accepts another number. After computation has been completed, the register should store the sum of the two numbers. Consider Fig. 10-4. At the start, assume that the ADD and CLEAR signals are 0. Thus, no clock pulses are supplied to the D flip-flops and the lower inputs of AND 0, 1 and 2 are 1's. The D flip-flops constitute the register. Their outputs are made available to the rest of the computer. Assume that there is a binary number stored in the register which will be supplied to the a_j inputs of each full adder. The b inputs to the full adder will be supplied from an external source. Thus, after sufficient time has elapsed, the full adder will produce the sum $s_2s_1s_0$. These digits will be applied to the inputs of AND 2, AND 1, and AND 0, respectively. Thus, these can be applied at the D_2, D_1 and D_0 inputs, respectively, of the flip-flop. However, the flip-flops will not change state since no clock pulse is applied to their clock pulse terminals. We assume that the clock is running and generating pulses. However, AND a prevents the clock pulses from being applied. Now suppose that the ADD signal is given(for one clock pulse). That is, a 1 is placed on the ADD line. After the next clock pulse is applied, the state of the flip-flops will become equal to their respective D inputs, which are s_2, s_1 and s_0. The number now stored in the register will be the sum of the two original numbers as desired.

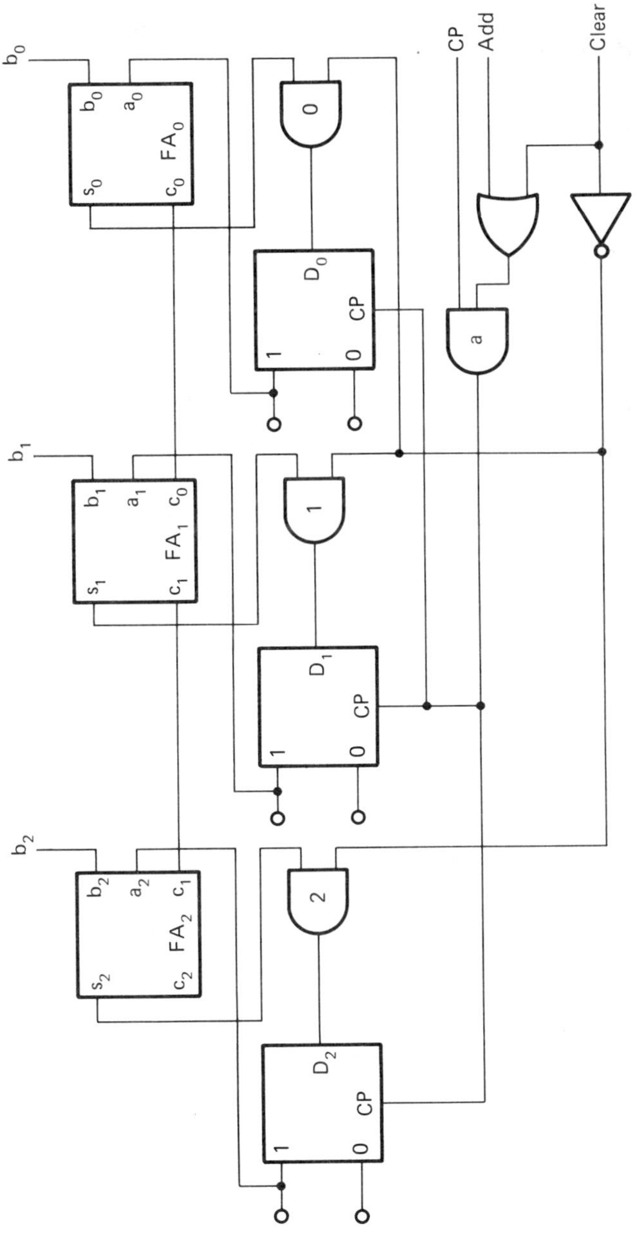

Figure 10-4. A simple accumulator.

If a signal is put on the clear line, then one input to each of AND 0, 1 and 2 becomes 0. Hence, D_0, D_1 and D_2 are 0. At the next clock pulse, all flip-flops will set themselves to state 0. Hence, we have cleared the register. Note that the next $b_2b_1b_0$ added will be added to the 000 in the register given a new stored number $b_2b_1b_0$.

The B register which supplies b_0, b_1, and b_2 is not in the accumulator. However, such a storage register is often associated with each accumulator.

Sequential Adders

We have discussed parallel adders. That is, information is inputted to each full adder simultaneously(with the exception of carry propagation). Thus, to add n bits we required n full adders. Now let us consider the serial or sequential adder. Only one full adder is used. One bit of each number is inputted at a time and the resulting sum and the carry bit are stored. Then, the next bit of each number is added using the last stored carry bit and the new sum and carry bits are stored. Note that the carry bit need only be stored until the addition using it is performed. Thus,only one register is needed to store the carry bit. Let us consider a simple acuumulator which performs sequential addition, see Fig. 10-5. The storage is a shift register of the type shown in Fig. 6-23b. AND gates 0,1 and 2 are included to allow the register to be cleared. This is essentially the same procedure as in Fig. 10-4. At the start, the shift register will store one of the numbers to be added. When the addition is completed, it will store the sum.

Now let us assume that an ADD signal is present and the CLEAR signal is 0. The b input to the full adder is also assumed to come from a shift register. Prior to the first clock pulse, a_0 and b_0 are supplied and $c_{j-1}=0$. The adder then computes s_0 and c_0. It is assumed that all this is completed before the next clock pulse arrives. Note that s_0 is calculated before this clock pulse. Thus, it has not as yet changed the state of flip-flop D_2.

After the next clock pulse, the following takes place: The shift register shifts to the right. Thus, a_1 is stored in D_0.

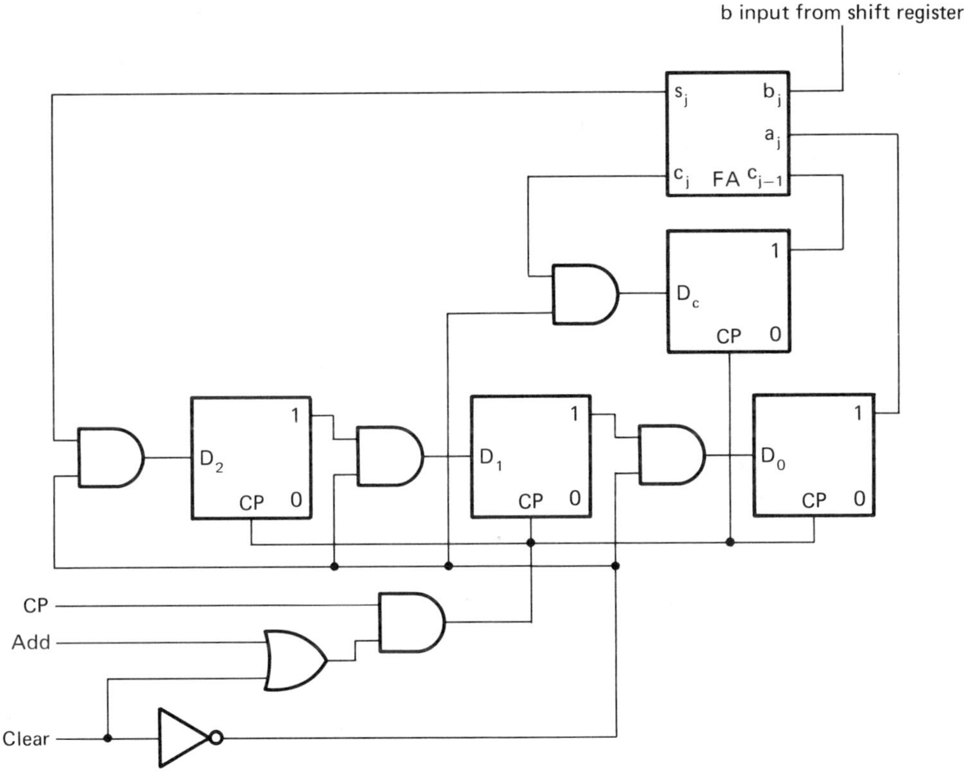

Figure 10-5. A serial adder.

Then, b_1 and a_1 are inputted to the adder and c_0 becomes the output of flip-flop D_c. The adder then computes s_1 and c_1. At this clock pulse, s_0 becomes the state of the leftmost flip-flop D_2 of the shift register prior to addition. Note that the spacing between clock pulses must also be long enough to allow all this to occur.

Proceeding in this manner, after two more clock pulses, the contents of the shift register will be $s_2s_1s_0$. The ADD signal must be removed now.

10.2 ACCUMULATOR OPERATIONS

We have considered simple accumulators and seen some of the operations that they can perform. In this section we shall discuss other operations which we would expect from accumulators. In the next section we shall see how these can be used to perform other arithmetic operations. Finally, we shall discuss a circuit for a complete accumulator. In Chapter 11 we shall see how this relates to the central processing unit.

In order to efficiently work with the operations, we shall also introduce notation that will be used to indicate them. We shall also introduce notation for arithmetic operations.

Registers will be indicated by capital letters. For instance, A represents register A. The number stored in a register, i.e., its *contents*, will be indicated by parentheses, i.e., (A) represents the contents of A.

If information in register A is transferred to register B, we write this as

$$(A) \rightarrow B \tag{10-4}$$

This means that, after the next clock pulse, the contents of B will be the same as the contents of A.

The operations of arithmetic are indicated by the following symbols

Addition +
Subtraction - (10-5)
Multiplication *
Division /

Thus, we can represent addition using an accumulator with register A which receives a number from register B as

$$(A) + (B) \rightarrow A \tag{10-6}$$

That is, the contents of registers A and B are added and the sum is placed in register A. If we want to divide the contents of register B by the contents of register C and place the result in register E, we would write

$$(B)/(C) \rightarrow E \tag{10-7}$$

In addition to the arithmetic operations, we can perform logical operations. To avoid confusion with the arithmetic symbols, we shall use the alternate notation

AND $\wedge$ (10-8a)

OR $\vee$ (10-8b)

EXOR $\oplus$ (10-8c)

Thus, we can write

$$(A) \vee (B) \rightarrow C \tag{10-9}$$

This means that the first locations of A and B are compared, using an OR gate, and if either or both is a 1, then a 1 is put in the first location of register C. Otherwise, a 0 is placed there. This is repeated with each location in turn.

Another notation we shall use is that (A_i) is the contents of the ith location of register A. (A_{ij}) represents the contents stored in the i to j locations of register A. For instance, if the contents of register A is 10011010, then $(A_0) = 0$ and $(A_{02}) = 010$. Then, we can also represent Eq. (10-9) as

$$(C_i) = (A_i) \vee (B_i) \quad i=0,1,2,\ldots,N \tag{10-10}$$

We can also speak of signals. For instance, in Fig. 10-5, the output signal could be the voltage on the 1 lead of flip-flop D_0. We shall denote this by a lower case letter, e.g., y. The signal at a particular time will be written as (y) or $y(t_0)$. If we want to designate that a particular signal is the result of the addition of the contents of two registers, we would write

$$(A) + (B) = y \tag{10-11}$$

For instance, if y were the output signal of a serial adder, then it would consist of an appropriate sequence of 0's and 1's. Note that y can be a series of bits or a single bit depending upon the circumstances.

In addition to the arithmetic and logical operations, there are certain functional operations that are used in many computer operations. For instance, suppose that there is a signal y and we want y to equal 1 if the contents of two registers, A and B are equal(e.g., they represent the same binary number), and a 0 otherwise. We would represent this by

$$y = EQ(A,B) \tag{10-12}$$

That is

$y = 1$ if $(A) = (B)$

$y = 0$ if $(A) \neq (B)$

Some other functional operations that are commonly used are

$$y = LT(A,B) \tag{10-13}$$

where

$y = (1)$ if $(A) - (B)$ represents a negative number

$y = (0)$ if $(A) - (B)$ represents zero or a positive number

Another function operation is

$$y = GT(A,B) \tag{10-14}$$

where

$y = (1)$ if $(A) - (B)$ represents a positive number

$y = (0)$ if $(A) - (B)$ represents zero or a negative number

Another functional operation is

$$y = SG(A) \tag{10-15}$$

where

y = (1) if (A) represents a negative number

y = (0) if (A) represents zero or a positive number

Still another function operation is

$$y = B_k(A) \tag{10-16}$$

where

$y = (A_k)$

We could consider other functions. However, these are the most important ones.

In order to implement the mathematical operations, we require the accumulator to perform certain operations, in response to control signals generated by the control unit. We have already discussed some of these operations in this chapter and in Chapter 6. However, we shall list a more complete set of them here.

TABLE 10-1
ACCUMULATOR OPERATIONS

Operation	Abbre-viation	Control designa-tion	Symbol
Addition	ADD	c_1	$(A) + (B) \rightarrow A$
Clear	CL	c_2	$0 \rightarrow A$
AND	AND	c_3	$(A) \wedge (B) \rightarrow A$
OR	OR	c_4	$(A) \vee (B) \rightarrow A$
EXOR	EXOR	c_5	$(A) \oplus (B) \rightarrow A$
shift right	SR	c_6	$(A_{j+1}) \rightarrow A_j \quad j=0,\ldots,n-1, (A_n)=0$
shift left	SL	c_7	$(A_{j-1}) \rightarrow A_j \quad j=1,\ldots,n, (A_0)=0$
complement	COMP	c_8	$(\bar{A}) \rightarrow A$
increment	INCR	c_9	$(A) + 1 \rightarrow A$
negative check	NC	c_{10}	y=1 if (A) represents a negative number
zero check	ZC	c_{11}	y=1 if (A) represents a 0

Most of these have been considered previously. For instance, the shift right and shift left cause the entire contents to be shifted as in a shift register. In the complement operation, each element of the register is replaced by its complement(e.g., 011011 is replaced by 100100). In the increment, 1 is added to the contents of the register.

The accumulator has a set of command leads, one for each of the desired operations. When a 1 is placed on that lead, then, after the next clock pulse, the accumulator performs in accordance with the given indtruction. A block diagram of the accumulator is given in Fig. 10-6.

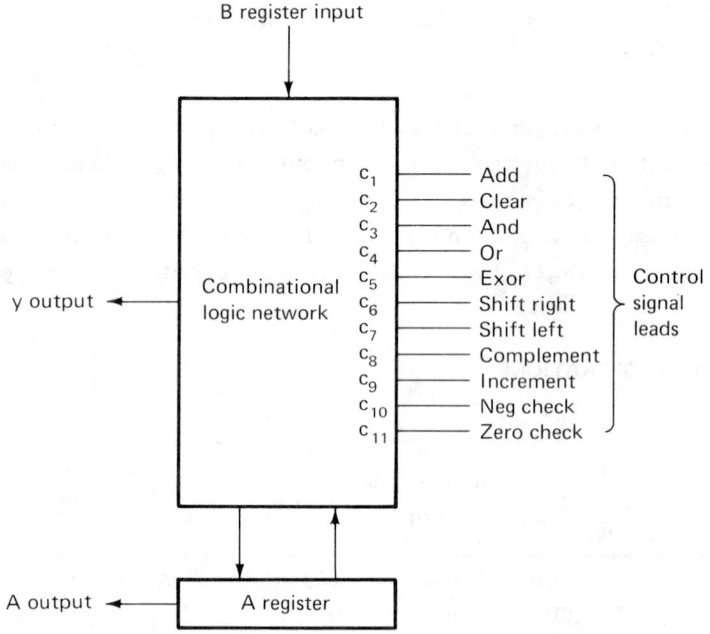

Figure 10-6. A block diagram of an accumulator.

Suppose that we want to perform the following operations which, as we shall see, are related to 2's complement subtraction. Take the complement of the contents of (A). Then, add 1 to this. Finally, to the resulting number, add the contents of another binary number in B and store the final result in A. The sequence of operations to be performed is

$$(\bar{A}) \rightarrow A$$

$$(A)+1 \rightarrow A \qquad (10\text{-}17)$$

$$(A)+(B) \rightarrow A$$

Suppose that, originally,

$$(A) = 0100$$

and

$$(B) = 0011$$

The results of the operations (10-17) are

$$(A) = 1011$$

$$(A)+1 = 1100$$
$$(A)+(B) = 1111$$

In order to perform the sequence of operations, 1's must be put on the proper control signal lines at the appropriate times. This is accomplished by a circuit called a *controller*, which could be part of the control unit. For instance, in order to carry out operations(10-17), we would have to use control lines c_1, c_8 and c_9. The first set of control signals would be (0,1,0). That is, c_1=0, c_8=1, c_9=0. The next signal would be (0,0,1) and then (1,0,0). An input signal starts the operation of the controller. A block diagram of this is shown in Fig. 10-7. Once the start signal is given, the controller produces the following outputs on successive clock pulses: (0,1,0),(0,0,1),(1,0,0) where the order of the outputs is (y_1,y_2,y_3). The controller is just a sequential circuit of the type realized in Sec. 6-5. The state transition diagram for this controller is given in Fig. 10-8. The state remains at q_0 with zero output signals until the input (START) becomes a 1. Then, on successive clock pulses, the desired sequence of output control signals is obtained. Note that we have chosen the transition so that once the cycle is started, it is completely independent of the input signal to the controller(i.e., whether it is a 0 or a 1). Of course, the input must be a 1 to initiate the action. Note that a ROM can also be used to store the sequence of control signals.

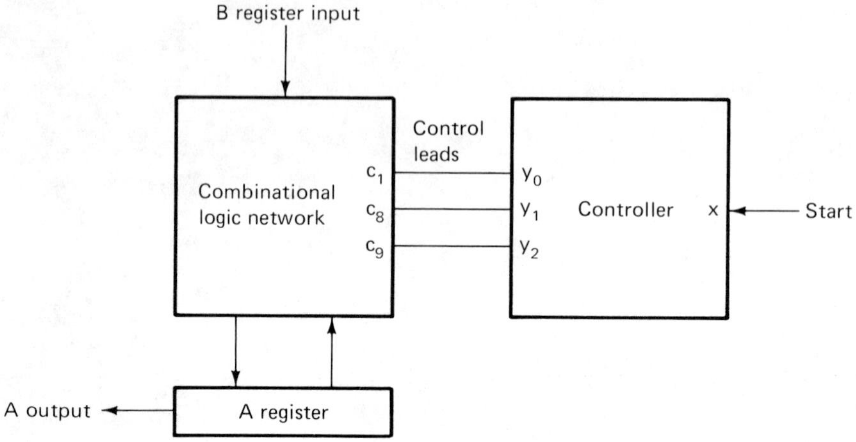

Figure 10-7. A block diagram of the register and controller that realize sequence (10.17).

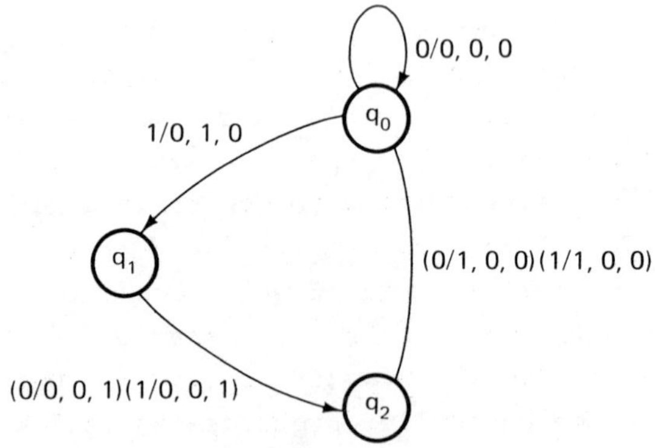

Figure 10-8. State transition diagram for the controller of Fig. 10.7.

In this section we have considered some elementary operations. In the next, we shall consider the basic arithmetic operations of subtraction, multiplication, and division.

10.3 ARITHMETIC OPERATIONS-SUBTRACTION, MULTIPLICATION AND DIVISION, BCD-BINARY CONVERSION

There are many techniques that can be used to implement the operations of subtraction, multiplication and division. We shall only consider a few of them to establish some fundamental ideas. The text by Stein and Munro cited in the bibliography contains many such procedures.

Subtraction

We shall use 2's complement arithmetic in subtraction. We assume that the reader is familiar with the discussion in Secs. 9-2 and 9-3.

Now suppose that we want to subtract (C)-(D). We form the 2's complement of (D) by complementing each of its bits and then adding 1 to the result. Then, the 2's complement of (D) and (C) are added. Their sum is the desired result of the subtraction. Then, using the notation developed in the last section, we have the following algorithms, or set of operations, that are used to obtain the subtraction (C)-(D).

$$0 \rightarrow A \qquad (10\text{-}18a)$$

$$(D)+(A) \rightarrow A \qquad (10\text{-}18b)$$

$$(\bar{A}) \rightarrow A \qquad (10\text{-}18c)$$

$$(A)+1 \rightarrow A \qquad (10\text{-}18d)$$

$$(C)+(A) \rightarrow A \qquad (10\text{-}18e)$$

We assume that the minuend is stored in the C register and that the subtrahend is stored in the D register. Let us consider these steps. In the first step, (10-18a), we clear the register of the accumulator. In (10-18b) we add the contents of D into the accumulator register. In (10-18c) we take the complement of the contents of the accumulator and in (10-18d) we add 1 to the contents of A. Now (A) is the 2's complement of the original (D). In the last step we take the sum (C)+(A) which is equivalent to (C)-(D).

Let us use Table 10-1 to obtain the command sequence that can be used to implement (10-18). This is

$$c_2 \quad (10\text{-}19a)$$
$$c_1 \quad (10\text{-}19b)$$
$$c_8 \quad (10\text{-}19c)$$
$$c_9 \quad (10\text{-}19d)$$
$$c_1 \quad (10\text{-}19e)$$

Note that we assumed that the minuend and subtrahend were stored in register C and D, respectively. At the appropriate time, the contents of D must be supplied to the input of the accumulator and, then, at a subsequent time, the contents of C must be so supplied. It is assumed that the computer has a multiplex circuit, which allows this to be done. In Figs. 3-20 and 3-21, a simple multiplexer is illustrated. At times, the input to an accumulator can only come from a single register, e.g., B. In this case, we would have to transfer the desired information into B. Thus, the first state of (10-18) would be

$$(D) \to B$$

and, proceeding, before (10-18e), we would have

$$(C) \to B$$

Then, (10-18e) would become

$$(B)+(A) \to A$$

A block diagram illustrating the implementation of sequence (10-18) and a state transition diagram for the controller are given in Fig. 10-9. The d indicates don't cares. That is, once the SUBTRACT signal has initiated subtraction, we don't care whether or not it is present. Note that the subtract signal must be synchronized with the information transfer from the C and D registers.

Multiplication

Now let us consider the multiplication of two numbers contained in registers C and D. We shall assume that they are both positive

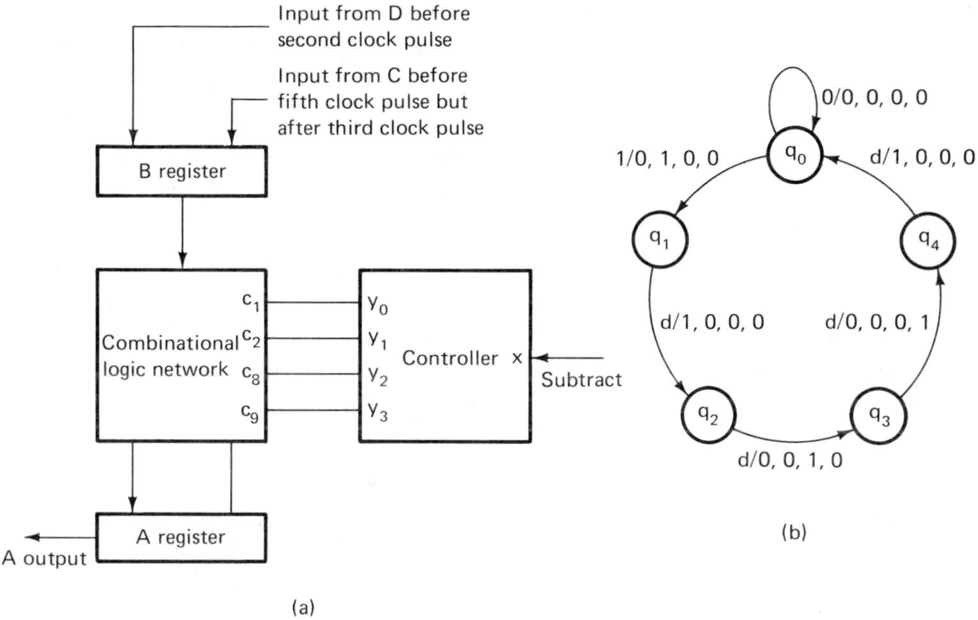

Figure 10-9. A simple subtractor. (a) block diagram; (b) controller state transition diagram. The order of the output signal is (y_0, y_1, y_2, y_3).

binary numbers. Multiplication algorithms exist for multiplication using 2's complement numbers. However, we shall discuss the multiplication of two positive binary numbers since it is simpler conceptually.

Let us start by considering the binary multiplication of two positive integers as would be done by pencil and paper. Then we shall see how this can be implemented by the computer. Suppose that we want to multiply 1101 by 1011. Then, following the usual rules for multiplication, we have

```
    1101    multiplicand
    1011    multiplier
   ------
    1101    first partial product
   1101     second partial product               (10-20)
  0000      third partial product
 1101       fourth partial product
---------
10001111    product
```

We multiply the multiplicand by the first digit to obtain the first partial product. Next, we multiply the multiplicand by the second digit of the multiplier, shift the result one digit to the left and add the result to the first partial produce,etc. When we implement this in the computer, we could shift the partial products to the left. However, it is more convenient if we consider the fourth(last)partial product as the reference. Then, the third partial product is shifted one digit to the right, the second partial product is shifted two digits to the right and the first partial product is shifted three digits to the right.

Now let us consider the digital computer implementation of this. In order to perform multiplication, we need more than an accumulator, although it is the central part of the multiplier. To conform with this example, we shall assume that we have a five bit accumulator with register A. Actually, the extra sum term can be obtained from the unused carry bit in the leftmost full adder of the accumulator. This is often not part of the accumulator and is called a *link* register, L. We must also have the usual register B, called a storage register, which supplies the input to the accumulator. In addition, we assume that we have one other register MQ, called the multiplier-quotient register. (This will also be used for division.) Both B and MQ will be assumed to store four bits.

The MQ, L and A registers are arranged into a single register such that data can be shifted from one to the other. If both registers are cleared, then we have

$$\overset{L}{0}\;\overbrace{0\;0\;0\;0}^{A}\;\overbrace{0\;0\;0\;0}^{MQ} \qquad (10\text{-}21a)$$

Now suppose that the multiplicand 1101 in example (10-20) is stored in the B register. Assume that the multiplier is also stored. For the time being, let us not say where it is stored, but we shall assume that we can use its digits for control. The multiplier is 1011. Now, since the rightmost digit of the multiplier is a 1, we add the contents of the B register to the A register, i.e., we multiply the contents of the B register by 1 and add. Doing this we obtain

$$\overset{L}{0}\;\overbrace{1\;1\;0\;1}^{A}\;\overbrace{0\;0\;0\;0}^{MQ} \qquad (10\text{-}21b)$$

Now shift the contents of the combined L-A-MQ register one digit to the right. This yields

```
L    A       MQ
0 0 1 1 0 1 0 0 0                    (10-21c)
```

The next digit of the multiplier(second from right)is 1. Thus, we again add the contents of the B register into the accumulator. This yields

```
L    A       MQ
0 0 1 1 0 1 0 0 0
  1 1 0 1                            (10-21d)
-----------------
1 0 0 1 1 1 0 0 0
```

Note that the accumulator register has five bits even though the number has only four. This allows for the "overflow" bit. Actually, this is not an overflow since we will, next, shift to the right.

Next, shift the L-A-MQ register to the right, to obtain

```
L    A       MQ
0 1 0 0 1 1 1 0 0                    (10-21e)
```

The next digit of the multiplier(third from right) is 0. Thus, we add 0 to the A register, leaving the results unchanged. Now shift to the right again. This yields

```
L    A       MQ
0 0 1 0 0 1 1 1 0                    (10-21f)
```

The last(fourth from the right end)digit of the multiplier is a 1. Thus, we add the contents of the B register to the A register to obtain

```
L    A       MQ
1 0 0 0 1 1 1 1 0                    (10-21g)
```

Finally, shift to the right. This yields

```
L    A        MQ
0 1 0 0 0 1 1 1 1                                        (10-21h)
```

which is the result of the multiplication.

The process of shifting to the right and adding that we have performed exactly follows the steps used in multiplying by pencil and paper.

We assumed that the multiplier was stored in some register. Actually, we can use the MQ register. Note that, from the start until after the first step is performed [(10-21b)], the MQ register is empty. After the first step, we no longer need the rightmost digit of the multiplier. Thus, we can shift the contents of the MQ register to the right as before. Therefore, we could store the multiplicand in the MQ register. The rightmost digit of MQ controls the multiplication that generates the first partial product of (10-21b). When we shift to the right, the rightmost digit of the multiplier is lost and the second digit of the multiplier becomes the rightmost digit. Now it can control the multiplication for the second partial product. Therefore, we can continue in this way. Then, for the example just considered, the sequence will be

```
0 0 0 0 0 1 0 1 1                                        (10-22a)
0 1 1 0 1 1 0 1 1                                        (10-22b)
0 0 1 1 0 1 1 0 1                                        (10-22c)
1 0 0 1 1 1 1 0 1                                        (10-22d)
0 1 0 0 1 1 1 1 0                                        (10-22e)
0 0 1 0 0 1 1 1 1                                        (10-22f)
1 0 0 0 1 1 1 1 1                                        (10-22g)
0 1 0 0 0 1 1 1 1                                        (10-22h)
```

The block diagram for a multiplier is given in Fig. 10-10a. Note that the L and MQ registers are interconnected with the A register as a single register. The same command shifts the entire L-A-MQ register, etc. The control has two inputs. One initiates multiplication. The other stops it. The stop signal is usually obtained from a counter, such as that in Fig. 6-37. Suppose that the counter counts clock pulses. The counter and its associated logic is such that a 1 is outputted after eight add and shift pulses. (Note that the multiplication by 0 is not shown in (10-22).) This then terminates computation.

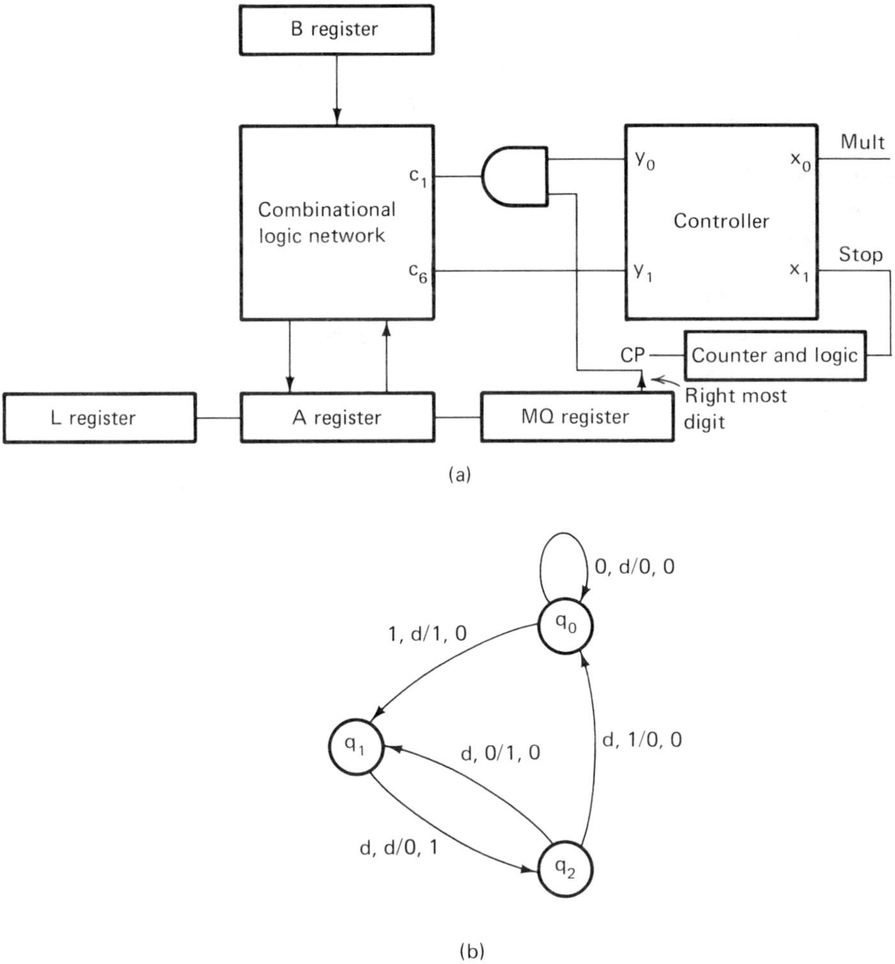

Figure 10-10. A simple multiplier. (a) block diagram. Note that all clock pulse connections are not shown; (b) state transition diagram for controller. The order of the output signals is y_0, y_1.

Note that, because of the AND gate, the ADD command c_1 can only be applied when the rightmost bit of the MQ register is a 1. Thus, we either multiply by 0 or 1.

A state transition diagram of the controller is shown in Fig. 10-10b. When the MULT signal is given, the first addition takes place. Then, the shift signal is generated, etc. Note that the counter and logic is such that the stop signal will be generated after the last shift command has been given.

To avoid complexity, we did not clear the A register at the start. However, the first step would be to clear all registers. Then, enter the multiplier into the MQ register and the multiplicand into the B register.

In the example of multiplication just performed, two four bit binary numbers were multiplied and the result was an eight bit number. It may seem as though this is an overflow. If the remainder of the computer circuitry is such that we can only work with four bit binary integers, then it is indeed an overflow. That is, the multiplier and/or multiplicand are too large. Now suppose that we multiply

```
      0 0 1 1
      0 0 1 0
-------------
      0 0 0 0
    0 0 1 1
  0 0 0 0
0 0 0 0
----------------
0 0 0 0 1 1 0
```

In this case the four rightmost digits, which will be stored in the MQ register after the multiplication is completed, constitute the complete answer and an overflow does not result even though there is an eight digit answer. Note that, in the case of integers, it is the rightmost digits, which are stored in the MQ register, with which we are concerned.

In the case of floating point numbers, the converse occurs. Suppose that we multiply 0.1101×2^a and 0.1011×2^b. The exponents a and b must be added. Now consider the actual multiplication of the fractional parts.

```
  0.1101
  0.1011
--------
    1101
   1101
  0000
 1101
---------
.10001111
```

Note that the steps are the same as for the integer multiplication. The leftmost digits represent the most significant figures. Thus, if we can only save four digits, roundoff error results, and we get

.1000

as the answer. This is stored in the A register. The rounded off answer would be

$$0.1000 \times 2^{a+b}$$

The details of the storage are further discussed in Sec. 9-7.

Note that, in the case of both integers and floating point numbers, the most significant digits lie to the left. However, in the case of integers, we usually do not tolerate *any* round off error. Thus, if any significant digits are in the A register, we can say that an overflow has occurred. Circuits are usually included to check for such overflows.

On the other hand,we do allow roundoff when we work with floating point numbers. Thus, we take the 4(or however many digits we are using)most significant figures. These are the leftmost digits which do lie in the A register.

We have considered that the extra digits constitute an overflow as in the case of integers or must be rounded off in the case of floating point numbers. However, the number stored in the combined

A-MQ register is correct. It is possible to control the computer (i.e., to write programs)which can use this data, in which case, overflow or roundoff need not occur. For instance, suppose that we have an 8 bit number and the accumulators can add 4 bit numbers. If we want to add the 8 bit numbers, then we can use two accumulators, one of which operates on the rightmost bits of each number and the other one on the leftmost bits. Of course, provision must be made for carries between the pairs, etc. Thus, the total number stored in the A-MQ register may or may not be usable. It is usually a function of how the computer is programmed.

We have considered parallel multiplication. Serial multiplication circuits also exist which utilize only one adder in the accumulator.

Division

Now let us consider the process of division. The operation is carried out using a minor modification of the circuitry used for multiplication.

Let us start by considering a pencil and paper division. This follows the ordinary rules for long division. We shall use 2's complement for subtraction. Let us divide 10001111 by 1011.

```
1011 |10001111
```

The first step would be to see if 1000 is greater than 1011. Obviously, it is not. However, the computer is not aware of this and would try it out. Thus, a first trial would be

```
         1
1011|10001111                                             (10-23a)
    -1011
```

In 2's complement we would have(the digits to the left of the dotted line represent "sign bits")

```
  0 | 1 0 0 0
 (1)| 0 1 0 1
 ------------
  1 | 1 1 0 1
```

The leftmost digit, to the left of the dotted line, is a 1. Thus, the result is negative. If we ignore the circled number, we can see that there is no carry into, nor existing 1 in, the fifth position. Thus, the result is negative. Then, the choice of the first digit of the quotient as 1 is too large. Thus, the next step in the division is

```
       1
1011|10001111                (10-23b)
    -1011
```

In the 2's complement, the subtraction becomes

```
  0 1 0 0 0 1
+(1 1)0 1 0 1                (10-23c)
(1 0 0)0 1 1 0
```

This is positive and, thus, the choice of quotient digit is correct. Note that, to be completely correct, we must include the encircled digits when we use the 2's complement. If we do not, then we have

```
1 0 0 0 1
  0 1 0 1
1 0 1 1 0
```

Now there is a 1 introduced into the fifth position of the answer. Thus, we know that the result will be positive. Note that, if we discard this 1, then we will have the correct answer. Thus far, then, we have

```
       01
1011|10001111
    +0101
     01101
```

Continuing this way, we obtain

```
       011
1011|10001111
    +0101
     01101
     +0101
    (1)0010
```

The fifth position has a 1 which indicates that the result of the subtraction is positive. As before, we discard it and proceed. Thus, we have

```
         01101
1011|10001111
     +0101
      01101
      +0101
      001011
        0101
      (1)0000
```

Again, the 1 in the fifth position is disregarded. Thus, there is no remainder.

In these cases, when there is a digit 0 in the quotient, the following operation would be performed by the computer. A 1 would be tried. The result of the subtraction would be negative. Thus, the 1 in the quotient would be rejected and the subtrahend would be added back. When humans divide, we can look and see if the result of the subtraction would be negative. However, the computer cannot do this.

Now let us consider the computer operation. The multiplier of Fig. 10-10a is used with a different controller. The combined A and MQ registers are used to store the dividend. The B register is used to store the divisor. (Note that the dividend can have essentially twice the number of bits as the divisor. The discussion of this in regard to multiplication applies here.)

After each subtraction, we shift in one more digit of the dividend. This is equivalent to shifting the dividend to the left. Note that this will also eliminate the encircled terms in the example.

As we shift to the left, 0's will be stored at the right side of the MQ register. We can use these "empty spaces" to store the digits of the quotient. Then, when the division is complete, the quotient will be stored in the MQ register.

Now let us carry out the previous example illustrating the contents of the L-A-MQ register. At the start, we have

```
L    A       MQ
0 1 0 0 0 1 1 1 1                                   (10-24a)
```

Note that the A register has 5 bits. Actually, the fifth bit will be in the link register. To avoid confusion, we shall show this bit circled. Note that it can be used to check that the results of each subtraction are positive. Thus, we shall only have four bits in the A register. After (10-23a), we have

```
L    A       MQ
(0) 1 1 0 1 1 1 1 1
```

Note that the link bit indicates a negative result. (Remember that we have omitted the sign bit from the B register.) Hence, the divisor is added back, giving

```
L    A       MQ
(1) 1 0 0 0 1 1 1 1
```

If we clear the link, we obtain the original expression

```
L    A       MQ
(0) 1 0 0 0 1 1 1 1
```

Now, the contents of the L-A-MQ register are shifted to the left. (Note that the above clearing of the link bit may seem a useless step, since the link bit is lost. However, in implementing the procedure, the link bit can be used to determine the next bit of the quotient. Thus, it is desirable that it be correct.) Then, after shifting, we have

```
L    A       MQ
(1) 0 0 0 1 1 1 1 0
```

Note that the previous link bit is 0, which is the correct rightmost bit of the MQ register. Next, subtraction is performed and succeeds. Thus, we have

```
L    A       MQ
(1) 0 1 1 0  1 1 1 0                                  (10-24c)
```

Since the subtraction succeeded, the term in the quotient is 1. This can be obtained from the link bit. Thus, a 1 is put at the right input of the MQ register. Now the contents of the L-A-MQ register are shifted to the left. This yields

```
L    A       MQ
(0) 1 1 0 1  1 1 0 1                                  (10-24d)
```

After the next subtraction, we have

```
L    A       MQ
(1) 0 0 1 0  1 1 0 1                                  (10-24e)
```

Since the subtraction succeeded, i.e., there is a 1 in the L register, a 1 is put at the right input of the MQ register. Thus, after shifting left, we have

```
L    A       MQ
(0) 0 1 0 1  1 0 1 1                                  (10-24f)
```

The next subtraction fails. Then, after adding back the divisor, clearing the link, and shifting, we have

```
L    A       MQ
(0) 1 0 1 1  0 1 1 0                                  (10-24g)
```

After the next subtraction, we have

```
L    A       MQ
(1) 0 0 0 0  0 1 1 0                                  (10-24h)
```

The subtraction succeeds. Thus, after the next shift, we have

```
L   A       MQ
0 0 0 0 0 1 1 0 1
```

(10-24i)

Thus, the quotient is in the MQ register and the remainder (0) is in the A register.

The block diagram is similar to that for the multiplier. However, the contents of the B register must be either subtracted from or added to A. Thus, some special provision must be made for this. For instance, the B register could be an accumulator. Alternatively, the variables can be continuously shifted into and out of the accumulator. Of course, there are different control functions. The control signal that indicates if the result of the subtraction is positive and that a 1 is to be entered into the MQ register can be obtained from the fifth position in the A register or the link register.

BCD-Binary Conversion

The numerical data supplied to a computer often comes from punched cards or a teletype. In this case, there is a sequence of decimal digits. These are often encoded by the associated electronic circuits in a BCD code, see Sec. 8-1. Arithmetic can be performed using numbers encoded in this way. At times this is done, for example, in desk calculators and in some small computers. However, for a digital computer, it is much more efficient if standard binary numbers are used. Thus, there must be a conversion from BCD to binary. This can be done by a simple combinational logic circuit. However, there would be millions of possible inputs, corresponding to each number that could be entered, and an equal number of binary outputs. Thus, the number of gates required would be prohibitive. We shall now discuss a procedure that can be used to convert without requiring extensive electronics.

In Eqs.(2-8) to (2-11) we illustrated a procedure for obtaining a binary number from a decimal number by successively dividing by 2. The remainders yielded the binary number. It is assumed that the reader is familiar with this procedure. Let us illustrate it by obtaining 43 in binary form. Then,

$$\begin{aligned} 43/2 &= 21 + 1/2 \\ 21/2 &= 10 + 1/2 \\ 10/2 &= 5 + 0/2 \\ 5/2 &= 2 + 1/2 \\ 2/2 &= 1 + 0/2 \\ 1/2 &= 0 + 1/2 \end{aligned} \qquad (10\text{-}25)$$

Now take the remainders in reverse order. Thus, we have

$$43_{10} = 101011_2$$

Now we shall consider how this can be implemented on the computer. We cannot simply divide since the arithmetic circuits are set up for binary numbers. Let us assume that an 8-4-2-1 BCD binary code is used. Then, *each* decimal digit will be encoded in binary. But the complete number is not. For instance, 43 will be encoded as

0100 0011

Since each digit is encoded in binary, we can use the division circuits to divide them, *each set of four bits*, by 2. Let us see how this would work. We shall use the decimal representation of the digits

$4/2 = 2 + 0/2 \qquad 3/2 = 1 + 1/2$

Note that the 4/2 actually represents 40/2. The next step is

$2/2 = 1 + 0/2 \qquad 1/2 = 0 + 1/2$

Next we have

$1/2 = 0 + 1/2 \qquad 0/2 = 0 + 0/2$

Thus, there is a remainder of 1/2 in the tens column. This is actually 10/2 = 5. This must now be added into the ones column.

Thus, we have 0 + 5 = 5.

0 5

The next steps are then

5/2 = 2 + 1/2
2/2 = 1 + 0/2
1/2 = 0 + 1/2

Again we consider the remainders in the ones column and obtain

$43_{10} = 101011_2$

A block diagram for the implementation of this is shown in Fig. 10-11.

We have illustrated this procedure using two digits. It can, of course, be extended. Remember that each carry adds 5 to the next "column" to the right. The electronics must account for this.

When the results of the computation are obtained, we often want the results to be printed out as a decimal number, that is, as a sequence of decimal numbers. In this case, the binary number is converted into a BCD number. Each digit is then used to activate a printer. The converse process can then be used to convert from binary to 8421 BCD. For instance, consider

101011_2

We perform the following operations. (Note that we work from left to right in this case.)

1 x 2 = 2	2 + 0 = 2	(Note 0 is second digit from the left)
2 x 2 = 4	4 + 1 = 5	(Note 1 is third digit from the left)
5 x 2 = 10	10 + 0 = 10	(Note 0 is fourth digit from the left)
10 x 2 = 20	20 + 1 = 21	(Note 1 is fifth digit from the left)
21 x 2 = 42	42 + 1 = 43	(Note 1 is sixth digit from the left)

Thus we can, by means of multiplication by 2 and addition, obtain the decimal number. The actual details of the implementation are the converse of the preceding ones. Remember that we must obtain each digit encoded in 8421 BCD. The details will be left to the reader.

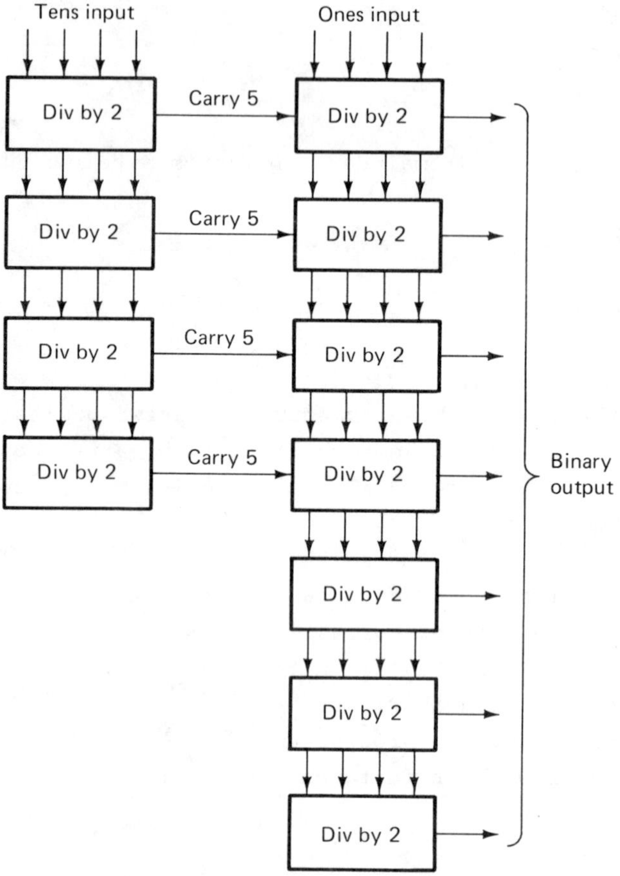

Figure 10-11. A block diagram which implements two digits of an 8421 BCD to binary conversion.

10.4 THE ACCUMULATOR

We shall now consider the circuit of an accumulator which performs the operations of Table 10-1. Some of these operations have already been considered. For instance, addition and clear have been implemented by the circuit of Fig. 10-4. To easily obtain more control, the accumulator registers are often J-K flip-flops instead of D flip-flops. To avoid excessive amounts of drawing, we shall consider only one stage of the accumulator. For instance, one stage of the accumulator using J-K flip-flops which performs the ADD operation is shown in Fig. 10-12. Note that the J-K flip-flop and the NOT gate function as the D flip-flop of Fig. 10-4. We introduce the ADD control signal c_1 differently here. When $c_1=0$, then

$$J_i = K_i = 0 \qquad (10\text{-}27)$$

and the flip-flop does not change its state. When $c_1=1$, then

$$J = S_i \qquad (10\text{-}28a)$$

$$K = \bar{S}_i \qquad (10\text{-}28b)$$

and the flip-flop sets itself to S_i. The point labeled c_1 is where the ADD command signal is applied.

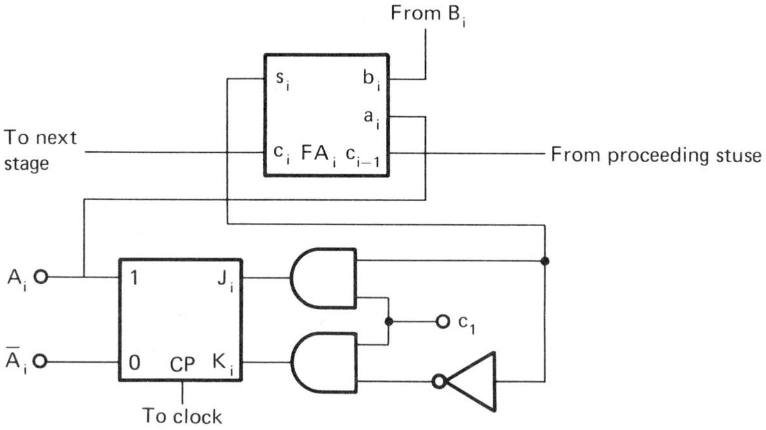

Figure 10-12. A portion of an accumulator which performs an addition when $c_1 = 1$.

To clear the flip-flop we use the circuit of Fig. 10-13. When the clear signal c_2 is given

$$J_i = 0 \qquad (10\text{-}29a)$$

$$K_i = 1 \qquad (10\text{-}29b)$$

Thus, the state of the flip-flop becomes 0 after the next clock pulse. We assume that only one command will ever be given at a time. Thus, we can use OR gates to apply the various signals to the flip-flops. We shall illustrate this subsequently. The β_{2i} is a connection point which we shall discuss later.

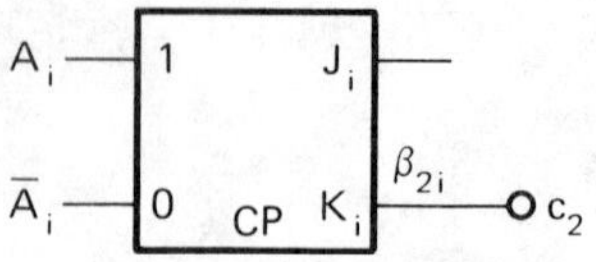

Figure 10-13. The circuit used to clear stage i of the accumulator when $c_2 = 1$.

Logical AND

To perform the logical AND we use the circuit of Fig. 10-14. Note that, if $\bar{B}_i$ is not available, then a NOT gate must be included to obtain it from B_i. Note that if c_3 is a 1, and A_i and B_i are not *both* 1's, then, after the next clock pulse, A_i=0. We could draw truth tables and design these sequential circuits using the procedures of Sec. 6-5. However, these circuits are so simple that this step can be omitted. Note that, in this case

$$J_i = 0 \tag{10-30a}$$

$$K_i = \bar{B}_i \tag{10-30b}$$

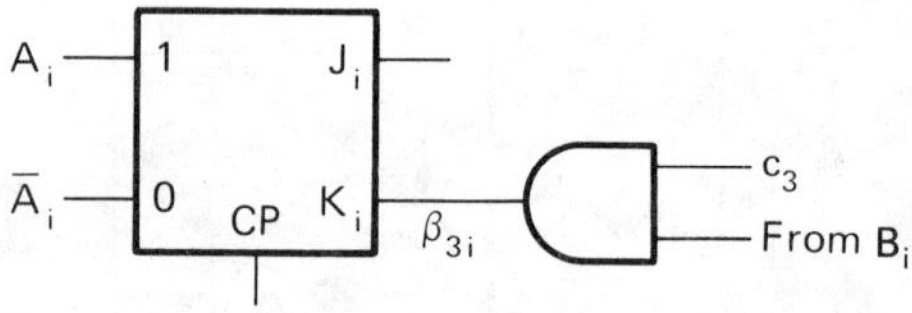

Figure 10-14. The circuit used to perform the logical AND operation with stage i of the accumulator when $c_3 = 1$.

Logical OR

The logical OR is implemented using Fig. 10-15. In this case, when c_4 is a 1, we have

$$J_i = B_i \tag{10-31a}$$

$$K_i = 0 \tag{10-31b}$$

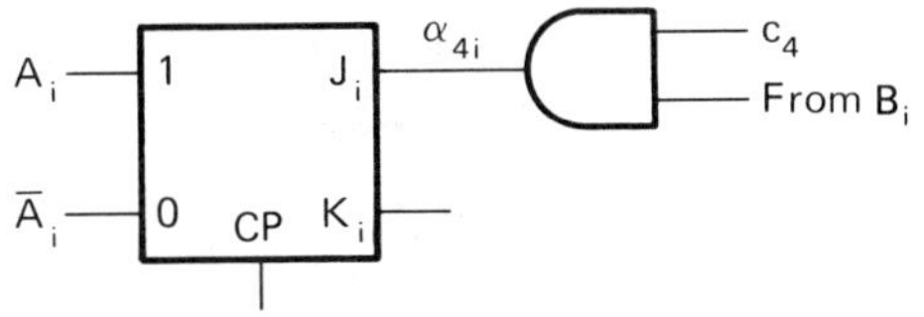

Figure 10-15. The circuit used to perform the logical OR operation with stage i of the accumulator when $c_4 = 1$.

Logical EXOR

The logical EXOR is implemented using Fig. 10-16. In this case, when $c_5=1$

$$J_i = K_i = B_i \tag{10-32}$$

Thus, the flip-flop functions as a T flip-flop. Hence, if $A_i=1$, and $B_i=0$ and if $A_i=0$ and $B_i=1$ then, after the next clock pulse, we will have $A_i=1$. If both A_i and B_i are 1's, then, after the next clock pulse, $A_i=0$. If both A_i and B_i are 0's, then the state will not change.

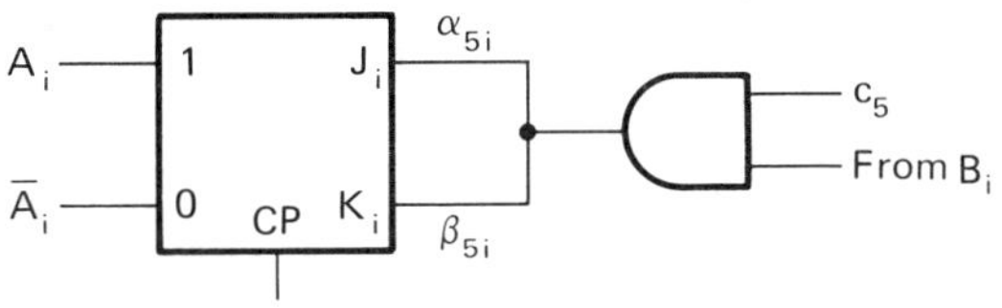

Figure 10-16. The circuit used to perform the logical EXOR operation with stage i of the accumulator when $c_5 = 1$.

Shift Right

To shift right we take the output from the flip-flop to the left(i.e., the $i+1^{st}$ flip-flop)and set the ith flip-flop to its state. To do this, see Fig. 10-17, with $c_6=1$, we set

$$J_i = A_{i+1} \tag{10-33a}$$

$$K_i = \bar{A}_{i+1} \tag{10-33b}$$

For the leftmost flip-flop we connect the A_{i+1} input to 0 and the $\bar{A}_{i+1}$ input to 1. This can be done by open circuiting the A_{i+1} input and connecting the $\bar{A}_{i+1}$ input to c_6.

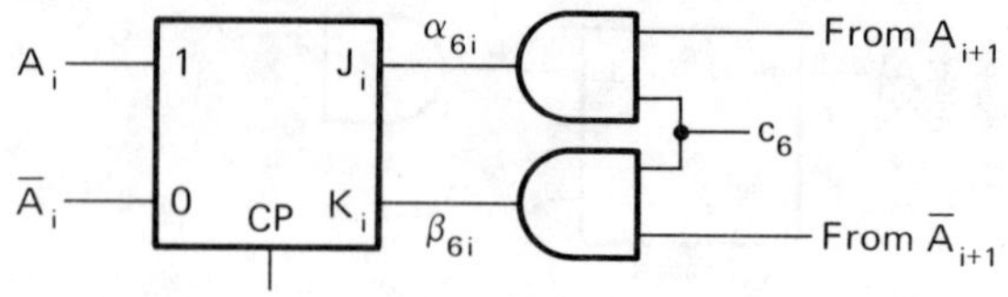

Figure 10-17. The circuit used to shift right with stage i of the accumulator when $c_6 = 1$.

Shift Left

This basically follows the shift right circuit, except A_{i-1} replaces A_{i+1}. In this case, it is the rightmost(i.e., first position) A_0 whose A_i lead is open circuited and whose $\bar{A}_{i+1}$ lead is connected to c_7. The circuit is shown in Fig. 10-18.

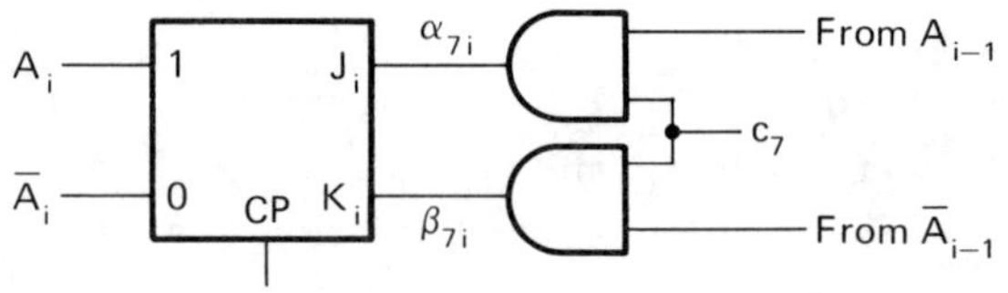

Figure 10-18. The circuit used to shift left with stage i of the accumulator when $c_7 = 1$.

Complement

To obtain the complement, when $c_8=1$, we change the state of the flip-flop. To do this, we connect it as a T flip-flop, and make its inputs equal to c_8. This is shown in Fig. 10-19.

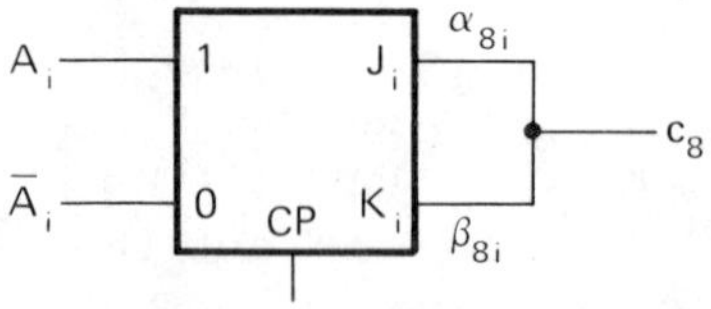

Figure 10-19. The circuit used to complement with stage i of the accumulator when $c_8 = 1$.

Increment

When we increment, 1 is added to the contents of the register. Thus, it is essentially an addition operation. In Fig. 10-20, we have combined both the operations of addition and incrementation. Note that when c_1 is applied, we have essentially the circuit of

Fig. 10-12. If $c_1=0$, then the B inputs are disconnected by the AND gate. If $c_9=1$, then the add command is essentially given except that now all the B inputs are cut off. Effectively (B)=0. However, $c_9=1$ and it is connected to the carry input of the first full adder. (We assume that all the adders are full adders.) Thus, we add 1.

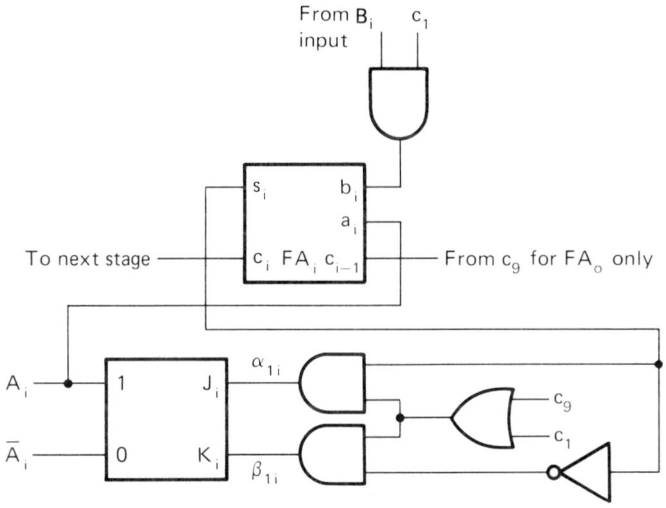

Figure 10-20. The portion of the accumulator which performs an addition when $c_1 = 1$ and increments when $c_9 = 1$.

Negative Check

We assume here that 2's complement arithmetic is being used. Then, the number in the accumulator is negative if the number stored in the leftmost register is a 1. Thus, in Fig. 10-21, when $c_{10}=1$, the output y will be 1 if the stored number is negative. If it is positive or 0, then y=0.

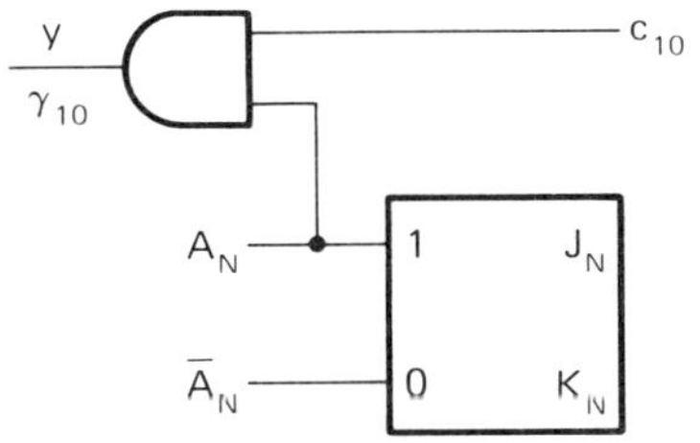

Figure 10-21. Negative check in accumulator. If stored number is negative then y = 1; when $c_{10} = 1$. This is leftmost stage.

Zero Check

If the stored number is 0, then

$$\bar{A}_0 = \bar{A}_1 = \ldots = \bar{A}_N = 1 \tag{10-34}$$

In the circuit of Fig. 10-22, when $c_{11}=1$, y will equal 1 if Eq. (10-34) is satisfied. If the accumulator has very many registers, then the fan in of the NAND gate can become excessive. In this case, several AND gates can be used in a cascade.

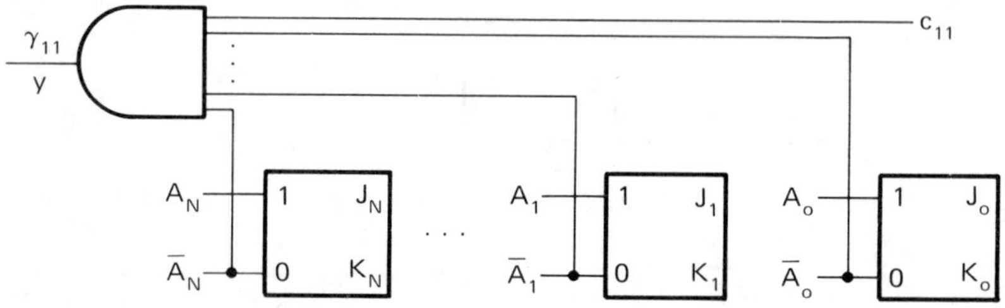

Figure 10-22. Zero check in accumulator. If stored number is 0, then $y = 1$ when $c_{11} = 1$.

Complete Accumulator

Now let us consider the complete computer. Note, we assume that the controller is such that only one control signal of c_1-c_{11} at a time will be a 1. Thus, to implement the complete accumulator, we can break all the inputs to the JK flip-flop at the points marked with an α or a β. Each set of α inputs for each stage is used as the input to an OR gate, which is connected to the J_i input. A similar statement can be made for the β points and the K inputs.

Thus, we break the α_{ni} and β_{ni} connections and insert the OR gates as shown in Fig. 10-23a. Note that each flip-flop of the accumulator has a pair of OR gates inserted as shown. (Note that there is no α_{9i} or β_{9i}, since two functions were realized in Fig. 10-20.) Note that there is one circuit of the type shown in Fig. 10-23a for *each stage* of the accumulator(i.e., a total of N+1).

There are two y outputs, one for zero check and the other for negative check. Since only one can be used at a time, we connect

γ_{10} and γ_{11} of Figs. 10-21 and 10-22 to the input of an OR gate as shown in Fig. 10-23b. Note that there is only one such circuit for the *entire* accumulator.

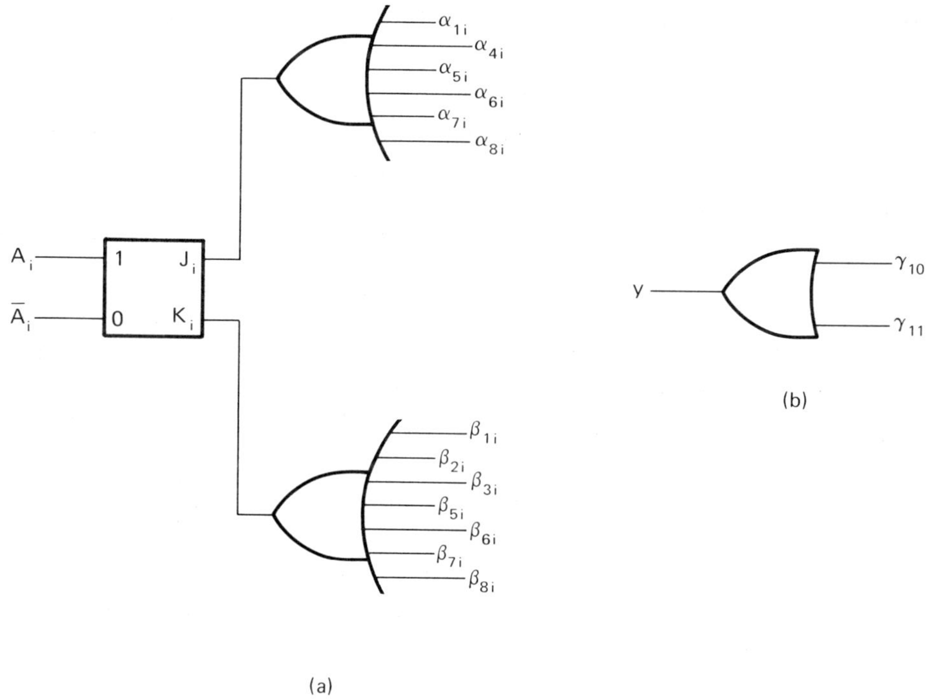

Figure 10-23. Diagram showing interconnection of Figs. 10-12 to 10-21 to produce complete accumulator. (a) insertion of OR gates in series with each flip-flop input. Note α_{2i}, α_{3i} and β_{4i} are missing; (b) connection of γ_{10} and γ_{11} to OR gate to produce y output. There is only one y output for complete accumulator.

Let us conclude by discussing some definitions. The device we have developed is called an accumulator, or sometimes a *true accumulator*. At times, the logic circuits are not wired directly to the register but the computer is made flexible so that the logic may be interconnected with different registers, by suitable multiplexing. This is still called an accumulator.

There is a much simpler device which will just add two numbers. In general, this is no longer used in digital computers. However, it is used in some desk calculators. Such a device is called an *accumulating register*, or a *working register*. At times, confusion results due to the similarity of the words accumulating register and accumulator. However, the accumulator is a much more versatile device.

10.5 INTEGRATED CIRCUIT IMPLEMENTATION OF COMPUTATION OPERATIONS

All of the computation operations that we have discussed,e.g., addition, multiplication, BCD-binary conversion can be implemented electronically on a single chip using integrated circuits. We shall consider some representative ones here. Remember that these are only a small sample of the available integrated circuits.

In Fig.10-24a, the logic circuit diagram of a type MC 1059 full adder is illustrated. This chip contains two full adder circuits. Thus, two bit numbers can be added. The carry output of the first adder is directly connected to the carry input terminal of the second adder. In addition to the sum and carry outputs, other logic outputs are given at appropriate terminals. These extra terminals increase the versatility of the device. In Fig. 10-24b, the truth table is given. This full adder is implemented using ECL. Its schematic diagram is shown in Fig. 10-24c. Only one half of the circuit is shown since the other half is essentially the same.

We have shown a two bit adder here. Actually, chips are available which will add two 16 bit(and larger)binary numbers.

As another example of an integrated circuit implemented computation operation, consider the type SN54145 BCD to decimal decoder/driver, see Fig. 10-25. The input to this device is a BCD number. This causes one of the appropriate output lines to be energized. The output lines are numbered 0 to 9. The energized one corresponds to the input BCD number. These outputs can actuate relays or light lamps for display purposes. The logic circuit diagram is shown in Fig. 10-25a. Figures 10-25b and c give the equivalent input and output circuits. The truth table is given in Fig. 10-25d. (See Sec. 6-8 for a further discussion of such figures.)

We have considered two examples of integrated circuits here. Actually, almost all computation circuits are implemented as integrated circuits. For instance, complete accumulators have been fabricated on a single chip.

In this chapter we have considered some ideas of computation circuits. In the next one we shall discuss their use in the complete digital computer.

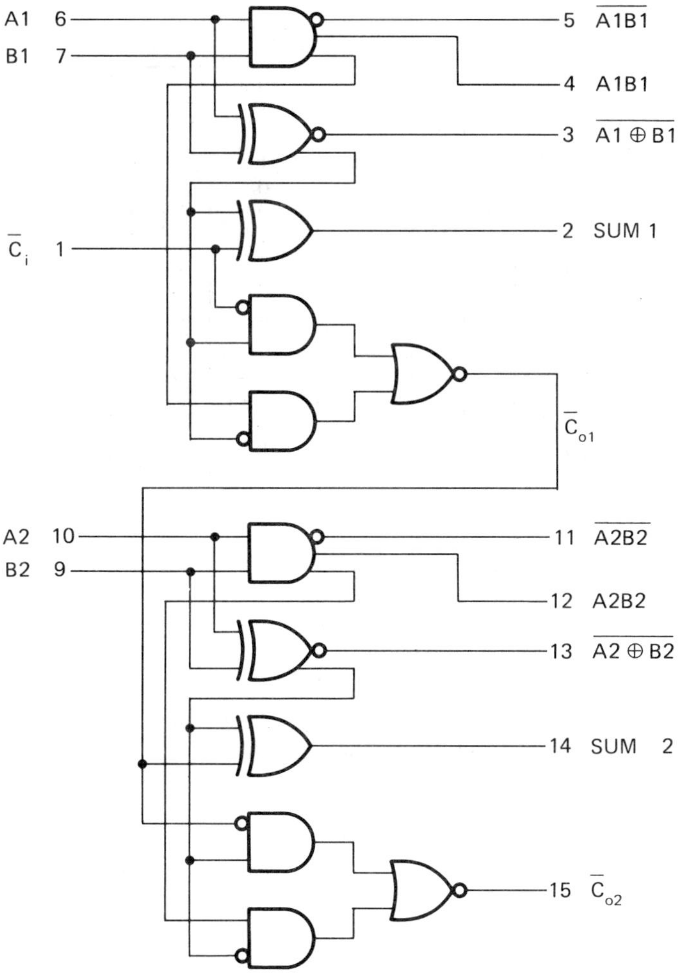

(a)

Figure 10-24. The type MC1059 binary adder. (a) logic circuit diagram; (b) truth table; (c) schematic diagram (Courtesy the Motorola Semiconductor Products, Inc.)

TRUTH TABLE

Inputs			Outputs			
A1	B1	$\overline{C}_i$	A1 · B1	$\overline{A1 \oplus B1}$	S1	$\overline{C}_{o1}$
0	0	0	0	1	1	1
1	0	0	0	0	0	0
0	1	0	0	0	0	0
1	1	0	1	1	1	0
0	0	1	0	1	0	1
1	0	1	0	0	1	1
0	1	1	0	0	1	1
1	1	1	1	1	0	0
A2	B2	$\overline{C}_{o1}$	A2 · B2	$\overline{A2 \oplus B2}$	S2	$\overline{C}_{o2}$
0	0	1	0	1	0	1
1	0	0	0	0	0	0
0	1	0	0	0	0	0
1	1	0	1	1	1	0
1	1	1	1	1	0	0
0	1	1	0	0	1	1
1	0	1	0	0	1	1
0	0	0	0	1	1	1

(b)

Figure 10-24. (Continued) part (b).

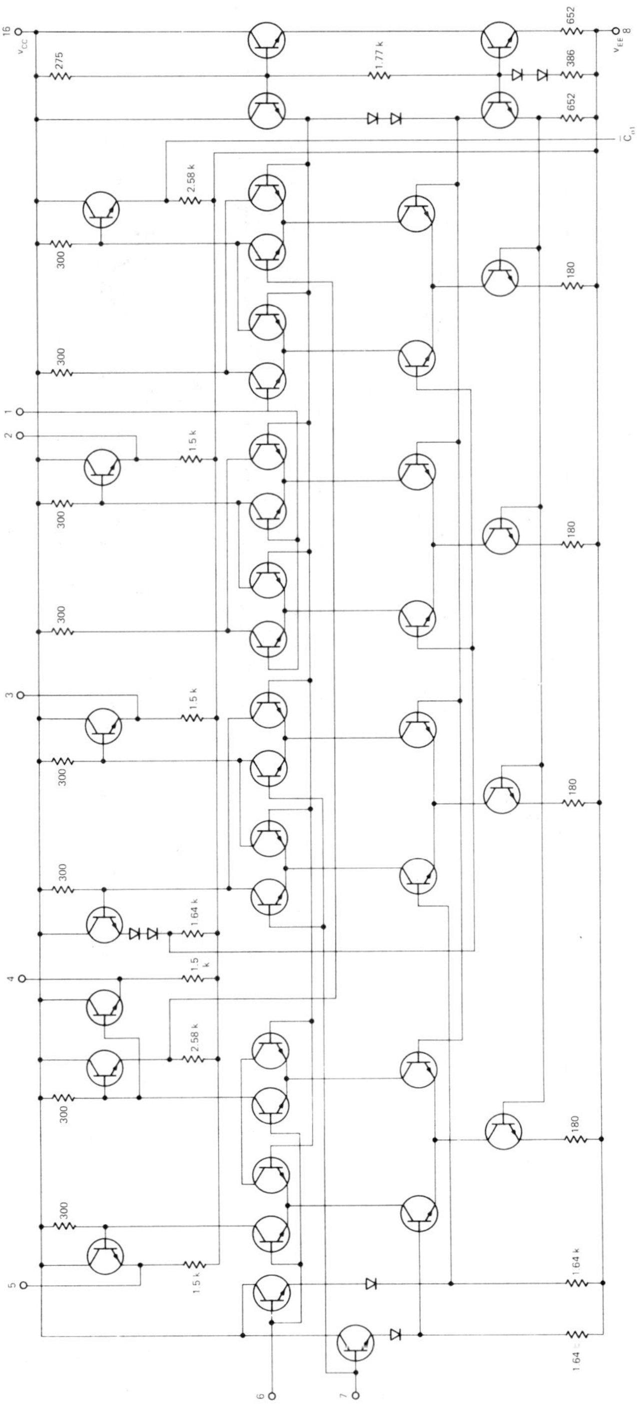

Figure 10-24. (Continued) part (c).

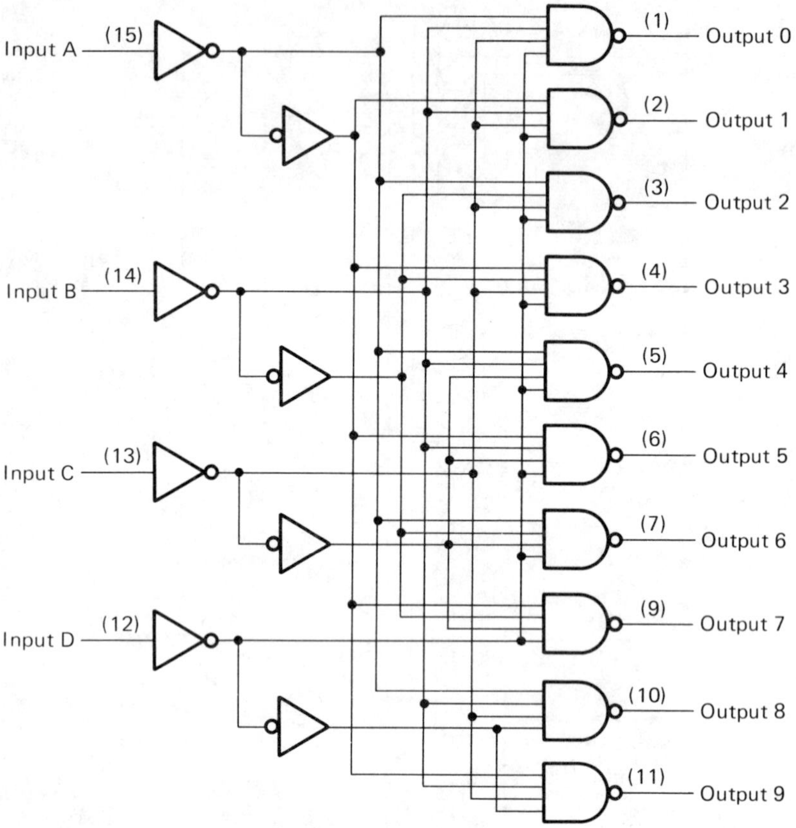

(a)

Figure 10-25. The type SN54145 BCD to decimal decoder/driver. (a) logic circuit diagram; (b) typical input circuit; (c) typical output circuit; (d) function table (Courtesy Texas Instruments, Inc.)

EQUIVALENT OF ALL INPUTS

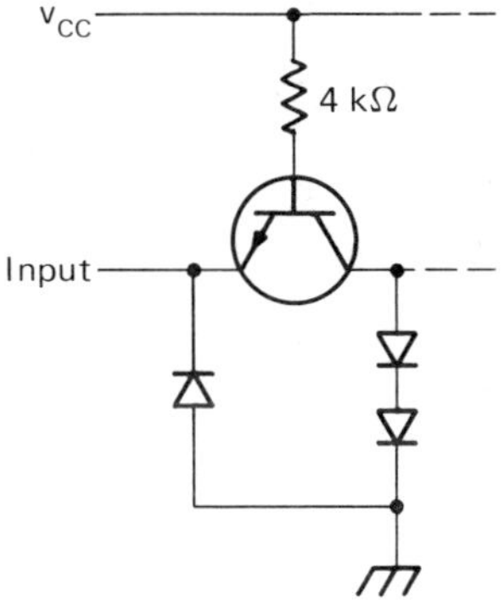

(b)

TYPICAL OF ALL OUTPUTS

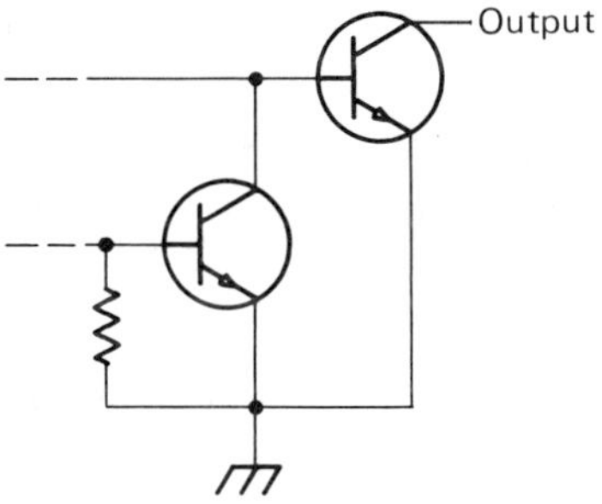

(c)

FUNCTION TABLE

No.	Inputs				Outputs									
	D	C	B	A	0	1	2	3	4	5	6	7	8	9
0	L	L	L	L	L	H	H	H	H	H	H	H	H	H
1	L	L	L	H	H	L	H	H	H	H	H	H	H	H
2	L	L	H	L	H	H	L	H	H	H	H	H	H	H
3	L	L	H	H	H	H	H	L	H	H	H	H	H	H
4	L	H	L	L	H	H	H	H	L	H	H	H	H	H
5	L	H	L	H	H	H	H	H	H	L	H	H	H	H
6	L	H	H	L	H	H	H	H	H	H	L	H	H	H
7	L	H	H	H	H	H	H	H	H	H	H	L	H	H
8	H	L	L	L	H	H	H	H	H	H	H	H	L	H
9	H	L	L	H	H	H	H	H	H	H	H	H	H	L
INVALID	H	L	H	L	H	H	H	H	H	H	H	H	H	H
	H	L	H	H	H	H	H	H	H	H	H	H	H	H
	H	H	L	L	H	H	H	H	H	H	H	H	H	H
	H	H	L	H	H	H	H	H	H	H	H	H	H	H
	H	H	H	L	H	H	H	H	H	H	H	H	H	H
	H	H	H	H	H	H	H	H	H	H	H	H	H	H

(d)

Figure 10-25. (Continued) part (b), (c) and (d).

BIBLIOGRAPHY

Hellerman,H., "Digital Computer System Principles", Chapter 1 McGraw-Hill Book. Co., New York, 1967

Kohonen, T., "Digital Circuits and Devices", Chapter 4, Prentice Hall, Inc. Englewood Cliffs, N.J., 1972

Mano,M.M., "Computer Logic Design", Chapter 9, Prentice Hall, Inc. Englewood Cliffs, N.J. 1972

O'Malley, J., "Introduction to the Digital Computer", Chapters 8, 11, Holt, Rinehart and Winston, Inc., New York, 1972

Rhyne, V.T., "Fundamentals of Digital System Design", Chapter 8 Prentice Hall, Inc., Englewood Cliffs, N.J. 1973

Stein,M.L. and Munro,W.D., "Introduction to Machine Arithmetic", Addison-Wesley Publishers, Inc., Reading, Mass. 1971

PROBLEMS

10-1. Derive Eqs.(10-1).

10-2. Describe the difference between a full adder and a half adder.

10-3. Design a four bit adder that does not suffer from the effect of carry propagation.

10-4. Modify the circuit of Prob.10-3 so that fewer gates are used. Carry propagation through one gate is now allowed.

10-5. Modify the circuit of Fig. 10-4 so that J-K flip-flops are used.

10-6. Repeat Prob. 10-5 but now use T flip-flops.

10-7. Repeat Prob. 10-5 for the circuit of Fig. 10-5.

10-8. Repeat Prob. 10-7 but now use T flip-flops.

10-9. Discuss and illustrate the notation introduced in Sec. 10-2.

10-10. An accumulator uses the control designations of Table 10-1. The initial values that are stored in the accumulator register A and the B register that supplies values to the accumulator are

(A) = 0000
(B) = 0010

What are the contents of the A register after the following sequence of commands is given?

c_1
c_1
c_1
c_8
c_9

Assume that the command signals are appropriately synchronized with the clock pulses.

10-11. Repeat Prob. 10-10 if the initial value of (A) is 0011.

10-12. Repeat Prob. 10-10 for the sequence

c_2
c_1
c_1
c_4
c_3
c_8
c_9
c_6

10-13. Design a circuit using JK flip-flops which realizes the controller whose state transition diagram is given in Fig. 10-8.

10-14. Modify Fig. 10-8 so that the A register can be cleared before the start of the operation.

10-15. Design a controller which realizes the control sequence of Prob. 10-10. Use JK flip-flops.

10-16. Repeat Prob. 10-15 using D flip-flops.

10-17. Repeat Prob. 10-15 for the control sequence of Prob. 10-12.

10-18. Repeat Prob. 10-17 using T flip-flops.

10-19. What modifications would have to be made in the subtraction algorithm if 1's complement subtraction were used?

10-20. Realize a controller which can direct the performance of 2's complement subtraction. Use J-K flip-flops.

10-21. Repeat Prob. 10-20 but now use different state assignments.

10-22. Describe the complete operation of the multiplier of Fig. 10-10.

10-23. Modify the multiplier and/or state transition diagram of Fig. 10-10 so that the A register can be cleared prior to the start of multiplication.

10-24. Obtain a block diagram and state transition diagram for the controller for a circuit that performs binary division.

10-25. Discuss the difference between integer and floating point multiplication.

10-26. Discuss the difference between integer and floating point division.

10-27. Discuss overflow and/or its lack in multiplication and division.

10-28. Extend the diagram of Fig. 10-11 to a three digit BCD conversion.

10-29. Obtain a block diagram for the decimal to 8421 BCD conversion. Assume two decimal digits.

10-30. Repeat Prob. 10-29 for three decimal digits.

10-31. Design a circuit which converts each digit of an 8421 BCD code to 2421 BCD code.

10-32. Implement the gate circuit of Fig. 10-12 using a D flip-flop.

10-33. Repeat Prob. 10-32 for the circuit of Fig. 10-13.

10-34. Repeat Prob. 10-32 for the circuit of Fig. 10-14.

10-35. Repeat Prob. 10-32 for the circuit of Fig. 10-15.

10-36. Repeat Prob. 10-32 for the circuit of Fig. 10-16.

10-37. Repeat Prob. 10-32 for the circuit of Fig. 10-17.

10-38. Repeat Prob. 10-32 for the circuit of Fig. 10-18.

10-39. Repeat Prob. 10-32 for the circuit of Fig. 10-19.

10-40. Repeat Prob. 10-32 for the circuit of Fig. 10-20.

10-41. Realize the circuit of Fig. 10-22 using AND gates with no more than three inputs and one NOT gate.

11

The Digital Computer

In this chapter we shall apply all of the material developed in the previous chapters to a discussion of the complete digital computer.

We shall start by considering the general configuration of the computer. The transfer of information among the parts of the computer and control of the sequence of operations that a computer must do when computation is performed are of fundamental importance. We shall consider them and discuss machine language and assembly language programming.

The design of computer circuits which perform the desired functions will be further discussed here. The interfacing of the computer with the "external world" will also be considered both here and in Chapter 12. We shall also describe a simple complete computer.

11.1 THE GENERAL ORGANIZATION OF THE DIGITAL COMPUTER

In this section we shall consider, in a rather broad, qualitative way, the general organization of the digital computer. We shall see that some of the most important aspects of the computer have already been considered in detail. We shall discuss the others in this chapter.

Fundamentally, a digital computer can be described by the block diagram of Fig. 11-1, which illustrates four basic units. There is a *central processing unit* (CPU), where all the computations are actually performed. In a simple computer, this could consist of an accumulator of the type discussed in Sec. 10-4, plus several additional registers. In large computers, there may be many accumulators and registers in the CPU. We have developed all the basic types of circuits which could be used in this unit.

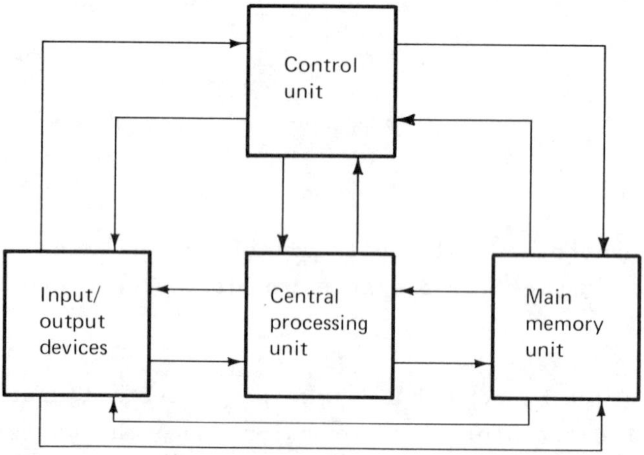

Figure 11-1. Block diagram of the basic computer units.

The operations of the central processor are controlled by a *control unit* (CU), just as a single accumulator was controlled by a control circuit, see Secs. 10-2 and 10-4. However, in a complete computer, the control unit must be more elaborate. For instance, in Sec. 10-3, we considered control circuits that controlled the operations of subtraction, multiplication, etc., in conjunction with an accumulator. Each control circuit that we discussed handled one operation, e.g., multiplication. In the complete computer, the control circuit may be capable of directing *all* these operations. In addition, it must control the transfer of information from the memory to the central processor and in the reverse direction. Transfer of data to and from the input and output devices may also be controlled by the control unit.

The *main memory* (MMU) is a random access memory of the type discussed in Chapter 7. It is often a magnetic core memory although magnetic film memories and semiconductor memories are also used. The main memory is capable of storing a large number of words. A single word can be selected and read or written in turn. These procedures were discussed in detail in Chapter 7. In this chapter, we shall be concerned with the details of data storage in the memory.

A computer is useless unless data and other information, such as programs, can be supplied to it and unless the computed information can be extracted from it. This is the function of the input/output devices. Some typical input devices are teletypes where the user types the information on a typewriter type keyboard. Contacts on the keys are connected with logic circuits which generate coded signals. For instance, the codes of Table 8-4 are typical of the ones generated. These are transmitted to the computer, sometimes over telephone lines. Further code conversion usually takes place in the computer. For instance, numerical data may be first encoded to BCD and then to binary, see Sec. 10-3. (Note that these codes are usually arranged so that the conversion to 8421 BCD is very simple.) Other forms of input devices are card readers, where the information is encoded by the location of holes punched in cards. A card reader senses the location of the holes and, from this information, generates the appropriate signals. There are also special purpose input devices. For instance, many "cash registers" now directly input information to a computer, which then can keep track not only of cash, but also of inventory, etc.

Output devices could be the same teletype that was used for the input. In this case, encoded signals are sent to the teletype which, with its logic circuits, decodes the signals and applies current to electromagnetic devices such that the appropriate character is printed. Line printers are also used. These have the advantage that an entire line can be printed at once. Thus, they are much faster than the teletype. Of course, the input/output devices are much slower than the computer itself.

At times, the computer does not receive or transmit printed information. For instance, the input may be electric signals from gyroscopes which indicate the position, altitude, etc., of an airplane. The output may be signals which control the airplane's throttle, airlerons, etc. There are many other forms of input-output devices. We have just discussed a representative sample.

The first large digital computers occupied entire rooms. However, now, using LSI(large scale integration), a small computer can be built on several chips. This is true of the computer that we shall discuss in this chapter. Of course, some of the peripheral devices such as magnetic tape drives, card readers and line printers are still large. However, the actual electronics of the computer can occupy a very small volume.

In this section we have, very briefly, discussed(or reviewed) the major components of a digital computer. In the next section we shall consider the details of information storage, transfer and control that are fundamental to the operation of the digital computer.

11.2 INFORMATION STORAGE AND TRANSMISSION-CONTROL IN DIGITAL COMPUTERS

In a digital computer, two types of information are stored in the memory, *instructions* and *data*. The data are the information acted upon by the computer. The instructions direct the actions of the computer. We could build a computer which did not store instructions. For instance, there could be buttons on the control panel corresponding to desired instructions and the operator could push a succession of them at the appropriate time. However, this would make the solution of complex problems, where many thousands of instructions are performed repetitively, almost impossible. The storage of instructions allows us to program the computer to perform the desired tasks.

Let us start by considering how data is stored in the memory. We shall assume that the memory organization discussed in Sec. 7-1 is used. In this case, the memory stores a word of fixed length(e.g., 4 bits, 16 bits, etc.). Each word has an address in the memory. We could use two devices of the type illustrated in Fig. 7-6 or 7-12 to select the coordinates of a word. There is a register which is associated with the main memory. The function of this register is to store the address of the current memory location(i.e., the location of the word we want to read or write). (Note that half of this word could be used to select one corrdinate and the other half the other coordinate.) The word can be designated by the one number stored in the register. This register is called the *memory address register*(MAR). There is also a *memory buffer register*(MBR). This stores a word prior to its being written into the memory, or it stores the word that has just been read from the memory, see Sec. 7-1. Remember that, with core memory, we must store the word since the readout is destructive. However, even with semiconductor memories, it is desirable to store this information in a separate register so that the word just read is always in the same place.

In order to talk about operations in a relatively compact form, we shall use the notation introduced in Sec. 10-2. Thus, the contents of the MAR register will be written as (MAR). Therefore, if the stored address is 1010, we would write

$$(\text{MAR}) = 1010 \qquad (11\text{-}1)$$

We shall also indicate the following notation. (M(1010)) is the word stored in memory location M(1010). In general, we can write M(MAR) which is the word whose location is stored in MAR. Note that, according to the procedures of Sec. 7-1, Eq.(11-1) would correspond to the location $x_2=10_2$; $y_2=10_2$. The control unit directs the operation of the main memory and its associated registers. There are relatively few commands that must be given to the memory

1 Write (MBR) into M(MAR), i.e.,(MBR) → M(MAR) (11-2a)

2 Clear M(MAR),i.e., 0 → M(MAR) (11-2b)

3 Read M(MAR), i.e., Write M(MAR) into MBR (11-2c)
{(M(MAR)) → MBR}

In Figs. 7-6 and 7-12 we illustrated control units for a main memory. These can be used for these operations. Slight modifications can be used to obtain the clear operation. Thus, in addition to generating signals to control the central processing unit, the control unit should be capable of supplying the memory control signals, indicated in (11-2).

Now let us consider the structure of the words which are stored in the memory.

Word Structure

The information stored in the main memory is always in the form of a binary number. This can actually represent a number. That is, it can be data. In addition, it can represent a stored instruction. In this case, the structure of the word must be considered since one word can convey several kinds of information. Actually, there is no general format for words in all computers. In some computers, a relatively small amount of information is supplied by a word while in others, much more information is supplied. We shall start with a discussion of simple word structures and then consider the more complex forms.

A relatively simple word usually contains an instruction(e.g., add)and an address in the memory. Words can thus be broken down into "subwords" which are called *fields*. For instance, in Fig. 11-2, we have illustrated a 15 bit word with two fields. We have used 6 bits for the instruction and 9 bits for the memory location. This means that there can be, at most, 64 instructions and 512 memory locations. Most computers have very many more memory locations. However, 512 is large enough to illustrate fundamental principles.

Let us demonstrate how the 15 bit word can be used to program the computer. We shall use Table 10-1 to indicate control signals. For instance, the c_1 control input to the accumulator causes it to add the contents of the B register to the accumulator register A. The binary number which is equal to the subscript of c will signify the operation. Thus, 000001 means add. c_8 commands complement. Thus, if the instruction field contained 001000, this would cause the $(\bar{A}) \rightarrow A$ to be performed, i.e., the

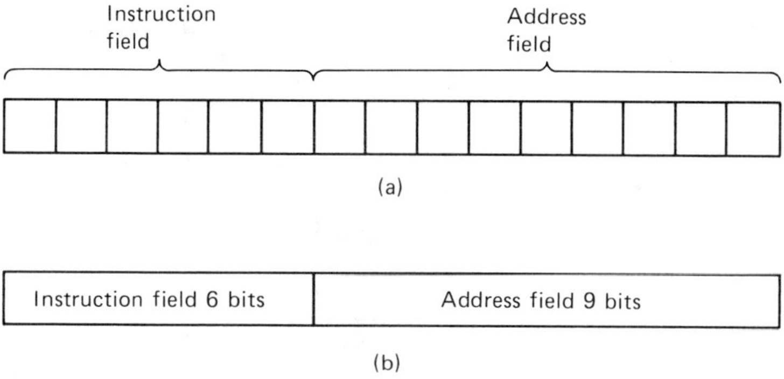

Figure 11-2. (a) A simple 12 bit memory work showing two fields; (b) a short hand representation of this word.

contents of A is replaced by its complement. Similarly, the command 000111 would initiate shift left. We shall also consider that 63_{10} = 111111 represents the command to stop computation. Now let us assume that, in some way, we have caused the following words to be written in the memory.

TABLE 11-1
A SIMPLE STORED PROGRAM

Memory location decimal numbers	Stored word
0	000000000000000
1	000011000000000
2	000001001000000
3	000001001000001
4	111111000000000
⋮	
64	000000000000011
65	000000000001100

We assume that the simple computer of Fig. 11-3 is used. Here this consists of just an accumulator, a control unit, and the memory. Note the block labeled PC. This is the *program counter*. Its purpose is to contain the address of the next memory instruction. For this simple computer, let us assume that it is set to

0 when the program starts running and then increments by 1 each time that an instruction is executed. We have not shown any input/output devices for this simple computer.

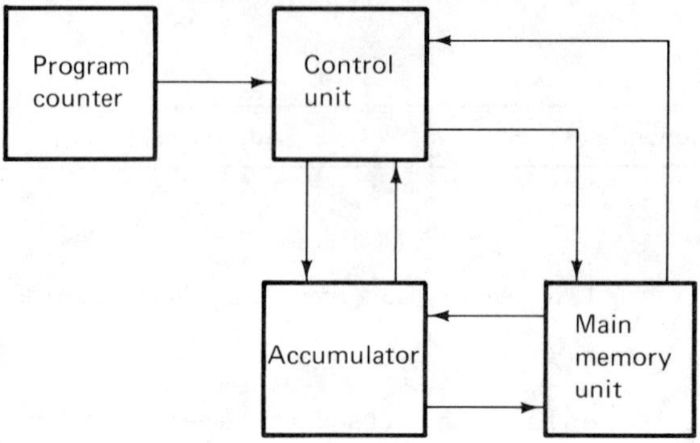

Figure 11-3. A very simple computer.

Now let us consider the program. When the start button is pressed, the value stored in the PC register is (PC)=0. The contents of memory location 0 is 0. Thus, no action will be taken. The value of PC is incremented by 1. Hence, (PC)=1. The word in memory location 1 is read into the MBR. The instruction code is 000011 which corresponds to clear accumulator. This is done. Next, (PC)=2. Thus, (MBR) = 000001001000000. 000001 is the command for ADD. Also, 001000000 = 64_{10}. Thus, the contents of memory location 64_{10} is added to the accumulator. This results in $000000000000011_2 = 3_{10}$ being stored in the accumulator register. The next word stored in the MBR is 000001001000001. The instruction code is 000001 and the memory location is $001000001_2 = 65_{10}$. Then, the contents of memory location 65_{10} is added to the accumulator. (The contents of memory location 65_{10} is $000000000001100_2 = 12_{10}$.) The contents of the accumulator is now $000000000001111_2 = 15_{10}$. The command in memory location 4_{10} now stops operation.

Note that there are many commands that are not given. For instance, suppose that the memory is a core memory. Then, the word would have to be read into the memory buffer register MBR, and then read back into the memory. However, this cycle can be incorporated in the control unit, or in the memory, and only a single read command need be given to the control unit and then it supplies the appropriate sequence of commands. We have seen, see Fig.10-9, that a single command can initiate a sequence of commands. This is the case here also.

A memory location just contains a number. The only way that the computer can tell that it is an instruction(and address)is by the time that it is called from the memory. That is, if the program counter initiates the extraction of information, then the resultant contents of the MBR are to be used as instructions. In this case, the contents of the MBR are stored in two other registers. The left six bits are stored in a register called the *instruction register*(IR). This register is connected to a set of gates that produces the sequence of command signals that correspond to the instruction. (The decoders of Sec. 7-5 can be used for this purpose.) Similarly, the right 9 bits of the MBR are stored in the memory address register (MAR).

It may seem as though the IR and MAR are superfluous since all their information is stored in the MBR. However, this is not the case. When the instruction is carried out, *data* from the memory is often written into the MBR. (Thus, any old information in the MBR is lost.) Hence, if the command were not stored in the IR, it could be lost before the execution of the instruction were completed.

Note that when an instruction activates the memory, the binary number transferred to the MBR is treated as data. When the PC activates the memory, the number transferred to the MBR is treated as an instruction and is transferred to the IR and MAR. We shall consider this subsequently. This multiplexing of the MBR can be accomplished using logic circuits of the type we have considered.

Let us modify the program of Table 11-1. This will illustrate some of the previously discussed ideas. Suppose that we replace the fourth statement and add a fifth so that we have

Memory location	Stored word
4	000001000000011
5	111111000000000

The first three steps will be the same so that, after their execution, the value of the quantity stored in the accumulator register is 000000000001111. (We assume a 15 bit accumulator.) The next instruction is 000001000000011 which causes the contents of memory location 3_{10} to be added to the accumulator register. The number stored in $M(3_{10})$ is 000001001000001. Thus, after addition, the contents of the accumulator is $000001001010000_2 = 592_{10}$. The next instruction terminates execution. Again, note that the number stored in the memory does not tell the computer what type of information it carries.

In order to work with complete programs, we require additional commands. However, we shall defer introduction of them until Sec. 11-3.

We have worked with a 15 bit memory word. It is tedious to keep track of binary numbers when they have this many digits. Many computers have 64 bit words. Thus, reading and writing them becomes tedious for the *people* involved. For this reason, most programmers replace the binary numbers with octal or hexadecimal ones. These bases are used since their conversion to binary is so simple, see Sec. 2-1. For instance, in octal, Table 11-1 would be

TABLE 11-2
TABLE 11-1 WITH THE BINARY REPLACED BY OCTAL NOTATION

Memory location		Stored word (octal)
decimal	octal	
0	0	00000
1	1	03000
2	2	01100
3	3	01101
4	4	77000
⋮		
64	100	00003
65	101	00014

The octal form is much more compact and easier to use. Remember that the computer works in binary . However, octal(or hexadecimal)is for the convenience of the users. Some computers are such that octal numbers can be inputted and read out. Thus, the programmer does not have to work with binary. However, the fundamental computer operations are carried out in a binary form.

We have considered that binary numbers of Table 11-1 are stored in the main memory. Let us see how they can be placed there. A set of instructions is called a *program*. The program and the data can be written on any appropriate input device, such as a card reader. The resultant electrical signals are encoded into binary. The information put into each card is read into a location of the memory. Often, the programmer can specify what memory locations are used. Let us assume that the computer logic circuits are such that the left 9 binary digits specify the memory location where the next 15 binary digits are to be stored. Then, the program for Table 11-1 would be

$$000000001000011000000000 \qquad (11\text{-}3a)$$
$$000000010000001001000000 \qquad (11\text{-}3b)$$
$$000000011000001001000001 \qquad (11\text{-}3c)$$
$$000000100111111000000000 \qquad (11\text{-}3d)$$
$$001000000000000000000011 \qquad (11\text{-}3e)$$
$$001000001000000000001100 \qquad (11\text{-}3f)$$

Again, the left nine digits constitute the memory location where the word in the next 15 digits is stored. (Note that the nine leftmost digits are not stored in memory.) In octal, this program becomes

$$00102000 \qquad (11\text{-}4a)$$
$$00201100 \qquad (11\text{-}4b)$$
$$00301101 \qquad (11\text{-}4c)$$
$$00477000 \qquad (11\text{-}4d)$$
$$10000003 \qquad (11\text{-}4e)$$
$$10100014 \qquad (11\text{-}4f)$$

The ideas of this program are very simple. In actual computers, the memory locations where each instruction word is stored are not specified by the programmer. There are special programs called LOADER programs, which direct the input of machine language (and other)programs into the memory. Thus, the 9 leftmost digits of program (11-3) would not be inputted. The programmer would not be concerned with all the details of the memory locations. Of course, he still must keep track of the location where data is stored. We shall discuss LOADER programs in Sec. 11-4.

We have considered a very simple and basic form of word which has one instruction and one memory location. However, it would often be convenient if we could write a single command which would do more than this. For instance, suppose that we want to add the contents of two memory locations. In Table 11-1 we accomplished this with two instructions. It would be more convenient for the programmer if he could do it with one command. We shall now consider how this can be done. Note that the type of command depends upon the computer hardware. That is, the type and interconnection of its logic circuits determine how complex its instructions can be. The larger computers often have much more complex commands. This does not necessarily make them more versatile since a series of simple commands can carry out the operations of a complex command. However, these computers are easier for people to use.

In the previous example, we wrote two instructions to add two numbers. This can be done with one. Suppose that we include in one memory location, an instruction and two addresses. The instruction could mean to add the contents of the first address to the accumulator and to then add the contents of the second address to the accumulator. Of course, the control unit would have to issue a longer sequence of commands(both to the memory and to the accumulator). The layout of the fields of such a word is shown in Fig. 11-4. There are two addresses here. We say that these are the addresses of *operands*, since the quantities stored in these memory locations are to be operated on.

Instruction code field	First address field	Second address field

Figure 11.4. A memory word with an instruction field and two addresses of operands.

There are other arrangements of fields that are used. Often, three addresses are included. The result of the operation performed on the contents of the first two memory addresses are stored in the third address. For instance, in program (11-3), the result remains in the accumulator register. It is often desirable to store it in a memory location. In this case, a three address command would be useful. Such a word would be divided into four fields.

Still another word has five fields. It has one instruction field and three address fields. The fifth field is for the specification of the address of the next instruction. This is illustrated in Fig. 11-5. If this is done, we do not need a program counter. Of course, memory space is occupied by the next instruction address which is a disadvantage. There is an advantage in that greater flexibility is introduced, however. Suppose that we have a sequence of instructions that are often repeated, for instance, the instruction required to multiply two numbers. It would be very tedious to write out the set of commands each time that multiplication was required. However, if, in the instruction just before multiplication, we made the next instruction address the address of the first word in the multiply sequence, then we would not have to repeat the multiply sequence each time that multiplication were needed. In this case, we would also have to record the last memory address, prior to multiplication, increment it by one, and store it. Subsequently, it would be placed in that part of the MBR where the next instruction address is stored.

These five field words can also be used in conjunction with a program counter. In this case, the multiply sequence would cause the numbers stored in the first two addresses to be multiplied and the result stored in the third address. The next instruction address would then have to be placed in the program counter. As the complexity of the stored word becomes greater, the associated logic of the controller becomes more complex.

Instruction code field	1st address field	2nd address field	3rd address field	Next instruction address field

Figure 11-5. A memory word with five fields.

We have discussed the accumulator register. In large computers, there may be many accumulators in the CPU. In this case, there may be a small field called the *tag* field added to the word. The tag specifies which register is to be used.

As the number of fields increases, the number of bits in the word usually increases. This is why 64 bit words are not uncommon in large computers.

The instructions that we have considered thus far involve one or more operands which are stored in the memory. This need not be the case. For instance, consider taking the complement of a number stored in the accumulator register. Here, these operations are performed on numbers which are stored in registers. Hence, an instruction word might only contain an instruction code and one or more register locations. In some cases, there might only be one set of registers to which the instructions would apply. In this case, the word might only contain an instruction.

At times, a memory word contains a sequence of other instructions. For instance, suppose we want to have a number stored in the A register and we want to subtract it from a number stored in the B register, see Sec. 10-3. Then, see Table 10-1, we want to have the control sequence

001000 (11-5a)
001001 (11-5b)
000001 (11-5c)

A single word would store all the information with the suitable operational code. For instance, we could have a word

110111001000001001000001

(This is somewhat larger than the words we have been considering.) 110111 is the operation code which means perform the following sequence of operations. This word is called *microprogrammed instruction*. To avoid excessively long commands, some computers utilize other facilities. For instance, ROMs, see Sec. 7-5, are used. The appropriate command then activates the desired word of the ROM which then produces an output which is the command word.

Some computers have provision for storage of words in several memory locations. Thus, we can form words that are longer than the length dictated by the physical layout of the memory. In this case, provision can be made for reading several words into one large memory buffer register (MBR) or several registers can be combined to form the MBR.

Memory words may also contain instructions for input/output devices. For instance, suppose we want information printed. In this case, the memory word would contain an instruction code field to indicate that the data is to be printed. A second field would indicate which device was to be used. A third field would give the address where the information is to be obtained. The information is supplied in binary form. Output devices often have circuits which convert this into a form suitable for printing. For instance, binary numbers are converted to decimal numbers.

The Execution of Instructions

Now let us consider the steps that are taken when an instruction is executed. Information from the program counter(actually part of the control unit)causes the next memory location to be read and stored in the memory buffer register(MBR). This is said to be the *fetch phase* of operation of the control unit, since it causes the information to be fetched from the memory. In this phase, the information from the program counter is put into the memory address register(MAR) and a read cycle is initiated. Material from the appropriate memory location is read into the MBR. Then, in the case of a core memory, a write cycle is initiated to place the information back in the memory. Remember that readout is destructive in core memories. The word remains stored in the MBR. The program counter is then incremented by the appropriate amount. This is usually one, unless there are other instructions. Now the computer transfers the commands from the MBR to the IR and the MAR. Now the computer is ready to execute the command stored in the IR. This is said to be the *execute phase*. The command portions of the word stored in the IR are now used to actuate the required control circuits, which then causes the memory, the accumulator, and other circuits of the central processor unit(CPU) to perform the required operation. When this is complete the program counter contents are put in the MAR and the procedure is repeated.

We have considered this operation in rather general terms. In the next section we shall discuss some of these ideas in detail.

11.3 THE COMPLETE SMALL DIGITAL COMPUTER

In this section we shall discuss a complete small computer and the various operations that it can perform. The basic ideas of machine language programming will be presented. To keep the procedures simple we shall perform all arithmetic operations with integers.

A block diagram of a computer is shown in Fig. 11-6. To avoid cluttering the diagram we have drawn only single lines, which may have one or two arrowheads, between components. Actually, the connection between components can represent many wires. For instance, if 16 bits are to be transferred in parallel, then 16 wires plus a common wire are needed. Such a set of wires is called a *bus*. We have assumed a word size and a memory size which are compatable with the previous examples of this chapter. This is a small computer in that the word size, number of words stored in the memory, size of registers and number of accumulators, etc. are small. However, most of the fundamental ideas can be conveyed with this computer.

Let us consider the components of the computer. The *central processing unit*(CPU) consists of an accumulator and its associated logic, see Sec. 10-4. The MQ and link(L)registers are included so that multiplication and division can be done as discussed in Chapter 10. The L register can also be used in the cases of addition and subtraction in conjunction with the rightmost bit of the A register, to determine if an overflow has occurred, see Sec.9-1. Although we do not show it explicitly, we assume that these last two bits in L and A can be sensed. It is assumed that 2's complement arithmetic is to be performed by this computer.

The storage register B supplies the input to the accumulator. The input to this register is multiplexed, see Sec. 3-4, through the main control circuit. The outputs of the L, A, and MQ registers are symbolically connected to the main control through the accumulator logic. This indicates that their values can be routed where desired. For instance, the contents of the A register could

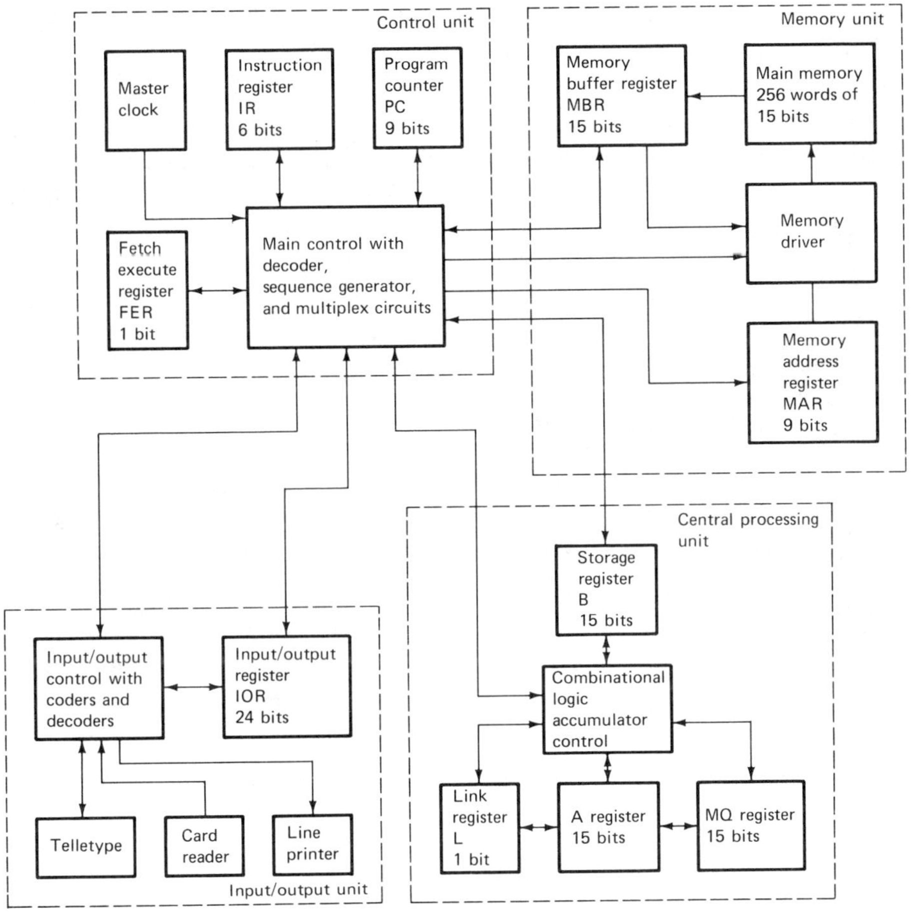

Figure 11-6. A block diagram of a small digital computer.

be put in the MBR and then stored in the memory. Alternatively, the result in the MQ register could be put into the MBR. We shall discuss this subsequently. The result in the A register could be used for other pruposes which we shall discuss.

In the *memory unit*, we have the main memory. Here we have assumed 512 words of 15 bits each. Note that this requires 512 x 15 = 7680 cores, or semiconductor memory units, etc. In Sec. 7-1 we indicated that this number should be a power of 2 for most efficient use of the interconnecting wires of the memory. Here we simply use the number 7680 for convenience. The memory driver is of the type discussed in Chapter 7. In accordance with signals it receives from the control unit, it causes the word in the MBR to be written into the memory or a word from the memory to be read into the MBR. The location of the word in question is in the MAR. The value stored in the MAR is obtained from the control unit. If the main memory is a core memory, with destructive readout, then it is assumed that the memory driver cycle is such that any word which is read is immediately rewritten.

The *control unit* has all the circuits which direct the operation of the computer. We shall discuss some aspects of this in great detail since we have not considered them before.

The *master clock* generates clock pulses which control the operation of all the synchronous circuits. If this memory is a core memory, then the time required for a core cycle will be considerably greater than the time required to cycle a semiconductor flip-flop. We could slow the clock so that it allowed for the slow memory cycle. However, this would be very wasteful of computer time. What is often done is that there are two sets of timing signals which are derived from the master clock. One set can be the master clock pulses; the second set of pulses can have a period which is an integral multiple of the master clock period. This is illustrated in Fig.11-7. A counter circuit such as that in Fig. 6-37 can be used, in conjunction with its associated logic, to produce the required signals from the clock pulses. Note that one flip-flop of a counter will change state with every input pulse, the next with every other input pulse, etc. Thus, using a counter and associated logic, we can generate pulses of the form of Figs.11-7b and c, which will be in *perfect synchronism* with the clock pulses. These slower pulses can be used to activate the memory cycle.

There is another reason that a memory cycle may have to be longer. Since the readout is destructive in a core memory, a write cycle must follow a read cycle. Then, two operations must be performed. The timing signal of Fig. 11-7 allows for this. The pulse of Fig. 11-7b can initiate the read cycle, and the pulse of Fig. 11-7c initiates the rewrite cycle. (The ratio of the master clock period and the memory cycle can differ from what is shown in Fig. 11-7.) Note that we assume that a single command causes both reading and rewriting.

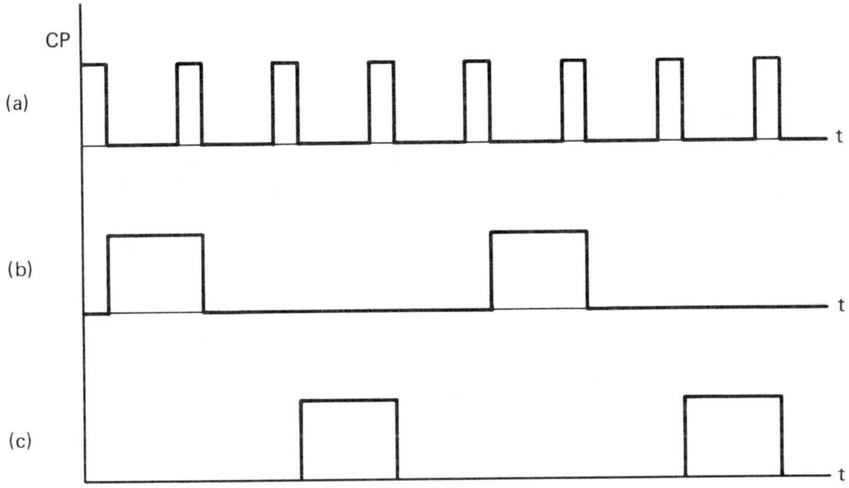

Figure 11-7. Timing signals: (a) master clock; (b) memory cycle initiate; (c) rewrite initiate.

The registers and other circuits are driven by the clock pulse from the master clock and operate at a faster rate. Of course, the logic must be such that if information must be extracted from the memory, then operations requiring this material must be delayed until the information is available.

The *instruction register* IR is used to store the current instruction (the one being executed). That is, after an instruction word is read into the MBR, the first 6 bits, containing the instructions, are read into the IR. It may seem that this is a duplication since the information is already stored in the MBR. However, carrying out the instruction may require that data be read out of and/or written into the memory. The MBR will be used in these operations. Thus, it cannot store the instructions.

The address bits of the instruction word will be transferred to the MAR. This completely frees the MBR.

The *program counter* PC contains the location of the next instruction. In the simplest mode of operation, the contents of the PC is increased by one each time that an instruction is executed. That is

$$(PC) + 1 \rightarrow PC \qquad (11\text{-}6)$$

The PC's function cannot be assumed by the MAR since its contents may be modified by the current instructions. For instance, in Table 11-1, the second instruction is to add the contents of memory location $64_{10} = 100_8$ to the accumulator. Thus, the contents of the MAR are changed from $000000010_2 = 2_8 = 2_{10}$ to $001000000_2 = 100_8 = 64_{10}$. Hence, the PC is needed to keep track of the address of the next instruction.

The next instruction may depend upon some value which must be computed(e.g., is y 0 or not?). In this case, the contents of the PC must be suitably modified. We shall consider instructions for this and discuss the logic involved subsequently.

The *fetch-execute register* (FER) is used to indicate if the control unit is fetching an instruction or is executing one. We have discussed, see Sec.11-2, that the computer cannot tell whether a number in the MBR is an instruction or just a binary number which is data. Thus, the control unit must keep track of whether it is obtaining instructions or executing them. The FER is a one bit register. If it is in state 1, then the sequence of commands generated causes the memory word specified by (PC) to be read from memory and placed in the MBR. That is

$$(M(PC)) \rightarrow MBR \qquad (11\text{-}7)$$

On the other hand, if the (FER) = 0, then the word in the MBR is treated as an instruction. Its first 6 bits are read into the IR, and its last 9 bits are transferred to the MAR. Any further instructions are also carried out.

The main control, as we have discussed, generates all the control signals needed to actuate the desired commands. We shall consider this subsequently.

The *input-output unit* contains all of the input/output devices. We have only shown a teletype, card reader and line printer. However, there are many others that can be used such as oscilloscopes, paper tape punches, and readers, magnetic tape units, disc pack, etc. In addition, there are signal processors that can input a continuous signal, such as a radar signal reflected from a distant planet, and produce a digital output which can be processed by the computer. There are output devices that produce a continuous signal from digital information. In the next section, we shall consider such devices.

In our small computer, we have assumed only a few input/output devices. The input devices produce a digital code, such as those in Table 8-3. For instance, if a key on a teletype is pushed, an electrical signal producing the code corresponding to the letter is generated. A ROM with a decoder can be used for this purpose. The input/output control then converts these codes to standard binary. In our small computer, we assume that the(decoded)binary word is stored in the input/output register IOR. We assume that this has 24 bits to accomodate programs of the type given in (11-3). In this case, the first 9 bits are placed in the MAR and the next 15 bits are then read into that memory location. As we have discussed, this represents a very simple approach to the input of programs, see Sec.11-2.

Note that we assume that there is a single input-output register IOR which handles all the transfer of information from the input/output unit and the computer. There is often a large amount of storage associated with the input/output devices, since these devices are much slower than any other computer element. Actually, this storage is part of the main memory. Such storage is termed a *buffer*.

We assume that, in the input/output control unit, there is a flag register which indicates if the device is ready to receive data, i.e., if it is in state 1; data can be imparted to the IOR since its previous data has been printed. If IOR cannot receive data because its present data has not yet printed out, then the state of the flag register is 0. We shall discuss how this is used later in this section.

Machine Language

Now let us consider the set of instructions that this computer uses. This depends upon the hardware of the computer. That is, corresponding to each instruction, there must be a logic circuit that produces the appropriate command sequence. In our small computer, we have a short list of commands. Of course, in larger computers, there would be more. The basic list of commands is derived from Table 10-1. This was discussed in Sec. 11-1. We shall add to this to produce a complete set. This is given in Table 11-3. (The input/output commands are not given. We shall discuss them subsequently.)

TABLE 11-3
INSTRUCTIONS FOR THE COMPUTER OF FIGURE 11-6

Instruction	Code binary	Code octal	Operation
Add	000001	01	$(A)+(B) \rightarrow A$
Clear	000010	02	$0 \rightarrow A$
AND	000011	03	$(A) \wedge (B) \rightarrow A$
OR	000100	04	$(A) \vee (B) \rightarrow A$
EXOR	000101	05	$(A) \oplus (B) \rightarrow A$
shift right	000110	06	$(A_{j+1}) \rightarrow (A_j)\ \ (A_n) = 0$
shift left	000111	07	$(A_{j-1}) \rightarrow (A_j)\ \ (A_0) = 0$
complement	001000	10	$(\bar{A}) \rightarrow A$
increment	001001	11	$(A) + 1 \rightarrow A$
negative check	001010	12	y=1 if (A) represents a negative number
zero check	001011	13	y=1 if (A) represents a 0
put in memory	001100	14	$(A) \rightarrow M(MAR)$, where (MAR) is specified in instruction
branch unconditionally	010000	20	$(M(MAR)) \rightarrow PC$
branch on zero	010001	21	$(M(MAR)) \rightarrow PC$ if $(A) = 0$ $(PC)+1 \rightarrow PC$ otherwise
branch on negative	010010	22	$(M(MAR)) \rightarrow PC$ if $(A) < 0$ $(PC)+1 \rightarrow PC$ otherwise
rotate right	010011	23	$(A_{j+1}) \rightarrow A_j\ \ (A_0) \rightarrow A_N$
rotate left	010100	24	$(A_{j-1}) \rightarrow A_j\ \ (A_N) \rightarrow A_0$
stop computer	111111	77	Stops computation

Many of these we have considered before. Let us discuss the new ones. Put in memory (14_8) is a command which puts the contents of the accumulator in the memory. For instance, the word

001100000000111 (11-8)

would cause the present contents of the accumulator to be placed in memory location 7_8.

For the moment, let us skip the next three commands. Instruction 23_8 is a right rotate. This is essentially the same as a shift right except that the values of the leftmost register A_N is not set equal to 0. In this case, $(A_0) \rightarrow A_N$. That is, the original rightmost bit is rotated around to the left input.

Instruction 24_8 is a left rotate. This is essentially the same as the previous one except that now the shift is to the left and $(A_N) \rightarrow A_0$.

Note that these rotations are especially helpful in division, see Sec. 10-3. If, in implementing that procedure, we rotated left instead of shifting left, the link bit would have been put into the rightmost bit of the MQ register. This would produce the desired result in a convenient way. If an instruction word does not have an address, then it refers to the A register. For instance,

001000000000000 (11-9)

causes the contents of the A register to be complemented.

Now let us consider instruction 20_8, 21_8 and 22_8 which have to do with branching. Here we cause the computer to stop taking instructions in order and to skip forward or backward to some other memory location. (That is, we do not just increment the PCR by 1.)

The *unconditional branching* is given by instruction 20_8. For instance, if we have the instruction

010000000000101

then the next instruction address will be taken from memory location $000000101_2 = 005_8$.

The other two branching instructions represent *conditional branching*. That is, the branching action depends upon some computed value. Instruction 21_8 causes the next instruction executed to be the specified one if the contents of the A register is 0. For instance, suppose we execute the instruction

010001000001001

which is in memory location 3_8. Then, if the contents of A is 0, then the address of the next instruction will be in $M(000001001_2) = M(007_8)$. If $(A) \neq 0$, then the next instruction executed will be that in $M(4_8)$. That is, if the contents of A are not 0, then the branching is ignored.

Instruction 22_8 performs the same type of operation. Now the branching depends upon whether or not (A) is negative. If $(A) < 0$, then the branching specified in the instruction occurs. If $(A) \geq 0$, then the next instruction executed is the next one in the memory(i.e., $(PCR) + 1 \rightarrow PCR$).

A Simple Program

Our program will be simple, since our objective is not to teach programming, but to obtain an understanding of the operation of the computer. We call such programs, written in binary(or octal, etc.)*machine language programs*.

Suppose that we want to subtract a number N_1 from another number N_2. If the result is negative, then we want to add 10 to the total; if the result is zero or positive, then we want to add 5 to the total. After doing all this, we want to add the result to another number N_3. The program is given in Table 11-4. Assume that the binary words of Table 11-4 are stored in the memory locations indicated. These words constitute the computer program.

TABLE 11-4
A SIMPLE PROGRAM

Memory location		Stored word	Comment
Octal	binary	octal	
1	000010000000000	02000	$(0) \rightarrow A$
2	000001001000000	01100	$(A)+(M(100_8)) \rightarrow A$
3	001000000000000	10000	$(\bar{A}) \rightarrow A$
4	001001000000000	11000	$(A) + 1 \rightarrow A$
5	000001001000001	01101	$(M(101_8))+(A) \rightarrow A$
6	010010000100000	22040	if $(A)<0; (M(040_8)) \rightarrow PC$
7	000001000100111	01047	$(A)+(M(47_8))+(A) \rightarrow A$
10	000000000000000	00000	CONTINUE
11	000001001000010	01102	$(M(102_8))+(A) \rightarrow A$
12	111111000000000	77000	STOP
40	000000000101000	00050	$(M(40_8)) = 50_8$
47	000000000000101	00005	$(M(47_8)) = 5_8 = 5_{10}$
50	000001001101000	01150	$(M(150))+(A) \rightarrow A$
51	010000001101001	20151	$M(151) \rightarrow PC$
100	000000000000111	00007	$M(100) = 7_8 \quad (7_{10})$
101	000000000000001	00001	$M(101) = 1_8 \quad (1_{10})$
102	000000000001000	00010	$M(102) = 12_8 \quad (10_{10})$
150	000000000001010	00012	$M(150_8) = 12_8 \quad (10_{10})$
151	000000000001000	00010	$M(151_8) = 10_8$

Now let us consider this program. The numbers N_1, N_2 and N_3 are stored in memory locations, 100, 101 and 102, respectively. In this case,

$$N_1 = 7_{10}$$

$$N_2 = 1_{10}$$

$$N_3 = 8_{10}$$

$M(1_8)$ contains a clear command. There is no address. Thus, the A register is cleared. M(2) results in the contents of M(100) being added to the A register. This is now

$$(A) = 000000000000111_2 = 7_{10}$$

$M(3_8)$ results in A's being replaced by its complement

(A) = 11111111111000

$M(4_8)$ results in A's being incremented by 1. Thus,

(A) = 11111111111001

$M(5_8)$ results in the contents of $M(101_8)$'s being added to A. This results in

(A) = 11111111111010

This is a negative number. Remember that we are using 2's complement representation. $M(6_8)$ contains a negative branch instruction. Since (A) is negative, $(M(40_8))$ is placed in the PC. Now, $(M(40_8))$ = 50_8. Thus, this number is placed in the PC register. The next command executed lies in $M(50_8)$. This causes $(M(150_8))$ to be added to A. $(M(150_8))$ = 10_{10}. Thus, we have

(A) = 000000000000100

Note that if the subtraction had not resulted in a negative number, then following the instruction in $M(6_8)$, we would have executed the instruction in $M(7_8)$ which, being executed, would have resulted in $(M(47_8))$ = 5_{10} being added to (A).

After carrying out the commands of $M(50_8)$, the PC is incremented by 1. Thus, the next command executed is in $M(51_8)$. This is an unconditional branch which causes the value stored in $M(151_8)$ to be put in the PC. $M(151_8)$ = 10_8. Thus, the next command executed is stored in $M(10_8)$. This is 000000000000000 and results in no action being taken. The PC is then incremented by 1. Then the instruction in $M(11_8)$ causes the contents of M(102) to be added to A, $(M(102_8))$ = 8_{10}. The resulting addition results in

(A) = 000000000001100 = 12_{10}

Then, the instruction stored in M(12) causes computation to stop. We have not considered the output of data. This shall be done later in this section.

Let us consider another example using branching where we make the program loop back on itself. Suppose that we want to obtain the sum 1+2+3+4+...+99+100. We could write a program which would do this directly. However, that would be a very lengthy program. We can use branching to eliminate much of the length. This can reduce much work for the programmer.

TABLE 11-5
PROGRAM FOR THE COMPUTATION OF $\sum_{k=0}^{100} K$ USING LOOPING

Memory location octal	Stored word octal	Stored word binary
1	02000	000010000000000
2	01151	000001001101001
3	11000	001001000000000
4	14151	001100001101001
5	10000	001000000000000
6	11000	001001000000000
7	01150	000001001101000
10	21160	010001001110000
11	02000	000010000000000
12	01151	000001001101001
13	01152	000001001101010
14	14152	001100001101010
15	20161	010000001110001
150	00145	000000001100101
151	00000	000000000000000
152	00000	000000000000000
160	00170	000000001111000
161	00001	000000000000001
170	77000	111111000000000

Now let us consider the program. All memory location references will be in octal. Thus, we shall omit the subscript 8. We start by clearing the accumulator. Next we add (M(151)) to A. At the start, there is a 0 stored in M(151). Thus, (A)=0. The instruction in M(3) causes the accumulator to be incremented by 1 and then the instruction in M(4) causes the new value stored in the accumulator to be put back in M(151). This value will be the number that is to be added on to the sum. That is, the desired sum is 1+2+...+100. Each cycle of the program will add on one more number. This number will be in M(151). Note that when we write over a location, the old value is destroyed and the new one entered. The command 14_8 not only puts the value stored in A in the specified memory location, but it also leaves it in A. Next, locations 5 and 6, we obtain the 2's complement of the value stored in A. Then, in step 7, we add the contents of M(150). This is the number 101_{10}. If the subtraction results in 0, it means that the term we are going to add has exceeded 100 and the computation must stop. In this case, the branch on zero(the instruction stored in M(10))causes the next command used to have the location stored in M(160). This location is M(170). The resulting command is 77000 which stops computation. On the other hand, if the check for branch on zero did not indicate a 0, then the next command will be in M(11). This clears A. In the next two steps, the contents of M(151) and M(152) are added to A. Next, this sum is put back into M(152). Then,there is an unconditional branch back to the instruction in M(1) and the cycle is repeated.

Note that the contents of M(151), i.e., (M(151)), consists of the last number added while (M(152)) is the accumulated sum. Then, when operation terminates, the answer is in M(152). We have not considered the output of data. This shall be done later in this section.

At times, a picture or flow chart is used to express the ideas of the program. Such a flow chart for this program is shown in Fig. 11-8. In our simple programs, we did not check for overflow.

Multiplication and Division

Let us consider the implementation of multiplication and division. The procedure of Sec. 10-3 can be implemented using the commands of Table 11-3. The multiplier must be placed in the MQ register. This can be implemented by putting it in the A register and then shifting it to the right sufficiently. The multiply routine would then involve shifting to the right and adding. Thus, the 06_8 and 01_8 instructions can be used. There is one

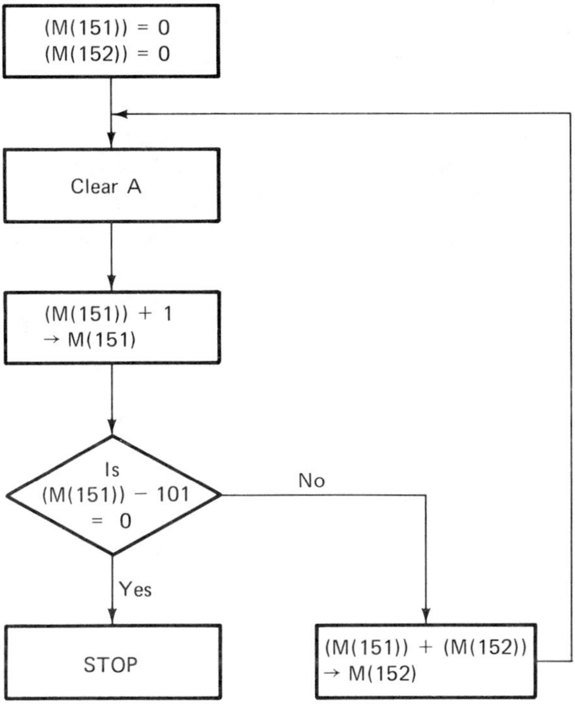

Figure 11-8. A simple flow chart for the program of Table 11-5.

minor problem. The rightmost bit of the MQ register must be sensed. This could be accomplished by adding a separate sense circuit and instruction. Instruction 22_8 branch on negative could also be used. Suppose we rotate the L-A-MQ register to the right twice. Now, the leftmost bit of the A register becomes the former rightmost bit of the MQ register. The branch negative instruction checks this bit to see if the result is negative. Thus, we can now branch to the appropriate commands. If the bit is a 1 (i.e., a yes answer for the branch negative), then we rotate left twice to restore the contents of the A-MQ register, and add the multiplier and then *shift* to the right. If the bit is a 0, then we just rotate left twice but do not add. The procedure is then cycled again. Multiplication thus seems to involve a lengthy program.

Actually, this can be avoided. We could have a single multiply instruction which would then actuate a microprogrammer or, alternatively, a sequence generator, which would carry out all the steps.

In a more complex computer, there might be a multiplier circuit, see Fig. 10-10, which automatically carries out the steps of multiplication.

In division, the ideas of Sec. 10-3 can be implemented using the instructions of Table 11-3. Note that the left rotate is of prime importance here. We would have to add an additional command which allows us to clear the link register. This procedure is discussed in detail in Sec. 10-3 and the details of the program will be left to the reader.

Indirect Address

There is another type of addressing, which gives the programmer more freedom. In this case, the address part of the instruction is not the address of the operand, but an address in the memory where the operand can be found. This is called an *indirect address* in contrast to the *direct addressing* that we have been using. Typically, the first bit of the address field is used to indicate whether direct or indirect addressing is used. If that bit is a 1, then indirect addressing is used. For instance, if we consider the 15 bit word that we have been discussing, then the first 6 bits are for the instruction code. The next bit would indicate the type of address and the last 8 bits would indicate the actual address. This format is indicated in Fig. 11-9. Let us illustrate this with an example. In Table 11-1 we have the word

000001001000000

Now, since the 7th bit is a 0, this is a direct address. Thus, as before, this means to add the contents of memory location $10000000_2 = 100_8$ to the accumulator.

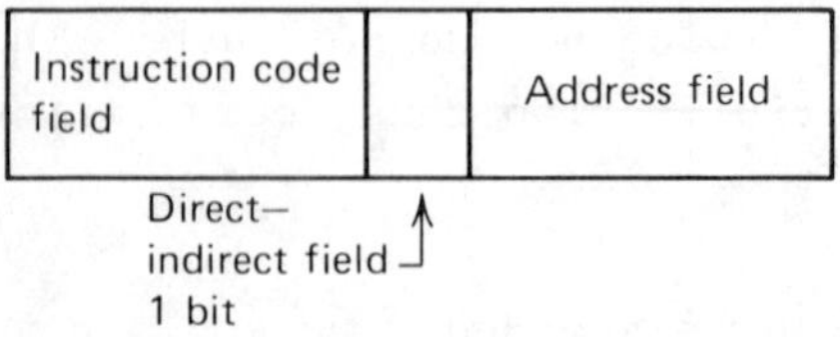

Figure 11-9. A memory field which is used with direct-indirect addressing.

On the other hand, if the 7th bit were a 1, then we would use indirect addressing. For instance, if we have the word

000001101000000

this would mean to add to the A register the number whose address is in the $01000000_2 = 100_8$ memory location. For instance, if we have the three words

Memory location	Stored word(binary)	octal
3_8	000001101000000	01500
⋮		
100_8	000000001101000	00150
150_8	000000000000111	00007

then, 3_8 instructs the computer to add the contents of the memory *location* to be found in $M(100_8)$ to the accumulator. The location in $M(100_8)$ is 150_8. The binary number stored in $M(150_8)$ is 7. Thus, this sequence of commands results in 7_{10}'s being added to the accumulator.

Indirect addressing allows us to specify the address of a memory location during computation. Advanced programming techniques make use of it. Note that when a direct address is used, the memory location in octal(in the last 3 digits)is the actual memory location. When an indirect address is used, then the address portion in octal is 400_8 too large. (Note that this number varies with the length of the word.)

If we use indirect addressing, then there is one less bit available for specifying an address. For instance, in our small computer, the memory would only contain 256 rather than 512 words.

Input-output Instructions

We did not include any provision for input to or output of data in the simple instruction set. Let us now do this. At the start, we shall discuss the outputting of data. The output of a computer consists of a set of wires called the input-output bus, I/O bus. For instance, in our simple computer, the I/O bus would consist of one wire from each flip-flop of the IOR plus a common ground wire. In addition, there are several wires that carry timing pulses to be used by the I/O devices.

Data to be outputted is placed on the I/O bus which is connected to all the I/O peripherals. These can include a teletype, line printer, magnetic tape drive, x-y plotter, etc. In general, at any one time, we only want to use one of these. For instance, we may want the data to be printed by the line printer. Alternatively, we may want it stored on magnetic tape. If there is a set of data which represents a graph, we may want it plotted by an x-y plotter, etc. Thus, the instructions must indicate which device is to be used and the data must be routed to that device only.

In addition, we must, at times, supply instructions to the input and/or output device(I/O device). That is, we must designate what *function* it is to perform. For instance, if the magnetic tape drive is to be used, then we may want to instruct it to wind the tape back to the beginning of the reel before data is inputted. If we use the x-y plotter, we must tell it if the data is the x variable or the y variable, etc.

The I/O instruction, thus, must have provision for both designating the device and the function. We shall use the following instruction for the output of data

$011000d_0d_1d_2d_3d_4d_5f_0f_1f_2$

The $d_0, d_1, d_2, d_3, d_4, d_5, f_0, f_1$, and f_2 represent binary digits. The first part of the instruction $011000_2 = 30_8$ instructs the computer to output the contents of the IOR to the I/O bus. The bits d_0, d_1, d_2, d_3, d_4, and d_5 designate the device to be used. Let us use

$000000_2 = 00_8$ teletype

$000001_2 = 01_8$ line printer

$000010_2 = 02_8$ x-y plotter

$000011_2 = 03_8$ magnetic tape drive

Then, if we want to supply the binary number stored in the IOR to the magnetic tape drive, the instruction would be

$$011000000011f_0f_1f_2$$

The $f_0f_1f_2$ indicates the particular function to be performed. For instance,

$001_2 = 1_8$ rewind tape to start $010_2 = 2_8$ skip to next block of tape	tape drive functions
$001_2 = 1_8$ data is x $010_2 = 2_8$ data is y $011_2 = 3_8$ print data	x-y plotter functions

Thus, if the data is to be plotted as the x variable on the x-y plotter, the complete instruction would be

$$011000000010001_2 = 30021_8$$

All of the I/O devices are connected to the same I/O bus. Let us discuss how the data is routed to only one device. When the output command is given, two sets of binary numbers are placed in the bus in sequence. The first number is

$$d_0d_1d_2d_3d_4d_5f_0f_1f_2$$

Then, subsequently, this is removed from the I/O bus and the actual data is transmitted. Associated with the peripherals, there is a logic circuit called a *device decoder*, which has as its input the bits $d_0d_1d_2d_3d_4d_5$. This device interprets these bits and only allows the designated I/O device to receive the subsequent data. In Sec. 12-7 we shall discuss the actual details of such a device. For the time being, let us just assume that it exists. For instance, if the instruction were

$$011000000010001_2 = 30021_8$$

then $d_0d_1d_2d_3d_4d_5$, which would be put on the I/O bus is

$000010_2 = 02_8$

Thus, the x-y plotter would be enabled to receive data while all other I/O devices could not. The last three bits $f_0f_1f_2$, i.e., 001, would now be supplied to the circuitry associated with the x-y recorder. This would indicate that the number which follows is an x coordinate.

At the next appropriate timing pulse, the data would be put on the I/O bus and it would be transmitted to the proper device(x-y plotter)and used properly(x variable). In Sec. 12-7 we shall discuss the interfacing of the computer with its peripherals and input and output in detail.

The instruction $011000_2 = 30_8$ transfers data to the IOR. When the instruction $011000d_0d_1d_2d_3d_4d_5f_0f_1f_2$ is given, the following sequence occurs: $d_0d_1d_2d_3d_4d_5f_0f_1f_2$ is put into the IOR and, then, onto the I/O bus. Next, the contents of the accumulator is put into the IOR and, at the appropriate time, this data is put on the data bus. *Thus, all data to be put on the I/O bus should be put in the accumulator.*

The input of data is similar to its output. Here we shall use the instruction

$011001d_0d_1d_2d_3d_4d_5f_0f_1f_2$

The first 6 bits $011001_2 = 31_8$ signify that data from a peripheral is to be read into the IOR. The $d_0d_1d_2d_3d_4d_5$ and $f_0f_1f_2$ are the same as an output. For instance,

011001000011001

would cause the magnetic tape drive to rewind to the start of the tape and then read the next binary number into the IOR.

Most input/output devices are very much slower than the logic circuits or even than the core memory. This could lead to problems if appropriate procedures are not used. Suppose that, in the program of Table 11-5, we print each partial sum. It might be very possible that the program could go through a complete cycle faster than the printer can print its number. This would be especially true in the case of the teletype. Thus, we would have to halt the computation to see if the flag(see the discussions at the beginning of this section)was a 1 before data was put in the IOR. Thus, we would also have to include a flag check command in our set of instructions. This would be a branching command which is like the branch on zero command. A flow chart for the flag check is shown in Fig. 11-10. Note that the flag check is repeated continuously until the printer is ready. (It is assumed that the printer can be controlled independently of the rest of the computer.) Thus, the flag check, in effect, stops the program until the printer is ready. In large computers, this would never happen. The data to be printed would be stored in a buffer memory, which is part of the main memory unit. In addition, these buffers would be coupled to a disc or drum memory. If the buffer memory became filled, then the data to be printed would be put on the disc or drum. Thus, all data to be printed would be stored. Now the input/output control would cause the material to be printed from memory independent of computation. Once the data is printed the memory space is made available(i.e., the memory space is erased). All this could be permanently wired and the programmer would not become involved with it. If this is done, the output information can be stored in the input/output memory and the computer can be released for other calculations. Actually, large computers are usually operated so that several tasks are performed at a time.

Input/output buffer memories are also used for input. Typing the information in, or even card reading, is also slow. Thus, the information can be entered into the buffer memory from which it can be inputted to the computer at a fairly rapid rate. Of course, in a small computer, if such buffers are not available, then operation is slowed by input and output.

There is often provision made to interrupt operation if a problem develops with an input or output(I/O)device. For instance, suppose that the output is being printed on the line printer and it runs out of paper. Operation should cease. There is a switch on the line printer which indicates that it is out of paper.

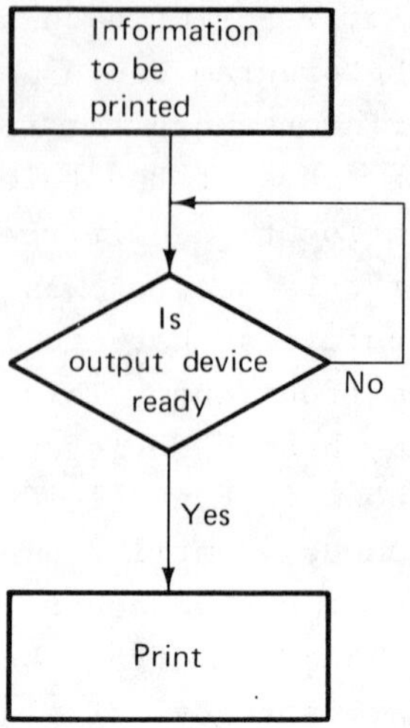

Figure 11-10. A flow chart for a printing operation.

This could be used to activate a flag which could be checked before each line is printed. However, this would be wasteful since very many lines are printed before the printer exhausts its paper supply. Thus, the following procedure is used. The signal from the out of paper switch initiates an interrupt sequence. This causes computation to temporarily stop without loss of information. When new paper is loaded, computation then can proceed. This is called an I/O *interrupt*. Such an I/O interrupt can be initiated by many of the problems associated with the I/O devices.

We assume, in our small computer, that there is a switch position on its panel such that programs of the form of (11-3) can be inputted. That is, each line typed on the teletype or punched on a card would be put into the IOR, and then into the appropriate memory location.

Design of the Computer Circuits

We have discussed the design of most of the components of the computer. The design of the control circuit must be considered. In general, even in a small computer, this circuit will contain a large number of gates. Thus, we will just consider some small aspects of the design since this will convey the general ideas without becoming reptitious .

Let us assume that a program and data are stored in the memory and that the program is to be executed. We assume that the PC, see Fig. 11-6, has been set to zero. When the start button is pressed, the fetch-execute register FER will be in the fetch phase. This will then cause the MAR to be set equal to the (PC) value. This can be done by having a set of AND gates interconnecting the outputs of each flip-flop of the PC to the input of the corresponding flip-flop of the MAR. One input of the AND gate is from each output of the FER. Thus, in the instruction fetch phase, the outputs of the AND gates will be equal to their PC inputs. Hence,

$$(PC) \rightarrow MAR \tag{11-10}$$

In Fig. 10-18 we illustrated how the state of one flip-flop is set equal to the state of another when a command signal is given.

The main control then activates the memory driver and the word, whose address is in the MAR, is read into the MBR. The fetch phase is then complete and the state of the FER is then set equal to zero. Again, transfer between registers takes place and the first 6 bits of the MBR are read into the IR. A decoder then generates a command signal from this signal. A section of a typical decoder is shown in Fig. 11-11. (Other forms of decoders are discussed in Sec. 7-5.) Here we assume that the IR gate has the bits:

$$p_5p_4p_3p_2p_1p_0 \tag{11-11}$$

Each instruction then produces an output on a separate wire. There is one wire for each of the computer's instructions. For instance, there is one wire for each instruction in Table 11-3. The decoder outputs are then applied to sequence generators. Some of these may be simple, some may be complex. For instance, 02_8 clears the accumulator. In this case, we need only put the control signal on the c_2 lead of the accumulator, see Sec.10-4.

The 01_8 command is more complex. Here, a value stored in the specified memory location is added to the contents of the accumulator. Let us consider the various steps that must be taken.

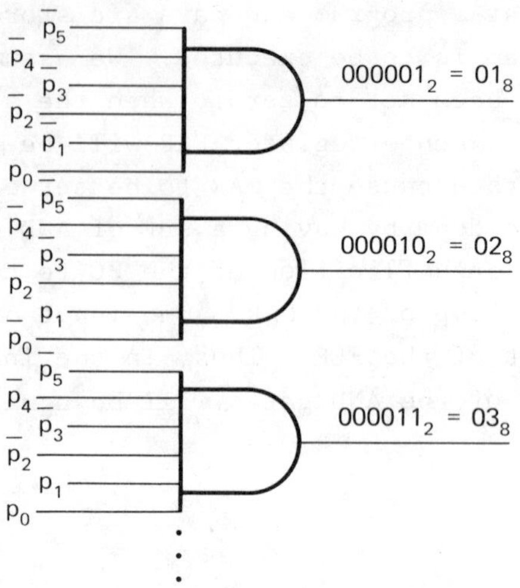

Figure 11-11. A portion of the instruction decoder.

The last 9 bits of the MBR must be transferred to the MAR. (We assume that there is no indirect addressing.) A read cycle must be initiated. After this is complete, the contents of the MBR must be placed in the B register. Then, a c_1 command, see Table 10-1, must be given to the accumulator to cause it to add. The value of the PC is increased by one, the FER is reset and another instruction is obtained. Note that, to increase the PC by one, it can be wired as a simple accumulator or its value can be transferred to the accumulator, incremented by one and then transferred back to the PC. Counters can also be used for the PC.

Thus, the control must generate signals which cause the following

$$(MBR_{8,0}) \rightarrow MAR \tag{11-12a}$$

$$(M(MAR)) \rightarrow MBR \tag{11-12b}$$

$$(MBR) \rightarrow B \tag{11-12c}$$

$$(B) + (A) \rightarrow A \tag{11-12d}$$

Remember that $(MBR_{8,0})$ represents the contents of the first 9 flip-flops of the MBR.

Now care must be taken. (11-12a), (11-12c) and (11-12d) are transfers between flip-flops and can take place at the master clock rate. (It is assumed that sufficient time is allowed to enable all gates to respond.) On the other hand, (M(MAR)) → MBR is a memory read cycle, which requires both a read and a write. The timing pulses of Fig. 11-7b are c are used here, e.g., 7 master clock pulses must elapse between (11-12a) and (11-12c). As was mentioned, a counter can be used to generate these slower pulses. The controller can use the slower pulses to change its states when the memory commands are given. Alternatively, the controller can have several state where the output does not change change. In the latter case, the controller changes state with each master clock pulse. In the former case, the controller has fewer states and may be simpler.

Each control signal must be applied to logic gates which perform the desired operation. We have considered each of them in detail.

In general, the ideas of each control sequence are similar. There is some difference in the ideas of conditional branching. Let us consider how they could be implemented. The conditions on which the branching depends is always the value of y, see Figs. 10-21 and 10-22, which are the zero check and negative check outputs. The appropriate value of y can be used as one of the inputs to the sequence generator of the controller for the branching sequence. For most of the control sequence, this y input can be a "don't care". However, at the time when the branching is to take place, the value of the y input determines the change of state. Thus, the branching is accomplished as desired.

Busses

We have shown interconnections on the block diagram of Fig.11-6 as single lines. Actually, this represents many wires. Several of the registers are 15 bit registers. Thus, to transmit information among them, 15 wires plus a ground are needed. In a large computer, 64 wires could be needed. (This assumes parallel transfer of information.) What is normally done is to form a set of wires called a *bus*(e.g., 64 wires)which connect all appropriate registers. These are connected through two input AND gates. Now suppose that we want to transfer information from register Q to register R(where Q and R are arbitrary designations). Then, a 1 could be applied to one input of all the AND gates at the outputs of Q and a 1 would be applied to the AND gates at the inputs to R. Thus, the output of Q would be connected to the input of R with no other registers affected.

11.4 ADDITIONAL ADDRESSING TECHNIQUES

We have discussed addressing procedures that were used in the simple computer developed in this chapter. Here we shall consider some additional addressing techniques.

Multiple Indirect Address

In indirect addressing, see Sec. 11-3, the address field did not contain the location of an operand but it contained the location of a memory word which contained the address of an operand. Use of indirect addressing gives us greater flexibility. It can be extended still further. For instance, the address field of the instruction could contain the location of a memory location which, in turn, contains the location of another memory location which is not the location of an operand, but of another memory location which, finally, contains the address of the operand. This procedure can be extended to an arbitrary number of steps. There must be a procedure for indicating when the memory location actually does express the location of the operand. In this type of address, there must be an additional bit(or bits)to be used as a *tag*. The tag bit indicates whether the contents of the next memory location is itself a memory location or an operand.

Relative Addressing

There are times, especially in small computers, when there are not sufficient address bits available. For instance, suppose that there are 1024 words but only 8 bits in the address field. Address bits could be obtained by taking them from the instruction code field. However, this reduces the size of the instruction code field and thus, limits the number of instruction codes that can be used. An alternative procedure, called relative addressing, is available. Suppose that there is a register called the *base address register*(BAR). Its bits, plus the address field bits, make up the total address. For instance, if the BAR is a 2 bit register and the address field has 8 bits, then 10 bits are available for addressing. Of course, instructions must be available for changing the base address register contents. However, this usually need not be done too often. For instance, in the previous example, the first 256 memory locations could be used with the (BAR) = 00; then the next 256 memory locations would be used with (BAR) = 01, etc. Thus, the BAR contents would only have to be modified after 256 locations were used.

Relocatability

At times, especially in large computers, it becomes convenient to put large blocks of stored information into specified areas. Relative addressing can be used to accomplish this. For instance, suppose that we have a program which, when entered, would occupy a sequence of memory locations. By setting the BAR, various starting locations can be picked for the sequence. In this way, the program will not be stored in locations reserved for data, etc.

Indexable Address

The previous idea can be extended so that the contents of the BAR(or similar)register can be determined during computation. This allows for increased versitility in branching type operations.

11.5 ASSEMBLER LANGUAGE AND HIGHER LEVEL LANGUAGES

Programming using machine language is a very tedious procedure. The programmer must remember numerical codes; see Table 11-3, which correspond to each instruction. In a large computer, there may be many more codes than are given in Table 11-3 and the job of remembering them can be very difficult. Another difficulty with machine languages is that the programmer must remember the numerical values of the storage locations of the variables. For instance, suppose that we want to obtain

$$b = x + y + z \tag{11-13}$$

We must choose and remember the memory locations for x,y,z, and b. This is not much of a problem in a simple program. However, if we have many variables, a table must be made and the procedures become very tedious.

To eliminate some of these difficulties, other languages have been developed. We shall start by considering *assembler languages.* Note that the objective here is not to teach programming but to discuss what these languages are and how the computer handles them.

When computation is performed, a machine language program always directs it. Thus, if an other form of language is used, it must be translated into machine language. This translation can be done by the computer. When we work with an assembler language, a computer program, called an *assembler* is written(in machine language). The assembler directs the computer which translates the assembler language program into machine language. Let us describe the assembler language. Then, we shall consider how the computer uses it.

In an assembler language, the numerical commands of the machine language are replaced by *mnemonics*. That is, instead of writing 001_8 = 000000001 for add, we would write ADD. Note that ADD is much easier to remember than 001_8. In Table 11-6 we will list some mnemonics that could be used for the instructions of Table 11-3. (Note that, like machine languages, assembler languages are *not* standard.)

We use three letter designations for our assembler language instructions although in more extensive languages, the number of letters need not be fixed.

TABLE 11.6
SOME MNEMONICS USED IN AN ASSEMBLER LANGUAGE

Instruction in octal	Assembler language
01	ADD
02	CLA
03	AND
04	ORR
05	XOR
06	SHL
07	SHR
10	COM
11	INC

12	NEC
13	ZEC
14	PIM
15	IIM
16	OFM
17	OFA
20	BRU
21	BRZ
22	BRN
23	ROR
24	ROL
77	STP

The second major difference between assembler languages and machine languages is that variables can be referred to by name instead of by memory locations. Usually, a variable can consist of up to six alphanumeric characters(letters or numbers). For instance, suppose we want to write the program of Table 11-1 in assembler language. It becomes

```
CLA                                    (11-14a)
ADD DATA1                              (11-14b)
ADD B                                  (11-14c)
STP                                    (11-14d)
DATA1:3                                (11-14e)
B:14                                   (11-14f)
```

DATA1 and B are names of variables. DATA1 is written instead of specifying memory location $M(100_8)$ and B is written instead of specifying memory location $M(101_8)$, as was done in Table 11-1. The programmer does not have to remember or even know the memory location of DATA1. Any time that he refers to DATA1, the correct memory location will be addressed. For instance, we refer to DATA1 in both (11-14b) and (11-14e). In each case, the proper memory location will be referenced.

We have considered a simple program but the same ideas hold in complex programs where there are many variables and many program statements.

Now let us consider how an assembler language program is used by the computer. A program, called an *assembler*, is written in machine language. This program need only be written once and most programmers are not concerned with it. The assembler is put into the main memory. The assembler language program is then inputted as *data*. The operation of the assembler is such that it translates the assembler language program into a machine language program. Let us see how this is done. Each statement of the assembler language program is scanned in turn. The first three letters are checked to see if they represent an instruction. (The number need not be three but it makes matters simpler if it is fixed.) The input/output devices are such that each letter generates a numerical code, see Table 8-4. Thus, each instruction has a unique code. The assembler program translates this into the corresponding machine language code. Let us see how this can be done with our simple computer. Note that there is a fixed total number of instructions. Thus, there is a fixed number of assembler codes, which uniquely correspond to the machine language codes. The assembler codes and their corresponding machine language codes are stored in the memory. The numerical code from a program statement is placed in the accumulator register. Then, it can be compared with each of the stored assembler codes in turn using the logical AND and then the complement operation. When (A)=0, then the mnemonic has been identified. Then, the corresponding machine language code can be generated. Thus, the mnemonics can be converted into their machine language equivalents. (We have done this using only the instructions of Table 11-3.)

Now let us consider the memory locations for variables. When a variable is used in assembler language, it follows an instruction, which,as we have discussed, can be identified. Thus, the alphanumeric characters following an instruction are identified as variable names. (An exception to this is data, which we shall discuss subsequently.)

Now let us discuss how the machine language program is formed. The assembler language program is scanned twice. In a simple computer, the program may actually have to be read in twice. In the first pass, all the variables are identified. The first time that a variable is encountered, its alphanumeric code is stored in a memory location. Subsequently, a number which gives the memory location which is assigned to store the data corresponding to this variable will also be picked. Thus, the variable name code and its corresponding memory location will be stored in a table called the *symbol table*. When pass 1 is completed, the symbol table will contain all the variable code names and their assigned memory locations. Note that a variable may be encountered many times in different instructions, but it only appears in the symbol table once.

Let us consider how the memory locations are assigned. During pass 1, the amount of memory needed for each statement is counted and stored cumulatively in a counter called a *program counter*. For instance, assume that the first statements are (11-14a) and (11-14b) and one word is used for each of the CLA and ADD commands, which may contain addresses. Then, two words would be required to store their binary representations. The second command is ADD DATA1. DATA1 is recognized as a variable and is listed in the symbol table. However, a memory location is not yet assigned. The scanning continues. The program counter is incremented at each step. Ultimately DATA1:3 is scanned. The colon (:) indicates that 3 is the numerical value to be stored in the memory location called DATA1. The symbol table is examined to find this variable. Now the memory location given by the program counter is reserved for it. This information is entered into the symbol table. The program counter is now incremented by the amount of memory needed to store the data. Next, statement B: 14 is encountered. The colon again indicates a symbolic address and B is found in the symbol table and the process is repeated. Note that during the assembler operation, words are not stored in their assigned locations. Only a machine language program is being written and stored.

Note that a data statement must be given for each memory location needed. For instance, suppose that the result of a computation is to be put into the memory in a location called SUM. We would have

```
PIM SUM
   .
   .
   .
SUM: 0
```

Then, a memory location would be reserved for SUM. Note that the value of 0 is arbitrary. When the value of SUM is computed and put into the memory, the 0 will be written over.

Now let us consider pass 2. (The symbol table was formed in pass 1.) Now the program is scanned a second time. The instruction mnemonics are converted into their binary codes. This forms the first part of the instruction word. The variable name is then found in the symbol table and its memory location is made the second part of the instruction word. The complete instruction word is then stored in a memory location. This procedure is repeated, step by step, until the entire machine language program is stored in the memory. Note that the instructions will be stored in sequence. The numerical values of the variables will be stored in locations following the instructions.

The numerical values of data are identified by writing the variable name followed by a colon. The appearance of a name followed by a colon indicates that the number following the colon is to be read into the memory location which has been previously assigned. (Assemblers may be such that the data is supplied in octal.)

The machine language program is then outputted as punched cards, a paper tape, or stored in disc or drum. The assembler is then erased from the main memory and the machine language program can then be read into the main memory and executed.

We assumed that the complete machine language was stored in the main memory. At times, there is not enough memory space to do this. Note that during pass 2, most of the core will be filled with the assembler and symbol table. There may not be enough space remaining to store the complete machine language program. This does not prevent the program from being assembled. As pass 2 proceeds, machine language statements are assembled and stored in the main memory. There is often an area of main memory called a *buffer* set aside for this. When the buffer becomes filled, pass 2 halts and the contents of the buffer containing the assembled machine language program are either stored in an auxiliary disc or drum memory, or are outputted on paper tape or magnetic tape or punch cards. This material is called a *block of data* or simply a *block*. After the block is transferred, the corresponding area of main memory is cleared and pass 2 continues. When the assigned block of main memory again becomes filled, the process is repeated and another block is put on disc, etc. When pass 2 is completed, there may be many blocks. These form the complete machine language program.

Note that, in the introduction to Chapter 7, we discussed the need for error prevention when data is transferred in and out of memories. This is also true here where blocks of data are transferred. Elaborate coding is included with these blocks for error correction and detection.

In the simple procedure we have discussed, instructions would be stored consecutively, starting with memory location 0. At times, when more sophisticated procedures are used, we may not want this. There are instructions called pseudo-ops which can be used to indicate such things as the starting location of the program. Pseudo-ops are also used to indicate to the assembler where the program ends(e.g., following B: 14), where the first executable statement is, or where a particular statement is to be stored.

Loaders

In Sec. 11-2 we discussed the entering of programs and presented a simple way of specifying the memory location for each command. In most computers, such simple procedures are not used. There is a program called a LOADER which removes the need for such things as specifying the memory location for each machine language instruction. The LOADER program is given the starting word address for each block of data and then allocates the memory locations according to a predetermined sequence. If the program is written in machine language, then the programmer must still specify the memory locations that are part of the program(i.e., the data locations). Thus, the starting addresses must be such that no blocks are written over each other.

Just as in the case of the assembler, data is transferred in blocks and can be stored in the auxiliary memory until loading is finished. Then, the LOADER is removed from the main memory and the program is then returned from the auxiliary memory to the main memory.

Higher Level Languages

Assembler language is considerably easier to use than machine language. However, it is still very tedious to use. For instance, suppose we wanted to execute the simple equation

$$b = (x+y)(z+x)/(y-z) \tag{11-15}$$

Very many machine or assembler language statement would have to be written to execute this. However, there are higher level languages where this is not the case. In one such language, FORTRAN, a program which would evaluate this for x=3, y=4 and z=5 is

```
X=3                          (11-16a)
Y=4                          (11-16b)
Z=5                          (11-16c)
B=(X+Y)*(Z+X)/(Y-Z)          (11-16d)
STOP                         (11-16e)
END                          (11-16f)
```

Note that * means to multiply. In the case of the higher level languages, the program is scanned as in an assembler language and memory locations are established. However, each single statement is scanned and, from this scan, a *sequence* of machine language commands are established. This involves much more sophisticated operations than those performed by an assembler. The programs which direct the converting of higher level languages into machine language are called *compilers*. Of course, even in higher level languages, the statements are highly structured and there are rules we must obey. However, these are very much easier to use than assembler language.

Note the instruction of STOP and END in program (11-16). The END signifies to the compiler that the program is finished. The STOP is compiled into machine language and is equivalent to 77_8 in Table 11-3.

Note that, in theory at least, the greatest amount of versatility is obtainable in a machine language program since all other forms of programs must be translatable into machine language. However, most of the programming is done in higher level languages because they are much easier to use and fewer mistakes are made by the programmers.

BIBLIOGRAPHY

Booth, T.L., "Digital Networks and Computer Systems", Chapters 9-13, John Wiley and Sons, Inc., New York, 1971

Flores, I., "Computer Organization", Chapter 1, Prentice Hall, Inc. Englewood Cliffs, N.J.,1969

Hill,F.J. and Petterson,G.R., "Digital Systems:Hardware Organization and Design", Chapter 6, John Wiley and Sons, Inc., New York, 1973

Mano,M.M., "Computer Logic Design", Chapters 10-12, Prentice Hall Inc., Englewood Cliffs, N.J., 1972

O'Malley,J., "Introduction to the Digital Computer", Chapters 10-12, Holt, Rinehart and Winston, Inc., New York, 1972

Rhyne,V.T., "Fundamentals of Digital System Design", Chapter 7, Prentice Hall, Inc., Englewood Cliffs, N.J. 1973

Souček, B., "Minicomputers in Data Processing and Simulation", Chapter 5, John Wiley and Sons, Inc., New York, 1972

PROBLEMS

11-1. Discuss the basic components of the digital computer.

11-2. A computer has a memory which stores 262,144 words. There are to be 300 machine language instructions. If the word structure of Fig. 11-2 is to be used(with a different number of bits), what is the minimum word length that can be used?

11-3. Repeat Prob. 11-2 for the word structure of Fig. 11-4.

11-4. Repeat Prob. 11-2 for the word structure of Fig. 11-5.

11-5. Write a simple machine language program which will perform 7+6-9+3.

11-6. Describe in your own words the operation of the memory unit of the simple computer of Fig. 11-6.

11-7. Repeat Prob. 11-6 for the control unit.

11-8. Repeat Prob. 11-6 for the central processing unit.

11-9. Repeat Prob.11-6 for the input/output unit.

11-10. If the computer of Fig. 11-6 were to be made more elaborate, what changes would be made?

11-11. Assume that a flip-flop can change its state in 100×10^{-9} sec. and a simple logic gate can change its state is 50×10^{-9} sec. A read or write from memory takes 1×10^{-6}sec. Discuss the timing signals that must be used.

11-12. Assume that N_1, N_2 and N_3 are three numbers which are stored in M(101), M(102) and M(103), respectively. Write a machine program which performs the following: It adds N_1 and N_2. If the sum is greater than 25, then N_3 is added. If the sum equals 25, then the operation stops. If the sum is less than 25, then $2N_3$ is added. Store the following values

$N_1 = 25$
$N_2 = 50$
$N_3 = 150$

11-13. Write a machine language program which adds $\sum_{k=51}^{250} 2k$.

11-14. Write a machine language program which performs

$$\sum_{k=1}^{100} (-1)^k k.$$

11-15. Discuss the operation that results when the 03_8 instruction of Table 11-3 is given.

11-16. Repeat Prob. 11-15 for 07_8.

11-17. Repeat Prob.11-15 for 10_8.

11-18. Repeat Prob. 11-15 for 22_8.

11-19. Write a machine language program that performs multiplication.

11-20. Write a machine language program that performs division.

1--21. Add output statements to the program of Prob.11-21.

11-22. Write the program of Prob.11-12 in assembler language.

11-23. Repeat Prob. 11-22 for the program of Prob. 11-13.

11-24. Repeat Prob.11-22 for the program of Prob. 11-14.

11-25. If you know FORTRAN, write the program of Prob.11-12 in FORTRAN.

11-26. Repeat Prob. 11-25 for the program of Prob. 11-13.

11-27. Repeat Prob. 11-25 for the program of Prob. 11-14.

11-28. Discuss the difference between machine language, assembler language, and higher level languages.

11-29. What is the difference between an assembler and a compiler?

11-30. What is the most versatile language in which to program? What is the most convenient?

12

Some Additional Circuits

In this chapter we shall discuss some additional circuits that are commonly used in conjunction with digital devices. Often, a digital computer is used to process or extract data from a continuous signal, e.g., see the discussion of digital filters in Sec.1-1. The result of this process may be another continuous signal. We shall consider analog to digital and digital to analog converters that can be used in these processes. Other associated circuits will also be considered. In addition, we shall discuss circuits used in computers for such things as generation of timing pulses. We shall conclude by discussing the interfacing of the computer with external devices.

12.1 THE OPERATIONAL AMPLIFIER

A basic building block that we can use in many digital devices is called the *operational amplifier*. This is a high gain amplifier which is commonly available in integrated circuit form. We shall start by considering some properties of these amplifiers.

The block diagram for an operational amplifier is shown in Fig. 12-1a. It is basically a high gain voltage amplifier with two inputs. (There is also a power supply input. However, it is conventional not to show it in the block diagram.) The circuits of the amplifier are such that the output voltage is

$$V_o = A_v(V_2 - V_1) \tag{12-1}$$

where the *voltage gain* A_v is a large positive number which may be 5000 or more. The input to which V_1 is connected is called the *inverting* input while the other input is called the *non-inverting* input. (Note that the output is proportional to $-V_1$ if $V_2 = 0$ and is proportional to V_2 is $V_1 = 0$.)

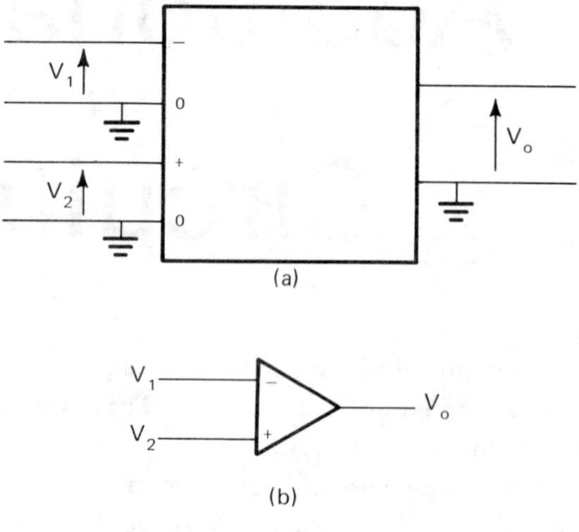

Figure 12-1. (a) The block diagram of an operational amplifier; (b) its symbol.

One terminal of each input and the output are connected to the common ground. Often, in drawing schematics, these grounded terminals are omitted from the diagram. Thus, a symbol that is used for the operational amplifier is shown in Fig. 12-1b.

A model that can be used to represent the operational amplifier is shown in Fig. 12-2. The voltage V_b is given by

$$V_b = V_2 - V_1 \tag{12-2}$$

The generator A_v represents the voltage gain. R_i and R_o represent the *input and output resistance*, respectively. R_o is usually a relatively small number and R_i a large one. In many circumstances, we can consider that R_i is infinite and that R_o represents a zero resistance.

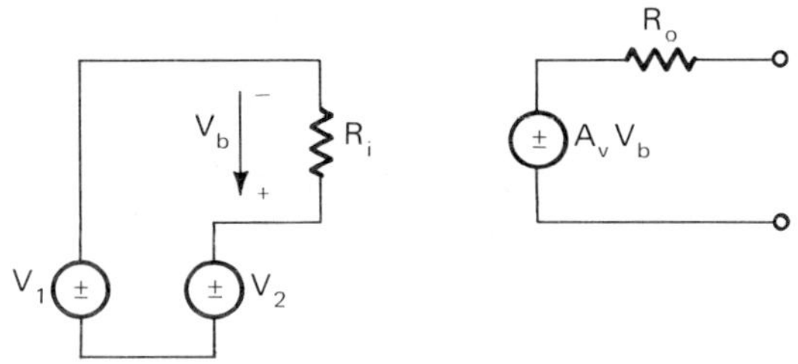

Figure 12-2. A model of the operational amplifier.

There are other assumptions that we shall make. One is that A_v does not vary with frequency. Also, if V_1=V_2, then V_o=0. Because of unbalance in the amplifier circuit, V_o may actually be a small voltage in this case. Also, if V_1=V_2=0, there may be a small direct voltage output. This is also due to unbalances which develop in the amplifier. This effect can be *offset* by applying a small direct voltage to one of the inputs. The characteristics of the amplifier can drift slightly with time and temperature. We shall, however, assume an ideal amplifier which is characterized by Eq. (12-1).

If we plot the output voltage of the amplifier versus V_i, a curve such as that in Fig. 12-3 results. The output will vary linearly until the curve levels off at $V_o = \pm V_A$. We say that, at these points, saturation is reached. Actually, the output and other stages are either at saturation or cut off and the output voltage does not change any further. We have idealized the characteristics. There is some rounding which is shown by the dotted lines. Because of the high gain of the amplifier, the value of V_2-V_1 which produces saturation is often a very small value (i.e., a fraction of a volt). In special circumstances, very high gain can be obtained by cascading amplifier stages, usually within the same integrated circuit chip. In such cases, the value of V_2-V_1 required to produce saturation has a magnitude of about a millivolt. That is, V_α, see Fig. 12-3, would be 1×10^{-3}volt.

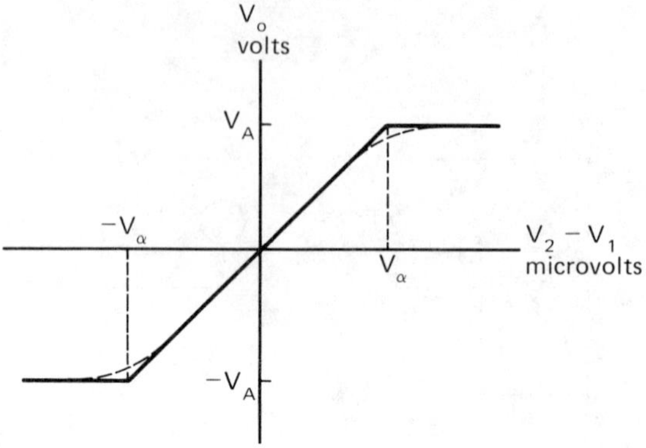

Figure 12-3. Idealized characteristics of an operational amplifier.

Now let us compute the voltage gain of the basic amplifier, shown in Fig. 12-4. We assume that the input impedance of the operational amplifier is infinite. Thus, the same current I will be in R_1 and R_2. Hence,

$$\frac{V_{i1}-V_1}{R_1} = \frac{V_1-V_o}{R_2} \tag{12-3}$$

Now

$$V_1 = -V_o/A_v \tag{12-4}$$

Then, manipulating, we have

$$V_{i1}/R_1 = -\frac{V_o}{R_2}\left(1 + \frac{1}{A_v}\left(1 + \frac{R_2}{R_1}\right)\right) \tag{12-5}$$

In general, A_v is very large, so that

$$1 >> \left|\frac{1}{A_v}\left(1 + \frac{R_2}{R_1}\right)\right| \tag{12-6}$$

Hence, we obtain

$$V_o = -\frac{R_2}{R_1} V_{i1} \tag{12-7}$$

Note that, in this case, the effective gain is essentially independent of the parameter A_v(as long as condition (12-6) holds).

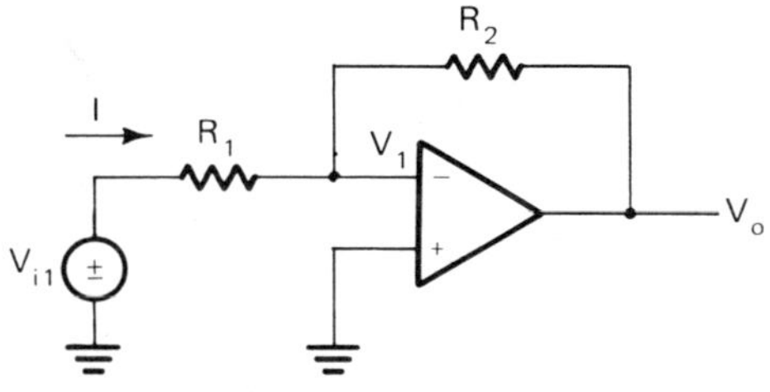

Figure 12-4. A basic operational amplifier configuration.

A_v is a very large number. Thus, V_1 will be very small, usually very much less than V_{i1} or V_o. Since the potential across the input of the amplifier is very small, and since the + terminal is grounded, we can say that the - terminal is also essentially at ground potential. This is called a *virtual ground*. (Note that we *cannot* state in general that V_1-V_2 is zero.)

Now let us consider the amplifier of Fig. 12-5. Proceeding as in the analysis of Fig. 12-4, we have

$$\frac{V_{i1}-V_1}{R_1} = \frac{V_1-V_o}{R_2} \qquad (12\text{-}8)$$

where V_1 is the voltage between the - terminal and ground. Thus,

$$V_1 = -V_b + \frac{R_4 V_{i2}}{R_3+R_4} \qquad (12\text{-}9)$$

Substituting and manipulating, we obtain

$$\frac{V_{i1}}{R_1} - \frac{V_{i2}}{R_1} \cdot \frac{1}{\frac{R_3}{R_4}+1} \cdot \frac{R_1}{R_2} + 1 = -\frac{V_o}{R_2}\left(1 + \frac{1}{A_v}\left(1 + \frac{R_2}{R_1}\right)\right)$$

If condition (12-6) holds, then we have

$$\frac{V_{i1}}{R_1} - \frac{V_{i2}}{R_1} \cdot \frac{1}{\frac{R_3}{R_4}+1} \cdot \frac{R_1}{R_2} + 1 = -\frac{V_o}{R_2} \qquad (12\text{-}10)$$

If

$$R_3/R_4 = R_1/R_2 \tag{12-11}$$

then

$$V_o = \frac{R_2}{R_1}(V_{i2} - V_{i1}) \tag{12-12}$$

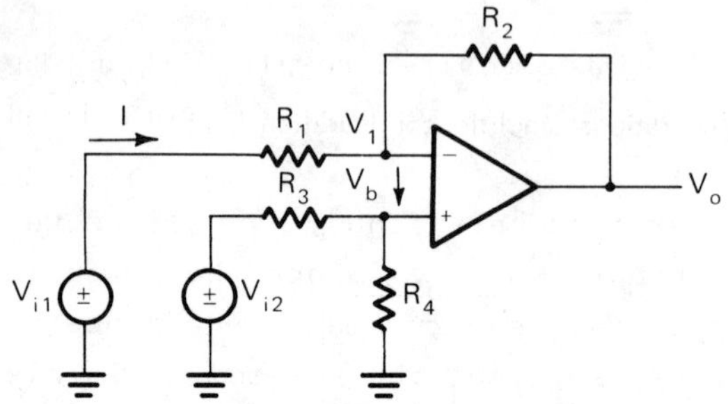

Figure 12-5. A two input operational amplifier circuit.

At times, we require a circuit wherein the output voltage exactly follows the input. This is called a *voltage follower*. We can obtain it from Fig. 12-5 by setting V_1=0, i.e., ground the input end of R_1, and setting R_1=R_2. Then, if Eq. (12-11) is satisfied, we achieve the desired result. However, this is wasteful of components. Consider the right side of Eq. (12-10). Suppose that we let $R_1 \to \infty$. Then, the right side becomes

$$\frac{V_{i2}}{R_2} \cdot \frac{1}{\frac{R_3}{R_4} + 1}$$

Thus, we have

$$V_{i2} \cdot \frac{1}{\frac{R_3}{R_4} + 1} = V_o$$

Since the value of R_2 is not included in this expression we can set it equal to any desired value. Let us use R_2 = 0 and also choose

$$R_3 = 0$$

$$R_4 = \infty$$

Then,

$$V_{i2} = V_o$$

and we have achieved the desired result with a minimum of components. The diagram for this voltage follower is shown in Fig. 12-6.

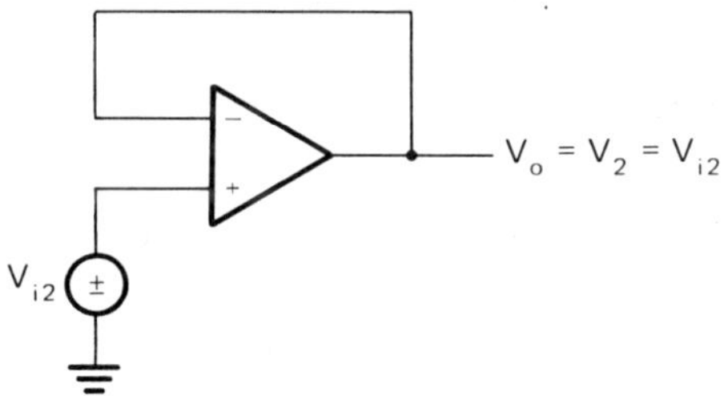

Figure 12-6. A voltage follower.

This result can be obtained by analyzing Fig. 12-6 directly. In this case, we have

$$V_o = A_v(V_{i2} - V_o)$$

or

$$V_o = \frac{A_v}{1+A_v} V_{i2}$$

If $A_v >> 1$, which is equivalent to condition (12-6), we obtain $V_o = V_{i2}$ as before.

Another useful circuit is shown in Fig. 12-7. Here we assume that there are many inputs, $V_{11}, V_{12}, \ldots, V_{1n}$ which are connected as shown. Then, proceeding as before, we have

$$\frac{V_{11}-V_1}{R_{11}} + \frac{V_{12}-V_1}{R_{22}} + \ldots + \frac{V_{1n}-V_1}{R_{nn}} = \frac{V_1-V_o}{R_2}$$

Manipulation yields

$$\frac{V_{11}}{R_{11}} + \frac{V_{12}}{R_{22}} + \ldots + \frac{V_{1n}}{R_{nn}} = \frac{-V_o}{R_2}\left(1 + \frac{1}{A_v}\left(\frac{R_2}{R_{11}} + \frac{R_2}{R_{22}} + \ldots + \frac{R_2}{R_{nn}}\right)\right)$$

$$\text{If } 1 \gg \left|\frac{1}{A_v}\left(\frac{R_2}{R_{11}} + \frac{R_2}{R_{22}} + \ldots + \frac{R_2}{R_{nn}}\right)\right| \tag{12-13}$$

then

$$V_o = -\frac{R_2}{R_{11}} V_{11} - \frac{R_2}{R_{22}} V_{12} + \ldots + \frac{R_2}{R_{nn}} V_{1n} \tag{12-14}$$

Thus, the output is the negative of the sum of the input voltages. This circuit is called an *analog adder*.

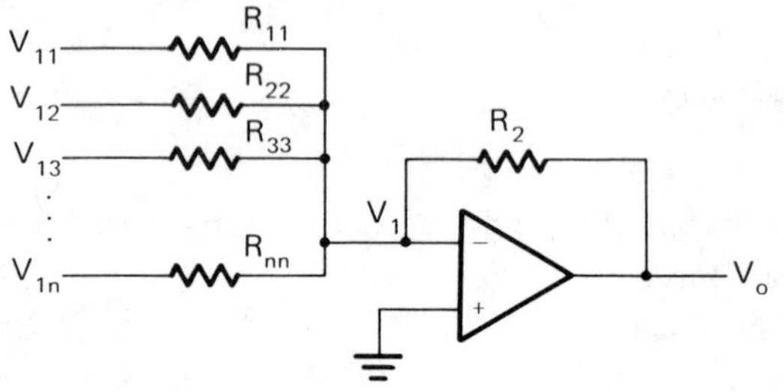

Figure 12-7. An adder circuit.

12.2 COMPARATORS-SCHMITT TRIGGERS

At times, we want to determine if a voltage is greater than or less than some reference level. We shall see that this is useful in analog to digital conversion. A circuit, whose block diagram is given in Fig. 12-8a, called a *comparator*, provides this information. The characteristics of a comparator are such that

$$V_o = V_C \text{ if } V_1 < V_R \tag{12-15a}$$

$$V_o = V_D \text{ if } V_1 > V_R \tag{12-15b}$$

A typical idealized characteristic for the comparator is shown in Fig. 12-8b. Thus, by sensing the output, we can tell if V_1 is greater than or less than the reference voltage V_R.

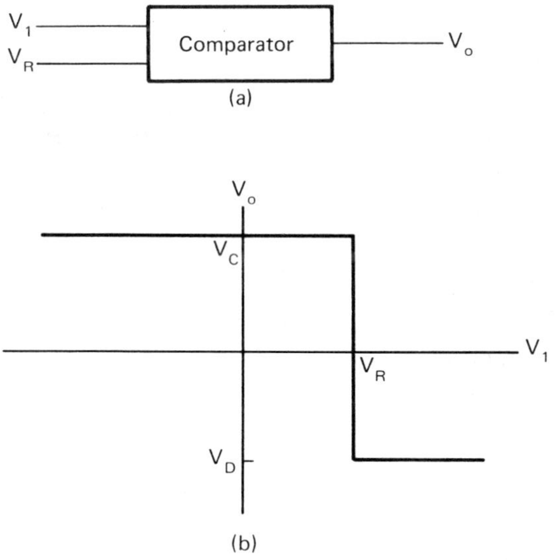

Figure 12-8. The comparator. (a) block diagram; (the ground terminals are omitted); (b) idealized characteristics.

A simple operational amplifier will act as a comparator. For instance, consider Fig. 12-9 in light of Fig. 12-3. If the amplifier is not to be saturated, then

$$|V_1 - V_2| \leq |V_1 - V_R| = V_\alpha \tag{12-16}$$

In general, V_α is a fraction of a volt. Thus, if $V_1 > V_r + V_\alpha$, then $V_o = -V_A$; if $V_1 < V_R - V_\alpha$, then $V_o = V_A$. Thus, the device functions as a comparator. Note that it is not an ideal comparator, since, when V_1 is close to V_R, the amplifier will be in its linear region and V_o will not be at either saturation level. There are special high gain operational amplifiers where V_α is a millivolt or less, in which case, the characteristics do tend to approach the ideal.

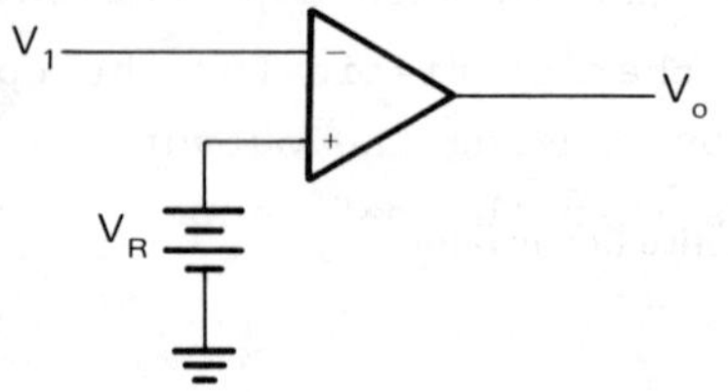

Figure 12-9. An operational amplifier connected as a comparator.

Let us consider one use of a comparator. Suppose that we have a periodic waveform like that of Fig. 12-10a. We want to determine its frequency. Since the waveform is arbitrary, we shall first convert it to a standard one. Suppose that we use the voltage as the input to a comparator, with $V_R=0$. Then, ideally, the comparator output would be of the form shown in Fig. 12-10b. This can be applied to a counter as discussed in Sec. 6-5, to determine the frequency of the waveform.

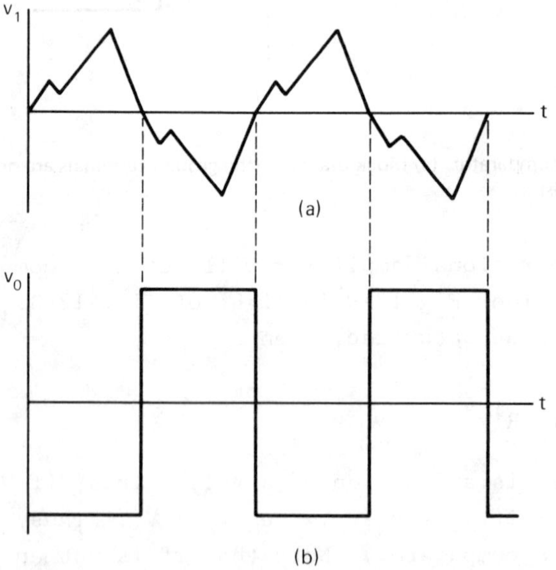

Figure 12-10. (a) A periodic waveform applied to the input of a comparator; (b) the idealized output of the comparator.

When the reference voltage $V_R=0$, then the comparator output voltage changes from V_C to V_D, or vice versa, see Fig. 12-8, each time that the input waveform passes through zero. This assumes that the voltage swing is greater than V_α, see Eq. (12-16). Then, the comparator can be used as a *zero crossing detector*. Such devices find extensive use in such applications as frequency measurement.

The Schmitt Trigger-Regenerative Comparator

The comparator circuits that we have discussed do not have the idealized responses of Figs. 12-8b and 12-10b. When $|V_1-V_R|$ is very small, the operational amplifier is not saturated and the output voltage varies linearly with the input signal.

Now suppose that the input waveform varies very slowly with time. Then, as it passed through the region where $|V_1-V_R| < V_\alpha$, the amplifier would not be saturated, and the output voltage would vary slowly with time. For instance, in Fig. 12-10a, if v_1 varied very slowly with time, then v_o would not appear as a square wave. The vertical sides would now follow the shape of v_1 as long as $|v_1-V_R| < V_\alpha$. However, we often require the transition of the waveform to be almost vertical, independent of the shape of the input signal and the rate at which it varies. In the simple comparator, we can only do this if we make V_α extremely small, ideally zero.

In Fig. 12-11 we illustrate a circuit called a *Schmitt trigger*, which makes the transition through the nonsaturated region very rapidly.

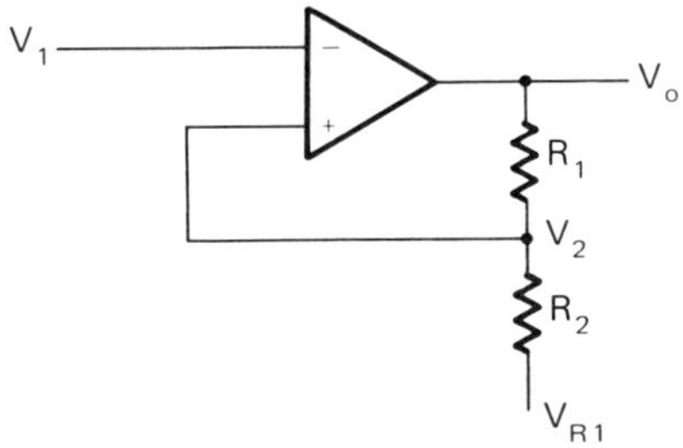

Figure 12-11. A Schmitt trigger.

The voltage V_2 of Fig. 12-11 is now the reference voltage of the comparator. Thus, we have replaced the fixed reference by a variable one. Let us consider this.

$$V_2 = V_o - \frac{(V_o - V_{R1})}{R_1 + R_2} R_1 \tag{12-17}$$

Manipulating, we obtain

$$V_2 = \frac{V_o R_2}{R_1+R_2} + \frac{V_{R1} R_1}{R_1+R_2} \tag{12-18}$$

(Note that the first component of the "reference"voltage is variable since it is a function of V_o.) Now let us consider the operation of this circuit. Suppose that $V_1 << V_2$ so that the amplifier is saturated. Then, $V_o = V_A$, see Fig. 12-3. Then, substituting in Eq.(12-18), we have

$$V_2 = \frac{V_A R_2}{R_1+R_2} + \frac{V_{R1} R_1}{R_1+R_2} \tag{12-19}$$

Now suppose that we slowly increase V_1. V_o will remain constant while the amplifier is saturated. When $V_i = V_2 - V_1$ becomes less than V_α, see Fig. 12-3, the magnitude of V_o will start to decrease. Now this decreases the magnitude of V_2, which makes $V_b = V_2 - V_1$ smaller. This change is amplified by A_v. Thus, V_o will decrease further, which further reduces V_2. This process then keeps repeating itself. Note that if V_2 were fixed, then as V_1 increased, $V_b = V_2 - V_1$ would decrease because V_1 was increasing. Now V_b also decreases because V_2 decreases. The change in V_2 is regenerative. That is, as V_2 decreases, $V_2 - V_1$ decreases which further decreases V_2. Thus, even if V_1 just starts the transition and then stops changing, the output of the amplifier will decrease until it becomes equal to $-V_A$, provided that A_v is large enough. Actually, if feedback theory is studied, we find that A_v must be large enough so that

$$\frac{A_v R_2}{R_1+R_2} > 1$$

If A_v satisfies this condition, then switching is very fast. Thus, we have obtained a rapid variation in output even if the input waveform varies slowly.

Once the transition is started, the regeneration carries it through to reverse saturation. The range of linear operation, see Fig. 12-3, is such that

$$-V_\alpha \leq V_b \leq V_\alpha \tag{12-20}$$

Once V_i enters this region, the transition will take place. Let us determine the location of this region in terms of V_1. If $V_2 > V_1$ and V_1 is increased, then V_o will equal V_A until switching takes place. This starts when $V_2 - V_1 = V_\alpha$. Substituting Eq. (12-19) we have

$$V_1 = -V_\alpha + \frac{-V_A R_2}{R_1+R_2} + \frac{V_{R1} R_1}{R_1+R_2} \tag{12-21a}$$

Now suppose that $V_2 < V_1$ and the amplifier is saturated. Now $V_o = -V_A$ and

$$V_2 = \frac{-V_A R_2}{R_1+R_2} + \frac{V_{R1} R_1}{R_1+R_2}$$

Switching will be initiated when V_1 is decreased until $V_2 - V_1 = -V_\alpha$. Thus,

$$V_1 = V_\alpha - \frac{V_A R_2}{R_1+R_2} + \frac{V_{R1} R_1}{R_1+R_2} \tag{12-21b}$$

Let us consider some typical values. Typical parameter values that are used are

V_A = 5 volts
R_1 = 10,000 ohms
R_2 = 100 ohms
V_{R1} = 3 volts
V_α = .01 volt

Substituting in Eqs. (12-21), we obtain

V_{11} = 3.0098 volts
V_{12} = 2.9308 volts

where V_{11} is the switching voltage when V_1 increases and V_{12} is the switching voltage when V_2 decreases. These are not the same.

The input-output characteristics of the Schmitt trigger are shown in Fig. 12-12. This is called a hysteresis curve because of its resemblance to the magnetic hysteresis curve. Note that, in contrast to the simple comparator, the Schmitt trigger achieves a very fast transition independent of the variation of the input signal levels.

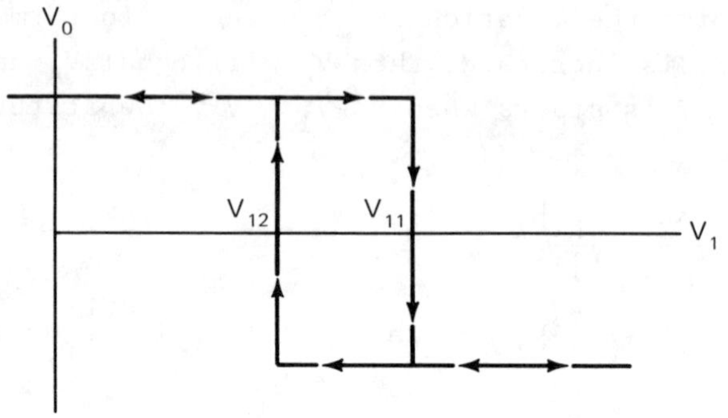

Figure 12-12. The input-output characteristics of the Schmitt trigger.

12.3 SAMPLE AND HOLD CIRCUITS-QUANTIZATION

When digital processes are applied to continuous waveforms, we sample the waveform at discrete times. These samples are used to determine information about the signal. Actually, it can be shown that, if a signal is bandlimited,i.e., all its signal components lie in a finite bandwidth B, then the signal can be uniquely determined by a set of samples spaced 1/(2B) sec. apart. This is illustrated in Fig. 12-13 where we first show a continuous waveform and then a set of samples which can be used to represent the waveform. The samples are the points shown.

We shall encode the levels of sampled voltages into a digital representation. Such signals can then be processed by a digital computer. The circuits that perform this operation require that the voltage remain constant for a short period of time. This would be satisfied if the voltage did not change between sample times as shown by the dashed curve. We can then process these constant voltages to obtain a digital representation. This will

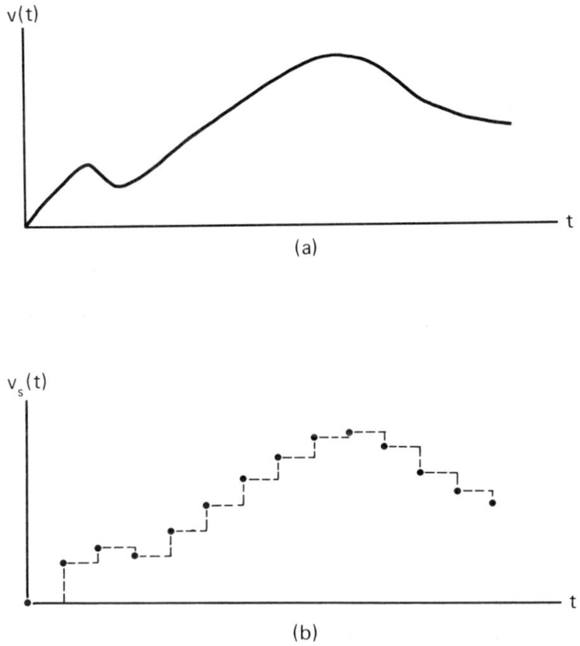

Figure 12-13. (a) A continuous function of time; (b) a set of samples used to represent this waveform. The dashed curve is the output of a sample and hold circuit.

be discussed in Sec. 12-5. The dashed curve is called a *sample and hold* representation of v(t).

A very simple sample and hold circuit is shown in Fig 12-14a. A switch SW is normally open and only closes briefly at the sample times. The capacitor will charge to the instantaneous value of v(t) when the switch is closed. (We assume that v(t) only changes by an insignificant amount during the time that SW is closed.) When SW opens, the charge will remain on the capacitor and $v_s(t)$ will not change.

This circuit has two disadvantages. First, it is difficult to operate a mechanical switch fast enough. Second, if the output is connected to a load which does not have a very high resistance, then charge will leak off the capacitor during the periods when the switch is open and the voltage will not remain constant.

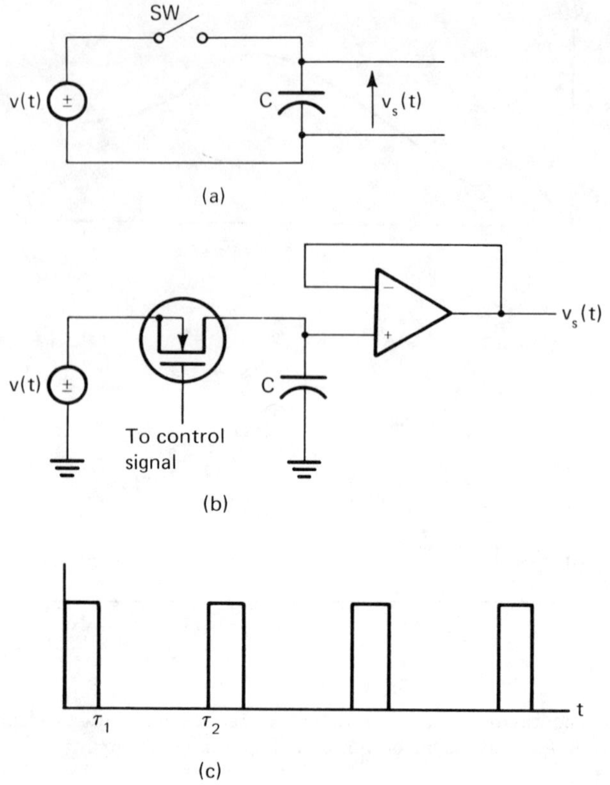

Figure 12-14. (a) A simple sample and hold circuit; (b) a more practical form of this circuit; (c) a control signal.

The circuit of Fig. 12-14b overcomes these difficulties. The mechanical switch is replaced by an enhancement MOSFET. When the control signal is cut off, the MOSFET acts as an open circuit. When the control signal is present, it acts as a very low resistance. The pulse of Fig. 12-14c can be used for the control signal. Then, the MOSFET acts as required. The voltage follower of Fig. 12-6 is used to obtain the output signal. The input impedance of the operational amplifier and the resistance of the cut off FET can be very large. Thus, the voltage across the capacitor will not change by an appreciable amount between samples.

If R_s represents the saturation resistance of the MOSFET, R_c represents its cutoff resistance and R_i is the input resistance of the operational amplifier. Then, during sampling, the time constant of the circuit is

$$T_1 = R_s C \tag{12-22a}$$

During hold, the time constant is

$$T_2 = \frac{R_c R_i}{R_c + R_i} C \tag{12-22b}$$

where we have assumed that $R_s << R_i$. Since R_s is very small and R_c and R_i are very large, we can make T_1 much shorter than the pulse length τ_1 and T_2 much longer than the period of the pulses τ_2, see Fig. 12-14c. Thus, the output of Fig. 12-14b will approach that of Fig. 12-13b.

In Sec. 12-5 we shall consider circuits that take the output from a sample and hold circuit and convert it into digital representation. Thus, we will obtain a sequence of binary representations for each sample. This involves *quantization*. For instance, suppose that the voltage can vary between 0 and 1 volt and we want to represent it by a four bit binary number. Then, there are only 16 different levels which can be represented. These represent the numbers 0,1,...15. For instance, we will have levels of 0, 1/16, 2/16,...,15/16 volts(i.e., 0, 0.0625, 0.125,...,0.9375). Then if

$$m/16 \leq v \leq (m+1)/16 \tag{12-23}$$

we would quantize it to m/16 volts. For instance, 0.0915 volts would be quantized as 0.0625. The levels must be taken close enough together so that errors due to quantization are small enough not to be troublesome. Thus, enough bits must be used to represent the number.

12.4 DIGITAL TO ANALOG CONVERSION

When a digital computer processes a continuous analog input signal, the signal is converted to a digital representation, i.e., a sequence of binary numbers. The processing takes place using these numbers. The result is another sequence of binary numbers. These may have to be converted back to a continuous analog signal. In this section we shall consider the digital to analog conversion. In the next, we shall discuss the analog to digital conversion.

Suppose that we have a four bit binary number which is to represent a voltage whose maximum value is V_M. If we have the four bits

$$a_3a_2a_1a_0 \tag{12-24}$$

then we can represent the voltage by

$$v = \frac{V_M}{16}(2^3a_3 + 2^2a_2 + 2a_1 + a_0) \tag{12-25}$$

Then, the binary number 0000 represents 0 volts, and 1111 represents 15/16 V_M volts, etc. This follows the rules for quantization discussed at the end of the last section.

Now let us consider the *digital to analog converter* (DAC) of Fig. 12-15 which can implement Eq. (12-25). We assume that the switches act in response to the bits of the number (12-24). That is, if a_0 is a 1, then SW_0 is up and $-V_M$ is applied to the left input of the $8R_1$ resistor. If a_0 is a 0, then SW_0 is down and the left input of the $8R_1$ resistor is connected to 0 volts (i.e., ground). The same idea is repeated with the other switches.

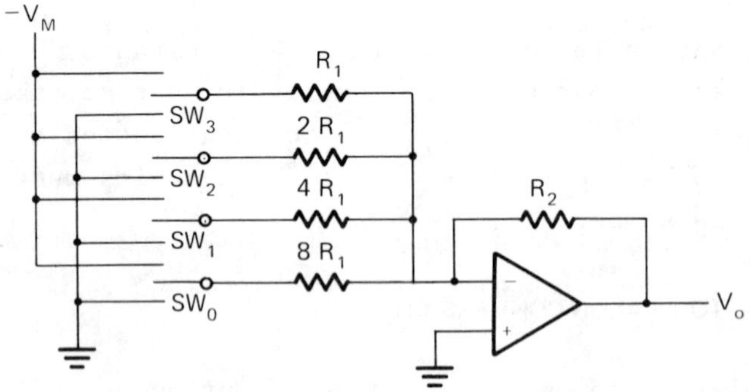

Figure 12-15. A simple digital to analog converter.

Figure 12-15 is basically the adder circuit of Fig. 12-7. Thus, the output voltage is

$$V_o = \frac{V_M R_2}{R_1}\left(a_3 + \frac{a_2}{2} + \frac{a_1}{4} + \frac{a_0}{8}\right) \tag{12-26}$$

Manipulating, we obtain

$$V_o = \frac{V_M R_2}{R_1}(8a_3 + 4a_2 + 2a_1 + a_0) \tag{12-27}$$

If we choose R_1 and R_2 such that

$$R_2 = 8R_1 \tag{12-28}$$

then the output voltage is the one specified by Eq. (12-25). Of course, we can set $V_M R_2/8R_1$ to give us any scale factor we desire. That is, we choose $V_M R_2/8R_1$ so that $15V_M R_2/8R_1$ yields the desired maximum voltage.

In order to practically implement the circuit of Fig. 12-15, we must replace the mechanical switches by electronic ones. One such switch is shown in Fig.12-16. One circuit is used for each switch of Fig. 12-15. (If more than four bits are used, the 2^{3-k} should be replaced by 2^{n-1-k}, where n is the number of bits.) Enhancement n-channel MOSFETs are used. We assume that a 1 is a large positive pulse and that a 0 is zero volts. Then, if a_k is a 1, the upper MOSFET will be saturated and the lower one will be cut off. The left side of the resistor will then effectively be connected to $-V_M$. On the other hand, if a_k is a 0, then the upper MOSFET will be cut off while the lower one is saturated. In this case, the left side of the resistor is effectively connected to ground. Thus, we must put four such switches into the circuit of Fig. 12-15. Note that we assume that the input comes from a register which can store the information. If this is not the case, then storage can be incorporated in the switch by using a D flip-flop. Then, the NOT gate of Fig. 12-16 is no longer needed.

In digital to analog conversion(or vice versa), there is always quantization error. The number of bits should be chosen so that an error in the last bit results in negligible error in the voltage.

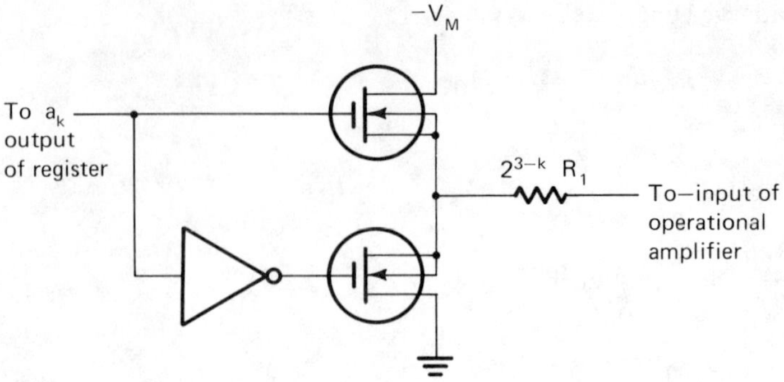

Figure 12-16. An electronic switch which can be used in conjunction with the digital to analog converter of Fig. 12-15.

The circuit of Fig. 12-15 presents some problems when there is a large number of binary digits. If there are n bits, then the ratio of the largest to the smallest input resistors is 2^{n-1}. For instance, if n=32, then 2^{n-1} = 2,147,483,648. Thus, the ratio of the largest input resistor to the smallest is very large. R_1 cannot be too small since it would tend to load down the electronic switch. Also, decreasing R_1 will result in a decrease in R_2 which will load down the operational amplifier. For instance, suppose R_1 = 5000 ohms. Then, the largest resistance, $2^{n-1} R_1$ would be 1.0737×10^{13} ohms. Such large resistances are difficult to obtain accurately. For one thing, the leakage resistance caused by contamination of their surface is often much less than the desired resistance.

In Fig. 12-17 we have another digital to analog converter that avoids this difficulty. Using the notation of (12-24), if a_j=1, then SW_j is thrown to the left. If a_j=0, then SW_j is thrown to the right. Analysis of this network shows that

$$V_o = \frac{V_M R_2}{2^4 \times 3R_1} (2^3 a_3 + 2^2 a_2 + 2^1 a_1 + a_0) \qquad (12\text{-}29)$$

Thus, by adjusting the value of

$V_M R_2 / 2^4 \times 3R_1$

we can obtain the desired digital to analog conversion.

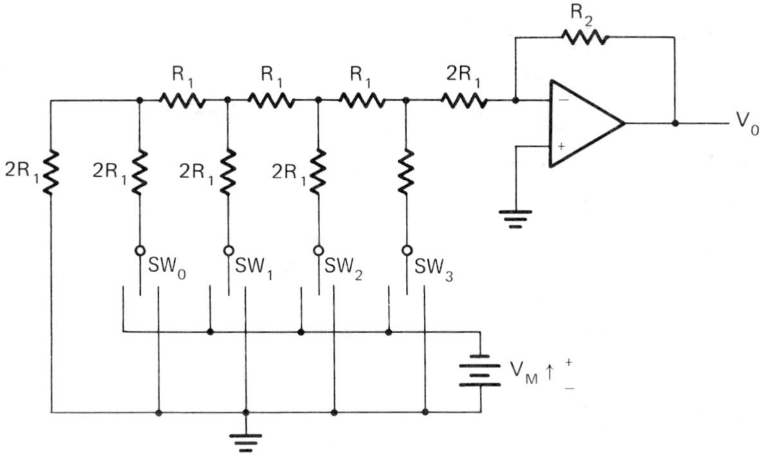

Figure 12-17. A digital to analog converter whose resistance values do not vary over a wide range.

BCD Digital to Analog Conversion

If the binary number uses a BCD code, then very similar digital to analog converters can be made. Let us assume that an 8421 BCD code is used. If we have only four bits, then the converter of Figs. 12-15 and 12-17 can be used directly. Suppose we have 8 bits representing a voltage whose maximum value is 99 volts.

$$b_{30}b_{20}b_{10}b_{00}b_3b_2b_1b_0$$

Then, the value of the voltage is

$$v = 10(8b_{30}+4b_{20}+2b_{10}+b_{00})+(8b_3+4b_2+2b_1+b_0) \qquad (12\text{-}30)$$

Again, the binary, analog to digital converter can be used with a suitable modification in the resistor values.

Alternatively, we can use two four bit 8421 converters, one of which converts the first set of bits and the second which converts the second set of bits. Then, see Fig. 12-18, the first output is multiplied by 10 and added to the second.

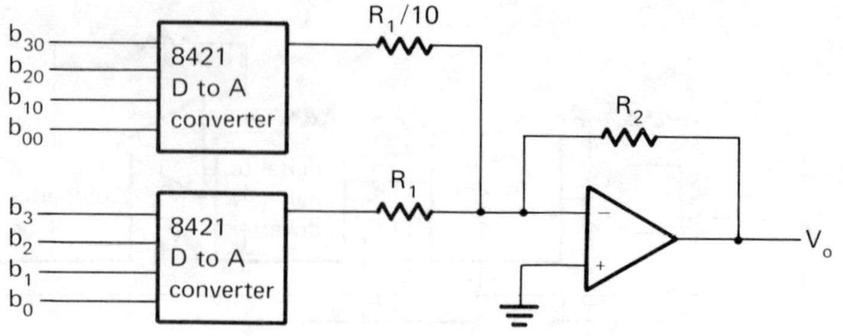

Figure 12-18. An 8421 BCD digital to analog converter.

12.5 ANALOG TO DIGITAL CONVERSION

Now let us consider the process of converting an analog signal to a digital representation. The device which performs this is called an *analog to digital converter* (ADC). We shall assume that the signal is sampled using a sample and hold circuit, see Sec. 12-3. Each sample is converted to a digital representation.

One form of analog to digital converter is shown in Fig. 12-19. Here we connect a counter to the input of a digital to analog converter. The clock pulses cause the counter to count in sequence. The various flip-flops of the counter then produce a binary number which drives the digital to analog converter. We call the output of the DAC V_R. As the output of the counter increases, V_R increases in discrete steps. V_R is connected to one input of a comparator whose other is from the input voltage V_{in}. The polarities of the output voltage of the comparator are such that when $V_{in} > V_R$, the output is a logical 1; when $V_{in} < V_R$, the output of the comparator becomes a logical 0. When this happens, the output of the AND gate becomes a 0. The counter stops advancing and the register is activated, and it stores the last output of the counter. This is the digital representation of the signal, within the quantization error.

The START signal resets the counter and the process repeats itself. The start pulses can also be used to activate the sample and hold circuit. The start and clock pulses can all be derived from the same master clock, see Fig. 11-7. The CP frequency must be such that the counter can be cycled through its complete range between successive START pulses.

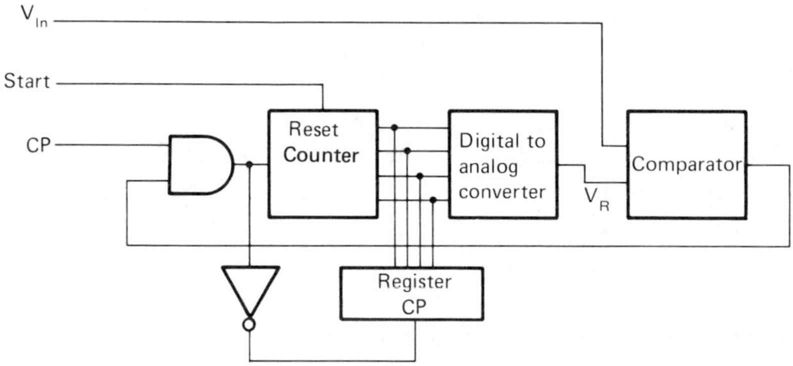

Figure 12-19. An analog to digital converter which uses a counter.

When analog to digital or digital to analog conversion is made, there may be an error in the least significant bit. For instance, suppose $V_{in}=0$. The circuit of Fig. 12-19 will give the digital representation of this as 0001. That is, the counter does not stop until $V_R > V_{in}$. The correct representation should be 0000. Enough bits should be used so that the quantization error of the last bit is negligible.

Note that this counter must always be such that V_R starts out less than V_{in}. We start with $V_R=0$. Then, V_R is successively increased. The conversion time then is larger with larger V_{in}. Now suppose that V_{in} changes at a rate which is slow in comparison with the sample rate. Successive samples of V_{in} will be relatively close together. Thus, it wastes time to reduce V_R to zero. To speed up the operation, the circuit can be modified. The counter is replaced by one which can count down as well as up. That is, when it is set in the up position, it counts in the usual way. When it is set to count in the down position, each count pulse reduces the count by one. Clock pulses are now continuously applied to the counter, independent of the comparator output. The output of the comparator is used to set the counter in its up or down mode. No reset signals are used in this case. The sampling rate can be increased with this circuit.

Suppose V_{in} is a constant; then V_R will be increased until $V_R > V_{in}$. Thus, V_R will be reduced by one "step". Now $V_R < V_{in}$ and V_R will be increased by one step. The output will then oscillate about the final value. This oscillation only occurs in the rightmost(least significant)bit. This bit can be ignored when the binary output is taken, e.g., it may not be supplied to the output register.

Now, if V_m changes slightly, the starting value of V_R will be close to it. Thus, the digital value will be obtained faster than if V_R started from zero.

Let us consider another form of analog to digital converter called the *successive approximation analog to digital converter.* The quantized voltage we are attempting to represent is of the form

$$V_{in} = \frac{V_M}{16}(8b_3 + 4b_2 + 2b_1 + b_0) \qquad (12\text{-}31)$$

We again assume four bits. These ideas are extended if there are more. That is, the voltage V_{in} is to be represented by the binary number

$b_3b_2b_1b_0$

Now consider Fig. 12-20. There are four digital to analog converters. Their parameters are adjusted so that their output voltages are

$$DAC_3: \quad V_3 = \frac{V_M}{16}(8) \qquad (12\text{-}32a)$$

$$DAC_2: \quad V_2 = \frac{V_M}{16}(8b_3 + 4) \qquad (12\text{-}32b)$$

$$DAC_1: \quad V_1 = \frac{V_M}{16}(8b_3 + 4b_2 + 2) \qquad (12\text{-}32c)$$

$$DAC_0: \quad V_0 = \frac{V_M}{16}(8b_3 + 4b_2 + 2b_1 + 1) \qquad (12\text{-}32d)$$

The comparators are such that if the voltage applied to their lower terminals is greater than that applied to their upper terminals, then their output will be a 1. Now suppose that $V_{in} > V_M/2$. Then, the output of comparator 3 will be a 1. Otherwise, it will be a 0. If $V_{in} > V_M/2$, then one input to comparator 2 will be $(V_M/16)(8+4) = 3/4V_M$. If $V_{in} > 3/4V_M$, then b_2, which is

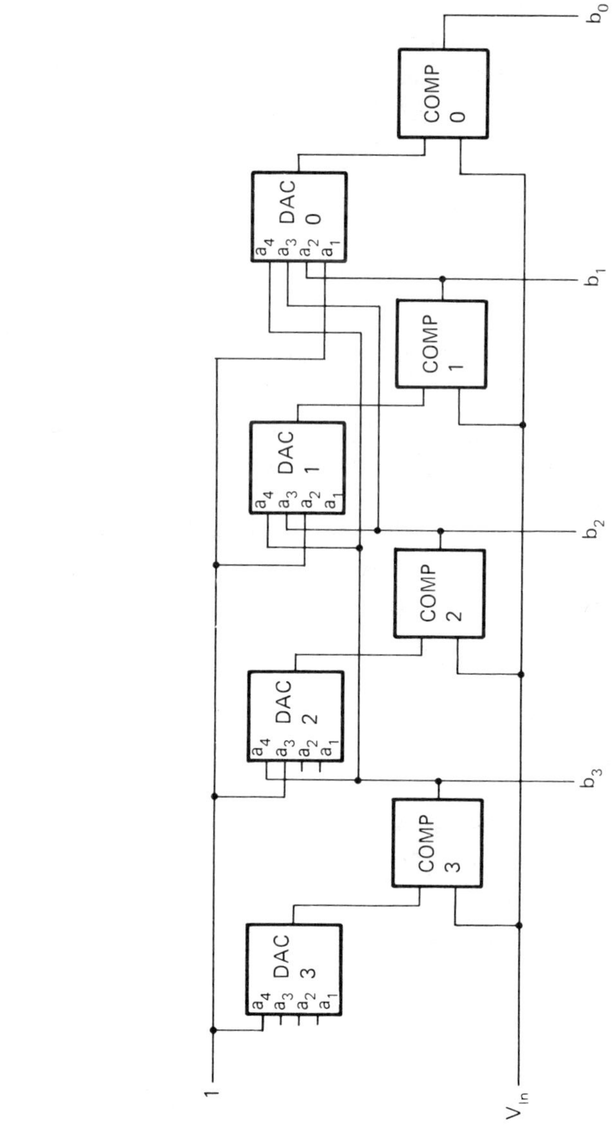

Figure 12-20. A successive approximation analog to digital converter.

the output of comparator 2, will be a 1; otherwise it will be a 0. On the other hand, if b_3 were a 0, then one input to comparator 2 would be $(V_M/16)(0+4) = V_M/4$. Now, if $V_{in} > V_M/4$, then b_2 will be a 1; otherwise it will be a 0.

In this way, b_2 and b_3 are correctly identified. Continuing the procedure all the binary digits can be obtained. Again, quantization error should be considered.

Analog to BCD Conversion

Using appropriate logic, we can design a converter which converts from an analog to a BCD representation. The basic ideas of the circuit used for analog to binary conversion are applied here.

Digital Voltmeters

Analog to digital converters are now finding extensive use as digital voltmeters. The circuits have no moving parts with their resultant friction and errors. Many significant figures can be read by using an ADC with enough stages. Voltmeters which can be read to four significant figures with an accuracy of 0.1 percent can be obtained without difficulty. Of course, this requires accurate reference voltages in the digital to analog converter. This can be obtained with zener diodes. The various resistances, see Figs. 12-15 and 12-17, in the digital to analog converter, must also be very accurate. Remember that there will be a quantization error equivalent to the least significant bit. This should be considered when the number of bits is chosen. We have considered one form of digital voltmeter. There are others.

The output of a digital voltmeter is often a series of lighted numbers, i.e., if the voltage is 123.4 volts, then the numbers 1, 2,3, and 4 plus the decimal point would appear.

In one type of number display, see Fig. 12-21, there are seven straight line segments, each of which can be lighted independently. (This is done electronically.) Note that, if we light the appropriate segments, all the numbers from 0 to 9 can be obtained. The binary output from the ADC in conjunction with decoders and read only memories cause the proper segments to be lighted. Thus, a very convenient and accurate voltmeter can be obtained.

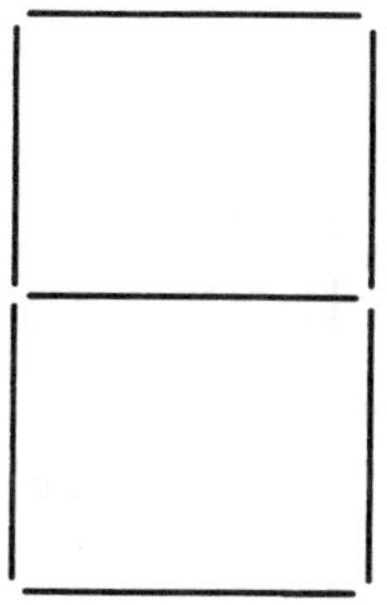

Figure 12-21. A display which can, by lighting the appropriate segments, produce the numbers 0 to 9.

12.6 PULSE GENERATORS

In a digital computer, we often have need for the generation of various types of pulses, e.g., a master clock. There are very many pulse generating circuits. In this section, we shall consider only a few of them. They shall utilize operational amplifiers. There are many pulse circuits that do not use them. The reader is referred to texts on pulse and/or electronic circuits for a more complete discussion.

Astable Multivibrators

The first circuit that we shall consider is called the *astable multivibrator*. It is illustrated in Fig. 12-22. Remember that if $|v_2 - v_1| > V_\alpha$, then the value of V_o will be $\pm V_A$, see Fig. 12-3.

Now let us assume that the device is initially turned on. Because of the capacitor, the initial value of $v_1(0) = 0$. Then,

$$v_2(0) = V_o \frac{R_2}{R_1+R_2} \tag{12-33}$$

Assume that the transient which results when the circuit is turned on, causes v_2 to be positive. Then, as in the case of the Schmitt trigger, we very quickly reach the case where $v_o = V_A$. Now the capacitor C begins to charge. When

$$v_1(t) = \frac{V_A R_2}{R_1+R_2} - V_\alpha \tag{12-34}$$

the circuit will switch as in the case of a Schmitt trigger.

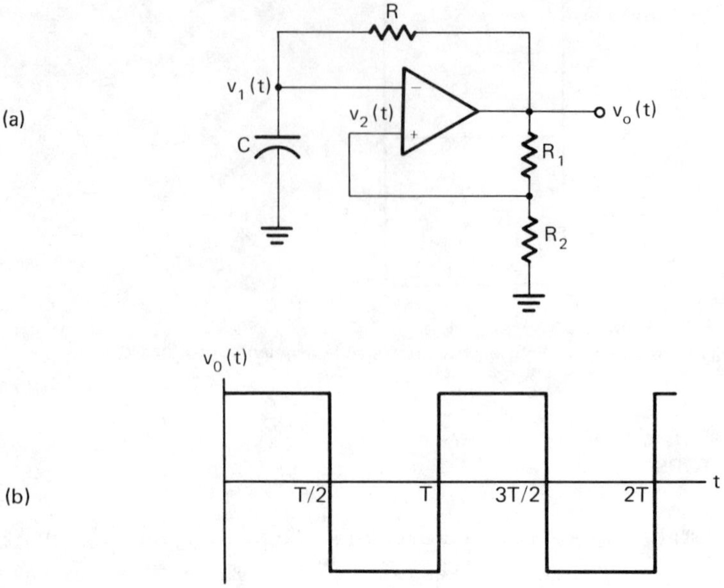

Figure 12-22. (a) An astable multivibrator; (b) the idealized output waveform.

Then, v_o very quickly becomes equal to $-V_A$ and the voltage across the capacitor falls exponentially from the value given in Eq. (12-34) toward $-V_A$. However, when

$$v_1(t_2) = \frac{-V_A R_2}{R_1+R_2} + V_\alpha \tag{12-35}$$

the circuit switches again and v_o becomes $+V_A$, and the cycle repeats itself. The idealized output waveform is shown in Fig. 12-22b. We assume that the switching action occurs in very much less time than the RC time constant. A simple circuit that can be used to calculate $v_1(t)$ is shown in Fig. 12-23.

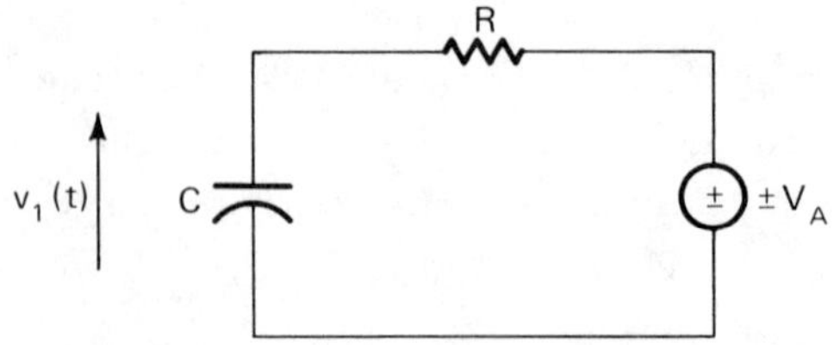

Figure 12-23. A circuit that can be used to calculate v_1 (t) in Fig. 12.22.

Let us obtain the period T of the oscillation, T/2 is the time required for the voltage across the capacitor to change from

$$\frac{V_A R_2}{R_1+R_2} - V_\alpha$$

to

$$-\frac{V_A R_2}{R_1+R_2} + V_\alpha$$

or vice versa. If V_0' is the initial voltage across a capacitor in a series RC circuit, and V_∞ is the final voltage across it, then the voltage across the capacitor as a function of time is

$$v_c(t) = V_\infty + (V_0' - V_\infty)e^{-t/RC} \qquad (12\text{-}36)$$

Then, substituting, we have

$$-\frac{V_A R_2}{R_1+R_2} + V_\alpha = -V_A + \left(\frac{V_A R_2}{R_1+R_2} - V_\alpha + V_A\right)e^{-T/2RC} \qquad (12\text{-}37)$$

Note the "effective voltage generator", or V_∞, is $-V_A$ since this is the value of V_o during the time that we are considering.

Manipulating, we obtain

$$T = 2RC \ln \left[\frac{R_1+2R_2 - \frac{V_\alpha}{V_A}(R_1+R_2)}{R_1 + \frac{V_\alpha}{V_A}(R_1 + R_2)}\right] \qquad (12\text{-}38)$$

Often, V_α is very small so that

$$R_1 >> \frac{V_\alpha}{V_A}(R_1+R_2)$$

Then,

$$T = 2RC \ln(1+2R_2/R_1) \qquad (12\text{-}39)$$

This circuit is called an *astable multivibrator*, since it has no stable states. Note that since flip-flops have two states, both of which are stable, i.e., they will not change unless input signals are applied, they are called *bistable multivibrators*.

We have drawn the response as an ideal square wave. Actually, during the time that $|v_2-v_1| < V_\alpha$, the amplifier is in its linear region and $v_o = A_v(v_2-v_1)$. During this time, the response will depart from the ideal square wave. However, V_α is usually very small and the regenerative switching is very fast. Thus, reasonably good waveforms do result.

Monostable Multivibrators

The astable multivibrator is used when we want to generate a sequence of pulses. For instance, it could be used as a master clock(if we did not require that the frequency remain very constant). Other times, we want to have a circuit which produces a single pulse in *response* to an input pulse. Such a circuit will remain in a stable state until an input pulse is applied. Then, it will switch producing an output pulse and finally return to its stable state. Such a circuit is called a *monostable multivibrator*, or *one-shot multivibrator*, since it has one stable state.

A simple monostable multivibrator is shown in Fig. 12-24a. Assume that $v_1(t) = 0$. Then, no matter what $v_o(t)$ is, eventually, the capacitor will act as an open circuit and $v_2(t) = -V_R$. Then, $v_o(t) = -V_A$. This is the stable state. We assume that V_R is large enough to cause saturation.

Now suppose that we put in a pulse $v_1(t)$ as shown in Fig.12-24b, where V_I is large enough to produce saturation, in the reverse direction. Then,

$$v_o = V_A$$

Assume that this pulse is $v_1(t)$ and that it is applied at t=0. In the first instant of time, the capacitor C has an initial voltage V_R-V_A, see Fig. 12-25. Note that the voltage across the capacitor cannot change instantaneously. Thus, at t=0+, after

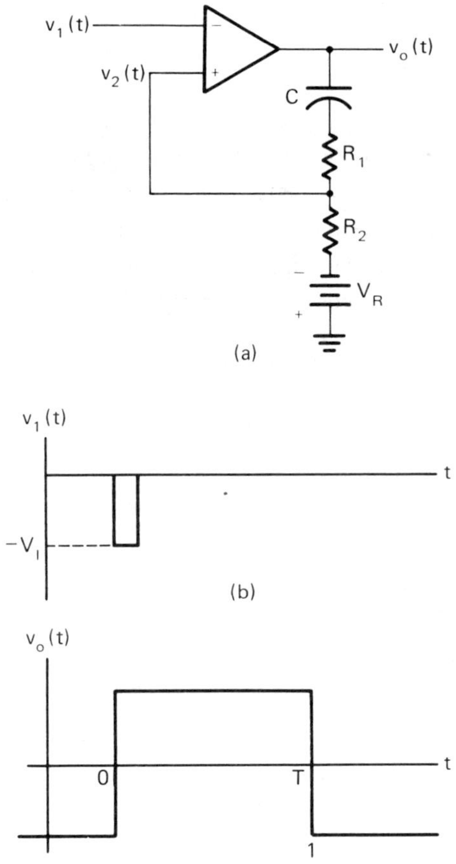

Figure 12-24. (a) A monostable multivibrator; (b) its input voltage; (c) its idealized output waveform.

v_o has switched to V_A,

$$V_2(0) = \frac{2V_A R_2}{R_1+R_2} - V_R \tag{12-40}$$

This will tend to keep $v_o = V_A$ even when the input pulse is removed. We assume that $2V_A R_2/(R_1+R_2) > V_R + V_\alpha$.

The voltage across the capacitor is given by, see Eq.(12-36)

$$v_C(t) = V_R + V_A - 2V_A e^{-t/(R_1+R_2)C} \qquad 0 \le t \le T$$

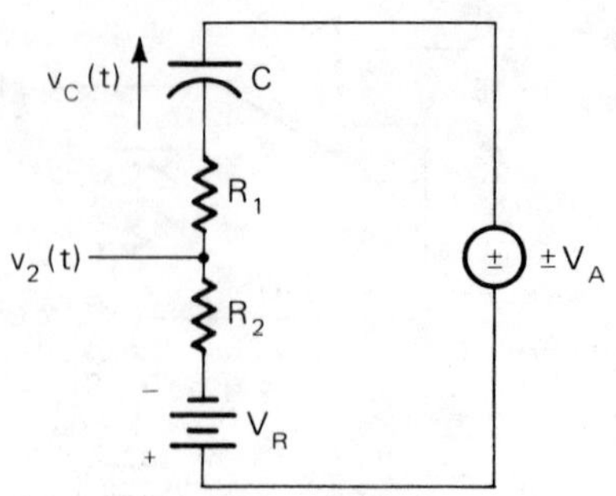

Figure 12-25. A circuit used to calculate the voltages of Fig. 12-24.

The current through the capacitor is

$$i_C = C\frac{dv_C}{dt} = \frac{2V_A}{R_1+R_2}e^{-t/(R_1+R_2)C}$$

Then,

$$v_2(t) = i_C R_2 - V_R$$

Substituting, we obtain

$$v_2(t) = \frac{2V_A R_2}{R_1+R_2}e^{-t/(R_1+R_2)C} - V_R \qquad (12\text{-}41)$$

Thus, at the first instant of time, the $v_o(t)$ is switched from $-V_A$ to V_A and $v_2(0)$ is given by Eq.(12-40). Now, as t increases, $v_2(t)$ falls off. When, at t=T, $v_2(t) = V_\alpha$, switching in the reverse direction occurs and $v_o(t)$ returns to $-V_A$ again. This reverses i_c and causes $v_2(t)$ to become more negative than V_R. The magnitude of $v_2(t)$ then falls off, finally reaching the value of $-V_R$. Of course, the output does not change during this time. Thus, after switching at t=T, the multivibrator returns to its stable state. Then, to find T, we solve Eq.(12-41) for the time when $v_2(t) = V_\alpha$. This is

$$V_R + V_\alpha = \frac{2V_A R_2}{R_1+R_2}e^{-T/(R_1+R_2)C}$$

Solving, we have

$$T = (R_1+R_2)C \ln\left[\frac{2V_A}{V_R+V_\alpha}\frac{R_2}{R_1+R_2}\right] \qquad (12\text{-}42)$$

Usually, $V_R >> V_\alpha$, so that this becomes

$$T = (R_1+R_2)C \ln \left[2 \frac{V_A}{V_R} \frac{R_2}{R_1+R_2}\right] \qquad (12\text{-}43)$$

We have used operational amplifiers in these circuits. Actually, their switching speed is slow in comparison with high speed logic. However, the switching speed is often adequate for these applications. There are special high speed integrated amplifiers which are used in computers which are sometimes used when high switching speeds are required.

Crystal Oscillators

Very often, we want to control the master clock frequncy very precisely. A monostable multivibrator's frequency will tend to drift with temperature since its parameters change. Oscillators with extremely stable frequencies of oscillation can be built. Usually these use *piezo-electric crystals*. Such a crystal will deform if a voltage is applied across its opposite faces. Similarly, if it is deformed, a voltage will be produced. The crystal acts as a resonant circuit with very low loss. The resonant frequency of the crystal is determined primarily by its mechanical resonant frequency and is extremely stable. Oscillators which can be constructed using these crystals are very stable. The reader is referred to an electronics circuits book for a discussion of crystal oscillators. When master clocks with very stable frequencies are constructed, they usually use crystal oscillators.

12.7 INTERFACING THE COMPUTER WITH ITS PERIPHERALS

The data calculated by the computer is of no use unless it can be supplied to the user. For instance, an answer stored in a register cannot be read by the programmer. It must be printed out. Similarly, data and programs must be supplied to the computer. The input and output of data involves the interconnection of the computer with peripheral input-output devices. This is called *interfacing* the computer with its peripherals. In Sec. 11-3, we discussed input-output instructions which were written in machine language. We assume that the reader is familiar with this discussion. We shall review its salient points. A typical machine language output command is

$011000d_0d_1d_2d_3d_4d_5f_0f_1f_2$

where the $011000_2 = 30_8$ signifies to the computer that data in the accumulator is to be outputted. $d_0d_1d_2d_3d_4d_5$ indicates the I/O device which is to perform the output and $f_0f_1f_2$ indicates a function which is to be performed.

In this section we shall draw some actual circuits. Each bit $d_0d_1d_2d_3d_4d_5f_0f_1f_2$ will require that a separate wire be drawn. These wires will clutter the diagram. To make it easier to read, we shall assume a much shorter word.

$011000d_0d_1f_0f_1$

Now d_0 and d_1 designate the I/O device and f_0 and f_1 designates the function. In this section, we shall use the following designations

00 teletype (12-44a)
01 printer (12-44b)
10 x-y oscilloscope plotter (12-44c)
11 magnetic tape drive (12-44d)

The following will define functions for the x-y oscilloscope plotter(which will be used as an example)

01 data is x (12-45a)
10 data is y (12-45b)
11 plot data stored in the x and y registers (12-45c)

We shall also assume that the IOR only has four bits. This is small but is again done to avoid cluttering the diagram since now the I/O bus need only have four data lines. It will also have control lines that carry clock pulses.

Now let us consider the consequences of the instruction(see Sec. 11-3 for a more complete discussion)

0110001001

This will cause the number

1001

to be placed on the data lines of the I/O bus. Now, the x-y plotter will be activated and be prepared to receive x variable data. Then, at the next appropriate timing pulse, the x data (the present value in the accumulator)will be placed on the data lines of the I/O bus.

Let us start by considering how the binary digits $d_0d_1f_0f_1$ can enable a particular device and function while inhibiting all others. We shall assume that the data lines of the I/O bus consists of four lines. In addition, we shall assume that two other lines are available for timing pulses.

The data lines of the I/O bus will have bits on them which either represent data or the lines will contain the $d_0d_1f_0f_1$ bits which represent a device address and a function. In order to enable the interface circuit to tell them apart, two different wires carrying timing pulses are used. This is illustrated in Fig. 12-26. There are two clock pulse lines CP_1 and CP_2. The bits on the data lines of the I/O bus which are present when there is a pulse on CP_1 are interpreted as data.

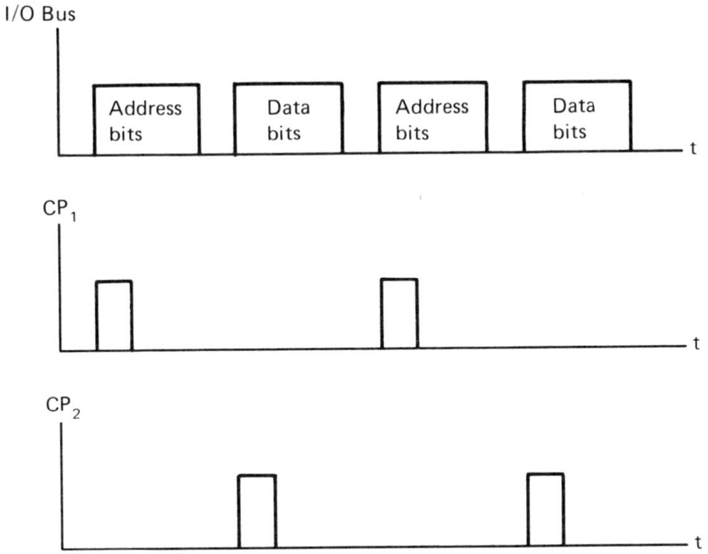

Figure 12-26. The timing pulses of an interface system.

Now let us consider an example of how device decoders operate. Consider Fig. 12-27. This shows four device decoders for the teletype, printer, x-y oscilloscope plotter and tape drive. The dashed lines separate the four decoders. The decoders' combinational logic is essentially the same as Fig. 7-22. Thus, when there are 0's on lines 1 and 2, the output marked TTY is a 1, while those marked Printer, x-y and Mag are 0. When there is a 0 on line 1 and a 1 on line 2, then the output marked Printer is a 1 while the other three outputs are 0. Note that the combinational circuit cannot distinguish between addresses and data on the I/O bus. However, the D flip-flops can only change state when CP_1 is present. Thus, the output of each D flip-flop will correspond to the signal which was on lines 1 and 2 when the last CP_1 pulse was present. For instance, if, when CP_1 is present, line 1 is a 0 and line 2 is a 1, then the output of D_p will be a 1, while the outputs of D_T, D_x and D_M will be 0. These outputs will remain unchanged until the next CP_1 pulse. Thus, the output of the flip-flop will always correspond to the addresses on lines 1 and 2 at the last CP_1 pulse. Now assume that each I/O device has a single input register(we shall subsequently modify this). The data is stored there so that the I/O device can operate on it. The data lines of the I/O bus are connected to the inputs of all of these registers. When a pulse is applied to the CP terminal of the register, then it will store the value that is present on the data lines of the I/O bus; *otherwise its value remains unchanged*. Each register receives its clock pulse through a different AND gate. One input of each of these AND gates is connected to the CP_2 line. The other input is connected to the input of the appropriate flip-flop, see Fig. 12-27. Let us consider an example. The only time that a pulse will be applied to CP_T(the clock pulse terminal of the TTY register)is when a pulse is present on the CP_2 line *and* when the output of D_T is also a 1. Similarly, the only time that a pulse will be applied to the CP_p terminal is when a pulse is present on the CP_2 line *and* when the output of D_P is a 1, etc.

Thus, when a pulse is present on line CP_1, the output of the flip-flop "addressed" by lines 1 and 2 becomes a 1. Next, when a pulse appears on line CP_2, the single "addressed" register sets itself equal to the data on the data lines of the I/O bus. All other registers remained unchanged.

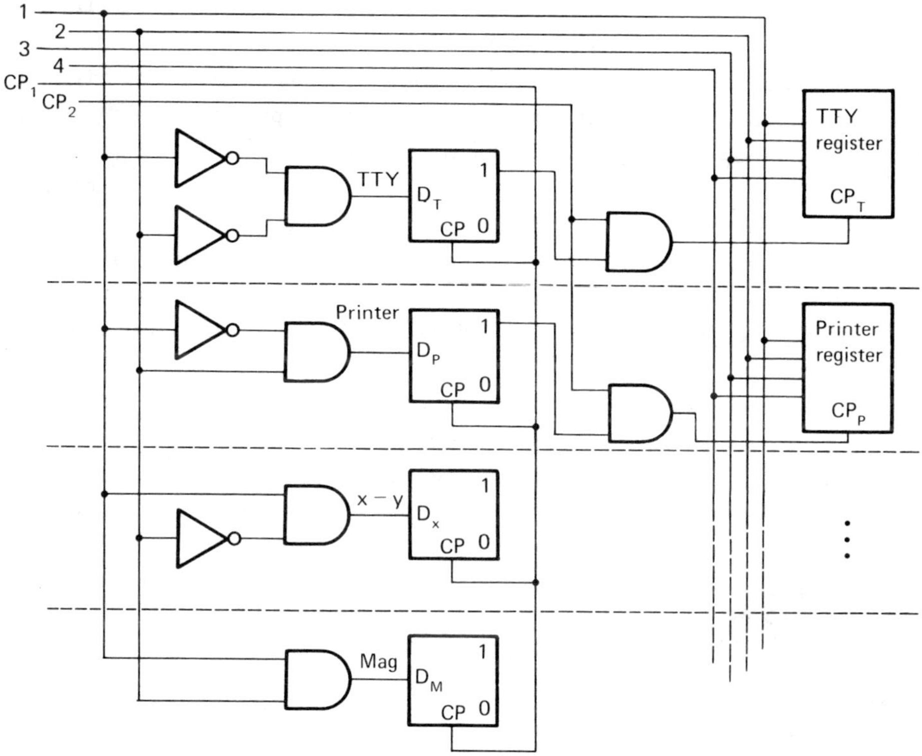

Figure 12-27. An illustration of a device decoder.

Now let us discuss a complete interface for an x-y oscilloscope plotter. We shall assume that an ordinary oscilloscope will be used to display the plot. As an introduction, we shall briefly describe the operation of the oscilloscope. There is a cathode ray tube which displays the curve. An electron gun within the cathode ray tube sends forth a narrow beam of electrons. These strike the face of the cathode ray tube and cause a spot to glow. The oscilloscope has two inputs called x and y. A voltage placed on the x input causes the beam to be deflected in the horizontal direction while a voltage placed on the y input deflects the input in the vertical direction. (To simplify the discussion, assume assume that the oscilloscope is adjusted so that when the x and y input voltages are zero, the beam is in the lower left hand corner of the screen.) Assume also that the x and y data are positive and that positive voltages cause right and upward deflection respectively.

There is a third input to the oscilloscope called the z input. This is used to vary the intensity of the electron beam. Assume that the oscilloscope is adjusted so that, if a positive signal is applied to the z input, then a spot will appear on the screen while, if a negative signal is applied, the screen will be dark. Note that very often we want the image to be made up of a series of dots and we do not want the beam displayed while it moves from point to point. For instance, such oscilloscope displays are often used to present alphanumeric data(e.g.,letters and numbers). Then, the lines made by the path which the beam takes between dots may obscure the character.

Now let us consider the interface for the x-y oscilloscope plotter. A circuit is shown in Fig. 12-28. Here we have modified the device decoder circuit so that we can also do function decoding without using too many elements. The combinational logic is the same as that enclosed in the dashed lines in Fig. 12-27. (i.e., one AND gate and one NOT gate). Now the output of the device decoder of the x-y plotter is applied to one input of each of three AND gates(AND x, AND y, AND z). The second inputs of each of these AND gates are connected to the CP_1 line. Then, AND gates x, y and z will have outputs of 1 only when the address of the x-y oscilloscope plotter is on lines 1 and 2 *and* when CP_1 has a pulse present.

Now consider the function decoder. Its output is in accordance with specifications (12-45). For instance, if line 3 has a 1 and line 4 has a 0, then the output line y of the function decoder will be a 1 while the output on the lines x and z will be 0's. Now consider the flip-flops D_x, D_y and D_z. If the address of the x-y oscilloscope plotter is on lines 1 and 2, then when CP_1 is present, the output of the flip-flops D_x, D_y and D_z will become equal to the values on the x, y and z lines, respectively. *These values will be stored until the next CP_1 pulse.*

Now consider the x and y registers. Their inputs are connected to the data lines of the I/O bus. The CP terminal of the x register is connected to the output of AND x_1 while the CP terminal of the y register is connected to the output of AND y_1. One input of both AND x_1 and AND y_1 is connected to the CP_2 line. The other input terminals of AND x_1 and AND y_1 are connected to the outputs

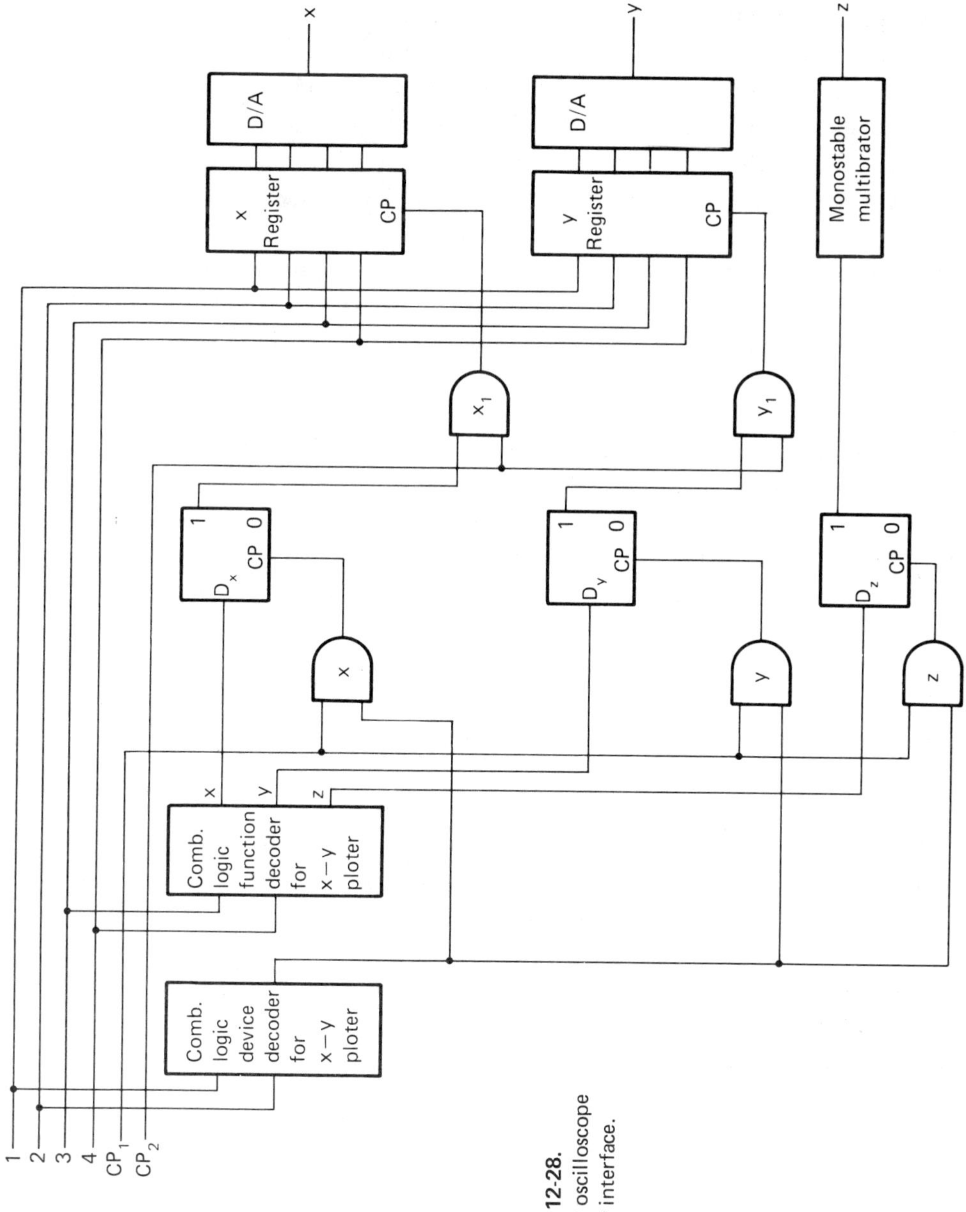

Figure 12-28.
An x-y oscilloscope plotter interface.

of D_x and D_y, respectively. Thus, if during the pulse of CP_1, 1001 is on the data lines of the I/O bus, then, after CP_1 there will be a 1 applied to the upper input terminal of AND x_1 and a 0 applied to the upper input terminal of AND y_1. Hence, when the next pulse of CP_2 occurs, the x register will be set to the value on the data lines of the I/O bus and the y register will remain unchanged. Similarly, if, during the pulse of CP_1, we had 1010 on the data lines of the I/O bus, the y register would be set to the value on the data lines of the I/O bus at the next CP_2 pulse and the x register would remain unchanged. The x and y registers are each connected to digital to analog converters. These produce an output voltage proprotional to the binary number stored in the x and y registers, respectively. These voltages can then produce the desired deflection of the spot on the face of the oscilloscope.

The output of D_z is connected to a monostable multivibrator that produces a pulse similar to that of Fig. 12-24c. Thus, whenever the address 1011 is present simultaneously with a pulse of CP_1, the monostable multivibrator will be triggered. The resulting positive z output results in a dot on the screen of the oscilloscope. It is assumed that the length of the pulse is such that the dot will fade before the next pulse of CP_1. Note that after a "z input", the data is ignored. That is, any data on the data lines of the I/O bus at the next pulse of CP_2 produces no effect.

Thus, in three consecutive pulses of CP_1 we can set the x value, set the y value and, finally, display the dot. If a curve is to plotted, then this sequence must be repeated for each point.

Let us consider a program which can be used to plot a curve. We shall use the assembler language mnemonic OUT to replace the digital code 011000. A typical program segment is

```
⋮  calculation of x value and storage in accumulator
OUT1001
⋮  calculation of y value and storage in accumulator
OUT1010
OUT1011
```

That is, the x value is calculated. Then, it is outputted to the x register of the x-y oscilloscope plotter interface, using the instruction OUT1001. Next, the value of y is calculated and it is outputted to the y register of the x-y oscilloscope plotter interface using the instruction OUT1010. Then, the beam is turned on using the instruction OUT1011.

The program segment results in the plotting of one point. Such a segment would have to be repeated for each point. Usually, the segment would not be written over and over, but looping, see Table 11-5, would be used. We assume that the computer is fast enough so that successive points would be plotted so rapidly that it would appear as though a continuous curve were being traced. (If this were not the case, then a pen type of x-y plotter can be used.) We also assume in this circuit that the times between pulses of CP_1 and CP_2 are long enough so that all circuits can change their states and that the plotter processes the data before the next data arrives.

12.8 SOME DIGITAL SYSTEMS

In this section we shall discuss some applications of digital devices. This shall be a qualitative discussion whose object is to illustrate a few of the myriad applications of digital devices. For instance, digital systems can be used to control lathes, drill presses, and other machines. Traffic flow on highways can be regulated by digital systems. Television signals can be transmitted using digital means. These are only a few examples of digital systems. Let us consider two of these.

A Traffic Control System

Suppose that we consider a busy highway which runs through a city. At rush hours, the highway becomes congested and slow moving. At such times, it would be desirable to restrict the number of cars entering the highway. Let us discuss a simple digital system that could be used to control the traffic flow. For instance, consider the highway section shown in Fig. 12-29.

There are sensors at points a_1 and a_2. Each time that a car passes these points, a pulse is generated. These pulses can be used to determine the total number of cars passing point a_2. In addition, since the distance between points a_1 and a_2 is known, the pulses can be used to compute the *average* velocity of the cars in the section between points a_1 and a_2. If the traffic flow past point a_2 becomes too great, then traffic jams will occur at subsequent portions of the highway if additional traffic is allowed to enter at the two entrance ramps. A traffic control system could sense the volume of the traffic flow at point a_2 and its average speed between points a_1 and a_2. This data could then be used to close the gates at the entrance ramps at appropriate times, preventing traffic from entering the highway below point a_2. On the other hand, if the volume of traffic exiting the highway through the two exit ramps is large enough, then it may be possible to keep the entrance ramps open even if the volume of traffic passing point a_2 is high. In general, the decision as to whether or not the entrance ramps should be closed depends upon a complex interrelation of all the measured data and the characteristics of the highway. In fact, relatively complex calculations may have to be made. If these are to be made rapidly, a computer type circuit such as that discussed in Chapter 11 must be used, to determine when the gates should be opened or closed.

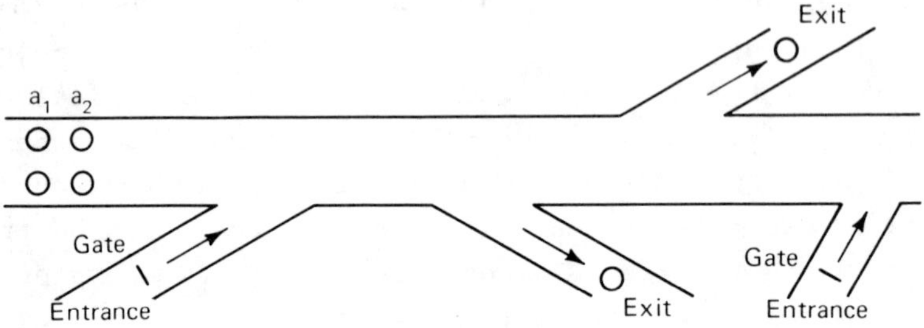

Figure 12-29. A section of highway. The circles are automobile sensors.

The sensors embedded in the roadway emit a pulse each time that a car passes. They can be activated by the pressure of the cars or they can be magnetic pick ups. The pulses are supplied as data to the computing circuit. Then, such quantities as average velocity and number of cars passing a point in a predetermined time interval can be calculated. A computation is performed and a signal is sent to the entrance ramp gates. Then, the computer updates its data and repeats the calculations.

Note that the calculations are not based on simple subtraction. For instance, the exit ramps may be some distance from point a_2 so that the data evaluated at different times may be compared. Thus, storage is needed. Actually, traffic flow is a complex problem and relatively sophisticated programs are used to direct the computer. In addition, other types of sensors can be used to obtain pertinant data. For instance, we must consider amount of day light and weather conditions. Sensors which detect light level, temperature and roadway moisture can also supply inputs to the circuit.

More elaborate systems have sensors such as Doppler radar which can detect the speed of individual cars rather than an average speed. When a car travels too slowly on a high speed highway, it is dangerous. Such sensors can be used, in conjunction with the highway computer to turn on hazard warning signal and to close appropriate entrance ramps when a dangerous situation occurs.

Digital Communication Systems

Let us now consider a communication system that functions digitally. Actually, the earliest communication systems were digital. For instance, the presence or absence of a puff of smoke could be represented by a 0 or a 1.

However, before the advent of large scale digital systems, voice or television signals were always transmitted by analog means. For instance, a voltage such as that of Fig. 12-13a could represent the signal, or, the waveform of v(t) would represent the sound levels that we want to transmit.

Suppose that such a v(t) is the output signal of a telephone. If the conversation is to be sent over many thousands of miles, ti becomes *attenuated*. For instance, suppose that the signal is transmitted along a coaxial cable. The signal strength will decrease exponentially as the distance from the transmitter increases. All devices generate random electric fluctuations called noise. If the signal level becomes small enough, the noise signals become comparable in magnitude with the signal, and begin to obscure it. If the ratio of the signal to the noise is not sufficiently large, the noise can render the signal unintelligible.

To prevent this from happening, amplifiers called *repeaters*, which increase the magnitude of the signal, are placed about every 50 miles, see Fig. 12-30. If the total length of the transmission system is long, then many amplifiers may have to be used. Suppose that the amplifier introduces some nonlinear distortion. Even if each amplifier only introduces a little distortion, the net contribution of all the amplifiers can produce enough distortion to make the signal unintelligible. Thus, very low distortion amplifiers must be used in such systems.

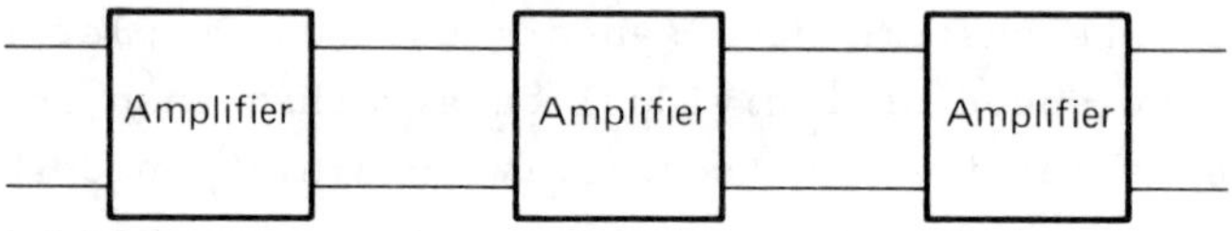

Figure 12-30. A transmission line with repeater amplifiers.

Now let us consider a digital system. Again, suppose that v(t) represents a telephone signal which contains frequency components up to 5000 Hz. It can be shown that this signal can be completely reconstructed by sampling it each 10^{-4} sec.(1/5000 x 2). Thus, if the signal is applied to a sample and hold circuit, see Sec. 12-3, a sampled signal such as that of Fig 12-13b results. If this signal is supplied to an analog to digital converter, see Sec. 12-4, which is clocked once for each sample, a sequence of binary numbers, equal to the quantized sampled values, will be obtained. Such signals can be transmitted down the transmission line of Fig. 12-30. Now, nonlinear distortion is no longer a problem since we are no longer concerned with the amplitude of a signal but only with its presence or absence (i.e., a 0 or a 1).

At the receiving end, the signal is supplied to a digital to analog converter which restores the signal. Such a system is shown in Fig. 12-31. The transmission channel represents the transmission line and amplifiers. Actually, the digital to analog converter produces a signal which is of the form shown in Fig. 12-13b. The output low pass filter is used to smooth out the signal (i.e., to remove the "steps").

Digital transmission systems are not only used because of nonlinear distortion in the amplifier. There are many other important reasons for their use. For instance, noise can be a problem in a digital as well as in an analog system. That is, the noise may make a 0 appear as a 1 and vice versa. However, special redundant codes can be used which reduce the effects of noise. For instance, we can repeat each pulse 5 times and then use a majority rule to determine if it is a 0 or a 1. For example, 10010, 00000, and 11000 all would represent a single 0 since three or more of their digits are 0's, see Prob. 6-49. This code reduces the likelihood of noise producing an error. There are more sophisticated noise reducing codes. One advantage of digital systems over analog ones is that such techniques can more easily be applied to reduce the effects of noise. The error correcting codes of Sec. 8-4 are an example of such redundant codes.

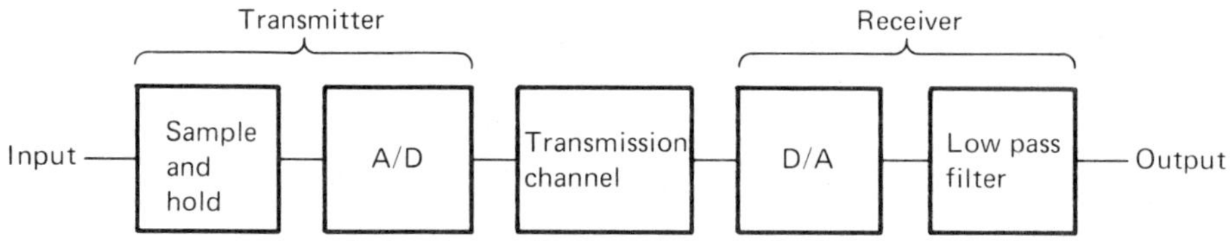

Figure 12-31. A digital transmission system.

Digital systems can be easily multiplexed(i.e., many signals can be transmitted simultaneously). For instance, consider Fig. 12-32, which represents a signal sampled every 10^{-4}sec. The crosshatched region represents the time range when the digital information is transmitted. Suppose that the bits that represent the sampled value can be transmitted in τ seconds where, in this case, $\tau = 10^{-5}$sec. Then, see Fig. 12-32, for much of the time, no signal will be transmitted. Let us call these times "empty times". Now suppose that a second signal were sampled. Its digital data could be transmitted during the empty times. Thus, both signals could be sent over the same channel. Then, the pulse for the original signal would be sent during the time durations

$$n \times 10^{-4} < t < n \times 10^{-4} + \tau \qquad n = 0,1,2,\ldots$$

The second signal's digital representation would be transmitted during the time durations

$$(n+0.5) \times 10^{-4} < t < (n+0.5) \times 10^{-4} + \tau \qquad n = 0,1,2,\ldots$$

Thus, these pulses would occur midway in time between the original ones.

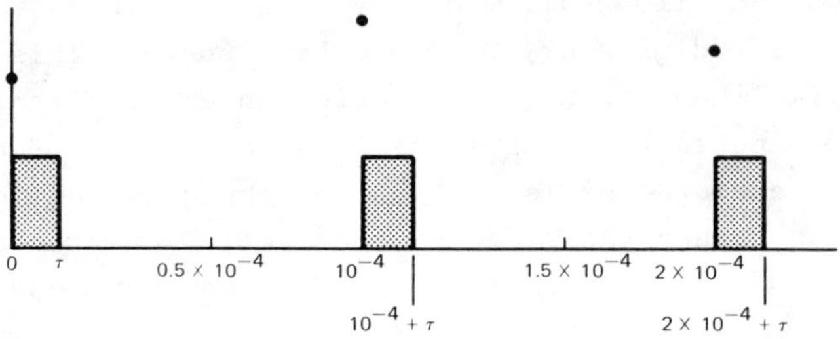

Figure 12-32. An illustration of the time required to send digital data. The digital data is transmitted during the time intervals shown shaded.

Now there would be two receivers, one for each signal. Each receiver would be gated by a sequence of clock pulses so that it would only receive its "own" signals.

Actually, this procedure can be extended so that many signals can be simultaneously transmitted over the same channel.

For the reasons we have discussed, digital transmission is becoming increasingly important. It is expected that, in some areas, it will eventually replace analog transmission entirely.

BIBLIOGRAPHY

Chirlian, P.M., "Electronic Circuits: Physical Principles, Analysis and Design", Sec.14-6, Chapter 15, McGraw-Hill Book Co., New York, 1971

Kohonen, T., "Digital Circuits and Devices", Chapter 8, Prentice Hall, Inc., Englewood Cliffs, N.J., 1972

Millman, J. and Halkias, C.C., "Integrated Electronics:Analog and Digital Circuits and Systems", Chapters 15-17, McGraw-Hill Book Co., New York, 1972

O'Malley, J., "Introduction to the Digital Computer", Chapter 12, Holt, Rinehart and Winston, Inc., New York, 1972

Rhyne, V.T., "Fundamentals of Digital System Design", Chapter 11, Prentice Hall, Inc., Englewood Cliffs, N.J., 1973

Sifferlen, T.P., and Vartanian, V., "Digital Electronics with Engineering Applications", Chapters 5 and 7, Prentice Hall, Inc., Englewood Cliffs, N.J. 1970

Strauss, L., "Wave Generation and Shaping", Second Ed., McGraw-Hill Book Co., New York, 1970

PROBLEMS

12-1. An operational amplifier has a gain A_v = 5000. Its maximum output voltage is ± 5 volts. What is the value of V_α? Assume the idealized curve of Fig. 12-3 can be used to represent the input-output characteristics of the amplifier.

12-2. The operational amplifier of Fig. 12-4 has a voltage gain A_v=5000, R_1=10,000 ohms and R_2=50,000 ohms. If V_o = 2 volts, what are the values of V_i and V_1?

12-3. The voltage follower of Fig. 12-6 is to be such that $V_o = (1-\delta)V_2$ where $0 < \delta < 0.01$. What is the minimum value that A_v can have if δ is to meet these specifications?

12-4. Design a circuit whose output is $v_o = 10\, v_1 + 20\, v_2 + 5\, v_3$, where v_1, v_2 and v_3 are continuously varying voltages.

12-5. The comparator of Fig. 12-9 is such that $V_C = -V_D = 5$ volts, see Fig. 12-8. Now consider the operational amplifier. If $V_1 > V_R + 0.001$, then $V_o = -V_A$; if $V_i < V_R - 0.001$, then $v_o = V_A$. What is the minimum value that A_v can have?

12-6. Discuss the advantages and disadvantages of using a Schmitt trigger as a comparator.

12-7. The Schmitt trigger of Fig. 12-11 has the following parameter values

V_A = 5 volts
R_1 = 15,000 ohms
R_2 = 100 ohms
V_{R1} = 2.5 volts
V_α = 0.002 volt

where V_A is the saturation voltage of the operational amplifier. Draw the input-output characteristics for this device.

12-8. Discuss the operation of the sample and hold circuit.

12-9. Discuss the need for quantization when analog signals are processed by a digital computer.

12-10. Modify the circuit of Fig. 12-15 so that an 8 bit binary number can be used. The maximum value of v_o is to be 1 volt. What should be the relation among V_m, R_1 and R_2? Assume that the operational amplifier does not saturate.

12-11. Redraw the circuit of Fig. 12-15 using electronic switches.

12-12. Repeat Prob. 12-10 for the circuit of Fig. 12-17.

12-13. Repeat Prob. 12-11 for the circuit of Fig. 12-17.

12-14. Design a 12 bit 8421 BCD to analog converter.

12-15. Draw the complete circuit for a four bit analog to digital converter using a counter.

12-16. Discuss the effect of quantization of analog to digital and digital to analog conversion. Why should both the least significant bit and the next one represent very small quantities?

12-17. Design an eight bit successive approximation analog to digital converter.

12-18. The astable multivibrator of Fig. 12-22a is to produce a 1 MHz square wave. If $R_1 = R_2 = 10{,}000$ ohms and $R = 20{,}000$ ohms, what should be the value of C? Assume $V_\alpha = 0$.

12-19. If, in Prob. 12-18, we want to double the frequency by changing R_2, what must its new value be?

12-20. Design a monostable multivibrator whose pulse length $T = 1 \times 10^{-3}$ sec. Assume that $V_A = 5$ volts and that $V_\alpha = 0$.

12-21. Describe, in your own words, the operation of the x-y oscilloscope plotter interface.

12-22. Design an interface which "connects" the computer with a pen x-y plotter. Assume that the plotter has a pen whose x and y positions are proportional to two input voltages. (If the inputs are zero, then the pen will be in the lower left hand corner of the page.) The input data can be assumed to be positive. There are two commands used to drive the pen. One is pen up, which lifts the pen from the paper. The other is pen down. Write the necessary machine language instructions.

Appendix — Semiconductor Devices

In this appendix, the basic physical properties of the semiconductor devices used in computer circuits will be described. We shall start with an elementary discussion of semiconductors. Next, the basic ideas of diodes, transistors, field effect transistors and MOS devices will be presented. The switching properties of these devices will be given. This appendix will conclude with a discussion of integrated circuits.

A-1 BASIC SEMICONDUCTOR PROPERTIES

In this section we shall discuss some basic ideas of semiconductor physics. In subsequent ones, we shall consider devices such as diodes and transistors, that are used in switching circuits. This discussion will be brief. Its purpose is to provide the reader with some insight into semiconductor circuits which will allow us to consider both the analysis and design of switching circuits.

Semiconductor Materials

Let us consider a crystalline structure, that is, one whose atoms are arranged in a regular array. Semiconductor crystals are, in general, made up of elements with a valence of 4. The valence electrons of an atom interact with the adjacent atoms. This is called a *covalent bond*. Such bound electrons are not free to wander throughout the material. Conduction can take place when these electrons break free from their covalent bonds and are free to travel through the crystal. The atomic structure of a crystal with four valence electrons is symbolically shown in Fig. A-1.

In order for the valence electrons to break their covalent bonds, they must acquire energy from an external source. To study this, we must consider energy in solids. A single isolated atom is said to possess a set of *discrete energy states*. That is, the electrons in the atom cannot possess just any arbitrary energy. They can only have the allowable energy of these states. Also, no more than two electrons of the atom can have the same energy state. There is a fixed number of states of specific energy levels. If two atoms are brought together, the energy levels of each shift slightly. Thus, each single level now becomes two *different* (but close) levels. Thus, in the two atom system, we have doubled the number of allowable energy levels. Then, in the two atom system, we have pairs of closely spaced allowable energy levels. These are separated by gaps of *forbidden energy levels*. Each time that another atom is brought into close proximity, the energy levels split again. Thus, in a solid where there is an extremely large number of atoms, the energy levels split so often that there are regions containing a great number of very closely spaced energy levels. It is often convenient to think of this as a *continuous band of allowable energy levels.*

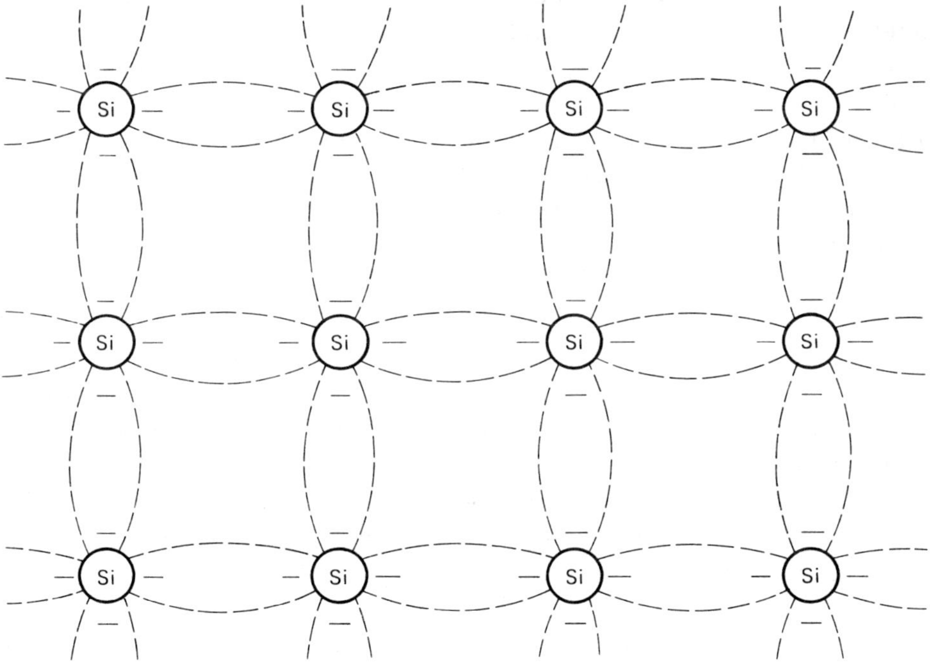

Figure A-1. A diagrammatic representation of the silicon crystal structure. Each of the four covalent electrons of an atom is shared with an adjacent atom.

The allowed energy bands are separated by prohibited energy bands. That is, there are regions of disallowed energy. In an insulator, or in a semiconductor at a temperature of absolute zero($T=0^{o}K$), the lowest allowed energy bands are filled. That is, every discrete state in the lower bands is occupied. This situation is illustrated in Fig. A-2a. Note that all the allowable bands are either completely filled or completely empty. The highest filled band is called the *valence band.* The next higher band is called the *conduction band.*

Now suppose that we try to obtain a current by applying an external voltage which sets up an electric field to accelerate the electrons. If an electron is accelerated, it must gain energy. Since the lower allowable levels are filled, it must gain at least a few electron volts of energy to reach the next allowable band if it is to be accelerated. (An electron volt is the amount of energy gained by an electron when it is accelerated through a potential difference of one volt.) It may seem as though this energy could be obtained easily if a voltage of several volts were applied. However, the electron must gain this increase in energy while being accelerated over distances approximately equal to the atomic spacing of the lattice. This would require the application of extremely large electric fields or, equivalently, in a material of reasonable dimensions, of extremely large voltages. Thus, at a temperature of absolute zero, under ordinary circumstances, there will be no current in an insulator.

Now consider Fig. A-2b, which represents the energy levels for a conductor at absolute zero. The valence and conduction bands are combined. That is, the valence electrons are in the conduction band which is only partially filled. Now, if an electron is accelerated, it need only gain the extremely small amount of energy between allowed levels. Thus, small applied voltages can result in conduction in a conductor.

Suppose that the temperature of an insulator is raised above absolute zero. The electrons will be excited thermally. At room temperature, some of them will attain enough energy to be raised into the conduction band. This is equivalent to saying that they have acquired sufficient energy to break their covalent bonds. They now become free electrons. Now, if a small voltage is applied, these free electrons can be accelerated by it and a current results. The current is small since only a(relatively)few electrons have sufficient energy to be raised into the conduction band.

In an insulator, the gap between the conduction and valence bands is relatively large(i.e., several electron volts). Thus, there are relatively few electrons whose energies lie in the conduction band. On the other hand, the energy barrier in a semiconductor is smaller(i.e., about one electron volt or less). Thus, at room temperature many more electrons can be raised into the conduction band. Hence, in a semiconductor, an applied voltage will result in much greater currents than it does in an insulator, although the current will still be very much less than in a conductor. Note that the energy gap difference between a semiconductor and an insulator does not appear large. However, the energy densities of the charges vary in an inverse exponential fashion. Thus, a small change in gap width can make a very large change in the densities of the free electrons.

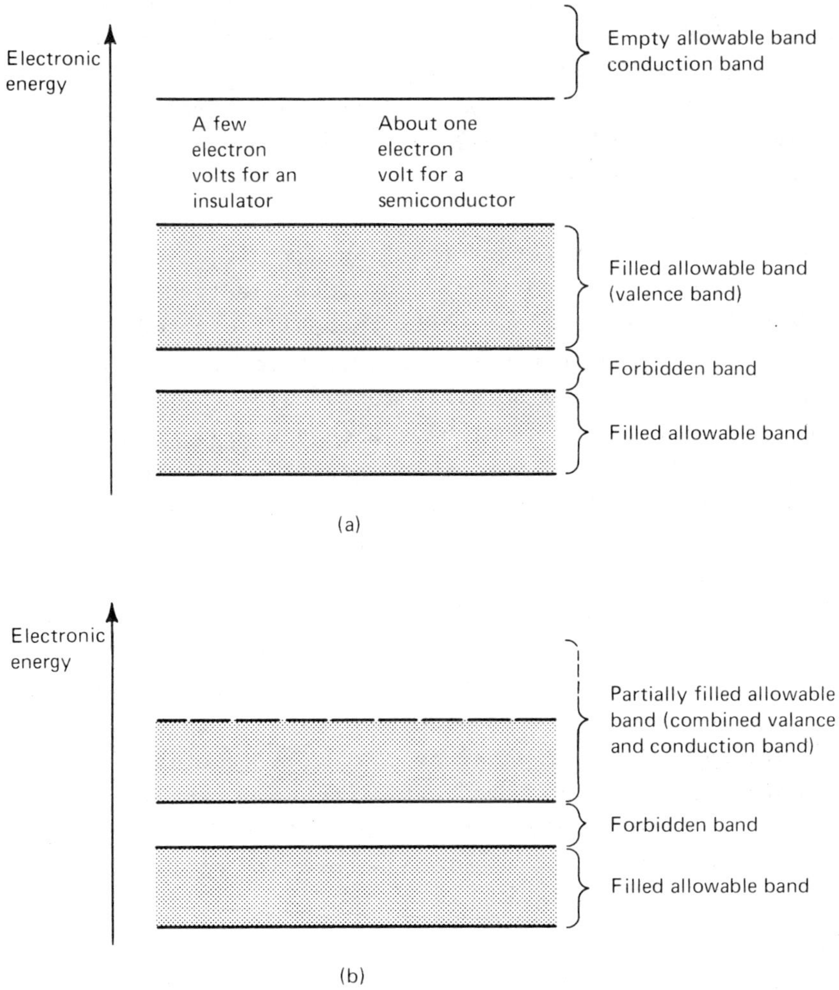

Figure A-2. Energy levels in a crystal at T = 0 ° K. (a) an insulator or semiconductor; (b) a conductor.

When the electrons in a solid are accelerated by an electric field, they do not gain velocity continuously. The free electrons interact with the nuclei of the atoms if they are not perfectly positioned in the crystal lattice. These interactions are spoken of as collisions and they tend to scatter the direction of the electrons as well as extract energy from them. On the average, when a potential is applied, the free electrons drift at an average(or drift)velocity under the influence of the electric field. Because of thermal effects, the atoms vibrate. These vibrations tend to increase the number of scatterings. In a conductor, an increase in temperature thus results in an increase in resistance. However, in a semiconductor or an insulator, this effect is more than offset by the increase in free electrons. Thus, the resistance of a semiconductor or of an insulator decreases with temperature.

We have thus far considered that, in a semiconductor or an insulator, conduction is carried on by free electrons. There is another mode. When an electron is thermally excited from the valence band into the conduction band, it leaves a vacant level in the valence band. It is possible for a valence electron from an adjacent atom, which does not have enough energy to move into the conduction band, to move into this vacant level. The vacant level thus moves to the adjacent atom. In this way, the vacant level can move throughout the crystal. This vacancy in the level is called a *hole*. Note that holes contribute to conduction. That is, as a hole moves to the left, there is a net motion of a free electron to the right. The motion of a hole throughout the crystal is actually the motion of many free electrons. However, this motion can be analyzed by considering that a hole is a positive charge with a mass about equal to that of a free electron. In a pure semiconductor, each time that a free electron is created, it leaves behind a hole. Thus, the number of free electrons and holes is equal. Such pure semiconductors are called *intrinsic semiconductors*.

A free electron can be captured by a vacancy in a covalent bond. That is, a free electron and hole can combine. This eliminates both the free electron and the hole. In a semiconductor, free electrons and holes are constantly recombining. However, other free electron-hole pairs are also constantly being thermally generated. Thus, there is an equilibrium density of free electrons and holes. These are designated by n and p. Their dimensions are number of charges/meter^3. Note that free electrons and holes have an *average lifetime*. This is, on the average, the time between the thermal creation of a free electron (hole)and the time that it recombines.

Impurities in Semiconductors

If very small quantities of impurities take the place of some of the atoms in a semiconductor crystal, the electrical properties may be greatly altered. This effect leads to useful electrical properties. The addition of impurities is called *doping*.

Donors

Suppose that a small quantity of an impurity which has a valence of 5 is introduced into a silicon crystal. Assume that this is done in such a way that these atoms will replace some of the silicon atoms in the lattice structure. The impurities have 5 valence electrons. Four of these will be closely bound to their neighbors in covalent bonds, see Fig. A-1. However, the fifth valence electron does not enter into this type of bond, and it is only weakly held to the atom. At room temperature, almost all such electrons acquire sufficient energy to become free electrons. The impurity atom which has supplied this free electron is called a *donor* atom since it has supplied a free electron.

When a donor atom supplies a free electron, a vacancy in a covalent bond is *not* created. Thus, a hole is not created when the free electron is produced. Instead, a *fixed positively ionized atom* is left behind. This ion does not enter into the conduction process since it is an atom of the crystal and, thus, fixed in place.

In semiconductor devices, the amount of impurities used is extremely small. For instance, it is measured in parts per million. However, at usual operating temperatures with typical impurity densities, the number of free electrons from donor electrons far exceeds the number of free electron-hole pairs created intrinsically. Thus, the density of free electrons far exceeds the number of holes. For this reason, such a semiconductor material is called an *n-type* semiconductor. In n-type material, the free electrons are called *majority* charge carriers while the holes are called *minority* charge carriers.

Acceptors

Now suppose that an impurity with a valence of 3 is added to a pure semiconductor. Each impurity atom can only form three covalent bonds. Thus, a valence electron from an adjacent atom will require very little energy to enter into the vacant covalent bond. Thus, a hole will be created. At room temperature, almost all acceptor atoms will cause holes to be created. Note that when a hole is formed in this way, a free electron is *not* created. Instead, a *fixed, negatively ionized atom* results.

For usual impurity densities, the number of holes far exceeds the number of free electrons and such material is called *p-type*.

In a p-type semiconductor, the *holes* are the *majority carriers* while the *free electrons* are the *minority carriers*.

In general, if a free electron encounters a hole, the two will recombine. Suppose that the material is n-type, then the number of free electrons will be much greater than the number of holes. The probability of a hole's being eliminated by combining with a free electron thus increases greatly. Thus, in n-type material, the hole density will be much less than it would be in intrinsic material. The hole lifetime is thus reduced. The converse effect occurs in p-type material.

It can be shown that the product of the free electron density and the hole density is independent of the impurity concentration. For instance, if n_i and p_i are the densities in the intrinsic materials, and n and p are the densities in the material after it is doped, then, assuming that the temperature is constant,

$$np = n_i p_i \tag{A-1}$$

For instance, in n-type material, if doping causes n to increase by a factor of 10, then p will decrease by a factor of 10.

A-2 p-n JUNCTION DIODES

By combining p-type and n-type regions in a single crystal, many important electronic devices can be obtained. We shall consider a basic one here. Assume that we fabricate a semiconductor crystal with p-type and n-type regions as shown in Fig. A-3a. The crystal is electrically neutral. However, the n-type region has very many more free electrons that holes and the p-type region has very many more holes than free electrons. (Note that there are fixed ionized atoms in each region.) Because of thermal agitation, the free electrons and holes will wander at *random* throughout the crystal. Such random motion is termed *diffusion*.

At the start, there are many more free electrons in the n-type region than in the p-type region. Thus, there will be a net movement of free electrons from the n-type region into the p-type region. Thus, the n-type region will acquire a net positive charge. Similarly, the p-type region will acquire a net negative charge. There is also a net flow of holes from the p-type region to the n-type region. This aids the effect just discussed.

These charged regions will set up an electric field. Its direction will be such that it opposes the flow of free electrons from the n-type to the p-type region and the flow of holes in the opposite direction. That is, the potential sets up a barrier and only those free electrons in the n-type region and those holes in the p-type region which have sufficient thermal energy to surmount the barrier can move from one region to the other. Finally, an equilibrium condition is established where the free electrons from the n-type region flowing into the p-type region equals the number of free electrons from the p-type region flowing into the n-type region. Note that free electrons flowing from the p-type region to the n-type region do not have to surmount a potential barrier. The electric field there actually sweeps them into the n-type region. However, there are far fewer free electrons in the p-type region than in the n-type region. Thus, the equilibrium in the flow of the free electrons is reached. A similar discussion applies to the holes.

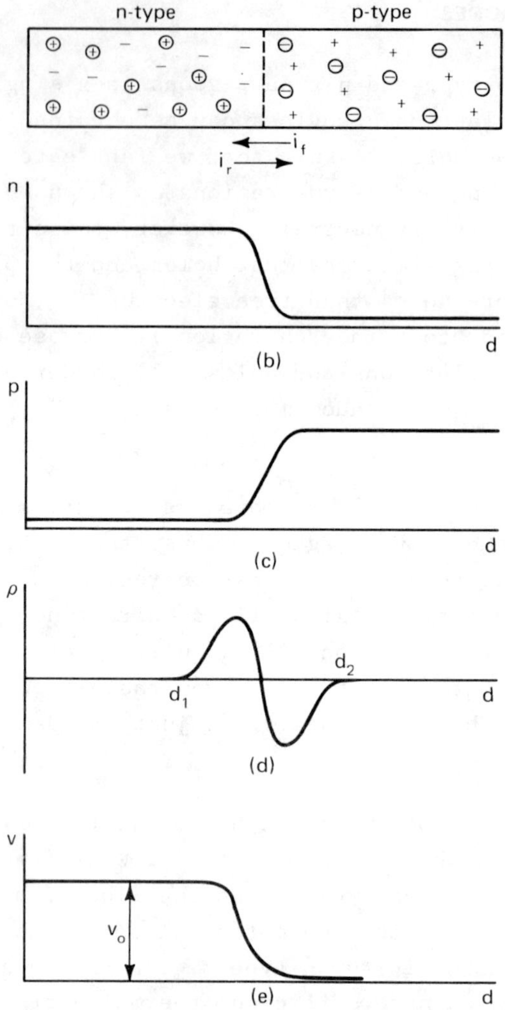

Figure A-3. The p-n junction. (a) diagrammitic representation; (b) free electron density, (c) hole density; (d) net charge density; (e) potential distribution.

Note that we have discussed two types of charge carrier motion. Diffusion is due to random motion. On the other hand, an electric field can cause the charge carriers to move in a specific direction. This is termed *drift*.

The electric field between the n-type and p-type regions tends to drive free electrons in the n-type region and the holes in the p-type region away from the junction. (There are the majority carriers in their respective regions.) Thus, there is a region on either side of the junction which is almost free of mobile charge carriers. This is called a *depletion region*. Since there are no mobile charges to neutralize the fixed donor and acceptor ions, there is a net charge density in the depletion region. This is illustrated in Fig. A-3d. The depletion region lies between d_1 and d_2 there. It can be considered that these fixed depletion layer charges set up the electric field and resulting potential, see Fig. A-3e, across the junction. The number of positive charges on one side of the depletion region is equal to the number of negative charges on the other side.

We can speak of two components of current in a p-n junction as shown in Fig. A-3a. One, from right to left, is composed of holes flowing from the p-type region to the n-type region and free electrons flowing in the opposite direction. This is called the *forward current*. The other component, called the *reverse current*, is composed of holes flowing from the n-type region to the p-type region and free electrons flowing in the reverse direction. Note that the forward current charges must surmount the potential barrier. The charge carriers constituting the reverse current "fall down" the barrier. If the p-n junction is not connected to any circuit, then, at equilibrium

$$i_f = i_r \qquad \text{(A-2)}$$

The p-n junction actually constitutes a device called the *p-n junction diode*. It is very important and we shall now consider its electrical properties. Assume that a p-n junction is connected to a battery as shown in Fig. A-4a. The polarity of the battery is such that it opposes the potential across the depletion layer. The net effect is that the depletion layer potential is reduced by the applied voltage. The distribution of thermal energies of the free electrons and holes is such that even a small decrease in the barrier results in a very large increase in i_f. Note that i_f is made up of charge carriers that *diffuse* across the barrier.

The reverse current is made up of carriers that diffuse to the barrier and then are swept across it. Thus, changing the barrier height will not change i_r substantially. The net current is $i_f - i_r$. Thus, applying a voltage of the polarity shown in Fig.

A-4a results in a very large current i_b. The diode acts as a very low resistance. A voltage of this polarity, which makes the p-type region positive with respect to the n-type region is called a *forward bias*.

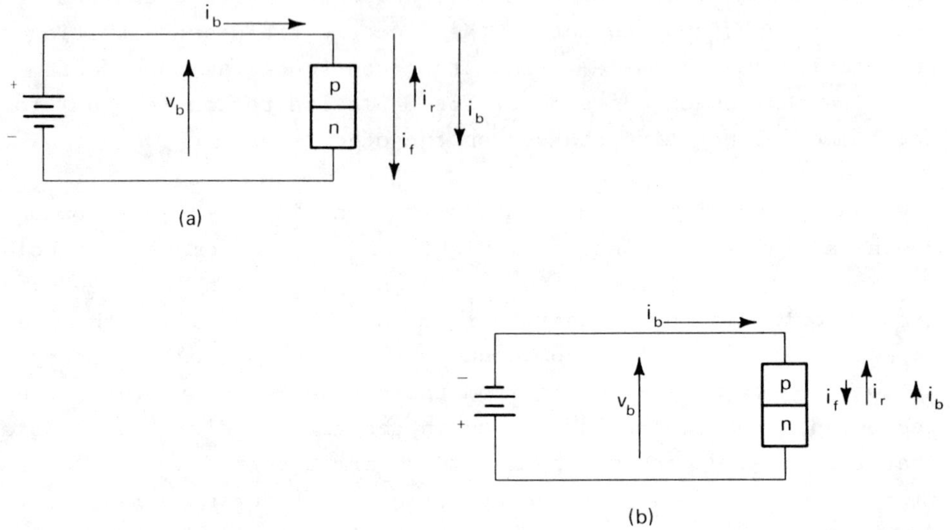

Figure A-4. (a) A forward biased p-n junction; (b) a reverse biased p-n junction.

If, on the other hand, the polarity of the applied voltage is reversed, see Fig. A-4b, then the height of the barrier is increased. This is termed a *reverse bias*. Now i_f will be reduced. As before, i_r will be essentially unchanged. The current is $i_b = i_f - i_r$. Thus, the largest that the current can become in the reverse direction is $-i_r$. In general, this is a very small value. Thus, the reverse biased diode acts as a very large resistance.

The p-n junction diode will, then, conduct a great deal of current in one direction but little in the other. An expression for the current can be found to be

$$i_b = I_s(e^{v_b/\eta kT} - 1) \tag{A-3}$$

where

k = Boltzmann's constant = 8.61×10^{-5} ev/^{o}K
T = temperature in ^{o}Kelvin
η = adjustment constant $\eta = 1$ for germanium
$\eta = 2$ for silicon
I_s = reverse saturation current = $-i_r$

Note that v_b is positive for forward bias and negative for reverse bias. If v_b is negative, the largest magnitude that i_b can have is I_s, which is small. However, if v_b is positive, then i_b can become very large.

The reverse saturation current $I_s = i_r$ depends on the thermally generated free electron-hole pairs and is very temperature dependent.

The symbol used for the p-n junction diode is shown in Fig.A-5. The arrow i_b points in the direction of forward current.

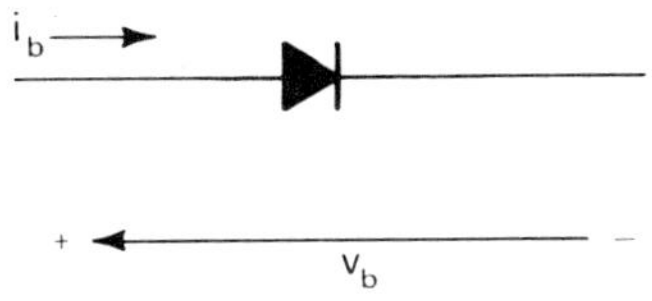

Figure A-5. The symbol for the p-n junction diode.

Equation (A-3) is an idealization and the actual current often differs from it somewhat. For this reason, measured curves of voltage versus current are often given. A typical one is shown in Fig. A-6. The reverse current is so small that is appears as essentially zero on the graph. If the reverse bias is made large enough, a large increase in current occurs. This is termed *reverse breakdown*. In some applications such as voltage limiters and regulators, this reverse breakdown is useful. However, in most applications, the breakdown region should be avoided.

When the diode is forward biased, the current increases exponentially with the applied voltage. There are values of voltage for which the current remains very small. It is convenient to think of the diode as an open circuit in this case, even though a small forward bias is applied. This is shown in Fig. A-6. We can assume that the current is essentially zero until $v_b = 0.6$ volts. This voltage is called the *cut in* voltage. It is typically 0.6 volts for silicon diodes. The symbol used for cut in voltage is V_γ

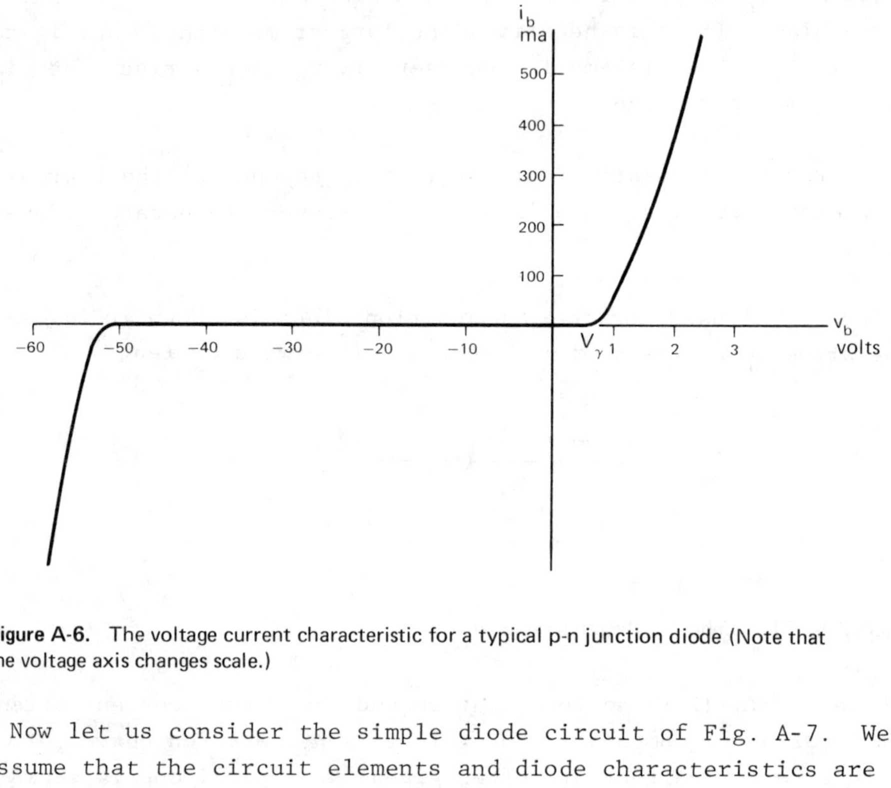

Figure A-6. The voltage current characteristic for a typical p-n junction diode (Note that the voltage axis changes scale.)

Now let us consider the simple diode circuit of Fig. A-7. We assume that the circuit elements and diode characteristics are known and we want to find v_b and i_b. The diode characteristics constitute one equation relating the two unknowns. To determine another equation we apply Kirchhoff's law to the circuit. This yields

$$v_b = V_{BB} - i_b R_L \tag{A-4}$$

This is the equation of a straight line, called the *load line*. It is plotted in Fig. A-7b. The intersection of the characteristics and the load line gives the actual currents and voltages i_{b0} and v_{b0}.

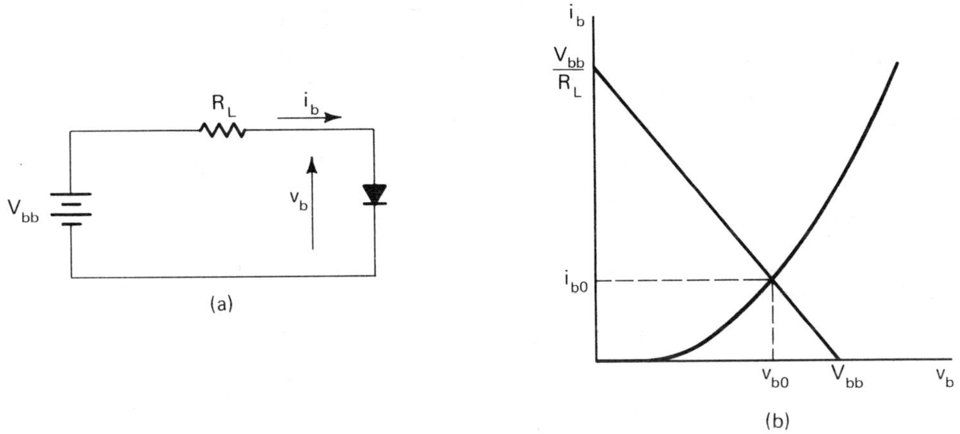

Figure A-7. (a) A simple diode circuit; (b) characteristics and load line for this circuit.

At times, when we deal with diodes, it simplifies the analysis, and is accurate as well, to consider them as ideal elements. In this case, the diode acts as a short circuit for current in the forward direction and as an open circuit in the reverse direction.

When a diode is used in a switching circuit, factors other than those discussed here are also important. We shall discuss them in Sec. A-5.

A-3 THE JUNCTION TRANSISTOR

The p-n junction diode is a useful device. However, it does not provide *amplification*. That is, a small input signal cannot result in a large output signal. In this section we shall consider an extremely useful two junction device, the *junction transistor*, which does provide amplification.

Consider the device shown in Fig. A-8. It is a diagrammatic representation of a *p-n-p junction transistor*. That is, it consists of a semiconductor crystal containing two p-n junctions. The p-type regions could be replaced with n-type and vice versa. In this case, we would have an n-p-n junction transistor. We shall discuss the theory of the p-n-p junction transistor. However, the theory for the n-p-n junction transistor follows this closely. It can be obtained by reversing the potential and reversing the roles played by the free electrons and holes.

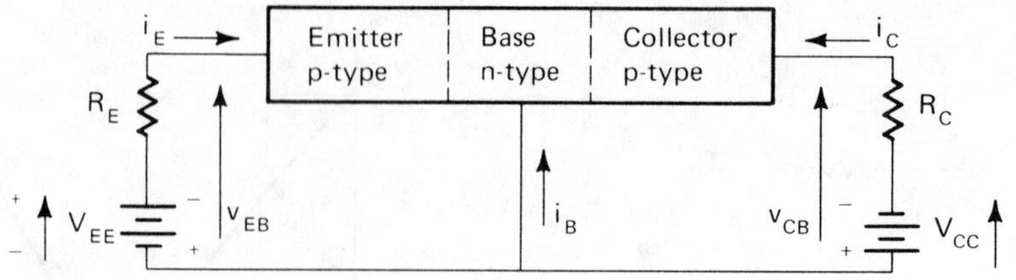

Figure A-8. A diagrammatic representation of a p-n-p junction transistor. Note that V_{CC} is a negative number.

The transistor is divided into three regions, the *emitter*, *base* and *collector*, see Fig. A-8. Actually, the base region is very narrow(i.e., the spacing between the emitter and collector is very small.) The base is much narrower than the diffusion length of the minority carriers there. (The diffusion length of a minority carrier is the average length that it will diffuse before it recombines with a majority carrier.)

Let us assume, at the start, that there are no connections made to the transistor. Then, as in the case of the p-n junction diode, potential barriers will be established at each junction. This is shown in Fig. A-9a. In the usual transistor, the emitter is more highly doped than the collector. Thus, v_{01} is larger than v_{02}.

Now consider that the transistor is connected into the circuit of Fig. A-8. The polarities of the applied voltages are such that the emitter-base junction will be forward biased while the collector-base junction will be strongly reverse biased. The potential distribution now becomes that of Fig. A-9b.

The large reverse bias across the collector-base junction causes the forward current(holes from collector to base and free electrons from base to collector)to become very small(i.e., the saturation current). This small current is called the *collector cutoff current* and is designated by I_{CO}.

The emitter-base junction is forward biased. Thus, its forward current is much greater than its reverse current. The forward current consists of two components; holes flowing from emitter to base and free electrons from base to emitter. The doping in the emitter is very much greater than the doping in the base. Thus, the hole component is very much greater than the free electron component. Those holes in the emitter, with sufficient energy to surmount the barrier, can diffuse to the base where they become minority carriers which diffuse throughout the base. Some of these holes will recombine with free electrons. However, the base width is very much less than the diffusion length of the holes so that almost all the holes injected to the base will diffuse to the collector-base junction. There they are swept into the collector by the strong electric field. Thus, these holes contribute to the collector current i_c, see Fig. A-8. Let us consider all the components of emitter current. The largest, and most important is the one we have discussed, holes from emitter to base. Almost all of these diffuse to the collector-base junction and contribute to collector current. Some of these holes recombine in the base. These contribute to base current but not to collector current. This is a relatively small number because the base is very narrow.

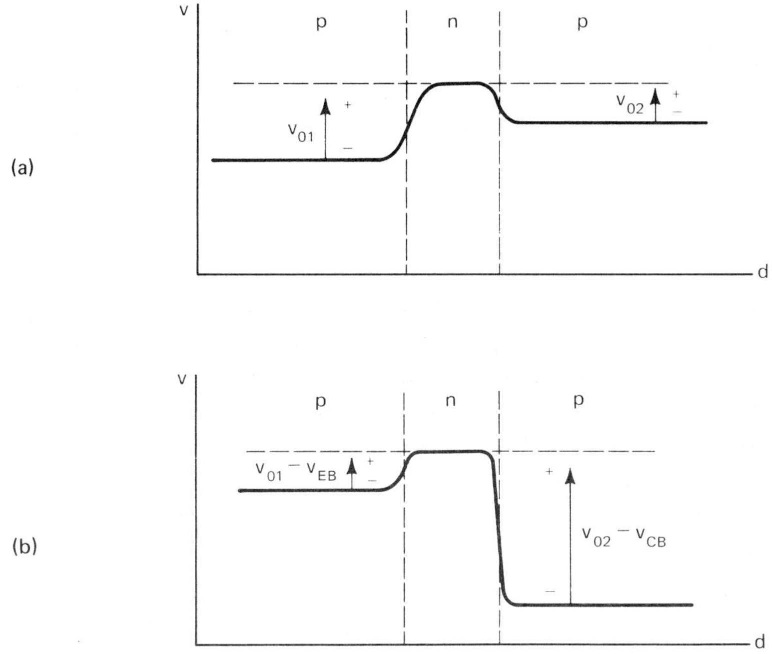

Figure A-9. Potential distributions in a p-n-p junction transistor. (a) unbiased case; (b) conventionally biased transistor. Note that v_{CB} is a negative number.

The free electrons traveling from the base to the emitter do not contribute to the collector current. Again, because of the doping levels, this current is small. However, this does contribute to the emitter and base currents. Note that, because of Kirchhoff's current law

$$i_E + i_C + i_B = 0 \tag{A-5}$$

The reverse component of collector-base junction current(i.e.,free electrons from emitter to base and holes in the opposite direction) contributes to emitter current and to base current but not to collector current. The hole current from emitter to base is by far very much greater than any other component of emitter current.

The collector current will thus have a magnitude that is only slightly less than the emitter current. This leads to the amplifying ability of the transistor. Let us explain this. The impedance of the forward biased emitter-base junction is low while the impedance of the reverse biased collector-base junction is very high. Thus, we can make the resistor R_C(see Fig. A-8)relatively large without affecting i_C to any substantial degree. Now consider that V_{EB} changes by a small amount, ΔV_{EB}. This will cause i_E to change by an amount Δi_E. Then, i_C, the collector current will change by almost the same amount

$$\Delta i_C \approx -\Delta i_E \tag{A-6}$$

This will produce a change in the voltage drop across R_C which is much larger than ΔV_{EB} because R_C is much larger than the resistance of the forward biased emitter-base junction. In general, the input to this circuit is between the emitter and the common base lead while the output is between the collector and the base lead. Thus, we can state that there is a voltage gain from input to output. Note that we have considered changes in the input voltage. In an actual case, this change could be brought about by a signal voltage applied at the input. Thus, there would be voltage amplification of this signal. Note that there is also power amplification. That is, the output signal power is greater than the input power supplied by the signal generator. Of course, this extra power is supplied by the power supply.

The changes in the emitter and collector currents will not be exactly the same. To characterize this, a parameter α, the *common base short circuit forward current transfer ratio*, is defined as

$$\alpha = -\left.\frac{di_C}{di_E}\right|_{v_{CB,\text{const}}} = -\frac{\partial i_C}{\partial i_E} .$$

The value of α is usually only slightly less than unity, in typical transistors.

The symbols for the p-n-p and n-p-n junction transistors are shown in Fig. A-10. Note that, in the remainder of this book, we shall call the junction transistor simply the *transistor*.

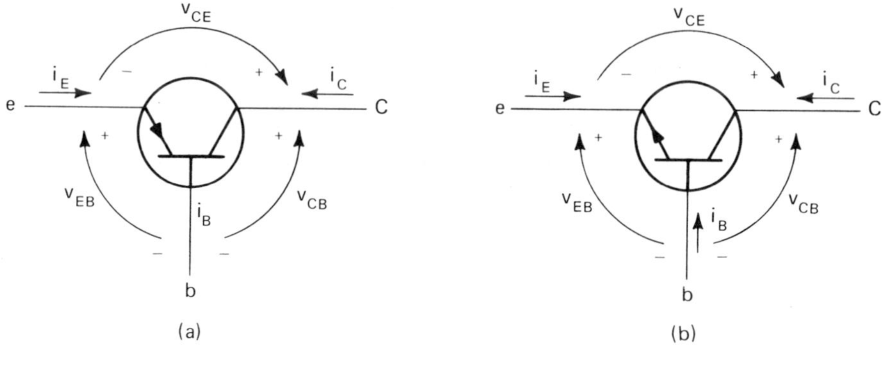

Figure A-10. Transistor symbols. (a) p-n-p; (b) n-p-n.

Ebers and Moll Equations

The equations that characterize the idealized behavior of the transistor are similar in form to those of the diode. They are

$$i_E = c_{11}(e^{v_{EB}/\eta kT} - 1) + c_{12}(e^{v_{CB}/\eta kT} - 1) \tag{A-8a}$$

$$i_C = c_{21}(e^{v_{EB}/\eta kT} - 1) + c_{22}(e^{v_{CB}/\eta kT} - 1) \tag{A-8b}$$

where the c_{11}, c_{12}, c_{21} and c_{22} are constants which depend upon the geometry of the transistor and η, k and T are as defined in

Eq. (A-3). v_{EB} and v_{CB} are the voltages defined in Fig. A-8. In these equations the voltages are considered positive if they forward bias their respective junctions and negative otherwise. To find the base current, we use Eq. (A-5). Thus,

$$i_B = -(i_C + i_E) \tag{A-9}$$

Equations (A-8) are called the *Ebers-Moll* equations.

Let us relate the constants of some previously determined parameters. Manipulating Eqs. (A-8), we have

$$i_C = \frac{c_{11}c_{22} - c_{12}c_{21}}{c_{11}} (e^{v_{CB}/\eta kT} - 1) + \frac{c_{21}}{c_{11}} i_E \tag{A-10}$$

Using Eq. (A-7), we obtain

$$\alpha = -\frac{c_{21}}{c_{11}} \tag{A-11}$$

The collector cutoff current I_{C0} is the collector current that results when $i_E=0$ and v_{CB} is a very large negative number. Substitution in Eq. (A-10) yields

$$I_{C0} = \frac{c_{12}c_{21} - c_{11}c_{22}}{c_{11}} \tag{A-12}$$

Substituting Eqs. (A-11) and (A-12) into Eq. (A-10), we have

$$i_C = I_{C0}(1 - e^{v_{CB}/\eta kT}) - \alpha i_E \tag{A-13}$$

We can proceed similarly and define a common base reverse current amplification factor as

$$\alpha_r = - \left.\frac{di_E}{di_C}\right|_{v_{EB},\,\text{const}} = - \frac{\partial i_E}{\partial i_C} \tag{A-14}$$

and an emitter cutoff current, I_{E0}, which is the emitter current when the collector current is zero and the emitter base junction is strongly reverse biased. Manipulating Eq. (A-8) and substituting we have

$$i_E = I_{E0}(1 - e^{v_{EB}/\eta kT}) - \alpha_r i_C \tag{A-15}$$

Static Characteristics of Transistors

Equations (A-8), (A-13) and (A-15) are used assuming that c_{11}, c_{12}, c_{21} and c_{22}, α and α_r are constants which do not depend upon the voltages and currents of the transistor. Actually, this is not the case and these equations are only approximately correct. Graphical curves, called static characteristics, obtained from experimental measurements, are often used to characterize transistors. Let us consider these now. Figure A-11 represents the common base output characteristics. There are three variables, i_C, v_{CB} and i_E. To obtain a two dimensional graph, we plot i_C versus v_{CB} for a fixed i_E and then repeat the procedure for a different i_E, etc. Note that we have divided the characteristics into three regions. The active region is where operation takes place when small signals are amplified and for some pulse applications. There, the collector-base junction is reverse biased and the emitter-base junction is forward biased. Note that the emitter and collector currents are almost equal in the active region. To illustrate the use of this region, let us consider the amplifier circuit of Fig. A-12. Suppose that $V_{CC} = -20$ volts and $R_C = 4000$ ohms. Then, a load line can be drawn as shown in Fig. A-11. Now assume that V_{EE} is such that $i_E = 3$ ma. The curve for $i_E = 3$ ma now characterizes the transistor. The transistor will operate at the point Q. This is called the *quiescent operating point*, that is, the operating point when no signals are present. Suppose that the signal voltage v_E causes the emitter current to vary sinusoidally about 3 ma. For example

$$i_E = 3 + \cos \omega t \text{ ma} \tag{A-16}$$

When $\cos \omega t = 1$, $i_E = 4$ ma. This curve (for $i_E = 4$ ma) characterizes the transistor at this instant of time. Thus, the instantaneous operating point at this time is a. As t varies, i_E will vary from 2 to 4 ma. Thus, the load line between points a and b will represent the locus of operation. The output will follow the input and linear amplification will occur. Note that this configuration does not produce current gain. That is, the output swing in current will actually be a little less than the input swing. We shall subsequently consider a circuit that removes this difficulty.

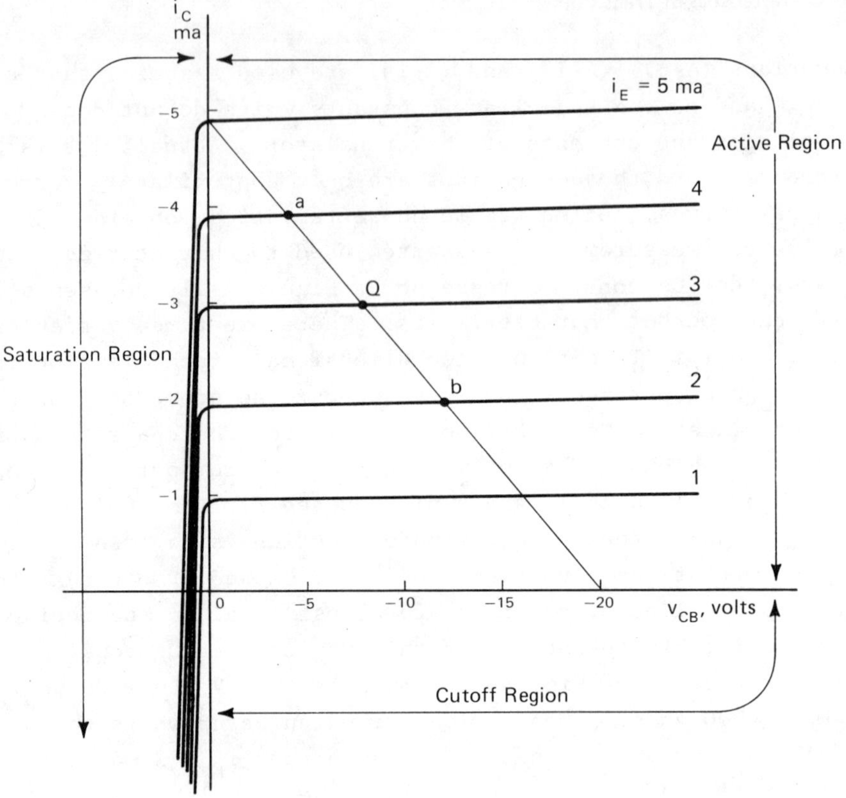

Figure A-11. Common base output characteristics for a typical p-n-p junction transistor.

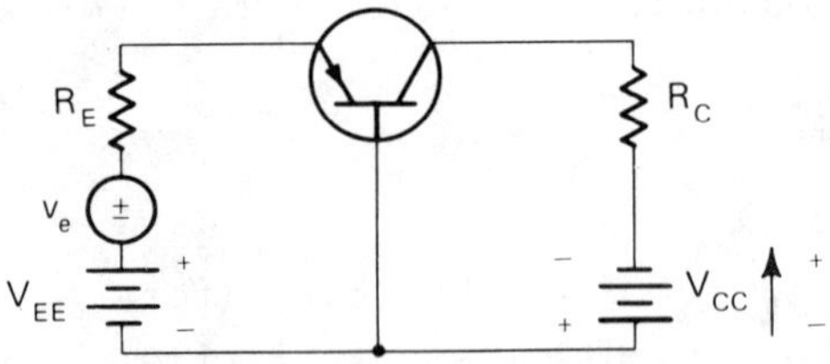

Figure A-12. A common base transistor amplifier. Note that V_{CC} is a negative number.

Transistors are always operated in the active region when they are to amplify linear signals. For instance, a high fidelity audio amplifier represents such amplification. However, we are primarily concerned with digital circuits. In digital circuits, at times, the operation is restricted to the active region. We shall see in Sec. A-5 that this can improve switching speed. However, pulse circuits are often not restricted to this region. Let us consider why this is done.

If $i_E = 0$ and v_{CE} is negative, then $i_C = I_{CO}$. This is a very small value and appears as zero on the graph. If we reverse bias the emitter-base junction, then i_E would become very small and negative. In this case, i_C would remain at very small values. Thus, if $i_E \leq 0$ and $v_{CE} \leq 0$, then $i_C \approx 0$. In this case, we say that the transistor is *cut off*. Note that both junctions are reverse biased and the impedance between the emitter and base leads and between the collector and base leads is very high, ideally approaching an open circuit. The transistor ideally acts as an open switch.

Now assume that we reverse the polarity of v_{CC}. Now the collector-base junction becomes forward biased. Assume that the potentials are such that the emitter-base junction also becomes forward biased. Note how i_C increases in positive value. When it is operated this way, the transistor is said to be saturated. Since both junctions are forward biased, the impedance between the terminals is low. Ideally, the transistor acts as a short circuit or a closed switch. Note that if v_{CE} is negative but very small, we can also consider that we are in the saturation region if the collector current is large.

The greatest swings in voltage and current occur when the transistor is operated between saturation and cutoff. Thus, in a digital circuit, if a 1 is represented by the voltage corresponding to cutoff and a 0 by the voltage corresponding to saturation. then there will be the largest possible difference between a 0 and a 1. This will make for the most reliable operation. Often, to make sure that cutoff or satuation is achieved, the transistor is overdriven. For instance, if $v_{EB} \approx 0$, then the transistor is probably cutoff. However, if we make $v_{EB} = -2$ volts, then the transistor will certainly be cutoff. Similarly, to make sure that the operation is well within the saturation region, the applied voltage which forward biases the collector-base junction may be made larger than is absolutely needed. This makes for reliable operation but, as we shall subsequently discuss, slows the speed of switching.

Now let us consider another configuration that is very often used, see Fig. A-13. Here, the emitter lead is common between the input and output. Thus, this is called a *common emitter* amplifier. The base lead is the input lead. Since $i_C \approx i_E$, then i_B will be very small. Thus, when the common emitter configuration is used, current gain can be obtained. In analyzing this circuit, it is convenient to plot i_C versus v_{CE} with i_B as a parameter. Note that this information can be obtained, in theory, from Fig. A-11 since $i_B = -(i_C+i_E)$. However, i_C and i_E are very close in magnitude and opposite in sign. Thus, the graph of Fig. A-11 does not provide sufficient accuracy to obtain i_B. However, we can measure the desired curves directly. A typical set is shown in Fig. A-14. A load line is drawn for $V_{CC} = -20$ volts, $R_L = 4000$ ohms. In the circuit, a signal of 10μa will produce a swing of about 1ma. This represents a current gain with a magnitude of about 100. The common emitter configuration is a very commonly used one because of its high current gain. It simultaneously provides voltage and power gains.

(a)

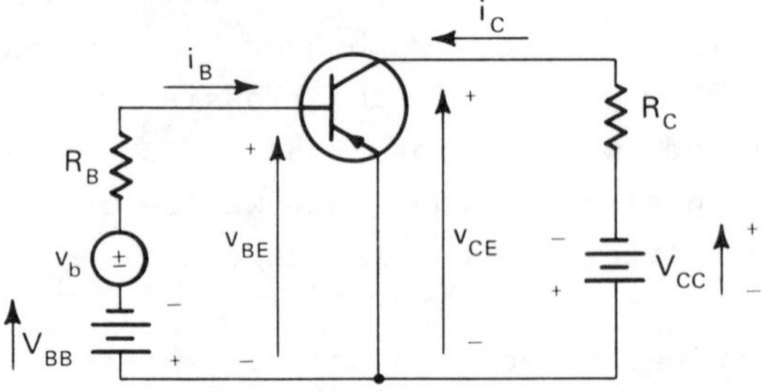

(b)

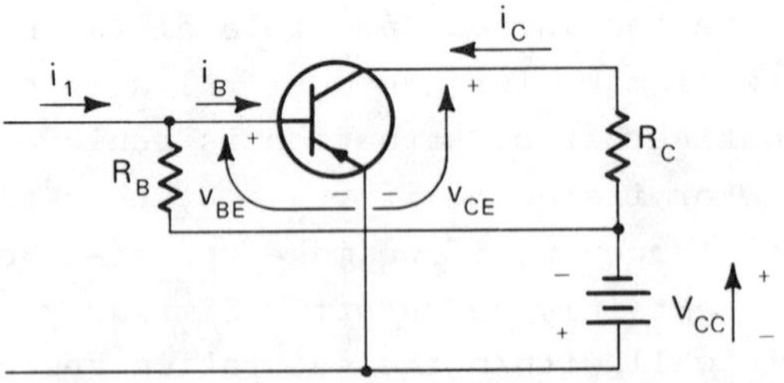

Figure A-13. (a) A common emitter transistor amplifier. Note that V_{BB} and V_{CC} are negative numbers; (b) a modification that uses only one power supply. i_1 represents the input signal current.

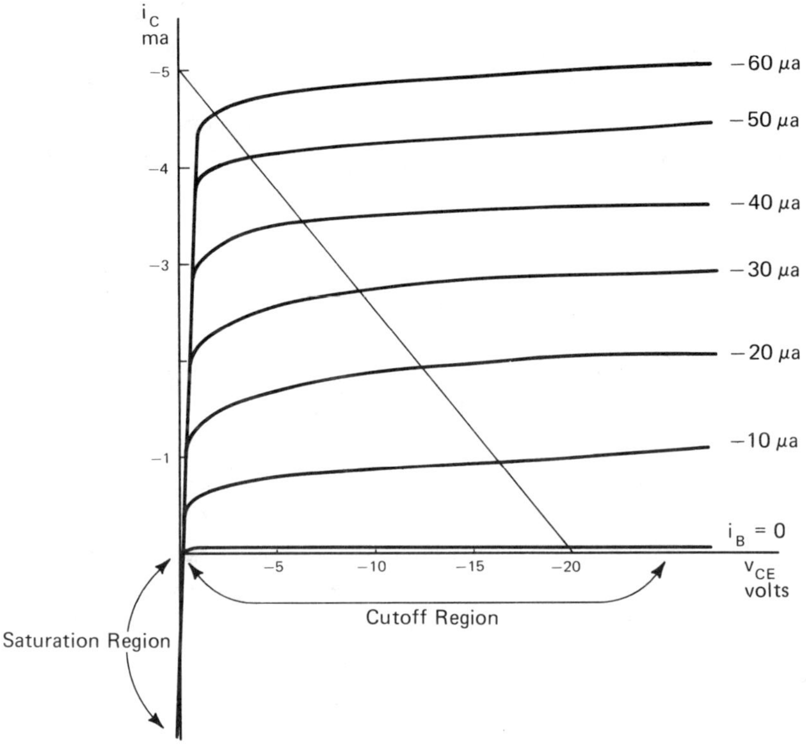

Figure A-14. Common emitter output characteristics for a typical p-n-p junction transistor.

The saturation region occurs at the left of the characteristics where all the curves come together and also where both junctions are forward biased. (Note that there is a small region where v_{CE} is very small but negative, i.e., where the base-collector junction is reverse biased.)

The cut off region is along the v_{CE} axis. Here i_C=0. Note that if $i_B = 0$, the transistor is essentially cut off. Actually, at cutoff, $i_E = 0$ and $i_C = I_{CO}$. Substituting in Eq. (A-9), we obtain

$$i_B = -I_{CO} \tag{A-17}$$

This corresponds to a reversed input bias. However, the value of i_C is often so small when $i_B = 0$ that this value of i_B can be considered to be the cutoff value of the base current. We shall often do so.

At times it is desirable to consider cutoff in terms of the base to emitter voltage. Note that when $i_B = 0$, we do not necessarily have $v_{BE} = 0$ since the currents through the transistor can establish a value of v_{BE} even if the base lead is open circuited. Actually, the collector current remains close to its cutoff value until v_{BE} has increased to about 0.5 volts for an n-p-n silicon transistor. This is analogous to the cut in voltage for a diode and is called the *cut in* value of v_{BE}, which is written as V_γ. Note that V_γ is a function of temperature.

In Fig. A-13a, we show two power supplies. However, they both have the same polarity with respect to the common lead. Thus, we need only use one power supply. Such a circuit is illustrated in Fig. A-13b.

We have not discussed the procedures for obtaining the input current. Actually, a set of graphical input characteristics can be drawn. However, this is often not necessary. For instance, consider Fig. A-13b. Applying Kirchhoff's laws, we obtain

$$i_B = \frac{V_{CC}-v_{BE}}{R_B} \tag{A-18}$$

However, v_{BE} is usually a fraction of a volt in magnitude, while V_{CC} has a magnitude of about 20 volts. Thus,

$$i_B \approx \frac{V_{CC}}{R_B} \tag{A-19}$$

and we do not need the graphical input characteristics.

We defined α as a quantity which related the output current change to the input current change for a common base circuit. Let us now define a parameter β which does this for the common emitter circuit. In this case

$$\beta = \left.\frac{di_C}{di_B}\right|_{v_{CE},\text{const.}} = \frac{\partial i_C}{\partial i_B} \tag{A-20}$$

Let us relate α and β. Substituting Eq. (A-9), we have

$$\beta = -\left.\frac{di_C}{d(i_C+i_E)}\right|_{v_{CE,\text{const.}}} \tag{A-21a}$$

Note that

$$\alpha = -\left.\frac{di_C}{di_E}\right|_{v_{CB,\text{const.}}} \tag{A-21b}$$

Different quantities are held constant in Eqs.(A-21a) and (A-21b). Thus, we cannot simply substitute one in the other. However, we can note that

$$v_{CE} = v_{CB} + v_{BE} \tag{A-22}$$

v_{CE} and v_{CB} are almost equal since v_{BE} is the small voltage across the forward biased emitter base junction. Thus, we assume that if v_{CE} is constant, then v_{CB} is also essentially constant. Thus, the restrictions on Eqs.(A-21) can be considered equivalent. Then, we can write

$$\frac{1}{\beta} = \left(-1 - \frac{di_E}{di_C}\right)_{v_{CB,\text{const.}}} = -1 + \frac{1}{\alpha}$$

Manipulating, we obtain

$$\beta = \alpha/(1 - \alpha) \tag{A-23}$$

Since $\alpha \approx 1$, then β can be very large. This tends to substantiate that the common emitter circuit can have high current gain.

Response to Low Level Pulses

The transistor behavior varies as the rate of change of the input signals increases. Let us discuss some important aspects of this here. We shall consider the response to small pulses here. Subsequently we shall discuss the pulse response when the transistor is driven between saturation and cutoff.

In the transistor there are capacitances across the two junctions which must be considered. These capacitances tend to "load down" the transistor circuit and reduce the gain when rapidly changing signals are encountered.

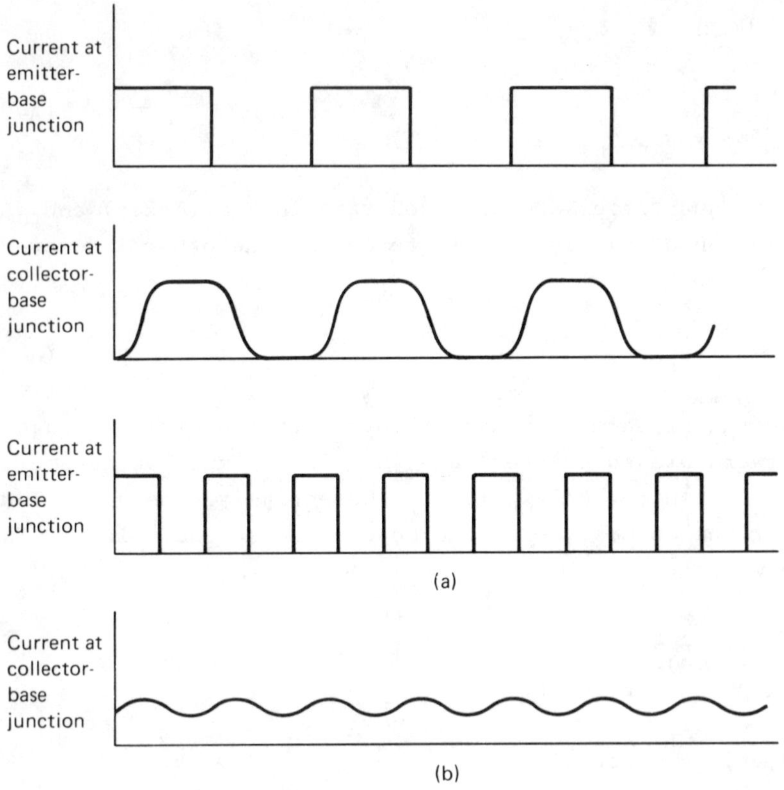

Figure A-15. An illustration of the transmission of a train of pulses across the base; (a) low frequency pulses; (b) high frequency pulses.

There is another effect which causes the high frequency response to deteriorate. Charge carriers are injected at the emitter-base junction. They then diffuse to the collector-base junction where they are swept into the collector to produce almost all of the collector current. Suppose that the injected carriers are of the form of a pulse train, that is, that the current across the emitter-base junction is in the form of the ideal pulse shown in Fig. A-15a. Hence, the injected minority carriers periodically vary with time in the form of a square wave. These carriers *diffuse* across the base. Diffusion is random motion. Thus, even if charge carriers are injected at the same time, they will not all arrive at the collector-base junction together. Thus, the current pulses at the collector-base junction will be rounded as shown. If the pulse duration is relatively long, then the rounding will only occupy a small part of the pulse and recognizable pulses will be obtained. However, if the pulses are short, see Fig. A-15b, then the rounding occupies much of the pulse. In this case, the size of the output pulses is greatly reduced and they are not of the form of a pulse. In digital work, this means that the time of the pulse must be large enough if satisfactory output pulses are to be obtained. This slows switching speed.

To improve the switching speed, we must reduce the time that it takes for charges to diffuse across the base. At times, the charge density of the impurities in the base is varied so that the ionized atoms set up a field that sweeps the minority carriers towards the collector. The greatest improvement in high frequency response is made by making the base very narrow. Modern manufacturing techniques are such that transistors with extremely narrow bases and, hence, extremely high switching speeds, on the order of 10^{-9}sec., can be obtained.

A-4 THE FIELD-EFFECT TRANSISTOR

Another semiconductor device that is used in switching circuits is the *field effect transistor*, commonly abbreviated FET. This device can be operated in switching modes which require very little power. This is important for several reasons. Modern integrated circuits, which will be discussed in a subsequent section, can incorporate many thousands of electronic devices, such as transistors, on a single silicon chip. Each device dissipates power and there are very many of them in very close proximity. This can cause the temperature to rise excessively. Thus, the number of devices that can be packed into a single chip is limited by power considerations. Reducing the power dissipated by each element thus increases the packaging densities that are allowed.

It is also desirable to reduce the power dissipated by the devices since this reduces the power that must be supplied to the computer. This is especially important in applications where the power is limited such as in space probes.

The input impedance(i.e., the impedance viewed into the input terminals)of the FET is very high, often greater than 10^9 ohms. In contrast, in the active region, the input impedance of a common emitter transistor amplifier is about 1000 ohms. This high input impedance is an advantage when one circuit "drives" another, that is, when the output of one circuit is used as the input of the next. The very high input resistance of the FET means that many circuits, in parallel, can be driven by a single one. This is often a great advantage in a digital computer.

The field effect transistor produces less noise, has better thermal stability, and is much less affected by radiation than the transistor.

From all of the above discussion, it might seem as though the FET would replace the transistor in most computer applications. However, the transistor has much faster switching speed. For this one reason, most large computers use a predominance of bipolar junction transistors. However, as we have discussed, there are many special applications where the FET is used.

The Junction FET

One form of the FET, called the junction FET is illustrated diagrammatically in cross section in Fig. A-16. This is not the actual construction of the FET but is one which is helpful in explaining it. Suppose that we have a cylinder of n-type material with a p-type region around its circumference. As in the case of the junction transistor, we can interchange the n-type and p-type regions.

One end of the FET is called the *source*, the other the *drain*. The p-type region is called the *gate*. A voltage is applied between the gate and source which reverse biases the p-n junction. A depletion region is set up. Now suppose that a voltage V_{DS} is applied as shown. A current I_D will result. Since the depletion region is essentially free of mobile charge carriers, the free electrons will flow in the space enclosed within the depletion region. This space is called the *channel*. In this case, we have an n-channel FET. Note that in the FET, the charge carriers are *majority carriers* and there is essentially only one type of carrier. For this reason, the FET is called a *unipolar* device while the junction transistor is called a *bipolar* device.

The depletion region is skewed, see Fig. A-16. This is because the potential difference between the gate and channel varies with position. The potential is larger near the drain side. This can be accounted for by the following: The current through the channel produces a voltage drop, which is a function of position. It is zero at the source end and it increases to V_{DS} at the drain end. Thus, the gate to channel potential varies from $-V_{GS}$ to $V_{DS}-V_{GS}$.

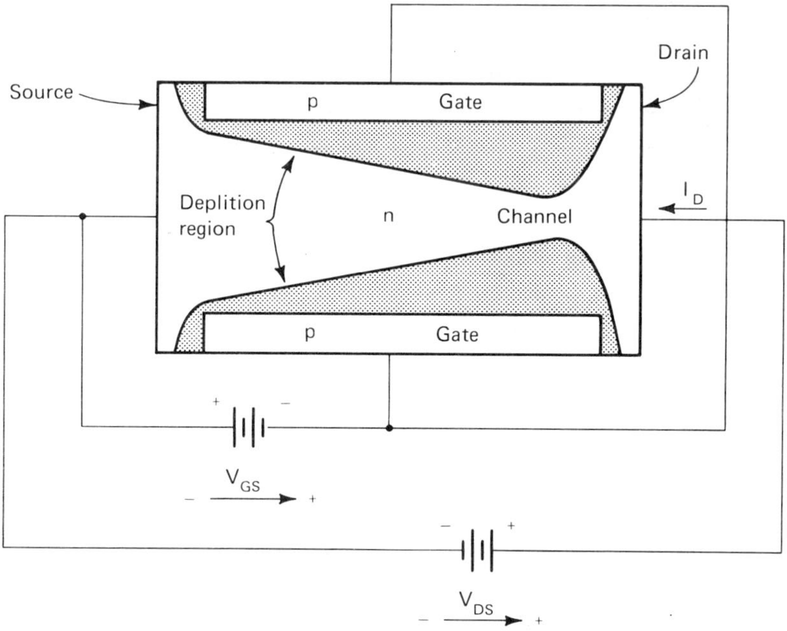

Figure A-16. A diagrammatic representation of a function FET. Note that V_{GS} is a negitive number.

Now suppose that V_{GS} is at some negative value(e.g.,-2 volts) and V_{DS} is increased. The resulting current produces a voltage drop which causes the width of the depletion region to increase. Thus, the effective resistance of the channel will increase. A typical set of characteristics is shown in Fig. A-17. Note that the curves become more horizontal as V_{DS} is increased. This is a manifestation of the increase in resistance with increasing V_{DS}. Note that we could reverse the polarity of V_{DS} and I_D would also reverse.

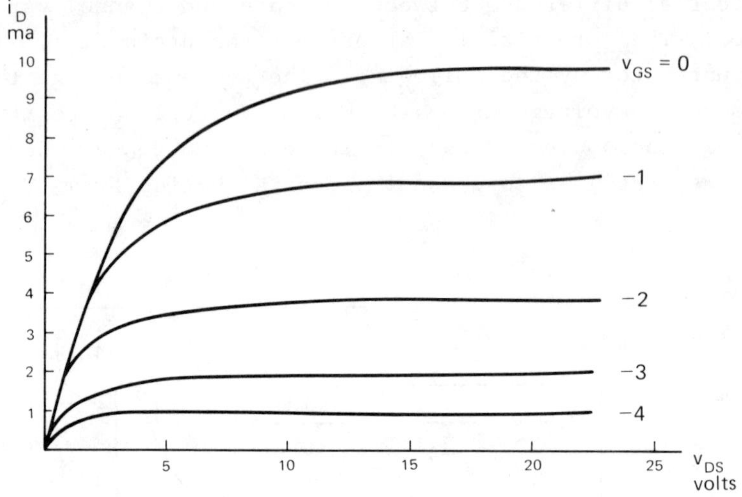

Figure A-17. Typical characteristics of a junction FET.

It might seem as though a sufficiently large V_{DS} would cause the channel width to reduce to zero, cutting off the current. However, this does not occur. The increased depletion region width is produced by the increased voltage drop in the channel, which, in turn, is produced by the current there. Since the current produces the effect, it cannot be exptected to fall off. What does happen is that the curves become nearly horizontal. This indicates that the resistance to *further* increases in current is high. Now the channel width is very small. The current densities there are very large. In this case, *pinch off* is said to have occurred.

The symbols for the junction FET are shown in Fig. A-18.

The construction of an actual typical junction FET is shown in Fig. A-19. The substrate is just a support for the FET. It can be isolated from the remainder of the FET by reverse biasing the p-n junction at the "top" of the substrate. The charge carriers must pass between the gate and the substrate. This is where the effective channel is.

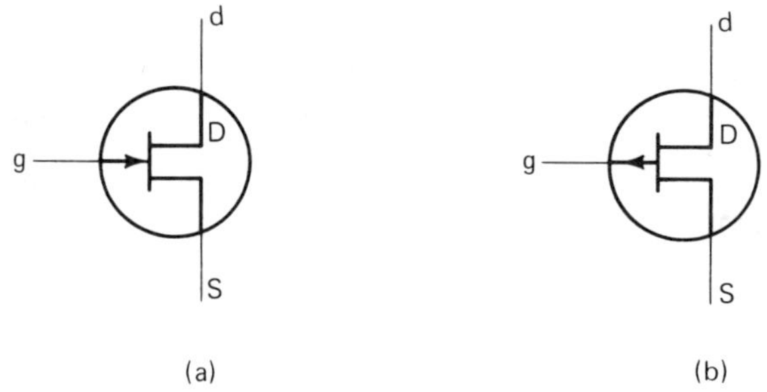

Figure A-18. Junction FET symbols. (a) n-channel; (b) p-channel.

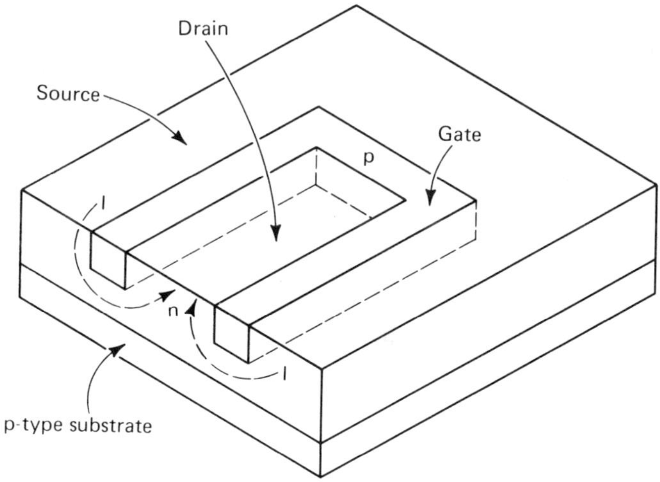

Figure A-19. A cross section of a more practical junction FET.

Before discussing the characteristics of the FET further, let us consider some other forms of the field effect transistor.

The Insulated Gate FET-IGFET-MOSFET

One disadvantage of the junction FET is that the gate-channel junction must be reverse biased to maintain the high input impedance. We shall now consider two devices that eliminate this difficulty. In each of these devices, the gate is insulated from the channel. Hence, the name insulated gate FET, which is abbreviated IGFET, is used.

These devices are fabricated using an integrated circuit technology which will be described in Sec. A-6. The gate consists of a metal layer which is insulated from the body of the FET by a layer of silicon dioxide(SiO_2). These devices are made of silicon. Thus, they are often called MOSFET(metal-oxide-silicon FET).

The Enhancement MOSFET

Consider the structure shown in Fig. A-20. The gate consists of a metal conductor separated from the p-type substrate by an insulator layer of quartz(SiO_2). The source and drain are n-type regions. Assume, for the moment, that $V_{GS} = 0$ and that V_{DS} is applied as shown. The drain current I_D will be very small since, no matter what the polarity of V_{DS}, there will be a reverse biased p-n junction. Also, the doping in the substrate is light so that there are not many mobile charges there.

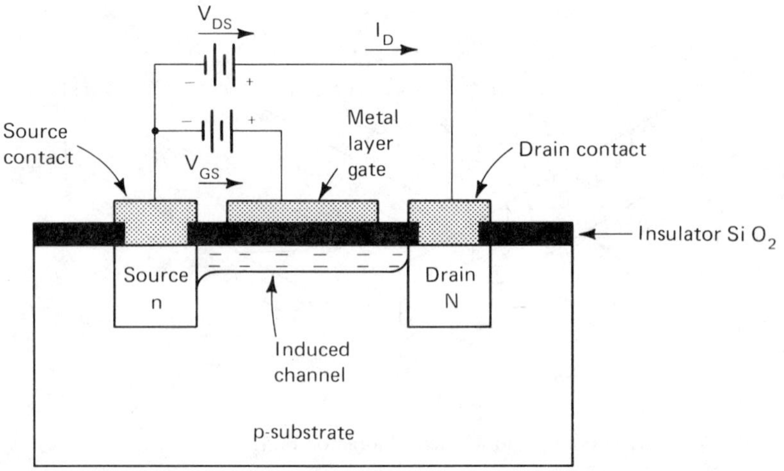

Figure A-20. An enhancement n-channel MOSFET. Note that the substrate is p-type.

Now suppose that a voltage is applied between gate and source with the polarity shown. The gate and substrate will act as plates of a capacitor and free electrons from the substrate will be attracted into the area indicated by the *induced channel*. The large number of free electrons present now convert this region into an effective n-type region. Thus, conduction can take place between drain and source. Note that the induced channel is skewed. This is for reasons similar to those discussed for the junction FET. This device is called an enhancement MOSFET or enhancement IGFET because the number of charge carriers in the induced channel is enhanced.

Typical output characteristics for an enhancement MOSFET are given in Fig.A-21. Note that, in the p-type channel case, if V_{GS} is 0 or negative, then the current is essentially zero.

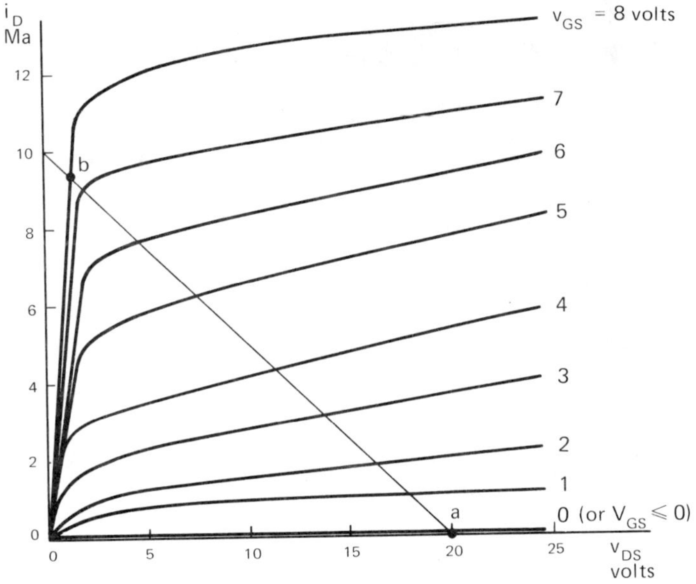

Figure A-21. Typical output characteristics for an enhancement MOSFET.

The Depletion MOSFET

We can construct another form of MOSFET as illustrated in Fig. A-22. The source and drain are the n+ regions. The superscript + indicates heavy doping. An actual n-type channel is fabricated in this device. Thus, conduction between the drain and source can take place even if $V_{GS} = 0$.

Now assume that V_{GS} is of the polarity shown. As before, the gate and channel will act as a capacitor. Now, however, the free electrons will be driven away from the gate. This will reduce the conduction between drain and source. This is called a depletion mode of operation since the free electrons are depleted from the region. Typical output characteristics for a depletion MOSFET are shown in Fig. A-23. Note that, for positive values of V_{GS}, enhancement type operation takes place.

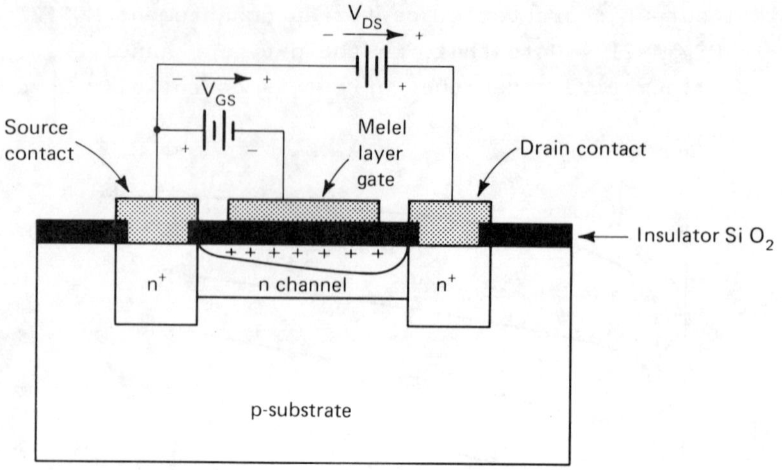

Figure A-22. A depletion MOSFET. Note that V_{GS} is a negative number.

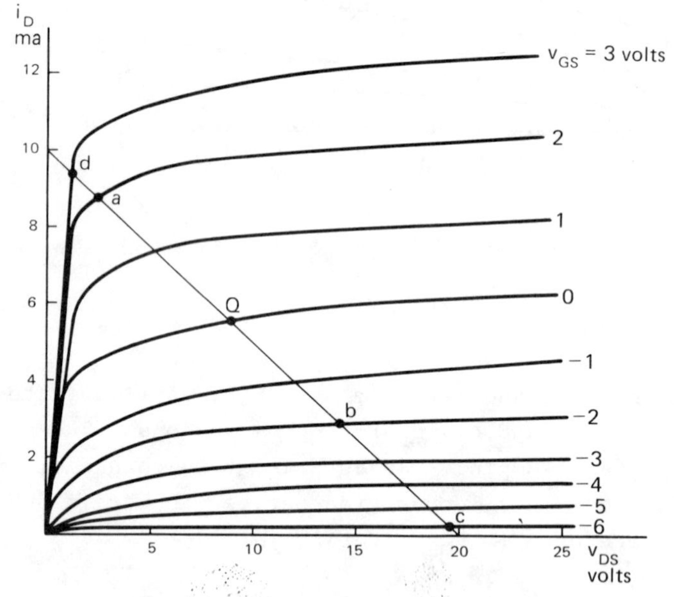

Figure A-23. Typical output characteristics for a depletion MOSFET. If v_{GS} is positive, enhancement operation results.

Some symbols used for MOSFETs are shown in Fig. A-24. Note that the substrate represents the bulk of the semiconductor in which the MOSFET is fabricated.

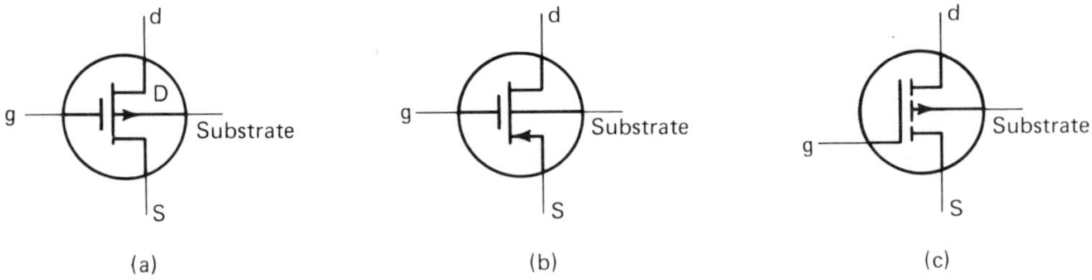

Figure A-24. Symbols for a p-channel MOSFET. For an n-channel case reverse the arrowheads. (a) and (b) are used for either enhancement or depletion operation; (c) is used only for enhancement MOSFETs. At times, a separate substrate lead is not used. In such cases, the substrate lead may be omitted from the symbol.

Now consider the circuit of Fig. A-25. Assume that $V_{DD} = 20$ volts and $R_L = 2000$ ohms. The load line for this circuit is drawn in Fig. A-23. If there is no signal, then the $V_{GS} = 0$ curve characterizes the FET. Thus, point Q is the quiescent operating point. Now suppose that the input signal is of the form

$$v_1 = 2 \cos \omega t \text{ volts}$$

The locus of operation will vary along the load line from point a to point b. Thus, the output signal will swing between 3 and 14 volts. If we neglect the curvature of the characteristics, the output will be

$$v_2 = -5.5 \cos \omega t$$

which represents an increase in the signal level. The minus sign indicates a phase reversal. Note that when v_1 increases, v_2 decreases.

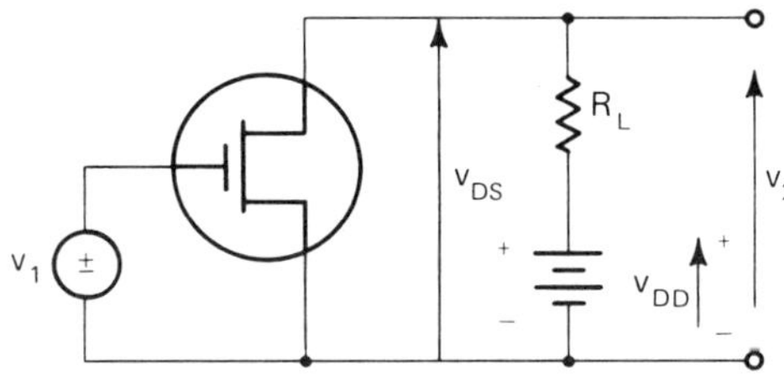

Figure A-25. A simple amplifier or inverter.

This phase reversal results in this circuit's being an inverter or NOT gate. For instance, suppose that v_1 represents a pulse which varies between +3 and -3 volts. Then, when $v_1 = +3$, $v_0 = +1$ volt and when $v_1 = -3$ volts, $v_0 = +15$ volts. This represents inversion of a sort. However, much better inversion takes place if we add a large negative voltage in series with v_1 as shown in Fig. A-26. Suppose that $V_{GG} = -6$ volts and v_1 varies between 0 and 9 volts. That is, if v_1 represents a 0, then $v_1 = 0$(actually 2 volts or less); if v_1 represents a 1, then $v_1 = 9$ or more volts. If $v_1 = 0$, then point C is the operating point and $v_2 = 19$ volts. If $v_1 = 9$ volts, then $v_{gs} = 9-6 = 3$ volts and point d is the operating point. Hence, $v_2 = 1$ volt. Thus, a 0 input results in a large positive output while a 1 input results in a very small output. Thus, inversion has been achieved.

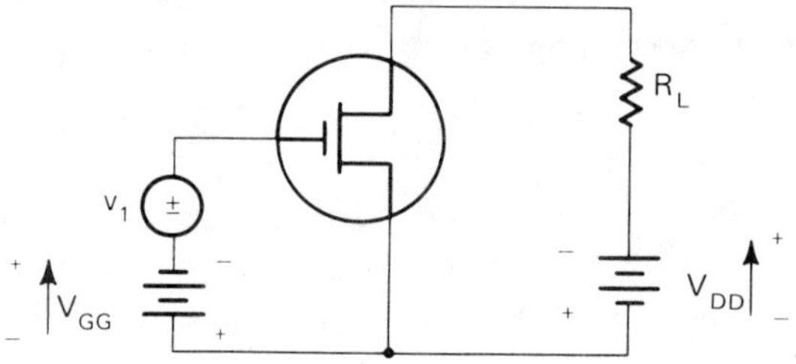

Figure A-26. An amplifier or inverter that uses a negative bias. Note that V_{GG} is a negative number.

A disadvantage of this circuit is that the negative voltage V_{GG}, called a *negative bias voltage* is necessary to obtain proper operation. If we use an enhancement MOSFET this is no longer needed. For instance, a load line for the circuit of Fig. A-25 is drawn on the characteristics of Fig. A-21. Now assume that v_1 varies between 0 and 8 volts. That is, suppose 0 or a small positive voltage represents a 0 while anything more positive than 5 volts represents a 1. When $v_1 = v_{GS} = 0$, then $v_2 = V_{DS} = 20$ volts(point a)and when $v_1 = V_{GS} = +8$ volts, then $v_2 = v_{DS} = 1$ volt (point b). Thus, we have achieved inversion. Note that if the output of this circuit is used to drive a succeeding circuit, a 1 will be a large positive number (≈ 20) while a 0 will be a small positive number(≈ 1).

In a FET, the cutoff region is "along" the v_{DS} axis where $i_D=0$. For instance, for an enhancement MOSFET, this occurs when the value of v_{GS} is zero or negative for a p-substrate n-channel MOSFET.

The saturation region occurs at the low value of v_{DS} where the curves all tend to come together. Note that, in this region, the value of v_D is very small in magnitude and almost independent of v_{GS}.

The active region comprises most of the curves. As in the case of the transistor, we use parameters to characterize the slopes of curves in this region. These parameters are most useful if we want to determine the response of the FET to small changes about a quiescent operating point. However, they are also used in studying the pulse response as it moves through the active region from saturation to cutoff or vice versa. The parameters are the *transconductance*, g_m

$$g_m = \left.\frac{di_D}{dv_{GS}}\right|_{v_{DS},\text{const.}} = \frac{\partial i_D}{\partial v_{GS}} \qquad \text{(A-24a)}$$

the *drain resistance*, r_d

$$r_d = \left.\frac{dv_{DS}}{di_D}\right|_{v_{GS},\text{const.}} = -\frac{\partial v_{DS}}{\partial i_D} \qquad \text{(A-24b)}$$

and the *amplification factor*, μ

$$\mu = -\left.\frac{dv_{DS}}{dv_{GS}}\right|_{i_D,\text{const.}} = -\frac{\partial v_{DS}}{\partial v_{GS}} \qquad \text{(A-24c)}$$

It can be shown that these three parameters are related by the equation

$$\mu = g_m r_d \qquad \text{(A-24d)}$$

Low Level Pulse Response of the FET

Just as in the case of the transistor, the speed at which an FET can respond to low level pulses is limited. In a FET, the primary causes of the fall of in high frequency response are the capacitance between the gate and source, C_{gs}, between the gate and drain, C_{gd}, between the drain and source, C_{ds}, and between the various elements and the substrate.

A-5 SWITCHING IN SEMICONDUCTOR DEVICES

In this section we shall consider the response of semiconductor devices to large pulses which drive them between saturation and cutoff. We shall start by discussing some simple switching models. Then, we shall consider switching speed.

Suppose that we have a simple transistor inverter, see Fig.A-27. The input is represented by the voltage generator v_1. Suppose that a 0 is represented by a small positive voltage while a 1 is represented by a large positive voltage. If v_1 is a 0, then the current through the transistor will be relatively small and v_2 will be almost equal to V_{CC}. On the other hand, if v_1 is a large positive voltage, then the transistor will conduct heavily and v_2 will be very small.

Let us now obtain models which ignore switching times and that can be used in the saturation, active and cutoff regions. That is, we shall obtain circuits which can be used to represent the transistor in these regions.

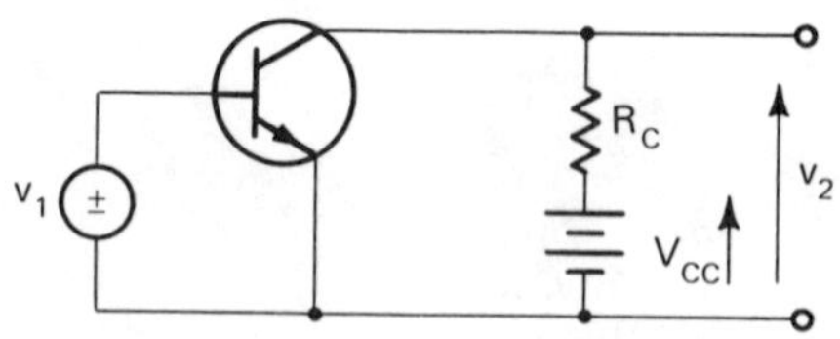

Figure A-27. A simple transistor inverter.

In the saturation region, both junctions are forward biased or the base current is so high that the collector base junction voltage is very low. Thus, the transistor can be modeled as shown in Fig. A-28a, where R_{is} and R_{os} are very small resistances.

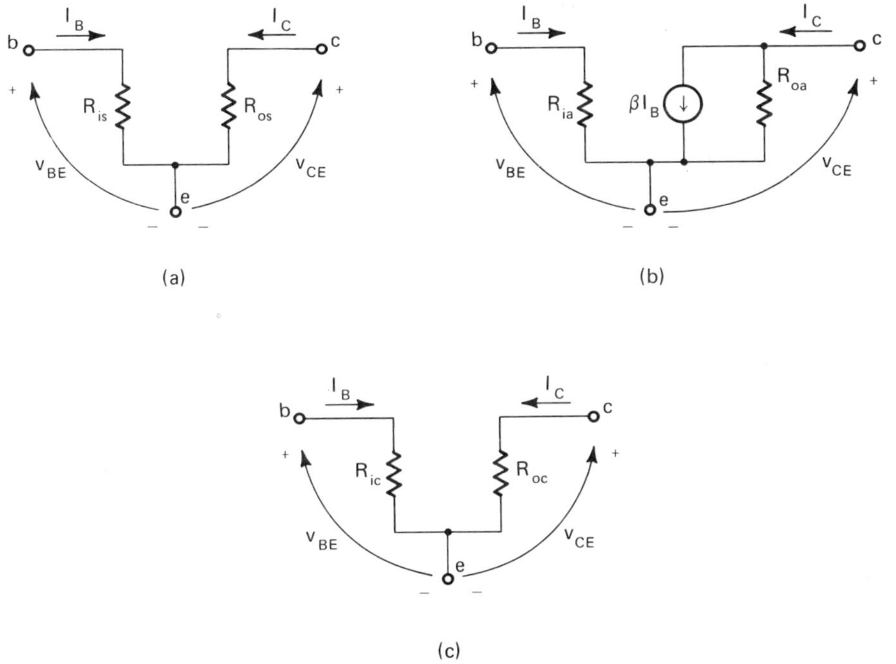

Figure A-28. Simple models for the junction transistor. (a) saturation region; (b) active region and (c) cut-off region.

Figure A-28b is a model that can be used in the active region. The generator βI_B takes the current amplifying ability of the transistor into account. (i.e., the generator current is β times the input current.)

Finally, in the cutoff region, both junctions are reverse biased. Thus, Fig. A-28c can be used to characterize the transistor in the cutoff region. Here R_{ic} and R_{oc} are large resistances. Note that the saturation and cutoff models have the same form. However, the magnitudes of the resistances are very different.

Suppose that the inverter of Fig. A-27 is switched from saturation to cutoff by the input signal. The models that we can use to analyze this are shown in Fig. A-29. Let us determine the output voltage. For the saturation model, we have

$$v_{2s} = \frac{V_{CC}R_{os}}{R_{os} + R_C} \tag{A-25a}$$

while for cutoff we have

$$v_{2c} = \frac{V_{CC}R_{oc}}{R_{oc} + R_C} \tag{A-25b}$$

Remember that $R_{oc} >> R_{os}$. In a proper inverter circuit we choose R_C such that

$$R_{os} << R_C \tag{A-26a}$$

$$R_{oc} >> R_C \tag{A-26b}$$

Then, substituting in Eqs.(A-25), we obtain

$$v_{2s} = V_{CC}\frac{R_{os}}{R_C} \approx 0 \tag{A-27a}$$

$$v_{2c} \approx V_{CC} \tag{A-27b}$$

Thus, the device function as an inverter. The output voltage for 0's and 1's can be computed from these relations.

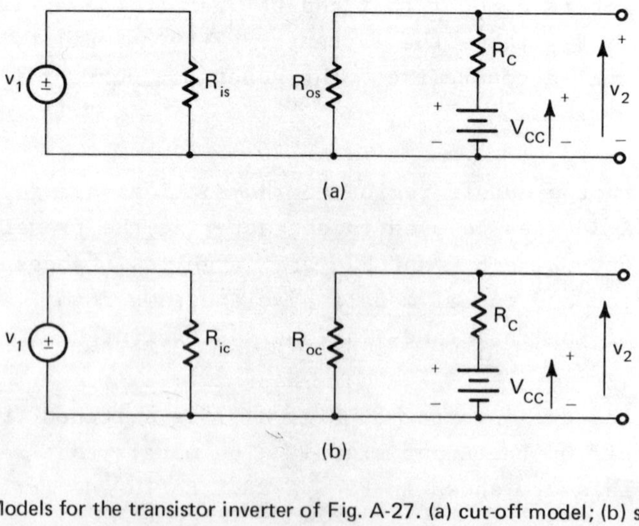

Figure A-29. Models for the transistor inverter of Fig. A-27. (a) cut-off model; (b) saturation model.

The models for the MOSFET are shown in Fig. A-30. These simple models are similar to those for the transistor except that, at low frequencies, we can consider that the input impedance is infinite.

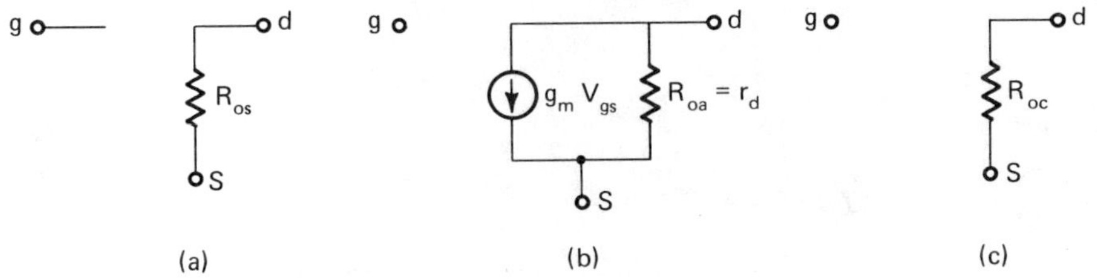

Figure A-30. Simple models for the MOSFET. (a) saturation region; (b) active region; and (c) cutoff region.

Switching in the p-n Junction Diode

Let us now study switching speed in semiconductor devices. We shall start the discussion by considering the p-n junction diode. This will provide us with insight into switching in a transistor.

In order to study switching in a p-n junction diode, we must consider the charge densities there. If a p-n junction is forward biased, then minority carriers will be injected across the junction. Thus, see Sec. A-2, the equilibrium value of charge densities of holes in the n-type region and of free electrons in the p-type region will be greater than the equilibrium value if there were no current. (This equilibrium charge density falls off exponentially with position as the distance from the junction increases.)

Let us define q_p as the total number of holes in the n-type region in excess of the equilibrium value when there is no current. Correspondingly we can define q_n as the total number of excess free electrons in the p-type region. Note that if the current varies with time, then q_p and q_n will be functions of time. We would then write them as $q_p(t)$ and $q_n(t)$.

Now suppose that we connect a current generator to the diode as shown in Fig. A-31a. Assume that, at t=0, the current of the generator switches from the forward value I_0 to $-I_S$, see Eq. (A-3). When there is forward current, $q_p(t)$ will be greater than the equilibrium value for zero current. We call this value Q_0. (To make the discussion simpler, we shall only consider $q_p(t)$. However, a similar discussion applies to $q_n(t)$.) When the current through the diode reverses and becomes equal to $-I_S$, the equilibrium value of $q_p(t)$ is less than the equilibrium value for zero current. This is given in $-Q_r$ by Fig. A-31c.

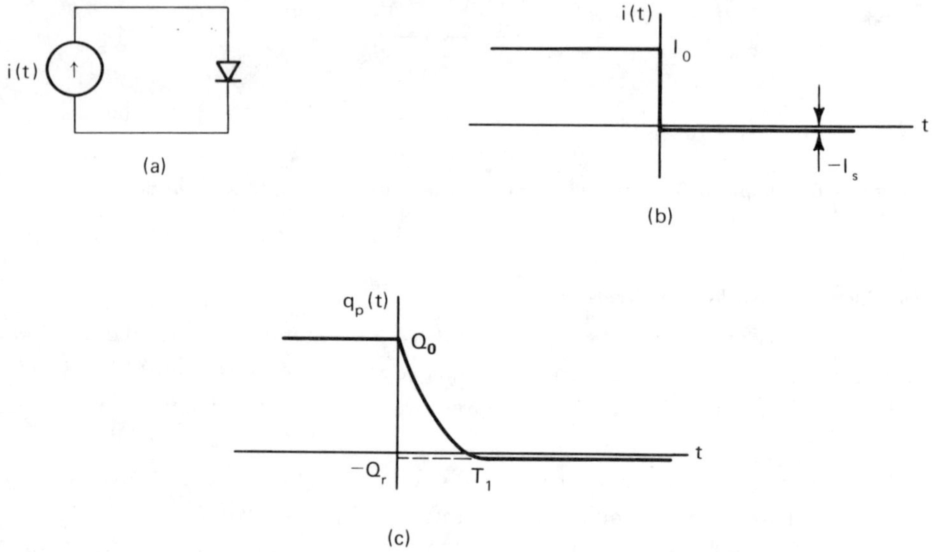

Figure A-31. (a) A p-n junction diode connected to a current generator; (b) the waveform of this generator; (c) the value of $q_p(t)$ or of $q_n(t)$.

Even though the value of current is forced by the generator to change instantaneously, the value of $q_p(t)$ does not change instantaneously from Q_0 to $-Q_r$. A finite time is required for the excess holes to combine with free electrons so that the new equilibrium can be established. Let us call T_1 the time that it takes $q_p(t)$ to fall to 0. Note that the larger I_0, the forward current is, the larger Q_0 will be. This, in turn, results in a larger T_1, the time for the charge to fall to zero.

Now assume that the diode is connected into the circuit shown in Fig. A-32a, which does not use an ideal generator. The voltage generator switches from forward to reverse bias at t=0. The charge density $q_p(t)$ falls off as before.

Now let us consider the current through the diode. For $t < 0$, $v = V_1$, the diode is forward biased. Then,

$$i = V_1/R \tag{A-28}$$

This assumes that the forward resistance of the diode is much less than R. At $t = 0$, $v(t)$ becomes $-V_2$ which is a reverse bias. The current reverses and becomes

$$i = -V_2/R \tag{A-29}$$

Note that *the diode acts as a small resistance.* The current essentially remains at $-V_2/R$, for T_1 seconds. That is, it remains there until $q_p(t)$ becomes zero. Note that while $q_p(t)$ is positive, the diode acts as though it is forward biased. After time T_1, the current falls off in magnitude and after an additional τ_f seconds, the current essentially reaches the reverse biased value of $-I_s$.

The time τ_f is called the *fall time* and can essentially be accounted for by the capacitance across the diode's junctions. The time $\tau_s = T_1$ is called the *storage time* and is the time required for the *excess stored charge* to be removed(i.e., to recombine).

The total time that it takes the diode to switch from its on to its off(i.e., forward to reverse)state is called the *switching time*. It is given by

$$\tau_T = \tau_s + \tau_f \tag{A-30}$$

Note that these times are net exactly defined since $i(t)$ approaches I_s asymptotically. Thus, manufacturers usually specify a 10-90 percent time, that is, the time that it takes for $i(t)$ to go from $0.9\ V_1/R$ to $-1.1\ I_S$.

Let us consider what can be done to reduce the switching time. If V_1 is increased, then V_1/R will be increased. This increases Q_0. Since there is greater excess stored charge, then τ_s will be increased. Similarly, reducing V_1 will reduce τ_s.

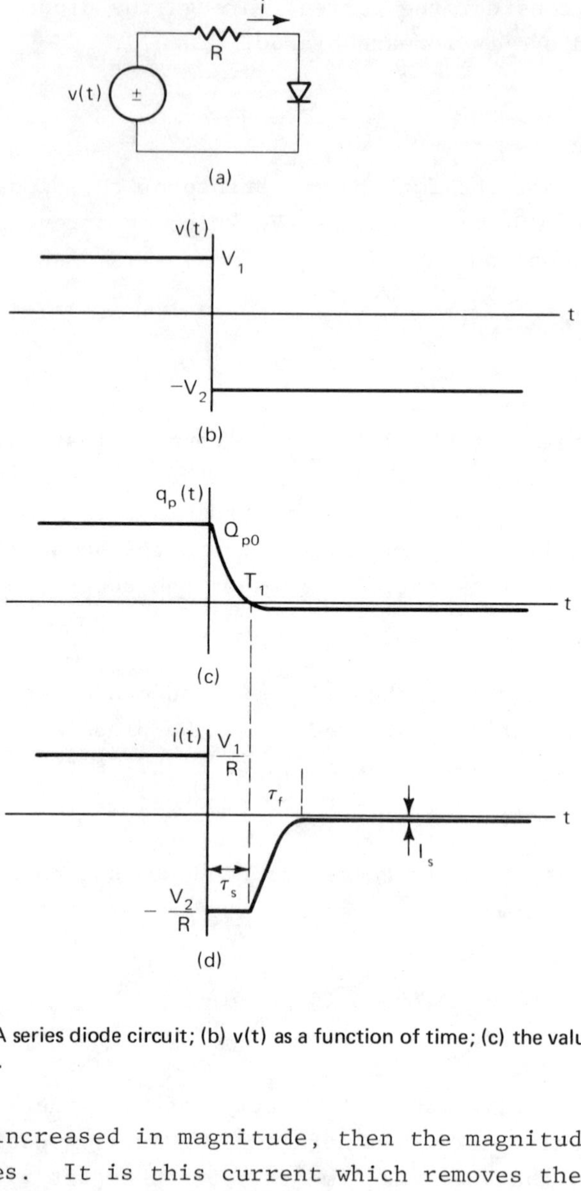

Figure A-32. (a) A series diode circuit; (b) v(t) as a function of time; (c) the value of q_1 (t); (d) the current i(t).

If $-V_2$ is increased in magnitude, then the magnitude of $-V_2/R$ also increases. It is this current which removes the excess stored charge. Thus, increasing V_2 reduces τ_s. Similarly, an increase in V_2 will reduce τ_f also.

Now let us consider the switching of a diode from reverse bias to forward bias. The phenomenon is similar to the previous case except that now the current prior to switching is extremely small. Hence, the excess stored charge will be very small and the storage

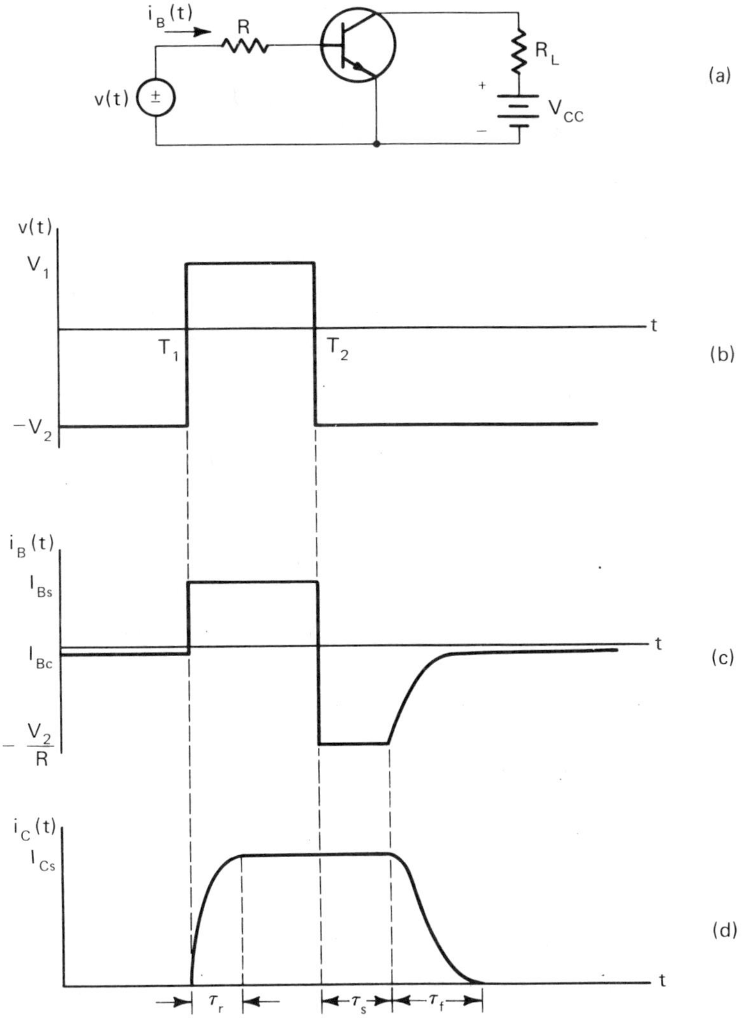

Figure A-33. (a) A transistor circuit; (b) the waveform of v(t); (c) the resulting base current; (d) the resulting collector current.

time effects will be negligible. Increasing the forward bias will reduce the switching time from reverse to forward bias. However, it will increase the storage time when the diode is subsequently switched to reverse bias.

Switching in a Transistor

The phenomena discussed in the switching of a p-n junction diode can be applied to the switching of a transistor. Consider the inverter of Fig. A-33a. We assume that the voltage v(t) switches as shown so that the base current is driven between a value which causes saturation I_{Bs} and a value which produces cutoff I_{Bc}. We have shown the base current changing almost instantaneously. Actually, this is not the case. However, the base current changes at a very much faster rate than the collector current. Hence, the time required for the base current to change is often ignored.

Now let us assume that i_B changes from the value I_{Bc} to I_{Bs}. The collector current does not change until the charge densities near the collector base junction change. This occurs some time after the charge density near the emitter base junction has changed. Thus, i_c rises from almost 0 to I_{cs} in the (10-90%)rise time τ_r as shown in Fig. A-33d.

When the base current is equal to I_{Bs}, the emitter base junction will be forward biased. Thus, there will be excess charge stored. When v(t) switches at time T_2, just as in the case of the diode, the current becomes $-V_2/R$ and essentially remains at this value until the excess charge is removed. Then it falls off to the small value I_{Bc}. As long as there is excess charge present, the collector current does not fall off. After τ_s seconds, the excess charge is removed and the collector current starts to fall to its cutoff value. This fall time takes τ_f seconds. Thus, on a 10-90% basis, the total time required to turn the transistor on is τ_r seconds and the time required to turn it off is $\tau_s + \tau_f$ seconds.

Let us see how we can reduce these switching times. If we increase I_{Bs}, then we drive the transistor into saturation in a shorter time. This consequently reduces τ_r. However, it also increases the excess charge stored, thus increasing τ_s. Thus, some compromise must be made. However, there is an alternative procedure. If we could make i_B very large for the first instant of time and subsequently make it very small, then τ_r would be reduced. However, excess charge could be small since i_B was small for most of the positive pulse. Thus, the storage time would be small. A circuit which accomplishes this, by adding a capacitor called a *speed up capacitor*, is shown in Fig. A-34. At the first instant of switching, the capacitor acts as a short circuit and the base current is large. However, after a short time, the capacitor acts as an open circuit and the base current becomes V_1/R. Thus, we can shorten both the rise and storage times by adding the capacitor and increasing R. The use of the speed up capacitor actually can greatly reduce the switching time. Integrated circuits usually do not make use of speed up capacitors, since it is difficult to obtain appropriate capacitance values.

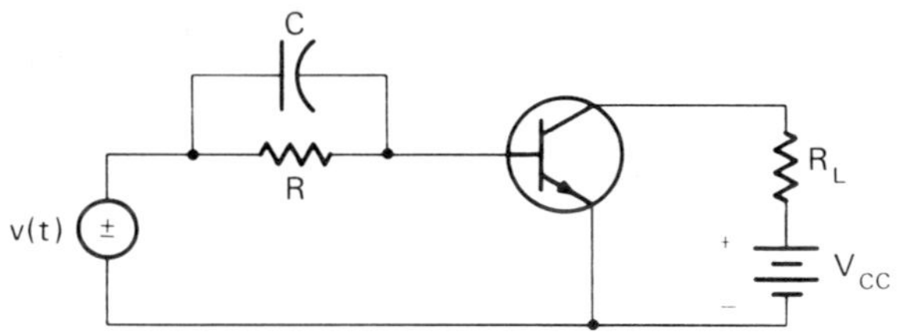

Figure A-34. A modification of Fig. A-33a that uses a speed up capacitor.

Increasing the voltage V_2(i.e., the magnitude of $-V_2$, see Fig. A-33b) reduces both the storage time and the fall time. Thus, it is desirable to make this voltage as large as practically possible.

Switching in an FET

When a junction FET is switched, the depletion region must be modified. The time required for this is similar to that for a diode. In an IGFET, somewhat similarly, a region of charge must be modified. In addition, the various capacitances tend to slow the switching times. Switching curves for the FET can be obtained. These are used in the design of switching circuits.

In general, switching times for "fast" transistors are much shorter than those for "fast" FETs.

A-6 INTEGRATED CIRCUITS

The modern digital computer utilizes electronic components which are manufactured as *integrated circuits*. This procedure allows thousands of elements such as transistors, FETs, resistors and capacitors to be incorporated on a single silicon chip. Integrated circuits as much as anything have allowed the construction of large complex computers. In addition, small minicomputers would not be practical without integrated circuits. In this section we shall briefly discuss some basic ideas of integrated circuits.

Introduction of Imputities by Diffusion

If a pure semiconductor is placed in an atmosphere containing impurities, these will diffuse into the bulk of the material. This process can be precisely controlled.

The diffusion process is speeded up as the temperature is increased. Thus, diffusion is performed in a *diffusion furnace* which is just an electric furnace whose atmosphere can be controlled precisely.

Integrated Circuits

The fabrication of integrated circuits, or even of single devices, using diffusion, make use of a property of silicon. When it oxidizes, it forms silicon dioxide(SiO_2) which is quartz. A thin layer of SiO_2 will effectively prevent diffusion of impurities from taking place.

Consider the following procedure. Suppose that a single crystal of silicon is placed in a diffusion furnace containing an atmosphere of water vapor and oxygen. The silicon will oxidize and become coated with a layer of SiO_2. Now suppose that we could remove the SiO_2 layer from a small portion of the surface. If the silicon crtstal is then placed in a diffusion furnace, whose atmosphere contained a donor type impurity, then a small n-type region will form under the area where the SiO_2 was removed. By repeating this procedure with different "openings" in the SiO_2 and with different impurities, a semiconductor device can be fabricated in the silicon chip.

We discussed removing the SiO_2 layer from a small region. This can be done photographically. There is a material called a *photoresist* which is polymerized by exposure to ultraviolet light and which also resists etching by hydrofluoric acid. The photoresist can be coated on the SiO_2 layer. Then, it is exposed to ultraviolet light at all points except where diffusion is to take place. The unpolymerized photoresist is then washed off this area. (Note that the polymerized photoresist is *not* removed by this process.) The silicon chip is then placed in a hydrofluoric acid atmosphere which removes the SiO_2 from those regions which are not covered by photoresist. Next, a solvent is used to remove all the photoresist and then diffusion takes place. The silicon chip can be reoxidized and the process repeated.

To illustrate this procedure we shall consider the formation of a transistor. This is called a *planar diffused transistor* since the entire fabrication process is done in a plane(i.e., the top surface)of the transistor.

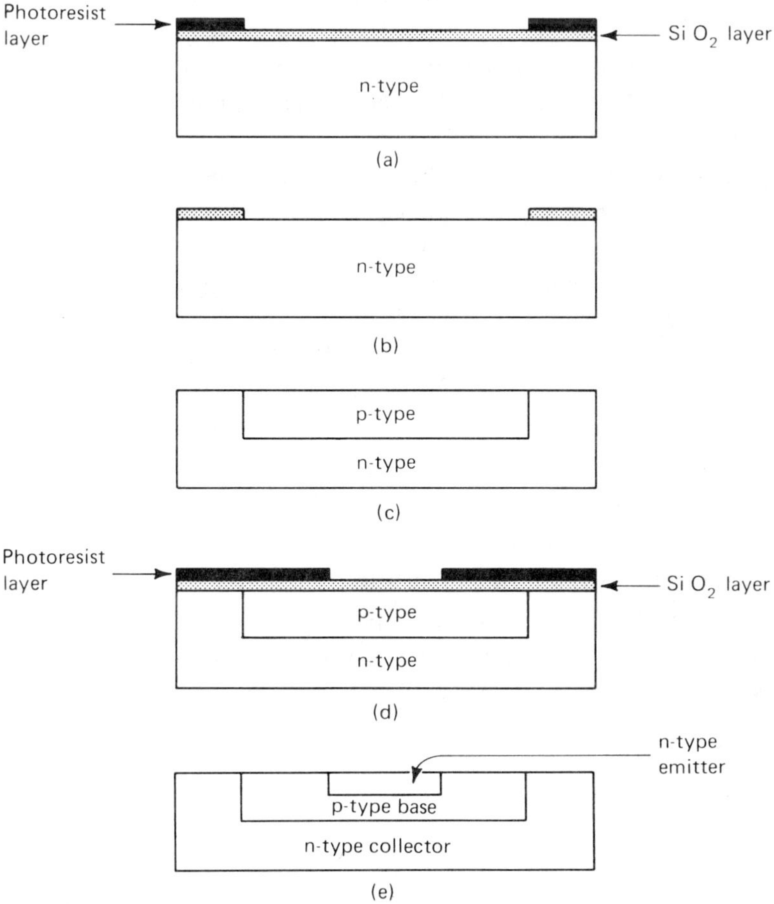

Figure A-35. Steps in the fabrication of a planar diffused transistor. (a) after oxidation, exposure and removal of unpolymerized photoresist; (b) after etching and subsequent removal of photoresist; (c) after diffusion of acceptor impurity; (d) after second oxidation and exposure and removal of unpolymerized photoresist; (e) final transistor.

Let us start with an n-type silicon crystal. It is oxidized and coated with photoresist. It is exposed to ultraviolet light at its edges. Thus, the photoresist can then be removed from a large region as shown in Fig. A-35a. After etching and subsequent removal of the remaining photoresist, the SiO_2 layer is as shown in Fig. A-35b. The crystal is put in the diffusion furnace where acceptors are diffused to form a p-type region, see Fig. A-35c. The process is repeated as shown in Fig. A-35d and the final transistor is shown in Fig. A-35e.

In general, we desire the impurity density in the collector to be large in regions far removed from the base. This makes the resistance low there. However, the impurity density near the base should be low since this reduces the junction capacitance. Diffusion produces just the opposite result. That is, the impurity density falls off as the distance from the surface increases. There is a procedure called *epitaxial growth* that can be used to eliminate this problem. Here, a crystal of silicon is placed in an atmosphere containing silicon vapor. The process is controlled so that silicon condenses on the original crystal. If this is done properly, the resulting silicon will still be a single crystal. We have oversimplified the discussion here. Actually, the atmosphere contains silicon compounds and chemical reactions take place which result in the condensation of pure silicon on the crystal.

The original crystal can be highly doped. Then, a lightly doped region can be epitaxially grown on it. In this case, we have the "lower region" with the heavy doping. This is just what was desired. In Fig. A-36 we show two such transistors fabricated on a single chip. Note that the highly doped and the lightly doped regions can both be epitaxially grown on a substrate.

When many transistors are fabricated on a single chip, they must be electrically isolated from each other. In Fig. A-36, the isolation is performed by reverse biasing the collector substrate p-n junctions. Other isolation schemes are used. In one, each transistor or group of transistors, or other elements are surrounded by SiO_2 layers (this is called EPIC isolation). In another scheme, regions of the substrate are physically separated and are connected and supported by their leads(beam lead isolation). An insulated base is also used for support.

There are many other aspects of integrated circuits which we shall not discuss here. Actually, integrated circuits have progressed to the point where large parts of a computer can be fabricated on a single chip. This is called *large scale integration*(LSI). If fewer elements are put on a single chip, this is called *medium scale integration*(MSI). In order to obtain the masks used to expose the photoresist, very large drawings can be made by draftsmen. These are then photographically reduced. Actually, these masks are now prepared almost completely by computer so that it has become an automated procedure.

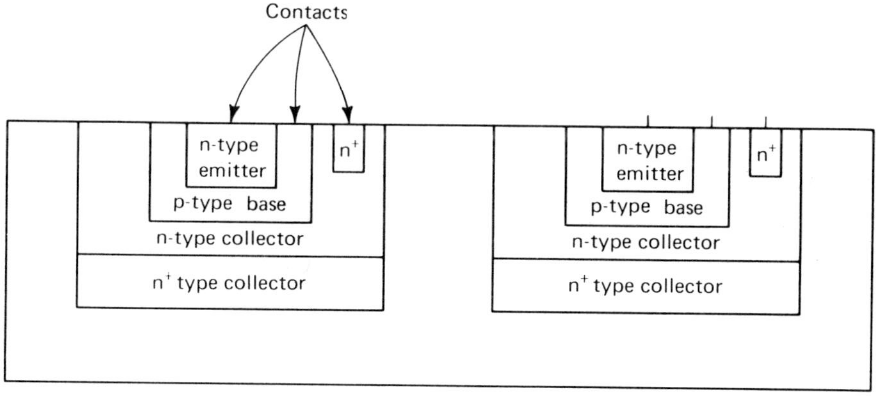

Figure A-36. Two transistors in a single substrate. The collectors in are epitaxially grown. The diagram is not to scale. Each transistor is typically 0.005 cm. long.

BIBLIOGRAPHY

Chirlian, P.M., "Electronic Circuits:Physical Principles, Analysis and Design", Chapters 3,4, and 15, McGraw-Hill Book Co., New York, 1971

Gray, P.E. and Searle, "Electronic Principles Physics, Models and Circuits", Chapters 4-10, 22, John Wiley and Sons, Inc., New York, 1969

Millman, J. and Halkias, C.C., "Integrated Electronics":Analog and Digital Circuits and Systems", Chapters 3-7, 10, McGraw-Hill Book Co., New York, 1972

PROBLEMS

A-1. Describe the difference between a conductor, a semiconductor and an insulator.

A-2. A silicon diode is characterized by Eq. (A-3). If $I_s = 10^{-7}$ amp., what value of forward bias is required to make the forward current 1000 times this value? Assume that the diode operates at a temperature of $20^oC(393^oK)$.

A-3. Describe the change in Fig. A-7b if the value of R_L is changed.

A-4. Repeat Prob. A-3 but now assume that V_{bb} is changed.

A-5. Describe the principles of operation of the junction transistor.

A-6. A transistor has the following parameter values

$I_{CO} = 10^{-7}$ amp.
$\alpha = 0.98$
$\alpha_r = 0.86$
$I_{EO} = 10^{-7}$ amp.

If the collector-base junction is strongly reverse biased and $i_E = 5$ ma, what will be the value of i_C?

A-7. If the transistor of Prob. A-6 is operated so that $i_E = 0$, what should the value of v_{CB} be to produce the same magnitude of collector current as in Prob. A-6?

A-8. The transistor whose characteristics are given in Fig. A-11 is operated in the circuit of Fig. A-12, where $V_{CC} = -15$ volts and $R_C = 3000$ ohms. If the input current is such that i_E is switched from 0 ma to 5 ma, what will the swing in v_{CB} be?

A-9. The transistor whose characteristics are given in Fig. A-14 is operated in the circuit of Fig. A-13b. The parameters of the circuit are $V_{CC} = -20$ volts, $R_C = 5000$ ohms, and $R_B = 666{,}667$ ohms. If there is no signal, what will be the values of v_{CE} and i_C?

A-10. Repeat Prob. A-9 but now assume that the input signal causes $i_B = 0$. Repeat the calculations for $i_B = 60\ \mu a$.

A-11. Discuss the relationship between transistor base width and switching speed.

A-12. Discuss the operation of the junction FET.

A-13. Repeat Prob. A-12 for the depletion MOSFET.

A-14. Repeat Prob. A-12 for the enhancement MOSFET.

A-15. For digital applications, why is the enhancement MOSFET the desirable one to use?

A-16. Why does it take a longer time to switch a transistor from saturation to cut off than from cutoff to saturation?

A-17. Discuss the reasons for using a speed up capacitor in switching circuits.

A-18. What property of silicon makes it especially useful for integrated circuits?

A-19. Describe the steps used to form two transistors in a single semiconductor chip.

Index

A

B

C

D

E

F

G

M

N

O

P

Q

R

S

T

U

V